Jens Kurreck and Cy Aaron Stein

Molecular Medicine

Related Titles

Meyers, Robert A. (eds.)

Translational Medicine

Cancer

2015

Print ISBN: 978-3-527-33569-5

Meyers, Robert A. (eds.)

Synthetic Biology

2015

Print ISBN: 978-3-527-33482-7

Vertes, Alain / Qureshi, Nasib / Caplan, Arnold I. / Babiss, Lee (eds.)

Stem Cells in Regenerative Medicine

Science, Regulation and Business Strategies

2015

Print ISBN: 978-1-119-97139-9

Pelengaris, S., Khan, M. (eds.)

The Molecular Biology of Cancer

A Bridge from Bench to Bedside

2 Edition

2013

Print ISBN: 978-1-118-00881-2; also available in electronic formats

Karp, G.

Cell and Molecular Biology

7 Edition

2013

Print ISBN: 978-1-118-20673-7; also available in electronic formats

Dickenson, J., Lloyd Mills, C., Freeman, F., Thode, C., Sivasubramaniam, S.

Molecular Pharmacology - From DNA to DrugDiscovery

2012

Print ISBN: 978-0-470-68443-6; also available in electronic formats

MacPherson, G., Austyn, J.

Exploring Immunology

Concepts and Evidence

2012

Print ISBN: 978-3-527-32412-5; also available in electronic formats

Acheson, N.H.

Fundamentals of Molecular Virology, Second Edition

2 Edition

2011

Print ISBN: 978-0-470-90059-8; also available in electronic formats

Pasternak, J.J.

An Introduction to Human Molecular Genetics

Mechanisms of Inherited Diseases, Second Edition

2 Edition

2005

Print ISBN: 978-0-471-47426-5; also available in electronic formats

Jens Kurreck and Cy Aaron Stein

Molecular Medicine

An Introduction

Verlag GmbH & Co. KGaA

Authors

Jens Kurreck
Berlin University of Technology
Institute of Biotechnology
Department of Applied Biochemistry, TIB 4/3-2
Gustav-Meyer-Allee 25
13355 Berlin
Germany

Cy Aaron Stein
City of Hope Medical Center
Department of Medical Oncology and Experimental Therapeutics
1500 E. Duarte Road
Duarte, CA 91010
USA

Illustrator

Anke Wagner
Berlin University of Technology
Institute of Biotechnology
Department of Bioprocess Engineering, ACK24
Ackerstr. 76
13355 Berlin
Germany

All books published by **Wiley-VCH** are carefully produced. Nevertheless, authors, editors, and publisher do not warrant the information contained in these books, including this book, to be free of errors. Readers are advised to keep in mind that statements, data, illustrations, procedural details or other items may inadvertently be inaccurate.

Library of Congress Card No.: applied for

British Library Cataloguing-in-Publication Data
A catalogue record for this book is available from the British Library.

Bibliographic information published by the Deutsche Nationalbibliothek
The Deutsche Nationalbibliothek lists this publication in the Deutsche Nationalbibliografie; detailed bibliographic data are available on the Internet at <http://dnb.d-nb.de>.

© 2016 Wiley-VCH Verlag GmbH & Co. KGaA, Boschstr. 12, 69469 Weinheim, Germany

Print ISBN: 978-3-527-33189-5
ePDF ISBN: 978-3-527-67508-1
ePub ISBN: 978-3-527-67507-4
Mobi ISBN: 978-3-527-67511-1

Cover Design   Adam Design, Weinheim
Typesetting   Thomson Digital, Noida, India

Printed on acid-free paper
Printed and bound in Singapore by Markono Print Media Pte Ltd

Dedicated to Ferdinand Hucho *and* John J. Rossi *for their contributions to the life sciences, and most importantly, for being extraordinarily well-respected, decent, honorable human beings.*

Contents

Preface

Molecular medicine is one of the hottest, most dynamic areas in all the life sciences and combines the disciplines of biology, biochemistry, human biology, and pharmacology, in addition to basic and clinical medicine. Knowledge of the molecular sources of disease has and will continue to pave the way for the development of novel diagnostic procedures and therapeutic strategies to treat inherited or acquired diseases. The expanding repertory of therapeutic approaches finally makes it possible to address diseases that hitherto have been untreatable. Molecular medicine will without doubt change our disease treatment paradigms; rather than employing standard therapies for every individual and adjusting treatment based on observed efficacy and side effects, personalized approaches based on specific constitutional features of the distressed individual will become possible.

In recent years, many universities worldwide have established bachelors and masters degree programs to study molecular medicine. In addition, molecular medicine has become an indispensible component of well-established curriculums. Courses of study in medicine, (molecular) biology, biochemistry, human biology, and biotechnology have frequently been offered. We thus had the sense that it was time to assemble an up-to-date introduction to the entire spectrum of molecular medicine, one written in a clear and uniform style. Numerous color figures and expositions of clinical relevance will, we hope, also ensure a systematic introduction to the topic.

This textbook primarily addresses undergraduates and lecturers. It is based on our experience in teaching courses in molecular medicine for many years, in laboratory research, and in medical practice. We assume that students who use this textbook have at least some basic knowledge of molecular and cell biology. If not, we recommend that students first consult appropriate introductory textbooks such as *Principles of Biochemistry* by Voet *et al.* or *Molecular Biology of the Cell* by Alberts *et al*. Although this textbook is primarily written for undergraduates, we also hope that medical and graduate

students and researchers in the field of molecular medicine will also benefit from it.

For many reasons, writing a book on molecular medicine is a challenging task. First, there is no common consent on just what molecular medicine actually is. We view molecular medicine as a discipline that investigates normal and pathological cellular processes at the molecular level. Such in-depth analysis will help us to better understand the causes of disease, improve diagnosis, and develop novel therapeutic strategies. This understanding of molecular medicine has led us to the selection of the topics that are covered in this book.

The broad range of topics poses a second challenge, as no two-author team can possibly be experts in all areas of molecular medicine. However, we chose not to edit a book composed of individual specific articles, each written by an expert in a particular field in his or her own style. Instead, we have attempted to present a homogeneous textbook in a uniform and consistent style.

Numerous cross-references will help the reader to understand the complex interdependencies between the different fields of molecular medicine. For example, the monoclonal antibody Herceptin not only provides a comparatively new treatment option for aggressive breast cancer but is also a recombinant protein drug that is challenging to produce. Its successful therapeutic use represents one of the paradigmatic applications of pharmacogenetic testing.

Finally, the dynamics of the field of molecular medicine required us to assemble a textbook with the most up-to-date information possible. An accompanying Web site – http://www.wileyvch.de/home/molecular_medicine – will contain figures and figure legends from the book, in addition to regular updates on the latest developments in the field.

This textbook is divided into chapters, each of which can be viewed as one lecture of a one semester course on molecular medicine. The sections are structured in a logical order. However, each chapter can also stand alone as an introduction to a single topic, for example, gene therapy or stem cell technology. Cross-references will help the reader to find sections in the book that should be consulted for an in-depth understanding of the topic.

The textbook commences with a short general introduction to molecular and cellular biology and then describes some selected methods widely used in modern life science research. Several chapters deal with the molecular causes of disease, and established as well as new diagnostic approaches are also described. Viral, bacterial, or eukaryotic pathogen infections are also covered, as these are a major cause of suffering and death worldwide, and require continuous improvements in therapeutics for the management of emerging pathogens and drug resistance. The outline of novel therapeutic approaches for the treatment of cancer and genetic disorders developed in the era of molecular medicine will certainly leave the reader impressed by the dynamism of this field: Drugs based on recombinant proteins, particularly monoclonal antibodies, have already become an important element in clinical practice. Timely topics such as the newest advances in gene therapy, stem cell research, and RNA technologies will be introduced with numerous helpful figures. The book will conclude with a chapter on the ethical dimensions of molecular medicine.

Every chapter is accompanied by recommendations for further reading. Rather than providing an exhaustive list of bibliographical references that would be of limited use for most students, we selected educationally valuable review articles for each topic. While we initially intended to choose current articles, we also found older reviews that can provide excellent introductions to various subjects. In addition, the reference lists are intended to help students to begin navigating through the jungle known as the modern scientific literature.

We hope that the readers will enjoy our journey through the field of molecular medicine and share our enthusiasm for this fascinating subject. We would be very pleased if our textbook helps prepare you for working in this exciting field.

Acknowledgments

This book would not have been possible without the help of many people. We are particularly thankful to Erik Wade for reading all the chapters and providing valuable suggestions. We also thank Anke Wagner for her comments on several chapters. We also want to thank our (JK) student Derya Günes for her chapter revisions. Additional thanks goes to Mary Houlemarde and Tonya Nickens for their support, and to Daniela Castanotto for her unfailing good humor, common sense, and help in probing some of the more complex issues of this narrative. A big thanks goes to Harry Kurreck for proofreading all the chapters and giving valuable comments.

We want to thank the team at Wiley, Waltraud Wüst, Anne Chassin du Guerny, Gregor Cicchetti, and Andreas Sendtko, for their advice and assistance. A special thank goes to our graphic designer Anke Wagner for converting our quirky sketches into illustrative figures.

It is impossible for two scientists to cover the whole field of molecular medicine in depth. We are, therefore, thankful to our expert colleagues for their valuable comments on specific topics: Thomas Bock, Toni Cathomen, Henry Fechner, Anja Pöhlein, Roland Lauster, and Daniela Castanotto. In addition, we want to thank Tatjana Schütze for support with the graphical representation of protein structures. We also want to thank our research groups for their enthusiasm about molecular medicine.

Last but not least, we express our most heartfelt appreciation and thanks to our families for their patience when we spent far too much time writing this book. Therefore, this book is for Harry, Marianne, Malte, Paul, and Anke; for Myra, Allison and Warren; Lauren and Brian; and Lily Leigh and Margot Nicole.

May 2015
Jens Kurreck, Berlin
Cy Aaron Stein, Duarte

Introduction

1

Summary

- Molecular medicine is a highly dynamic field of life science research that uses interdisciplinary approaches to understand normal and pathological cellular processes at the molecular level. The findings of basic research have entered clinical practice, as new diagnostic assays and novel therapeutic strategies focus not only on the symptoms but also on the causes of disease.
- The development of drugs is a long-term and expensive process that starts with basic and preclinical research. A candidate drug must then successfully pass through three types of clinical trial in humans before a novel agent can be approved for therapeutic purposes.
- The eukaryotic cell is compartmentalized into several cellular organelles by intracellular membranes. The nucleus harbors the genetic material, mitochondria are the cellular power plants, and the endoplasmic reticulum and the Golgi apparatus are responsible for the glycosylation and sorting of proteins.
- Cells follow a tightly regulated cycle of four phases. These include the two gap phases G_1 and G_2, the S phase in which new DNA is synthesized, and mitosis, during which the cell divides.
- Apoptosis is the process of programmed cell death, which is important as a normal physiological mechanism and for protection against infections and cancer. Apoptosis can be triggered by extrinsic or intrinsic signals.
- Genomic DNA is amplified by DNA polymerases in a process known as replication. The synthesis occurs in a semiconservative and semidiscontinuous way.
- Expression of genes requires two steps. In the first step, the DNA is transcribed into RNA. Most primary transcripts are posttranscriptionally processed. For mRNAs, this step includes the addition of a cap at the 5′ end and of a poly(A) tail at the 3′ end. Introns are spliced out to link the exons together. Several bases are modified in various types of RNAs. The second step in gene expression is the translation of the genetic information into proteins. This process is carried out by ribosomes. Posttranslational modifications of proteins include activation by proteolytic cleavage and covalent modification of amino

Contents List

acid side chains. This can occur, for example, by glycosylation or reversible phosphorylation.

- Sophisticated communication between cells is essential for the functioning of a multicellular organism. Neurons transmit signals at synapses. Hormones are molecules that induce physiological responses over a long distance or in adjacent cells. The extracellular signals are transmitted into the cell by cell surface receptors and induce a signaling cascade that leads to a biological response.
- The immune system protects an organism against (infectious) disease. The innate immune response recognizes general patterns of pathogens, while the adaptive immune system is directed against specific targets. The adaptive immune system involves a cellular immune response (T cells) and a humoral immune response (B cells that produce antibodies).

1.1
The Basics of Molecular Medicine

1.1.1
Topics of Molecular Medicine

Molecular medicine is a discipline dedicated to understanding normal and pathological cellular processes at the molecular level. This approach requires the use of many physical, chemical, biological, biochemical, and medical techniques (some of which are introduced in Chapter 2) to understand fundamental molecular mechanisms and how they go awry in disease. Molecular medicine combines classical disciplines such as cell and molecular biology, biochemistry, and medicine. Knowledge is often acquired via interdisciplinary investigation and can be used to develop new forms of molecular diagnosis and therapeutic intervention.

Molecular medicine can be divided into a basic research and an applied clinical discipline. The basic research component investigates molecular and genetic mechanisms of cellular function and identifies pathological processes. In many cases, this addresses a specific question with a hypothesis-driven approach, and can lead to large-scale investigations of whole genomes and proteomes (Chapter 7). The discipline known as translational research then tries to apply the findings from basic science to the clinic, where it may provide new forms of diagnosis and therapy.

A report published by Linus Pauling in 1949 laid the basis for the establishment of the field of molecular medicine. In his seminal paper, he showed that hemoglobin from patients suffering from sickle cell anemia had a different electrical charge than hemoglobin from healthy individuals. This study demonstrated that a disease could be traced to an alteration in the molecular structure of a protein. This novel perspective opened the possibility of establishing novel forms of diagnosis and therapy at the molecular level. Sickle cell anemia is not the only case in which a detailed understanding of the molecular etiology of the disease (e.g., of inherited genetic disorders, Chapter 3) has led to new diagnostic options (Chapter 8), although with only a modestly improved therapeutic outcome.

The field of oncology also illustrates the paradigm shift caused by a molecular perspective. While cancer treatment is still largely based on removal of the tumor by surgery (followed by chemotherapy and/or radiation therapy), molecular oncology (Chapter 4) tries to elucidate those pathways that lead to cellular transformation. This knowledge helps to produce a comprehensive molecular diagnosis of the disease basis in a single patient so that the treatment can be adjusted accordingly, an approach that has come to be known as "personalized medicine." Many modern anticancer drugs block specific pathways that lead to uncontrolled cellular proliferation. Similarly, elucidation of the life cycles of pathogens has helped develop new drugs for the treatment of infectious diseases (Chapters 5 and 6). For example, advancements in virus biology have led to the identification of novel targets for antiviral agents.

Most drugs belong to the class of small molecular compounds. To achieve oral bioavailability and to promote rapid diffusion across cell membranes and intracellular trafficking to their sites of action, the majority of (oral) drugs have molecular weights below 550 Da (although some antibacterial agents fall in the 700–900 Da range). A prominent example is acetylsalicylic acid (trade name Aspirin, Figure 1.1a), a drug mainly used as an analgesic.

Molecular medicine has broadened the spectrum of entities used as drugs. New medications are now often based on large molecules of biological origin (known as "biologics"). These include, for example, recombinant proteins such as monoclonal antibodies (Figure 1.1b) (Chapter 10), short pieces of DNA or RNA (Chapters 13 and 14), entire genes that can be delivered by viral vectors (Chapter 11), or even complete cells (Chapter 12). Pharmacogenetic investigations aim at discovering why the efficacy and toxic side effects of a drug at a given dose vary between individuals (Chapter 9). However, molecular medicine not only develops new diagnostic and therapeutic approaches but also poses heretofore unknown ethical issues, some of which will be introduced in Chapter 15.

(a)   **(b)**

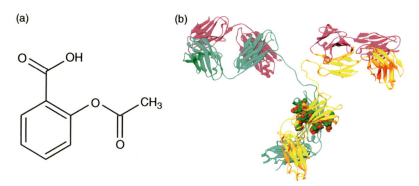

Fig. 1.1 Small molecular drugs and biologics. (a) The chemical structures of acetylsalicylic acid (Aspirin) and (b) the crystal structure of an antibody, shown for comparison. The two structures are not drawn to scale. (Part (b) adapted from Ref. [1] with kind permission from John Wiley & Sons, Inc.)

1.1.2
Stages of Drug Development

The development of a new drug is a time-consuming and expensive process (Figure 1.2) that may take 12–15 years (and in some cases even longer). The cost calculus of developing a new drug is complex and controversial, but the average cost to bring a new molecular entity (NME) to the market has been estimated to be as high as $1.8 billion. Drug development usually starts with the identification of a new target, which, for example, may be a proliferative factor that causes tumor growth. The next step in the process is to characterize the target, its location (extracellular, membrane-bound, cytosolic, and nuclear), and its function. Confirmation that the potential drug target fulfills the expected function is known as *target qualification* or *validation*. One way to identify a new compound is to perform what is known as a high-throughput screen (HTS). This approach allows testing a large number (up to millions) of compounds to identify an active molecule that modulates a particular target

(e.g., inhibits a proliferative factor). In almost all cases, the primary hit must be optimized by chemical modification to obtain higher binding affinities, better solubility, and so on. The efficacy and toxicology of the substance are then investigated in *in vitro* studies and animal experiments. The process of drug development may deviate substantially from this path depending on the type of drug being developed. Biologics, for example, are usually not obtained by HTS. Any substance will only be tested in humans after having passed extensive toxicological examination.

Clinical research is usually divided into three main phases. However, these phases are sometimes preceded by an exploratory trial (frequently called as phase 0) in a small number of subjects with a very small, subtherapeutic dose designed to gather data on the agent's basic properties in humans. This trial does not produce data about safety or efficacy. The actual clinical research starts with a phase I trial, usually carried out with a small number (20–100) of subjects. The main purpose of a phase I trial is to assess the safety of the drug. The

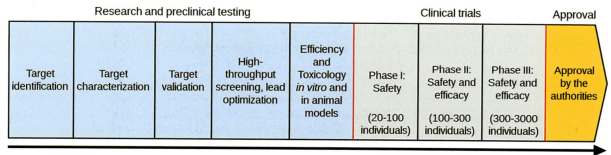

Fig. 1.2 Stages of drug development. The preclinical stages comprise the identification, characterization, and validation of a target and the identification and optimization of a compound, in addition to toxicological evaluation. The drug then undergoes three main phases of clinical testing before it is approved by the regulatory authorities.

trial is frequently designed to include a dose escalation to determine the optimal dose and the dose at which unacceptable toxicity supervenes. Although phase I trials are often carried out with healthy volunteers, under some circumstances, ill patients are enrolled. This is done most often with cancer patients, as the drugs to be evaluated are likely to make healthy individuals ill or may carry a significant risk of long-term toxicity.

Phase II trials are carried out with a larger number (100–300) of individuals. The central aim of the phase II study is to evaluate the efficacy of the drug. The trial is, therefore, usually performed in sick patients. Phase II studies are sometimes divided into phase IIA and phase IIB. While a phase IIA trial assesses the optimal dosing of the drug, a phase IIB trial is designed to study the efficacy of the drug at the prescribed dose. Another important goal of a phase II trial is to assess drug safety in a larger group of individuals.

Phase III trials are designed to assess the effectiveness and safety of a new drug in clinical practice in a large patient group (300–3000 or more individuals). These studies are carried out in randomized, controlled, multi-center trials. Phase III trials are usually designed as double blind studies, that is, the patients are randomly assigned to an experimental and a control group (in some trials, the control group may either receive a placebo or standard of care treatment). Neither the patients nor the physicians monitoring the outcome know which patient is receiving which treatment. Phase III trials aim at assessing the efficacy of a drug in comparison to placebo or the current standard of care treatment.

If drug safety and efficacy have been demonstrated in multiple phase III trial(s), approval for marketing can be applied for from the appropriate regulatory agency such as the Food and Drug Administration (FDA) in the United States or the European Medicines Agency (EMA) in the European Union. These agencies may request postmarketing monitoring, which is sometimes referred to as a phase IV trial. A phase IV trial involves safety surveillance after the drug has received permission to be marketed. In principle, it is designed to detect rare or long-term adverse effects in a much larger patient population and over a longer time period than was possible during the earlier phase of clinical trials. A phase IV trial may also identify interactions with other marketed drugs. Even after marketing, if harmful effects are discovered, any drug may be withdrawn at any time or its use restricted only to certain conditions.

The term phase V is increasingly used to describe studies that determine whether the therapeutic effect of a new drug is realized in day-to-day clinical practice. Community-based research is employed to survey whether the effects under typical (and somewhat variable) clinical contexts are similar to those that were found in the controlled efficacy studies. A phase V trial may also analyze the cost-benefit ratio of a drug or therapeutic intervention.

1.2
The Human Cell

Despite the extreme complexity of living organisms and the myriad number of functions that each constituent organ must carry on, only a surprisingly limited set of molecules are commonly employed. Typical biomolecules found in living organisms include nucleic acids, proteins, polysaccharides, and lipids. These macromolecules are composed of a relatively limited number of monomeric building blocks such as DNA and RNA nucleotides, amino acids, monosaccharides, and fatty acids. In addition, inorganic ions, organic acids, and a variety of metabolites are also important constituents of cells. While the basic features of biomolecules are extensively covered in the general textbooks of biochemistry listed at the end of this chapter, the cell as the basic functional unit of an organism and the major intracellular processes relevant to human physiology and pathology will be outlined here.

The human body consists of approximately 100 trillion (10^{14}) cells. Although all cells of a given organism carry the same genome, these cells have different functions and are highly specialized. A typical eukaryotic (animal) cell is illustrated in Figure 1.3. The most prominent characteristic that distinguishes eukaryotic cells from prokaryotes is its compartmentalization. The main membrane-bound organelles of animal cells are the nucleus, the endoplasmic reticulum (ER), the Golgi apparatus, the mitochondrion, the lysosome, and the peroxisome. Each of these organelles contains a specific set of proteins that fulfill a specific function. These organelles are embedded in a gelatinous fluid called the cytosol.

1.2.1
Organelles

1.2.1.1 The Nucleus
The central organelle of a eukaryotic cell is the nucleus, which contains the genetic material. The nucleus is surrounded by the nuclear envelope consisting of two membranes: the inner and the outer nuclear membranes (Figure 1.4a). The outer nuclear membrane is continuous with the rough endoplasmic reticulum (RER). Proteins referred to as nucleoporins form aqueous channels, called nuclear pores, through the envelope. These pores allow small water-soluble molecules to pass into and out

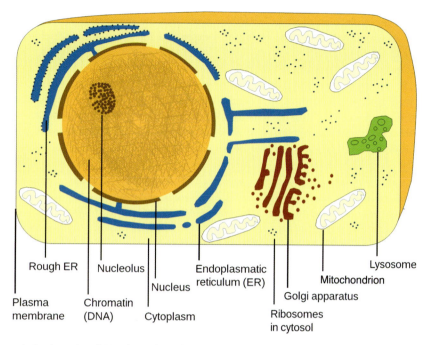

Fig. 1.3 Diagram of a typical eukaryotic cell. Membrane-bound organelles include the nucleus, endoplasmic reticulum (ER), Golgi apparatus, mitochondrion, lysosome, and peroxisome. (Reproduced with permission from Ref. [2]. Copyright 2008, Garland Science/Taylor & Francis LLC.)

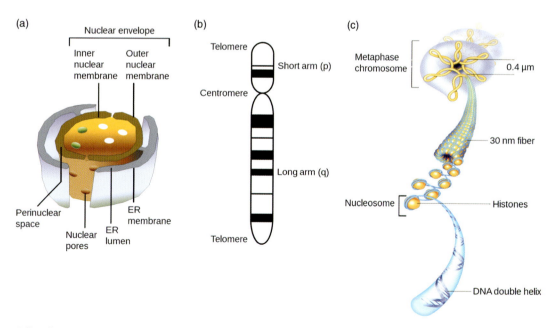

Fig. 1.4 Cell nucleus and chromatin organization. (a) The nucleus is surrounded by the nuclear envelope consisting of two membranes. Pores in the envelope allow the exchange of small water-soluble molecules. The outer nuclear membrane is continuous with the membrane of the endoplasmic reticulum (ER). (b) Schematic representation of a eukaryotic metaphase chromosome. Each chromosome has a specific banding pattern after staining. (c) DNA winds around histone proteins to form nucleosomes that fold into a 30 nm fiber. Loops of chromatin are then attached to a protein scaffold to form the metaphase chromosome. (Part (a) reproduced with permission from Ref. [2]. Copyright 2008, Garland Science/Taylor & Francis LLC. Part (c) adapted from Ref. [1] with kind permission from John Wiley & Sons, Inc.)

of the nucleus, while larger molecules must be actively transported in or out. A filamentous network of lamin proteins provides mechanical support, but is also involved in the regulation of replication and cell division. A distinct structure found in the cell nucleus is the nucleolus (indicated in Figure 1.3). The nucleolus occupies up to one fourth of the volume of the nucleus and forms around specific chromosomal regions. In the nucleolus, ribosomal RNA is transcribed and assembled with proteins to give (incomplete) ribosomes.

The DNA together with DNA-binding proteins and RNA molecules is organized in chromosomes. The human genome consists of 46 chromosomes that can be identified based on the specific banding pattern after staining. Figure 1.4b shows a schematic representation of a condensed metaphase chromosome. The ends of the chromosome are called *telomeres*. The *centromere* is the point where the two identical chromosomes touch and the microtubules attach for separation during mitosis. Most chromosomes are asymmetric; the centromere separates a short arm (p for the French word *petit*) and a long arm (q, chosen as next letter in the alphabet after p). The position of a gene or a DNA sequence in the human genome (denoted as the *locus*) is indicated by the chromosome number, the arm, and three numbers that refer to the region, the band and the subband. For example, the locus 11p15.3 indicates that a DNA sequence is located on the short arm of chromosome 11, in region 1, band 5 and subband 3.

If the DNA of every cell in the human body were lined up end to end, it would stretch from the Earth to the Sun and back 100 times. This means that a single cell must package DNA strands with a combined length of approximately 2 m into a nucleus with a volume in the cubic micrometer (μm^3) range.

Chromosomes consist of the DNA complexed with proteins, together known as chromatin (Figure 1.4c). The first step in chromatin condensation is achieved by winding the DNA around so-called histone proteins. These proteins are positively charged (i.e., contain many arginine and lysine residues) and can bind negatively charged DNA by electrostatic interactions. Two copies of each of the histones H2A, H2B, H3, and H4 form a core around which the DNA is wound twice. The fifth histone protein, the linker histone H1, binds to the middle of the DNA and to its two ends, locking the DNA into place. This structural unit, called the nucleosome, allows packaging in higher ordered structures, such as the so-called structural element "30 nm fiber."

The interaction between proteins and DNA has a major impact on gene expression (Section 1.3.3). Much of a cell's DNA is not expressed, as it does not encode a gene product. In addition, the expression of a particular gene may be inappropriate for a certain cell type. Non-expressed DNA is typically highly condensed; it is called *heterochromatin*. In contrast, lightly packed DNA (*euchromatin*) facilitates access to those enzymes required for gene expression and is, therefore, transcriptionally active. Nucleosomes also undergo remodeling as proteins required for transcription must bind to their target DNA. The *chromatin remodeling complex* disrupts the interaction between histones and DNA in the nucleosomes in an ATP-dependent manner so that the DNA becomes accessible. Histones undergo extensive covalent posttranslational modification (Section 1.3.5), which, together with DNA methylation, are two of the main mechanisms of the epigenetic regulation of gene activity (Section 1.3.4).

1.2.1.2 Mitochondria

Mitochondria (Figure 1.5) are organelles with double membranes that function as "cellular power plants" since they are the site of the oxidative metabolism that leads to the production of ATP. A eukaryotic cell typically contains 800–2000 mitochondria. In the internal matrix, they contain all the enzymes required for the citric acid cycle and for fatty acid oxidation. Redox proteins involved in the electron transport chain and oxidative phosphorylation are embedded in the inner membrane, which invaginates into cristae to expand its surface area.

A special feature of mitochondria is that they carry a genome that encodes several (but not all) mitochondrial proteins. The human mitochondrial genome is a circular molecule of approximately 17 000 base pairs encoding 13 proteins and 2 ribosomal and 22 transfer RNAs. Each mitochondrion contains multiple copies of the mitochondrial genome. Additional mitochondrial proteins are encoded in the nucleus. The mitochondria and their

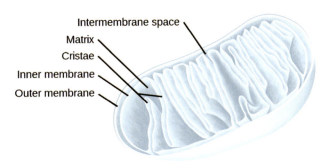

Intermembrane space
Matrix
Cristae
Inner membrane
Outer membrane

Fig. 1.5 The mitochondrion. The mitochondrion is bounded by two membranes. The inner membrane harbors the electron transport chain bearing the respiratory enzymes of the cell. It is extensively invaginated, forming the so-called cristae. The internal matrix contains high levels of enzymes involved in oxidative metabolism, and also holds the mitochondrial DNA. (Adapted from Ref. [1] with kind permission from John Wiley & Sons, Inc.)

genetic material are exclusively inherited from the mother. The *endosymbiotic hypothesis* is an attempt to explain the origin of mitochondria (and plastids such as chloroplasts in plants). According to this theory, mitochondria and plastids evolved from formerly free-living bacteria that were engulfed by another cell as an endosymbiont through endophagocytosis. The mitochondrial DNA (mtDNA) represents the remnant of the bacterial genome, while the remaining genetic material moved into the nuclear genome.

Mitochondrial dysfunction may result in mitochondrial diseases, which may be caused either by mutations in mtDNA or in nuclear genes that encode mitochondrial proteins (Section 3.1.4). Mitochondrial dysfunction may have severe consequences, due to their critical function in cellular energy supply. These effects usually vary from organ to organ and also depend on other variations in the genome. Mitochondrial diseases often lead to neuro-muscular disease symptoms (myopathy), but also include diabetes mellitus, deafness, visual loss, and dementia.

1.2.1.3 Endoplasmic Reticulum and Golgi Apparatus
The endoplasmic reticulum and the Golgi apparatus play an important role in the posttranslational modification and sorting of proteins (Figure 1.6). The ER is a network of interconnected, flattened, membrane-enclosed sacs known as cisternae. The enclosed space is called the lumen (or cisternal space). The main function of the ER is to support the synthesis and export of proteins and membrane lipids. Two types of ER can be distinguished, the rough ER (RER) and the smooth ER (SER). While the SER is involved in lipid and carbohydrate metabolism and in detoxification, the RER plays an important role in posttranslational modification of proteins. The rough appearance is caused by ribosomes that bind to the cytosolic side of the ER membrane and translocate newly synthesized proteins into the lumen of the ER during the translation process. The main functions of the RER include the support and control of protein glycosylation and of correct protein folding, as will be described in more detail in Section 1.3.5. The RER carries out initial N-linked glycosylation, which is terminated in the Golgi apparatus. Proteins are shuttled from the RER to the *trans* Golgi network in membrane-bound vesicles.

The main functions of the Golgi apparatus include the termination of *N*-glycosylation and the initiation of *O*-linked protein glycosylation. Vesicles from the RER carrying proteins fuse with the membrane on the *cis* face of the Golgi apparatus and release their cargo. Proteins are modified in the lumen and leave the Golgi apparatus from the *trans Golgi network*, which sorts proteins for extracellular release or for transport to the lysosome, according to the markers they carry.

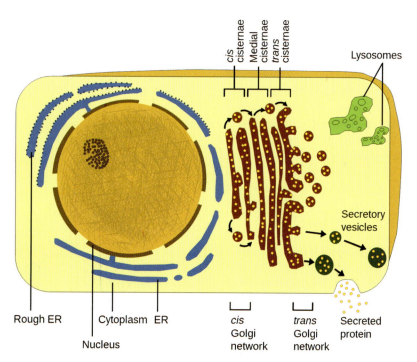

Fig. 1.6 Endoplasmic reticulum (ER) and Golgi apparatus. The endoplasmic reticulum and the Golgi apparatus are two organelles that are important for the correct folding, posttranslational modification, and sorting of proteins. *N*-Glycosylation of proteins is initiated in the ER and is terminated in the Golgi apparatus, while *O*-glycosylation exclusively occurs in the Golgi apparatus.

The importance of correct protein sorting is illustrated by the hereditary disease mucolipidosis II, which is characterized by psychomotor retardation, skeletal deformities, and death in childhood. The molecular cause of the disease is the lack of a phosphotransferase that normally phosphorylates mannose residues to mannose-6-phosphate on *N*-linked glycoproteins. Without this marker, enzymes that normally degrade glycosaminoglycans are transported from the Golgi apparatus into the extracellular space, rather than into the lysosome. This defect in protein sorting eventually leads to the accumulation of glycosaminoglycans and glycolipids in large inclusions in lysosomes that cannot be degraded. The disease is thus also known as Inclusion-cell (I-cell) disease.

1.2.1.4 Peroxisome and Lysosome

Peroxisomes and lysosomes are eukaryotic organelles with important metabolic functions, which consist mainly of degradative processes. Peroxisomes are involved in the catabolism of very long-chain and branched fatty acids and in the oxidative catabolism of amino acids. In addition, they contain enzymes of the pentose phosphate pathway. Peroxisomal disorders typically affect the nervous system, since peroxisomes also play a role in the biosynthesis of ether phospholipids, which are critical for the normal functioning of the mammalian brain.

Lysosomes are the cell's waste disposal and recycling system, as they contain enzymes that digest dispensable cellular components and ingested material. Lysosomes employ approximately 50 hydrolytic enzymes to break down cellular waste products, fats, carbohydrates, proteins, and other macromolecules into simpler compounds, which are returned to the cytoplasm as building blocks for biosynthetic processes. A variety of proteases, known as *cathepsins*, degrade proteins in a nonselective manner. Lysosomes maintain an interior pH of approximately 5, and their enzymes function optimally at acidic pH. The cytosol, with its higher pH (slightly above 7), is thus protected from the degradative activity of lysosomal enzymes that may leak into the cytosol, since they are inactive in this more alkaline environment.

Approximately 50 rare inherited metabolic disorders are known to result from malfunction of the lysosome. They are collectively referred to as *lysosomal storage diseases* (LSDs). The lack of a particular lysosomal enzyme may result in accumulation of substances destined for breakdown. LSDs have a variety of symptoms, but typically affect children, who die at a young age. Lysosomes also have a major influence on the pharmacokinetics of several drugs. Uncharged, weak bases may cross lysosomal membranes and accumulate in lysosomes. This phenomenon is called lysosomotropism and may explain the high tissue concentration and long elimination half-lives of some drugs. Accumulation in lysosomes may result in the inhibition of the activity of lysosomal enzymes.

1.2.2 Cell Cycle

The cell cycle is the sequence of events that leads to cell division into two daughter cells. It can be divided into three main periods: the interphase, the mitotic phase, and cytokinesis (Figure 1.7). Mitosis and cytokinesis together define the mitotic (M) phase of the cell cycle. A complete cell cycle in cultured cells takes approximately 16–24 h. In contrast, the doubling times of a cell in a multicellular organism can vary from 8 h to >100 days. Some terminally differentiated cells, such as muscle cells or neurons, never divide. This quiescent state is called the G_0 phase.

Interphase is characterized by cell growth, the accumulation of nutrients, preparation for mitosis, and the replication of DNA. It can be subdivided into three phases: The longest segment of interphase is the G_1 phase (G stands for gap), in which cells grow in size, have elevated biosynthetic activity, and produce enzymes required for DNA replication. DNA replication itself takes place in the synthesis phase (S phase), during which time all the chromosomes are replicated. S phase is followed by the second gap phase (G_2) during which time cells continue to grow. Several control mechanisms ensure that the cell is ready to enter mitosis.

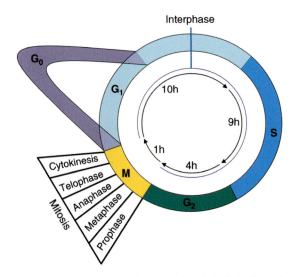

Fig. 1.7 The cell cycle. The eukaryotic cell cycle is divided into three main phases: (1) interphase consists of the G_1 and G_2 gap phases as well as the synthesis phase (S); (2) mitosis comprises pro-, meta-, ana-, and telophase; (3) in the final phase, cytokinesis, the cell divides.

Mitosis starts with *prophase*, which is characterized by spindle formation and condensation of the duplicated chromosomes. The nuclear membrane then breaks apart and the nucleus disappears (this process is sometimes called prometaphase). In the subsequent *metaphase*, the duplicated chromosomes move to positions midway between the spindle poles. The next step is the separation of the sister chromatids of the duplicated chromosomes and their movement to opposite poles of the cell. This occurs in *anaphase*. The last phase of mitosis is called *telophase* and involves decondensation of the chromosomes and restoration of the organelles and the nucleus. Although cytokinesis is not part of mitosis in the strict sense, it directly follows mitosis and involves division of the cytoplasm into two daughter cells.

As uncontrolled proliferation is a main characteristic of cancer, strict control of cell cycle is essential for the organism. The cell cycle has several checkpoints that monitor its progress and arrest the cell cycle if certain conditions have not been fulfilled. For example, the cell cycle only progresses to mitosis after the cell's DNA has been completely replicated. Another checkpoint ensures that all chromosomes are properly attached to the mitotic spindle.

Two classes of proteins, known as cyclins and cyclin-dependent kinases (Cdks), are the main regulators of progression through the cell cycle. Different types of cyclins are specific for each phase of the cell cycle. Cyclins are synthesized during one cell cycle phase and subsequently degraded during the succeeding phase. A cyclin forms a complex with its corresponding Cdk, which leads to the activation of the Cdk. Cdks are serine/threonine protein kinases that phosphorylate nuclear target proteins involved in the various processes of the cell cycle. Cdks are also regulated by cyclin-dependent kinase inhibitors, which arrest the cell cycle under certain conditions. These conditions include contact with adjacent cells, DNA damage, terminal cell differentiation, and cell senescence. Alterations of the inhibitory activity of Cdks are frequently found in cancer. Leland H. Hartwell, R. Timothy Hunt, and Paul M. Nurse were awarded the 2001 Nobel Prize in Physiology or Medicine for their discovery of the cyclin/Cdk system.

1.2.3
Apoptosis

Apoptosis (the Greek word for "falling off", for example, of leaves from a tree) is the process of programmed cell death. Apoptosis is a normal, physiological mechanism found in multicellular organisms that is important for the development and maintenance of proper physiological functioning and for the protection against cancer and

other diseases. An example of the essential function of apoptosis in developmental processes is the formation of fingers in vertebrates. Initially, the digits are fully connected by webbing. During development, the web is eliminated by programmed cell death. An estimated 10^{11} of the $\sim 10^{14}$ cells of an adult human die every day through apoptosis and are replaced by new ones. Apoptosis is also important for the elimination of virus-infected or malignantly transformed cells. Abnormal apoptosis may result in neurodegenerative diseases such as Alzheimer's, Parkinson's, and Huntington's diseases, as well as in damage caused by stroke or heart attack.

The apoptotic program includes a sequence of cellular morphological changes: As a first step, the cell begins to shrink. Its chromatin becomes condensed at the nuclear periphery and the cytoskeleton collapses. Subsequently, the nuclear envelope disappears, the DNA is fragmented, and the plasma membrane forms irregular bulges (blebs). Eventually the cell disintegrates into numerous membrane-enclosed vesicles called apoptotic bodies. Apoptotic bodies are phagocytosed by neighboring cells and macrophages. Apoptosis should be differentiated from necrosis, which is a form of cell death that results from trauma, such as acute cellular injury. In this case, water rushes into necrotic cell due to the loss of cell membrane integrity. The cell swells, its organelles lose their function, and the cell eventually bursts, spilling its contents into the extracellular space.

The processes of apoptosis are mainly controlled by *caspases*, a family of *c*ysteinyl *asp*artate-specific prote*ases*. Caspases are heterotetramers, consisting of two α- and two β-subunits, which are expressed as inactive, single-chained zymogens (procaspases). Upon activation, the N-terminal prodomain is excised and the α- and β-subunits are proteolytically separated to yield the active $\alpha_2\beta_2$ caspase. Two types of apoptotic caspases can be distinguished: initiator and effector (executioner) caspases. Initiator caspases cleave inactive zymogen precursors of effector caspases, thereby activating them. The effector caspases then cleave other cellular proteins, triggering the apoptotic process.

Apoptosis may be induced in two ways (Figure 1.8): External signals trigger the *extrinsic pathway* (death by commission), while the absence of external signals that inhibit apoptosis can lead to the activation of the *intrinsic pathway* (death by omission). The extrinsic pathway is induced by binding of a so-called death ligand (e.g., the Fas ligand) on an inducing cell to the death receptor (Fas receptor) on the surface of the cell destined to undergo apoptosis. This leads to activation of the initiator caspase-8 (and possibly caspase-10), which triggers executioner caspases to cleave their substrate proteins, driving the cell into apoptosis. The internal pathway is

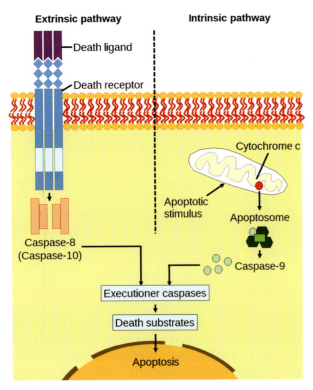

Extrinsic pathway

Intrinsic pathway

Death ligand

Death receptor

Cytochrome c

Apoptotic stimulus

Apoptosome

Caspase-8 (Caspase-10)

Caspase-9

Executioner caspases

Death substrates

Apoptosis

Fig. 1.8 Extrinsic and intrinsic pathways of apoptosis. Programmed cell death can be induced by extrinsic signals from a death ligand or by an apoptotic stimulus that leads to the release of cytochrome c from mitochondria. In both cases, an initiator caspase is activated that activates executioner caspases, which cleave their substrates. This program eventually leads to cell death. (Adapted from Ref. [3] with kind permission from Macmillan Publishers Ltd., Copyright 2002.)

activated, which then triggers executioner caspases as per the program described for the extrinsic pathway.

In 2002, the Nobel Prize in Medicine was awarded to Sydney Brenner, Robert Horvitz and John E. Sulston for their discoveries concerning the genetic regulation of organ development and programmed cell death.

1.3
DNA Replication and Gene Expression

(Anke Wagner, Berlin University of Technology, Berlin, Germany)

Cellular DNA must be copied before a cell can divide into two daughter cells. This occurs via a process known as *replication*. In addition, the genetic information stored in the DNA must be transferred into proteins, the actual primary effectors of cell function. This process of gene expression comprises two steps: the *transcription* of DNA into RNA and the *translation* of the RNA sequence into the corresponding amino acid sequence necessary for protein formation. This flow of genetic information, formulated in brief as "DNA makes RNA makes protein," was summarized by Francis Crick in 1958 in what is known as the *central dogma of molecular biology* (Figure 1.9). Crick himself acknowledged that the term dogma may be misunderstood as a belief that cannot be doubted. In fact, we now recognize several exceptions to the original model. The process of reverse transcription from RNA to DNA in retroviruses, or the replication of RNA molecules into new RNA molecules by other viruses (Section 5.1.1) are such examples. The dogma also does not cover processes such as epigenetic regulation of gene expression by DNA methylation (Section 1.3.4), posttranscriptional gene silencing by the RNA interference pathway (Section 13.3), or posttranslational protein modification. However, the assumption that proteins cannot serve as the template for the

induced in the absence of signals from the environment, for example, neighboring cells, which normally prevent the cell from committing suicide. Proapoptotic members of the Bcl-2 protein family are important mediators of these signals. They activate mitochondria to release cytochrome c into the cytosol, where it binds the apoptotic protease activating factor 1 (APAF1) to form the *apoptosome*. In the apoptosome, the initiator caspase-9 is

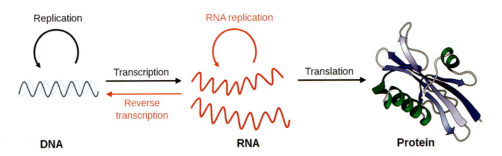

Replication

RNA replication

Transcription

Translation

Reverse transcription

DNA

RNA

Protein

Fig. 1.9 The central dogma of molecular biology. DNA is copied in a process known as replication. The genetic information stored in the DNA is expressed by transcription into RNA, followed by translation into an encoded protein. Several exceptions to the model (marked in red) are known, including the replication of RNA into RNA and the reverse transcription of RNA into DNA.

synthesis of new proteins or for the transfer of sequence information into RNA or DNA is still valid.

1.3.1
DNA Replication

Replication is the process of copying the genetic information contained in cellular DNA. The double-helix model of DNA first proposed by Watson and Crick suggested a molecular mechanism for the transmission of hereditary information: Each strand of the double-stranded DNA molecule can serve as a template for the synthesis of a complementary daughter strand. Since the newly synthesized DNA double strands consist of one parental strand and one daughter strand, this mode of replication is termed *semiconservative*.

The process of prokaryotic and eukaryotic replication is similar, although it is more complex in the eukaryotes. Altogether, replication involves the concerted action of more than 25 proteins. DNA synthesis occurs at the so-called replication fork, a simplified scheme of which is depicted in Figure 1.10.

In eukaryotic cells, the initial step of replication is the separation of the two strands of the DNA by a hetero-hexameric helicase called minichromosome maintenance (MCM). The unwinding process, which is driven by the hydrolysis of ATP (or another NTP), causes supercoiling

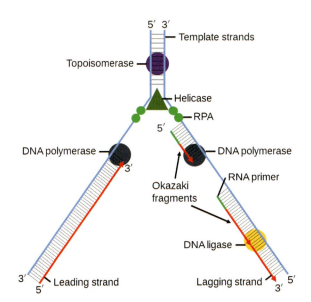

Fig. 1.10 Replication fork. Two DNA polymerase enzymes synthesize the new DNA strands, which are called the leading and lagging strands, respectively. A helicase unwinds the double-stranded template molecule and topoisomerases prevent supercoiling of the DNA. The replication protein A (RPA) stabilizes the unwound DNA single strands. DNA polymerase requires RNA primers to initiate synthesis of the Okazaki fragments, which are finally joined together by DNA ligase.

of the DNA ahead of the replication fork, which would eventually halt replication. To overcome these topological problems, the enzyme topoisomerase catalyzes the relaxation of DNA by cutting one or both strands of the DNA, rotating it, and reannealing the cleaved strand(s).

The two strands separated by the helicase have a strong tendency to reanneal to form a DNA duplex again. To prevent reannealing, the replication protein A (RPA) coats the single-stranded DNA. The central enzymes of DNA replication are the DNA-dependent DNA polymerases, simply known as DNA polymerases. Animal cells contain at least 13 distinct types of DNA polymerases. Like bacterial DNA polymerases, they can extend DNA strands only in the $5' \rightarrow 3'$ direction. The two strands of parental DNA, therefore, have to be replicated in different ways. While the *leading strand* is synthesized continuously in the $5' \rightarrow 3'$ direction of the replication fork movement, the other strand, known as the *lagging* strand, is also synthesized in the $5' \rightarrow 3'$ direction, but can only be made in small segments known as Okazaki fragments. The overall process is, therefore, described as a *semidiscontinuous* mode of replication.

Another common feature shared by all known DNA polymerases is that they cannot synthesize a new strand *de novo*, but can only add nucleotides to the free $3'$-OH group of a base-paired polynucleotide. The replication of DNA is, therefore, initiated by the synthesis of a short RNA primer. In eukaryotes, a primase synthesizes a 7- to 10-nt RNA primer. DNA polymerase α is associated with the primase and extends the primer by approximately 15 DNA nucleotides. The chain is then further extended by DNA polymerases δ and ε. Both polymerases are highly processive, meaning they can polymerize long stretches of the DNA without dissociating from the strand. DNA polymerase δ needs a sliding clamp protein named proliferating cell nuclear antigen (PCNA) to maintain its processivity. PCNA is also required for the synthesis of the lagging strand. In contrast, DNA polymerase ε is highly processive in the absence of PCNA and is involved in leading strand synthesis.

Both polymerases δ and ε carry out an additional *proofreading* step, for which they possess $3' \rightarrow 5'$ exonuclease activity: In case the DNA polymerase incorporates a mismatched nucleotide into the newly synthesized strand, its polymerase activity is halted and the mismatched nucleotide is hydrolytically excised by the exonuclease activity. The enzyme can then incorporate the correct nucleotide and resume its normal polymerase activity. DNA polymerases with proofreading activity are characterized by high fidelity, that is, their error rates are only about one in every 10^6–10^{11} nucleotides, depending on the type of polymerase.

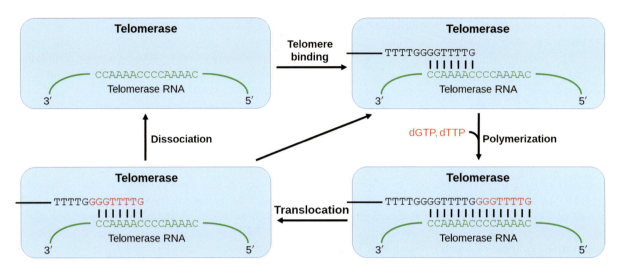

Fig. 1.11   Telomerase. Telomerase is a cellular reverse transcriptase that extends the ends of linear chromosomes (telomeres). It contains an RNA component to which the single-stranded 3′ end of the chromosome hybridizes. Telomerase then adds several nucleotides. The telomere can translocate and undergo several rounds of extension before it dissociates from telomerase.

Eventually, the RNA primers must be removed from the DNA and the Okazaki fragments must be joined together. In eukaryotes, this step is initiated by RNase H1 (H stands for hybrid, since the enzyme recognizes a DNA–RNA hybrid and hydrolyzes the RNA; this catalytic activity is also used in antisense technologies, as will be described in Section 13.1). The RNase removes most of the RNA and leaves only one RNA nucleotide adjacent to the DNA, which is then removed by the flap endonuclease-1. The excised nucleotides are replaced by DNA polymerase δ and the nick is eventually sealed by DNA ligase.

Eukaryotic chromosomes are large: The longest human chromosomes consist of more than 200 million base pairs. To achieve replication in a few hours, eukaryotic chromosomes contain multiple origins of replication, one every 3–300 kb depending on the species and tissue. Clusters of 20–80 adjacent replicating units (*replicons*) are activated simultaneously. Several sets of these clusters must be activated so that the entire chromosome can be replicated.

The linear ends of eukaryotic chromosomes, called telomeres, present a particular problem for the replication machinery. As already outlined, an RNA primer is required for the initiation of DNA replication. This primer can be removed but not replaced by the DNA polymerase. To prevent shortening of the linear chromosomes with each round of replication, eukaryotes have developed a mechanism to maintain the telomeres. The enzyme telomerase is a ribonucleoprotein that consists of a protein and an RNA component (of 451 nucleotides in humans). Since it functions as a reverse transcriptase, as illustrated in Figure 1.11, the protein component is called telomerase reverse transcriptase (TERT). The RNA component of telomerase is a complementary sequence to the repeating telomeric sequence and directs the addition of nucleotides to the 3′ end of the DNA. This process of polymerization followed by translocation can be repeated several times before the telomere dissociates from the enzyme. The complementary strand can then be synthesized by the normal cellular machinery for lagging strand synthesis.

Somatic cells of multicellular organisms lack telomerase activity; thus, the telomeres shorten with each round of replication. Cells in culture can only undergo a certain number of cell divisions before they reach senescence and die. The absence of telomerase in somatic cells is at least part of the basis for aging in multicellular organisms. The presence of shortened telomeres in somatic cells used for the cloning of animals is also believed to be one of the reasons for the health problems found in these clones (Section 10.1.5, Box 10.1). Telomerase is active mainly in two types of cells: germ cells that must maintain intact telomeres and cancer cells that divide rapidly and would stop growing without a mechanism to prevent shortening of the telomeres. The senescence of somatic cells and the process of aging due to lack of telomerase activity can be viewed as mechanisms that protect multicellular organisms from cancer. Telomerase is an attractive target for new anticancer drugs, since its inhibition should prevent the uncontrolled growth of tumor cells.

1.3.2
Mutations

Mutations are changes in the genetic information. They may cause cancer or genetic disorders. A large number of human cancers (estimates reach up to 80%) arise from substances referred to as *carcinogens*, which induce mutations. Over 70 000 man-made chemicals are currently of commercial importance and 1000 new ones are introduced every year.

Germinal mutations occur in germline cells and are transmitted through the gametes to the progeny. In contrast, *somatic mutations* occur in somatic cells; thus, the mutant phenotype will occur only in the descendants of that cell and will not be transmitted to the progeny. While *spontaneous mutations* occur without any known cause, perhaps due to inherent metabolic errors or to unknown agents in the environment, *induced mutations* result from exposure of organisms to mutagens. The degeneracy and order in the genetic code outlined below help prevent many mutations from affecting the phenotype of the organism. These changes are called *neutral* or *silent mutations.*

Several types of mutations are known: These include substitutions of one base pair for another and deletions and insertions of one or more base pairs (Figure 1.12). The most common type of mutation is the substitution of one base pair for another. A *transition* replaces a pyrimidine with another pyrimidine or a purine with another purine. A *transversion* replaces a pyrimidine with a purine or a purine with a pyrimidine. Altogether, four different types of transitions and eight different types of transversions are possible in DNA.

An example of a germline mutation is a single base substitution in the ß-globin gene that results in sickle-cell anemia in homozygous individuals (Sections 3.1.2 and 6.2). This disorder is caused by an A to T substitution, causing the hydrophilic amino acid glutamic acid to be replaced with the hydrophobic amino acid valine at the sixth position of the ß-globin protein. The exchange promotes the noncovalent polymerization (aggregation) of hemoglobin, which distorts red blood cells into a sickle shape and decreases their elasticity.

Base pair insertions or deletions within the coding regions of genes usually lead to so-called *frameshift mutations* (except if three nucleotides or multiples thereof are added or removed), because they alter the reading frame of all the base pair triplets. Since the triplets specify codons in mRNA and amino acids in the encoded protein (Figure 1.12), frameshift mutations mostly result in the synthesis of nonfunctional protein products.

Mutations can be induced physically (e.g., by UV radiation), chemically, or biologically (e.g., by transposons or viruses). Chemical mutagens may be mutagenic for replicating DNA only (e.g., acridine dyes) or mutagenic for both replicating and nonreplicating DNAs (e.g., alkylating agents and nitrous acid). For example, treatment of DNA with nitrous acid causes point mutations by oxidative deamination. Thus, cytosine and adenine can be converted to uracil and hypoxanthine, respectively.

Insertions and deletions can arise from the treatment of DNA with intercalating compounds such as acridine orange. Intercalation of such an agent into DNA almost doubles the distance between two consecutive base pairs. As a consequence, the replication of the distorted DNA occasionally results in the insertion or deletion of one or more nucleotides.

Environmental factors, such as UV or ionizing radiation, and some chemical compounds can also cause physical damage to the DNA. For example, UV light promotes the formation of a cyclobutyl ring between adjacent thymine residues that leads to the formation of

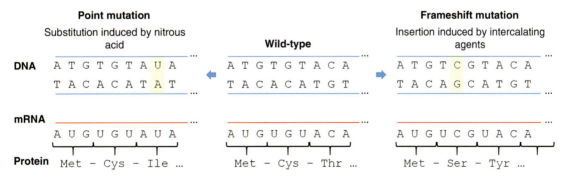

Fig. 1.12 Types of mutations. Substances such as nitrous acid can induce point mutations, that is, the substitution of one base by another. This may lead to a change of the codon and the incorporation of an incorrect amino acid into the protein. Frameshift mutations, induced, for example, by intercalating agents, occur by the introduction (or deletion) of one or multiple nucleotide(s). A reading frameshift changes all the codons and results in the synthesis of a nonsense protein.

intrastrand thymine dimers. Cytosine or cytosine–thymine dimers can also be formed, but at lower rate. These pyrimidine dimers all result in errors during transcription and replication. Treatment of DNA with ionizing radiation may lead to strand breakage. DNA damage may either be caused directly by radiation or by the formation of free radicals in the surrounding aqueous medium. Cells have developed several very effective mechanisms for the repair of DNA damage. Clinically, individuals with inherited disorders in the repair systems can develop disease such as *xeroderma pigmentosum*, which can lead to the early development of cancer through exposure to sunlight (Section 3.1.2).

The *Ames test* is a widely employed method to investigate the mutagenic potential of a substance. It is a quick and convenient bacterial test that serves as an alternative to time-consuming carcinogenesis tests on mice and rats; such evaluation in animals may take up to 3 years. In the assay, a special tester strain of *Salmonella typhimurium* that cannot synthesize histidine (denoted as his^-) is used. These bacteria are unable to grow in the absence of the amino acid. Bacteria of the tester strain are spread on a culture plate lacking histidine. In the presence of a mutagen, some bacteria revert to the his^+ phenotype, that is, they regain the ability to synthesize histidine and can grow and form visible colonies on the histidine-deficient medium. When this happens unusually frequently, it is an indication that the tested material is mutagenic.

1.3.3
Transcription

Transcription is the process of copying of DNA into RNA by the enzyme RNA polymerase (RNAP). Similar to replication, transcription relies on the transmission of genetic information by the pairing of complementary bases; however, transcription usually involves only copying of one of the two DNA strands, while both strands are doubled during replication. The DNA strand that serves as a template for the synthesis of RNA is called the noncoding or antisense strand, while the strand with the same sequence as the RNA is known as the coding or sense strand. As will be outlined in various chapters of this book, it has become apparent only recently that a great diversity of RNA molecules exists (Sections 7.2.4 and 13.4.2), all of which are generated by RNAPs. Animals contain three distinct types of RNAPs in the cell nucleus (as well as a separate mitochondrial RNAP), which transcribe different types of RNAs (Table 1.1).

Eukaryotic RNAPs are large complexes consisting of up to 14 subunits. They possess DNA unwinding activity, synthesize the new RNA molecule in the $5' \rightarrow 3'$

Table 1.1 Eukaryotic RNA polymerases.

Polymerase	Location	Transcribed RNAs
RNA polymerase I	Nucleolus	rRNA precursors
RNA polymerase II	Nucleoplasm	mRNA precursors; U1-, U2-, U4-, and U5-snRNAs
RNA polymerase III	Nucleoplasm	5S-rRNA, tRNAs, U6-snRNA, most miRNAs, and various other small RNAs

The table summarizes the location of the three eukaryotic RNA polymerases in the nucleus and the types of RNAs they transcribe.

direction, and carry out a proofreading step similar to the one described for DNA polymerases. The mode of action of the three RNAPs is somewhat different, and the following paragraphs will focus on the synthesis of mRNA precursors by RNAP II.

Transcription is initiated at specific sites of the genome known as promoters. RNAP II promoters contain certain sequence motifs, sometimes referred to as core promoter elements (Figure 1.13), such as a GC-rich region or the so-called TATA box (thymidine–adenosine–thymidine–adenosine) located 25–31 nucleotides upstream of the transcription start site (it is often overlooked that approximately two thirds of protein-encoding genes lack the TATA box!). The promoters of housekeeping genes that are expressed in all tissues have a GC box upstream of their transcription start site. However, this GC box is missing in structural genes selectively expressed in certain cell types. About half of the protein-coding genes have a conserved 7-nucleotide initiator element (Inr) that includes the first nucleotide of the transcribed RNA (+1). None of the core promoter elements are present in all promoters. Further elements, such as the CCAAT box located in the region −50 to −110 to the transcriptional start site, are indicated in Figure 1.13.

Additional elements, called enhancers, are required for the full activity of their cognate promoters. These control elements do not have fixed positions and orientations relative to their corresponding transcribed sequences. Enhancers are recognized by specific transcription factors that stimulate RNAP II to bind to the corresponding, but often distant, promoter. This requires that the DNA between the enhancer and the promoter forms a loop that permits simultaneous interaction of the transcription factors with the enhancer and the RNAP at the promoter site.

While a large number of promoters are ubiquitously active in all cell types, others are tissue specific. For example, α_1-antitrypsin is a protease inhibitor that is expressed only in the liver. Its promoter, abbreviated

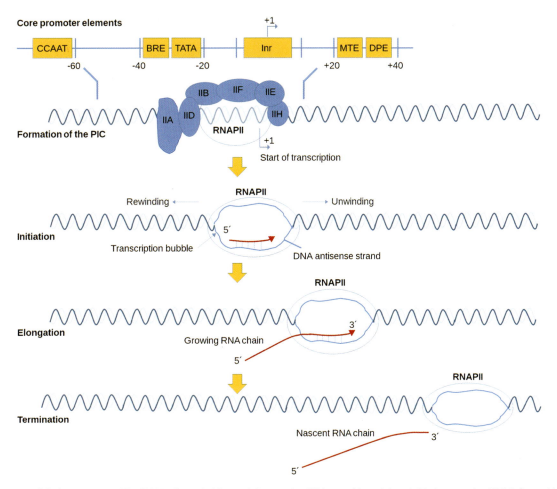

Fig. 1.13 Model of transcription. The DNA is shown in blue and the growing RNA strand in red. A preinitiation complex (PIC) is formed by binding of RNA polymerase II (RNAP II) and general transcription factors (blue circles; IIA, B, D-F, H) to the core promoter sequences. The initiation step includes unwinding of the double-stranded DNA to form an "open transcription complex." Elongation of the RNA chain occurs in the 5′ → 3′ orientation. During termination, the nascent RNA chain is released. CCAAT, CCAAT box; BRE, TFIIB recognition element; TATA, TATA box; Inr, initiator element; MTE, motif ten element; DPE, downstream promoter element.

as hAAT (human alpha antitrypsin), is active only in the liver and can be used for the tissue-specific expression of a transgene in gene therapeutic applications (Section 11.3).

The process of transcription can be divided into three stages: initiation of the synthesis of a new RNA chain, elongation of the chain, and termination of the transcription and release of the newly synthesized RNA molecule (Figure 1.13). During initiation, RNAP associates with the promoter sequences near the transcription start site. A so-called closed promoter complex is formed as the DNA at the transcription start site is still double stranded. Initiation of transcription is mediated by several proteins called *transcription factors*. Accurate transcriptional initiation of most genes requires the presence of six general transcription factors (GTFs). The designation of GTFs starts with the initials TF (for transcription

factor) followed by a Roman numeral (e.g., II indicates involvement in transcription catalyzed by RNAP II). The properties and functions of these multiprotein complexes are summarized in Table 1.2. Together with the RNAP, they form the so-called preinitiation complex (PIC).

Transcription factors are proteins that bind to specific DNA sequences, thereby controlling transcription. They perform this function alone or with other proteins in a complex by promoting (as an activator) or blocking (as a repressor) the recruitment of RNAP to specific genes. All transcription factors contain one or more DNA-binding domains (DBDs), which attach to specific sequences of DNA adjacent to the genes that they regulate. Approximately 2600 proteins (~10% of all protein-coding genes) encoded in the human genome contain DNA-binding domains, and most are presumed to

Table 1.2 Functions of the eukaryotic general transcription factors.

Factor	Number of unique subunits in humans	Function
TFIIA	3	Stabilizes TBP and TAF binding
TFIIB	1	Stabilizes TBP binding; recruits RNAP II; influences start site selection
TFIID:		Recognizes TATA box; recruits TFIIA and TFIIB
TBP	1	
TAFs	16	
TFIIE	2	Recruits TFIIH and stimulates its helicase activity; enhances promoter melting
TFIIF	2	Facilitates promoter targeting; stimulates elongation
TFIIH	10	Contains an ATP-dependent helicase that functions in promoter melting and clearance

TF: transcription factor; TBP: TATA-binding protein; TAF: TBP-associated factor.

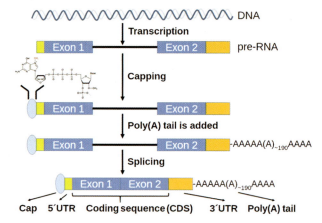

Fig. 1.14 Posttranscriptional processing of pre-mRNA. The products of transcription, the *primary transcripts*, can be altered by appending nucleotides to the 5′ end (capping) or the 3′ end (poly(A) tailing). Introns are removed during the splicing process. Posttranscriptional modifications lead to the formation of typical human mRNAs, including the untranslated regions (UTR).

function as transcription factors. Transcription factors are involved in numerous important biological roles, such as basal transcription regulation (GTFs TFIIA, B, D, E, F, H), the differential enhancement of transcription, developmental processes, response to intercellular signals (e.g., estrogen receptor transcription factors) or environmental factors (e.g., hypoxia inducible factor), and cell cycle control (oncogenic factors (e.g., myc) or tumor suppressor proteins).

RNA synthesis, such as DNA synthesis, occurs in the 5′ → 3′ direction (Figure 1.13). In the region being transcribed, the DNA double helix is unwound by RNAP so that the template strand can be transcribed into its complementary RNA strand. This complex is referred to as an "open promoter complex." During transcription, the DNA antisense strand transiently forms a short RNA–DNA hybrid with the nascent RNA chain. The transcription bubble travels along the DNA.

After initiation of transcription and the production of a short transcript, RNAP II changes into an elongation mode and leaves (clears) the promoter. A complex composed of six proteins known as "elongators" binds to the polymerase and assists with the synthesis of the mRNA. These proteins are dispensable, but transcription is accelerated in their presence.

In eukaryotes, a termination sequence for transcription has not been identified. A defined termination site is not required, since the transcript undergoes posttranscriptional processing (Figure 1.14): The major steps of maturation of the *primary transcript* are the addition of a "cap" at the 5′ end, appending a poly(A) tail to the 3′ end, and the removal of noncoding introns by a process denoted as splicing.

Eukaryotic mRNAs have a peculiar 7-methylguanosine at the 5′ end that is linked to the transcript via a 5′-5′ triphosphate bridge. This structure is called the "cap." It is added to the growing transcript before it is 30 nucleotides long. After formation of the 5′-5′ bridge, the guanine is methylated to produce 7-methylguanosine. Further nucleotides may be methylated at the 2′-OH-group of the ribose, generating Cap1 (first nucleotide of the transcript is methylated) or Cap2 (first and second nucleotide of the transcript are methylated).

Almost all mammalian mRNAs have 3′-poly(A) tails consisting of ~250 nucleotides. The poly(A) tail is generated from ATP by poly(A) polymerase, a template-independent RNA polymerase. The main function of the poly(A) tail is protection against nucleases. Poly(A) tails are added in a two-step reaction: In the first step, the transcript is cleaved 15–25 nucleotides downstream of an AAUAAA sequence, and less than 50 nucleotides before a U- or U- and G-rich sequence. In a subsequent reaction, the poly(A) tail is generated from ATP in a stepwise action by poly(A) polymerase. The cleaving enzyme and poly(A) polymerase are located within a large protein complex. Hence, the cleaved mRNA cannot dissociate before polyadenylation. In the cytoplasm, the poly(A) tail enhances the stability of the mRNA by interacting with the poly(A)-binding protein (PABP).

In eukaryotic genes, coding sequences (exons) are interspersed with noncoding sequences (introns). The primary transcripts initially transcribed are heterogeneous in length (*heterogeneous nuclear RNA (hnRNA)*). After capping and polyadenylation, the introns are excised and the exons are spliced together to form the mature mRNA (Figure 1.15). On average in humans, exons are 150 nucleotides long. Most of them are <300 nucleotides in length, but the largest exon (from the muscle protein titin) is 17 106 nucleotides in length.

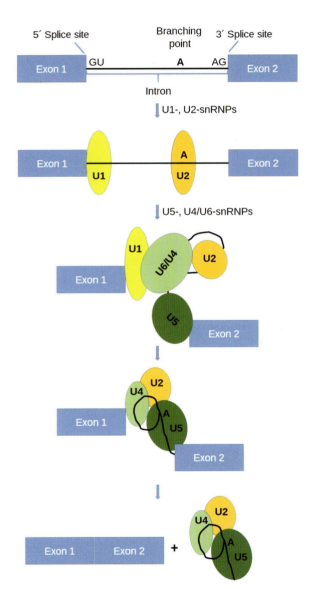

Fig. 1.15 The splicing reaction. In the first step, U1-snRNP binds to the 5′ splice site and U2-snRNP attaches to the branch site. Afterward the complete spliceosome assembles and cleaves the transcript at the 5′ splice site. The 5′ end of the intron is joined to the adenine in the branch position to form a lariat. U1- and U4-snRNP leave the complex. Finally, the 3′ splice site is cleaved. The 5′ end of exon 2 is joined to the 3′ end of exon 1 and the lariat structure is released.

Introns are usually much longer. On average, they are 3500 nucleotides in length, but they may be as long as 2.4 million nucleotides. The number of introns in a gene in the human genome averages ∼8 and can vary from none to more than 200.

Splicing occurs via two transesterification reactions. The first one yields a 2′,5′-phosphodiester bond between an adenosine at the branch point and the 5′ splice site. This process results in the formation of a lariat structure, as schematically shown in Figure 1.15. The second transesterification reaction splices the two exons together and releases the intron.

The splicing reactions are mediated by the *spliceosome*, which consists of the pre-mRNA, small nuclear ribonucleoprotein complexes (snRNPs) (commonly pronounced "snurps") and a number of pre-mRNA-binding proteins. In total, the spliceosome contains more than 100 proteins and five RNA molecules, called small nuclear RNAs (snRNAs). The snRNP complexes are named U1, U2, U4, U5, and U6, where U indicates that the RNAs are uracil-rich.

The composition of a protein encoded by a gene can be modulated by the selection of alternative splice sites. The most common mode of alternative splicing in mammalian cells is exon skipping. This is where an exon may either be spliced out of the primary transcript or preferentially retained (Figure 1.16a). Further modes of alternative splicing are the mutual exclusion of exons, the use of alternative splice donor or acceptor sites, and the retention of an intron. For the muscle protein α-tropomyosin, alternative splicing leads to the generation of seven tissue-specific variants encoded by a single gene (Figure 1.16b).

Alternative splicing is found in all multicellular organisms and is especially prevalent in vertebrates. It has been estimated that up to 95% of all human genes undergo alternative splicing. On average, each pre-mRNA is spliced to give three different variants. Splice variants may control the subcellular localization of a protein or whether it is soluble or hydrophobic. The selection of splice sites is tissue- and developmental stage-specific and, therefore, tightly regulated. Several studies have stated that 30–60% of the genetic diseases in humans are pre-mRNA splicing defects caused by point mutations. In addition, tumor progression is correlated with changes in the protein levels implicated in alternative splice site selection.

After completion of the various steps of post-transcriptional processing, a typical, mature eukaryotic mRNAs consists of I: the cap at the 5′ end; II: a short 5′ untranslated region (UTR); III: the coding sequence (CDS); IV: the 3′ UTR; and V: the poly(A) tail (Figure 1.14).

(a)

(b)

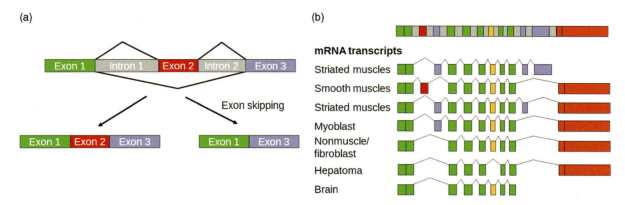

Fig. 1.16 Alternative splicing. (a) Exon skipping. Exon skipping is the most common mode of alternative splicing in mammals. In this case, an exon may be spliced out of the primary transcript or retained. (b) Alternative splicing of rat α-tropomyosin. Seven α-tropomyosin variants are produced in a tissue-specific manner by alternative splicing. The thin lines indicate the positions of introns before splicing. Tissue-specific exons are indicated with a color code: Green boxes represent exons expressed in all tissues; the exon indicated by a brown box is expressed in smooth muscles (SM); purple boxes depict exons expressed in striated muscles; variably expressed exons are depicted as yellow boxes. (Part (b) adapted from Ref. [1] with kind permission from John Wiley & Sons, Inc.)

Similar to mRNAs, rRNAs and tRNAs undergo post-transcriptional processing. The eukaryotic genome typically has several hundred tandemly repeated copies of rRNA genes that are contained in small, dark-staining nuclear bodies, the nucleoli (see nucleolus above). The nucleolus is a nonmembrane-bound substructure, in which rRNA transcription and processing and ribosomal subunit assembly take place. The primary rRNA transcript is about 7500 nucleotides in length and contains three rRNAs separated by spacer sequences.

In the first stage of processing rRNAs, the 45S RNA precursor is specifically methylated at numerous sites (106 sites in humans). In addition, many pre-rRNA uridines (95 in humans) are converted to an isomeric nucleoside called pseudouridine, which is characterized by a C-glycosidic bond instead of the normal *N*-glycosidic bond. The subsequent cleavage and trimming of the 45S RNA is catalyzed by enzymes possessing RNase activities. Some bases in rRNAs are methylated, a process that is guided by members of the large family of small nucleolar RNAs (snoRNAs). The length of these snoRNAs varies from 70 to 100 nucleotides. They contain segments of 10–21 nucleotides that are complementary to fragments of the mature rRNAs that contain the nucleotides to be methylated at the 2′ OH position. Methylation is mediated by a complex of at least six nucleolar proteins, including *fibrillarin*, the likely methyltransferase. Eukaryotic ribosomes contain four different rRNAs. While the 28S, 18S, and 5.8S rRNAs are processed as described, the fourth type, the 5S rRNA, is separately processed in a manner similar to tRNAs.

tRNAs deliver amino acids to the ribosome, the protein synthetic machinery. They consist of ~80 nucleotides and are commonly represented as a two-dimensional *cloverleaf structure*, although their three-dimensional structure was determined to be L-shaped. tRNAs contain a substantial number of modified bases (Figure 1.17). Mature tRNAs have the highest content of covalent modification diversity of all classes of RNA molecules. Up to 25% of the nucleotides of a tRNA are covalently modified. For accurate translation, the correct amino acid needs to be attached to its tRNA; the amino acid is linked to the 3′-terminal CCA sequence. This reaction is catalyzed by *aminoacyl-tRNA synthetases* that specifically link amino acids to the 3′-terminal ribose residue of their respective tRNAs to form aminoacyl-tRNAs. The *anticodon* that is complementary to the mRNA codon specifying the tRNA's amino acid is located in the loop of the cloverleaf structure opposite to the stem containing the terminal nucleotides.

Primary tRNA transcripts contain one to five tRNAs. The 5′ end of the tRNA is processed by RNase P, which consists of an RNA and a protein component. An important feature of RNase P is that the RNA component is the enzyme's catalytic subunit, that is, it functions as a ribozyme (Section 13.2). Endonucleases cleave the tRNA precursors close to the 3′ end. Subsequently, exonucleases such as RNase D remove nucleotides stepwise until they reach the 3′ end of the tRNA. Finally, CCA ends are posttranscriptionally added to the tRNA by the enzyme *tRNA nucleotidyltransferase*, which sequentially adds two Cs and an A using CTP and ATP, respectively, as substrates. Many eukaryotic primary tRNA transcripts contain introns adjacent to their anticodons that must be removed during the processing steps.

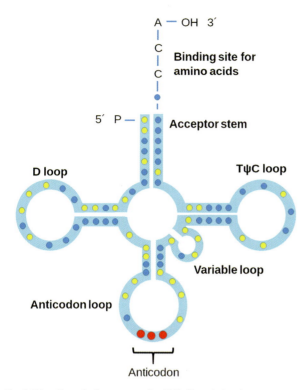

Fig. 1.17 Cloverleaf structure of a tRNA. Blue circles show nonmodified RNA nucleotides. Red circles reflect nucleotides forming the anticodon sequence. The average tRNA has 13 modified residues. The putative positions of modified nucleotides are highlighted by yellow circles. These modifications are highly conserved in eukaryotes, including humans.

1.3.4
Epigenetic Regulation of Gene Expression

The cell uses several mechanisms to regulate gene expression, for example, by the actions of transcription factors. A recent focus of intensive research is the *epigenetic* modulation of gene activity. The term "epigenetics" describes heritable changes in the level of gene expression that are not caused by changes in the DNA sequence. In a broader sense, epigenetics also refers to stable, long-term alterations in transcriptional levels that are not necessarily heritable. The term epigenetics (*epi*, Greek: outside of, around) was coined in 1942 by C.H. Waddington and describes phenomena other than those determined by the DNA sequence. Epigenetic processes are crucial for cellular differentiation. All cells of an organism contain the identical genetic information, but the selective activation and inhibition of genes determines their differentiation into a specific cell type, for example, a liver cell, a muscle cell, or a neuron.

Two main mechanisms produce epigenetic alterations: DNA methylation and histone modification (Figure 1.18), each of which influences gene expression without altering the DNA sequence. In addition, positioning of the nucleosomes relative to the transcription start sites and additional regulation by small noncoding RNAs such as miRNAs (Section 13.4) are sometimes considered mechanisms of epigenetic regulation.

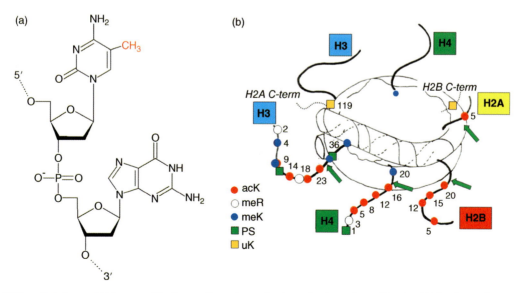

Fig. 1.18 DNA methylation and histone modification are the main mechanisms of epigenetic modification. (a) The chemical structure of a methylated CpG dinucleotide. The methyl group is marked in red. (b) The properties of histones are altered by posttranslational modifications. The figure shows the N- and C-terminal domains of histone proteins in the nucleosome core particle. Sites marked by green arrows are susceptible to cutting by trypsin in intact nucleosomes. Modifications are indicated by colored symbols: acK: acetyl lysine; meR: methyl arginine; meK: methyl lysine; PS: phosphoryl serine; uK: ubiquitinated lysine. (Part (b) adapted from Ref. [4] with kind permission from Elsevier.)

DNA methylation refers to the well-characterized chemical modification of nucleotides that influences the expression levels of certain genes. The only methylated base found in eukaryotic DNA is cytosine. This form of methylation occurs largely in dinucleotides of cytosine and guanosine (abbreviated as CpG dinucleotides, where p indicates the phosphate group connecting the nucleosides (Figure 1.18a)). CpG dinucleotides tend to cluster in regions called CpG islands. DNA methyltransferase (DNMT) enzymes transfer a methyl group to the C5 position of cytosines. The methyl group can be removed by demethylases.

CpG islands are frequently found in the 5′ region of genes, where promoters are situated. Methylation at cytosine switches off gene expression. In contrast, genes that are transcriptionally active will be hypomethylated at the CpG islands in their upstream regions. Methylation represses gene expression by recruiting chromatin corepressors and inhibiting the binding of transcription factors.

DNA methylation is stably maintained during cell division: The parental DNA strand directs the generation of the identical methylation pattern in its daughter strand. This process, also known as *maintenance methylation*, is mediated mainly by the DNA methyltransferase DNMT1 and results in the stable inheritance of methylation patterns in a cell line. Thus, all the cells will have the identical differentiated phenotype.

While methylation patterns are maintained during somatic cell division, they vary during the stages of embryologic development. Initially, the level of DNA methylation is high in mature gametes, but it is nearly eliminated in the blastocyst stage of development (i.e., approximately 200–300 cells, when the embryo embeds itself into the endometrical layer of the uterine wall). The level of DNA methylation then rises again. This *de novo* methylation is mediated by two specific DNA methyltransferases (named DNMT3a and DNMT3b). At the gastrula developmental stage (when the three germ layers ectoderm, mesoderm, and endoderm are formed), the level of DNA methylation rises to the adult level, where it remains for the lifetime of the individual. However, the CpG islands of germline cells are an exception, as they remain largely unmethylated throughout life.

There is a clear difference in the levels of DNA methylation between embryonic and somatic cells. This has consequences for patterns of gene expression. These differences probably explain (at least in part) the high failure rates of cloning experiments in mammals by somatic cell nuclear transfer (SCNT) (Section 10.1.5, Box 10.1). There, a nucleus of a differentiated somatic cell is injected into a denucleated egg. However, the epigenetic pattern of the transferred genomic material differs from that present naturally at this early stage of embryological development. As a consequence, only a small fraction of cloned embryos develop normally and even those that survive to birth usually have a variety of abnormalities. For example, the live newborn may be unusually large, a condition known as the large offspring syndrome.

DNA methylation has also been found to play an important role in cancer development, because of the induction of classic genetic mutations. Methylated cytosines mutate to thymidines at a comparatively high frequency, a mutational change frequently found in human cancers. These mutations may convert proto-oncogenes to oncogenes or inactivate tumor suppressor genes. In addition, many cancer cells also have a distorted epigenetic landscape. Typical changes include a profound global loss of DNA methylation, which occurs concomitantly with specific patterns of extended methylation at CpG islands of certain gene promoters. Hypermethylation at the CpG islands of tumor suppressor genes further favors tumor initiation and progression, whereas hypomethylation can induce aberrant expression of oncogenes and induce loss of imprinting (LOI) in some loci (see below).

Drugs are available that affect the epigenetic modulation of CpG motifs. An example is decitabine (5-aza-2′-deoxycytidine, trade name Dacogen) that hypomethylates DNA by inhibiting the DNA methyltransferase. This drug is indicated for myelodysplastic syndrome (MDS) and acute myeloid leukemia (AML). However, since the unspecific inhibition of DNA methylation has widespread consequences, the drug has many toxic side effects in treated patients.

After DNA methylation, posttranslational histone modification is the second major mechanism of the epigenetic fine-tuning of gene expression (Figure 1.18b). Histones are positively charged proteins that package and order genomic DNA. Modifications at the N-terminal tails of the histones influence the ability of genes to transcribe DNA into RNA and also affect replication, DNA repair, and chromosomal condensation. Posttranslational modifications include methylation, acetylation, phosphorylation, and ubiquitination. These modifications (except for methylation) produce negative charge, thus weakening the interaction of the histone with the negatively charged DNA. As a consequence, the chromatin becomes decondensed. In contrast, methyl groups increase the hydrophobicity of the side chain and tend to stabilize chromatin structure. The histone tails also interact with specific proteins, which further change the transcriptional accessibility of the associated genes. The different combinations of modifications are sometimes regarded as a *histone code*, where a certain modification pattern evokes a specific biological effect.

Posttranslational modification of histone proteins is carried out by specific enzymes, including histone acetyltransferases (HATs) and histone methyltransferases. These processes are reversible as histone deacetylases (HDACs), and the more recently discovered demethylases, may remove the substituents.

Aberrations of the histone-modifying proteins may have severe clinical consequences. An example is the Rubinstein–Taybi syndrome, which is characterized by short stature, distinctive facial features, and mental retardation. This condition is caused by a mutation in the CREB-binding protein (CBP), which binds to the cAMP-response element-binding protein (CREB). CBP has histone acetyltransferase activity and opens the chromatin structure. It also interacts with additional proteins of the transcription complex and controls the activity of many genes. CBP has an important role in regulating cell growth and division and is essential for normal fetal development.

Alterations in histone modification have also been found to promote tumor growth. A common alteration is the global reduction of an acetylated lysine residue in histone H4. The loss of acetylation is mediated by histone deacetylases that are overexpressed or mutated in different tumor types. Vorinostat (trade name Zolinza), used for the treatment of cutaneous T-cell lymphoma, was the first HDAC inhibitor that was approved by the FDA. In some cases, chromosomal translocations modulate the activity of histone acetylases and affect the global balance of histone acetylation. Altered methylation of histones is primarily found in hormone-dependent types of cancer, such as breast and prostate cancers.

Another important function of epigenetic regulation is *genomic imprinting*, a process by which certain genes are expressed in a manner dependent on the parent from which they are inherited. Since mammalian embryos carrying two copies of either only female or only male origin do not develop into viable offspring, the diploid genome always consists of genetic material obtained from both a male and a female parent. The vast majority of genes are expressed from both alleles simultaneously. However, a small proportion of genes are imprinted, that is, their expression originates from only one allele. In humans, only roughly 200 genes are known or predicted to be expressed in a manner dependent upon their parental origin. An example is insulin-like growth factor 2, which is only expressed from the paternal allele. Imprinted genes lie within clustered regions of the genome; the two largest regions are located on chromosomes 11 and 15 (11p15 and 15q11). Genomic imprinting employs DNA methylation and histone modulation to achieve monoallelic gene expression without altering the genetic sequence. Several imprinted genetic

disorders have been described. For example, the Prader–Willi syndrome (PWS) and the Angelman syndrome (AS) are both associated with genetic defects in the 15q11-13 chromosomal region (Section 3.1, Box 3.1). However, the phenotypic consequences depend on whether the mutation is inherited from the father or from the mother. Paternal inheritance of the alteration is associated with the PWS, while inheritance of the same mutation from the mother leads to the development of AS.

Epigenetic modifications are also important for a process called X-inactivation (or lyonization after the discoverer of this phenomenon, Mary Lyon). As females carry two X chromosomes, one of them must be inactivated to prevent them from having twice as many X chromosomal gene products as males, who only posses a single copy of the X chromosome. This process is also known as *dosage compensation*. Although most of the X chromosome is inactivated, some segments escape this process because there are comparable genes on the Y chromosome.

X-inactivation is initiated from a specific site on the chromosome, which expresses a long noncoding RNA denoted as the X-inactive specific transcript (Xist). The inactive X chromosome is coated by this RNA, whereas the active X chromosome is not. It becomes highly condensed and visible under a microscope in the nucleus as the so-called Barr body. Once an X chromosome is inactivated, it will remain silenced throughout the lifetime of the cell and in all of its descendants in the organism. The inactivation occurs as a random process in the human embryo approximately 16 days after fertilization. Therefore, female cells are mosaics; some cells have an inactive maternal X chromosome, while others have an inactive paternal chromosome. The process of X-inactivation is important for the symptomatic occurrence of X-linked recessive disorders in females (Section 3.1.3).

In addition to the inherited disorders described above and in the development of cancer, epigenetic alterations also play an important role in some neurological and autoimmune diseases. The findings that epigenetics are highly relevant to health and disease have triggered large-scale studies, sometimes referred to as epigenomics. These initiatives have succeeded in characterizing genome-wide DNA methylation patterns at single-nucleotide resolution (known as the human DNA methylome), and have also found new variants and modifications of histones.

1.3.5
Translation

The genetic code is a set of rules that defines how the information encoded within genetic material is translated into proteins. As only four bases must code for 20 amino acids, multiple bases (*codons*) are required to specify a single amino acid. Cells use a three-base code that can theoretically code for $4^3 = 64$ amino acids. Therefore, most of the amino acids are specified by more than one codon, that is, the code is *degenerate*: While most of the amino acids are specified by two, three, or four codons, the amino acids arginine, leucine, and serine are each specified by six codons. Only methionine and tryptophan are represented by a single codon. Codons that specify the same amino acid are termed *synonyms*.

Table 1.3 summarizes the standard genetic code. While 61 triplets correspond to particular amino acids, 3 triplets indicate chain termination. The stop codons UAG, UAA, and UGA are referred to as amber, ochre, and opal, respectively. The codon AUG (and less frequently GUG) serves as the initiation codon of translation. However, the codons also specify the amino acid residues methionine and valine, respectively, at internal positions in the polypeptide chains.

The standard genetic code is (almost) universal, that is, it is used by bacteria as well as eukaryotes. This phenomenon forms the basis of genetic engineering. The genetic code not only facilitates research but also permits the production of human proteins in bacteria (Chapter 10). It came as a surprise when research in the early 1980s revealed that certain mitochondria and ciliated protozoa employ a slightly different genetic code.

The code as depicted in the table is nonrandom. Changes in the first position tend to specify similar amino acids. Most synonyms differ only in their third nucleotide, which minimizes the deleterious effects of mutations. Many point mutations at the third codon position do not change the encoded amino acid.

While transcription and RNA processing take place in the nucleus, translation occurs in the cytoplasm (Figure 1.19). The primary sites of biological protein synthesis are ribosomes, which decode an mRNA and catalyze the formation of peptide bonds between amino acids that are delivered by tRNAs. Ribosomes are large and complex particles consisting of two subunits composed of protein and RNA molecules. The RNAs account for approximately two thirds of the total mass. The large subunit is mainly involved in mediating biochemical tasks such as catalyzing the peptidyl transferase reaction. The small subunit is responsible for recognition processes such as mRNA and tRNA binding. Messenger RNAs are read in the $5' \rightarrow 3'$ direction and encode the polypeptide, which is synthesized from the N-terminus to the C-terminus. For protein synthesis, the proper tRNA is selected only through codon–anticodon interactions, while the aminoacyl group does not participate in the process.

In 2000, the X-ray structure of the 50S ribosomal subunit at atomic resolution was published. Soon thereafter, the X-ray structure of the 30S ribosomal subunit was solved. For elucidating the structure of the ribosome, Ada Yonath, Thomas Steitz, and Venkatraman Ramakrishnan were awarded the Nobel Prize in Chemistry in 2009.

Their structure analysis revealed that the distribution of the proteins in the two ribosomal subunits is not uniform. The vast majority of the proteins are located on the surfaces of their subunits. The interface between the two subunits, particularly those regions that bind the tRNAs and mRNAs and form the new peptide bonds, is largely devoid of proteins. Thus, rRNA rather than the ribosomal proteins has the major functional role in ribosomal processes. Ribosomes have three functionally distinct tRNA-binding sites, known as A (aminoacyl) site, P (peptidyl) site, and E (exit) site (Figure 1.20). Ribosomes are tandemly arranged on mRNAs. The individual ribosomes in these *polyribosomes* (*polysomes*) are separated by gaps of 50–150 Å, producing a maximum density of ~1 ribosome per 80 mRNA nucleotides.

Protein biosynthesis requires a number of protein factors in addition to the ribosomal proteins. These soluble proteins can be grouped into three classes: initiation factors, elongation factors, and release factors (RFs).

Eukaryotic initiation is a complex process that involves at least 11 initiation factors (abbreviated as eIF*n*; e for

Table 1.3 The genetic code.

First position (5′ end)	Second position				Third position (3′ end)
	U	C	A	G	
U	Phe	Ser	Tyr	Cys	U
	Phe	Ser	Tyr	Cys	C
	Leu	Ser	Stop	Stop	A
	Leu	Ser	Stop	Trp	G
C	Leu	Pro	His	Arg	U
	Leu	Pro	His	Arg	C
	Leu	Pro	Gln	Arg	A
	Leu	Pro	Gln	Arg	G
A	Ile	Thr	Asn	Ser	U
	Ile	Thr	Asn	Ser	C
	Ile	Thr	Lys	Arg	A
	Met	Thr	Lys	Arg	G
G	Val	Ala	Asp	Gly	U
	Val	Ala	Asp	Gly	C
	Val	Ala	Glu	Gly	A
	Val	Ala	Glu	Gly	G

The amino acids are given by their three-letter codes. The stop codons and the start codons encoding methionine and valine, respectively, are marked.

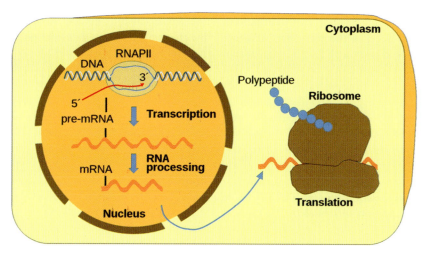

Fig. 1.19 Subcellular localization of transcription, RNA processing, and translation in a eukaryotic cell. RNAs are transcribed and processed in the nucleus. After their export to the cytoplasm, ribosomes translate the genetic information into proteins.

eukaryotic), including the Met-tRNA$_i$^{Met} (initiator tRNA) that delivers a methionine that is never formylated in eukaryotes (as it is in the case of prokaryotes). Eukaryotic mRNAs also lack the Shine–Dalgarno sequence that is present in prokaryotic mRNAs, but nearly all eukaryotic mRNAs have the m^{7}G cap described above. Furthermore, they are almost always monocistronic, that is, they contain the genetic information for a single protein only, so that they initiate translation at their leading AUG. The AUG is embedded in the consensus sequence GCCRCCAUGG, known as the Kozak sequence (R denotes one of the purines adenine or guanine). Initiation factor eIF4E binds the cap by intercalating between two tryptophan residues. In addition, eIF4G interacts with the poly(A)-binding protein bound to the mRNA's

poly(A) tail, circularizing the mRNA. Additional initiation factors subsequently bind to the eIF4F–mRNA complex.

The resulting complex then associates with the ribosomal 43S preinitiation complex. This complex translocates along the mRNA in an ATP-dependent process called *scanning* until it encounters the AUG initiation codon, thereby yielding the 48S initiation complex. Hydrolysis of GTP results in release of all the initiation factors. Finally, the large 60S subunit joins the mRNA-bound Met-tRNA$_i$^{Met}–40S subunit complex in a GTPase-driven reaction to yield the 80S ribosomal initiation complex.

Ribosomes elongate polypeptide chains in a three-stage reaction cycle that adds amino acid residues to a growing polypeptide's C-terminus: These three stages

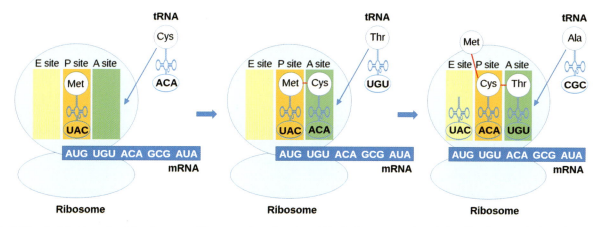

Fig. 1.20 Protein biosynthesis by ribosomes. Ribosomes elongate polypeptides in the process of translation. The ribosome harbors three functional sites, the E(xit) site, the P(eptidyl) site, and the A(minoacyl) site. tRNAs deliver the amino acids. The interaction between the codon of the mRNA and the anticodon of the tRNA determines the incorporation of the correct amino acid. A peptide bond is formed between the growing polypeptide chain and the incoming amino acid.

are decoding, transpeptidation (peptide bond formation), and translocation of tRNAs. The process occurs at a rate of 10–20 residues/s and requires several elongation factors.

In the decoding stage, the ribosome selects and binds an aminoacyl-tRNA whose anticodon is complementary to the mRNA codon at the A site (Figure 1.20). A complex of GTP and the elongation factor EF-Tu is associated with an aminoacyl-tRNA. The resulting ternary complex binds to the ribosome. GTP is hydrolyzed and while the aminoacyl-tRNA is bound to the A site, EF-Tu-GDP and P_i are released. The GDP in EF-Tu is replaced by GTP in a reaction mediated by the elongation factor EF-Ts, which functions as a guanosine exchange factor (GEF).

The next step of the elongation cycle is the formation of the peptide bond through the nucleophilic displacement of the P-site tRNA by the amino group of the 3′-linked aminoacyl-tRNA in the A site (Figure 1.20). The nascent polypeptide chain is thus lengthened by one residue at its C-terminus and transferred to the A-site tRNA.

In the final translocation stage of the elongation cycle, the now deacylated, uncharged P-site tRNA is transferred to the E site and expelled. Simultaneously, the peptidyl-tRNA in the A site, together with its bound mRNA, is moved to the P site. The translocation process requires the participation of the elongation factor EF-G, which is bound to the ribosome along with GTP, and is released only after the hydrolysis of GTP, which generates the free enthalpy for the translocation of the tRNA.

The three stop codons UAA, UGA, and UAG do not normally have corresponding tRNAs. Thus, they indicate that the newly synthesized polypeptide chain should be released. In eukaryotes, all three stop codons are recognized by a single *release factor*, the eukaryotic release factor 1 (eRF1). This protein functions as a tRNA mimic. Through its binding to the A site of the ribosome, eRF1 triggers the hydrolysis of the peptidyl-tRNA, resulting in the release of the peptide chain. eRF3, a second eukaryotic release factor, accelerates this process in a GTP-dependent manner. Finally, ribosome recycling occurs. This process is mediated by components of the initiation complex (eIF3, eIF1, and eIF1A).

Antibiotics are substances produced by bacteria or fungi that inhibit the growth of other organisms. They have many metabolic targets, but most antibiotics, including many used as drugs, block translation. Antibiotics are specific for bacterial ribosomes and bind to different regions of the ribosome, and block different ribosomal processes such as peptidyl transferase activity, translocation, and elongation. Antibiotic mechanisms of action are extensively discussed in Section 6.1.2.

Virtually all proteins are modified after their synthesis by the ribosome. Posttranslational protein processing may be reversible or irreversible. A common irreversible modification is the proteolytic cleavage of peptide bonds. The first step, occurring shortly after release of the newly synthesized protein from the ribosome, is the removal of the leading methionine. In addition, many proteins are synthesized as inactive precursors that are activated by limited proteolysis (examples are digestive enzymes such as trypsinogen, factors of the blood coagulation pathway, and insulin). The inactive proteins are called *proproteins* or *zymogens*.

In addition to activation by limited proteolysis, many proteins are processed by covalent modification of their amino acid residues. An example of this is the glycosylation of proteins; this type of modification promotes several processes, such as correct protein folding, enhancement of protein stability, and cell–cell adhesion. Glycosylation occurs in transmembrane proteins, lysosomal proteins, and proteins that are destined to be secreted. Proteins are initially synthesized with an N-terminal *signal peptide*, which is recognized by the signal recognition particle (SRP), a ribonucleoprotein. SRP mediates translocation of the nascent protein into the ER. Proteins bearing a signal peptide are called *preproteins* (or, if they also contain propeptides, as prepro proteins). The signal peptide is cleaved off once the peptide chain is translocated through a protein pore known as a *translocon*.

Proteins may be modified either by *O*-glycosylation at serine or threonine residues or by *N*-glycosylation at asparagine residues. The latter is initiated in the ER and continued in the Golgi apparatus. In contrast, *O*-glycosylation occurs only in the Golgi apparatus. Based on the glycosylation pattern, the proteins are sorted for their final destination, which may be the plasma membrane, secretory vesicles, or lysosomes.

In addition to glycosylation, many other types of posttranslational modifications have been described. Among them are the acetylation of the N-terminus of the protein or of lysine or serine residues, hydroxylation, methylation, or the addition of myristic acid, a common saturated fatty acid. Another common posttranslational modification is the reversible phosphorylation of serine, threonine, and tyrosine residues. Phosphorylation status modulates the activity of many enzymes.

1.3.6
Protein Degradation

Proteins are continuously synthesized and degraded. This turnover eliminates abnormal proteins whose accumulation would be harmful to the cell; it also permits the regulation of cellular metabolism by eliminating superfluous enzymes. The half-lives of enzymes vary greatly. While short-lived enzymes (e.g., ornithine decarboxylase and phosphoenolpyruvate (PEP) carboxylase) have half-lives of 0.2–5 h, long-lived enzymes (e.g., aldolase and lactate dehydrogenase) can have a half-life of more than 100 h.

Eukaryotic cells have two systems for protein degradation: a lysosomal mechanism and a cytosolic ATP-dependent system. As already described, lysosomes are membrane-encapsulated organelles that maintain an acidic internal pH and contain more than 50 hydrolytic enzymes, including a number of proteases collectively known as *cathepsins*. Lysosomal protein degradation is normally nonselective.

In addition to the lysosomal system, an ATP-dependent, targeted proteolytic system exists in the cytosol. The protein *ubiquitin*, consisting of 76 amino acid residues, is the marker for those proteins to be degraded. Ubiquitin, as its name suggests, is a ubiquitous, highly conserved abundant protein in eukaryotes. Proteins that are selected for degradation are initially marked by covalent linkage to ubiquitin. This process occurs in a three-step pathway (Figure 1.21a). At first, ubiquitin's terminal carboxyl group is conjugated to the *ubiquitin-activating enzyme* (E_1) in an ATP-depending reaction. Humans have only one type of E_1 protein. The ubiquitin is then transferred to one of the *ubiquitin-conjugating enzymes* (E_2). Mammals have about 20 E_2 proteins. Afterward, *ubiquitin-protein ligase* (E_3) transfers the activated ubiquitin from E_2 to an ε-amino group of a lysine residue of a target protein, forming an isopeptide bond. Cells contain many different E3 proteins, each of which mediates the ubiquitination of a specific set of proteins.

For efficient degradation, the target protein must be covalently linked to more than one ubiquitin. Typically, a chain of at least four tandemly linked ubiquitin molecules is attached to the target protein. Lys-48 of each ubiquitin peptide forms an isopeptide bond with the C-terminal carboxyl group of the next ubiquitin. The poly-ubiquitin chains can consist of 50 or more ubiquitin molecules.

A ubiquitinated protein is proteolytically degraded to short peptides in an ATP-dependent process. This process is mediated by a large multisubunit protein complex called the *26S proteasome* (Figure 1.21b). The 26S proteasome is composed of the 20S proteasome, which forms a hollow barrel and carries out the proteolysis of condemned proteins, and the 19S caps. The 19S cap consists of approximately 18 subunits. These caps may be attached to one or both ends of the 20S proteasome. The caps recognize ubiquitinated proteins, unfold them, and feed them into the 20S proteasome in an ATP-dependent manner.

1.4
Biological Communication

Intercellular communication is an essential prerequisite for the functioning of multicellular organisms. The

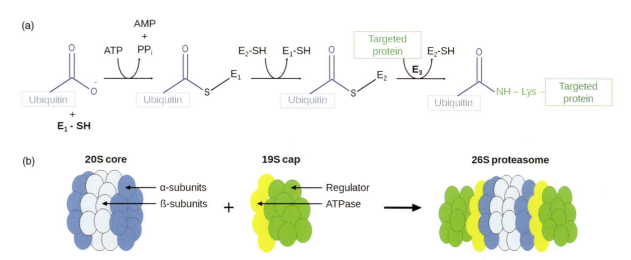

Fig. 1.21 Protein degradation by the proteasome. (a) Attachment of ubiquitin to a protein targeted for degradation. In the first step of the process, ubiquitin is covalently linked to the E_1 protein in a reaction driven by ATP hydrolysis. Subsequently, the activated ubiquitin is transferred to the sulfhydryl group of the E_2 protein. The final step, catalyzed by E_3, is the transfer of ubiquitin from E_2 to the target protein, which is then marked for degradation by the proteasome. (b) Structure of the proteasome. The proteasome consists of a central 20S barrel and 19S caps. Together, these components form the complete 26S proteasome. E_1: ubiquitin-activating enzyme; E_2: ubiquitin-conjugating enzyme; E_3: ubiquitin ligase.

following section will introduce the concepts of signal transmission between neurons, long-range and short-range intercellular signaling by hormones, and transduction of an extracellular signal across the cell membrane.

1.4.1
Neurotransmitters

The nervous system is characterized by a special mechanism that permits passage of a signal from a neuron to another cell, which may be another neuron, or an altogether different cell type. This occurs at a specific structure called a *chemical synapse* (Figure 1.22). When a nerve impulse in a presynaptic cell reaches the synapse, it triggers the fusion of synaptic vesicles, which contain neurotransmitters, with the presynaptic membrane. The neurotransmitter is then released via exocytosis into the synaptic cleft, and diffuses across it in less than 0.1 ms. The neurotransmitter then binds to specific receptors on the postsynaptic membrane, which trigger the continuation of the impulse in the postsynaptic cell. Due to the complexity of receptor signal transduction, chemical synapses can have many effects on the postsynaptic cell.

The known neurotransmitters are mainly classified into three groups (Table 1.4). The common transmitters

Table 1.4 Characteristics of major neurotransmitters.

Class	Neurotransmitter	Function	Secretion sites
Biogenic amines	Acetylcholine	Excitatory for skeletal muscles; excitatory or inhibitory at other sites	CNS, PNS, neuromuscular junction
	Norepinephrine	Excitatory or inhibitory	CNS, PNS
	Dopamine	Generally excitatory	CNS, PNS
	Serotonin	Generally inhibitory	CNS
Amino acids	GABA	Inhibitory	CNS
	Glycine	Inhibitory	CNS
	Glutamate	Excitatory	CNS
	Aspartate	Excitatory	CNS
Neuropeptides	Substance P	Excitatory	CNS, PNS
	Met-enkephalin	Generally inhibitory	CNS

CNS: central nervous system; PNS: peripheral nervous system; GABA: gamma aminobutyric acid.

glutamate, gamma aminobutyric acid (GABA), and glycine belong to the amino acid class. Serotonin, dopamine, or acetylcholine are biogenic amines. The third group includes neuropeptides, of which more than 50 are known today. In addition, nitrous oxide and ATP are involved in neuronal communication, but are not a member of any of the three groups.

The duration of action of neurotransmitters is usually a few milliseconds; their effects are strictly restricted to the synaptic area. However, cells that are far away from the source of the neurotransmitter can be affected by diffusion. Neurotransmitter activity is terminated either by enzymatic cleavage or by uptake of the neurotransmitter into the presynaptic cell. Neurotransmitters can act in an inhibitory or excitatory manner depending on the postsynaptic receptor and the intracellular signal transduction pathway. Multiple types of neurotransmitters can coexist in a single neuron.

Physiologically important and well-studied examples of neurotransmitters are acetylcholine, serotonin, and dopamine. A special feature of these three transmitters is that they have a relatively limited origin, that is, they are only produced by certain defined groups of neurons. However, their influence extends to over 100 000 synapses per neuron in many different sites in the brain. In addition, acetylcholine, serotonin, and dopamine have a longer lasting effect than, for example, glutamate.

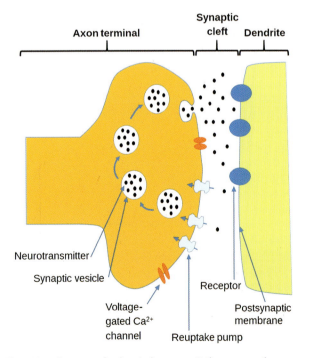

Fig. 1.22 Structure of a chemical synapse. At the synapse, the presynaptic neuron releases a neurotransmitter into the synaptic cleft. Binding of the neurotransmitter to receptors in the membrane of the postsynaptic neuron leads to its excitation (or inhibition).

Therefore, they play an important role in the regulation of conditions such as sleep or mood and in the development of disease.

Because it plays a crucial role in the vegetative nervous system and is also active at the interface between motor neurons and skeletal muscles, acetylcholine was the first neurotransmitter discovered. Neurons that mainly use acetylcholine as neurotransmitter are referred to as cholinergic neurons. They are present in the brain and participate in the control of attention and in the excitability of the brain during the sleeping and waking circadian rhythm. In addition, they are believed to play a crucial role in plasticity and learning. The hippocampus, the neocortex, and the olfactory bulb are innervated from the cerebrum. Cells of these areas are among the first to die in Alzheimer's disease. Among the approved drugs that aim to delay the progression of Alzheimer's disease are those that slow the cleavage of acetylcholine in the brain.

In the brain, the neurotransmitter serotonin is detectable only in neurons whose cell bodies are located in the *raphe nuclei* in the brain stem. These neurons innervate virtually all regions of the brain and affect pain, sleep and wake rhythms, and mood. The raphe nuclei are very active in a state of heightened vigilance and have the lowest activity during sleep. Studies have shown that agitation and hallucinations may occur with excessive brain levels of serotonin. Serotonin deficiency can lead to depression, anxiety, and aggression.

Dopamine-containing cells are found in many areas in the central nervous system. One group of dopaminergic neurons in the *substantia nigra* of the midbrain is of particular medical relevance. These neurons send their axons into the striatum and are essential for the control of volitional movement. The degeneration of dopaminergic cells in the substantia nigra triggers the motor dysfunction seen in Parkinson's disease.

1.4.2
Hormones

Hormones mediate another important mechanism of intercellular signaling. They are typically secreted by one cell type and act on another, inducing physiological responses such as cell growth or the activation of metabolic processes. Hormones can be classified in different ways. One way is to group them according to the distance over which they act (Figure 1.23a). *Endocrine* hormones act over a long distance, that is, on cells far away from the site of their release. They are produced in

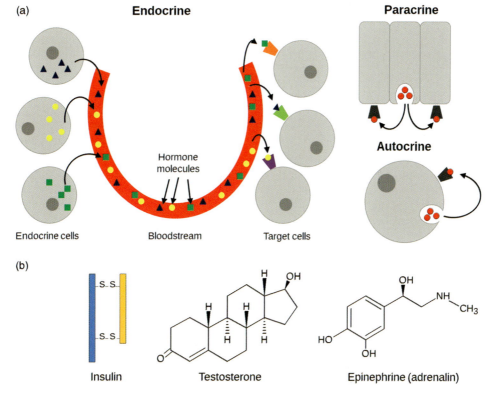

Fig. 1.23 Classification of hormones. (a) Hormones can be grouped into endocrine, paracrine, and autocrine hormones according to the distance over which they act. (b) Hormones belong to three different major chemical classes: peptides, steroid hormones (derived from cholesterol), and derivatives of aromatic amino acids. (Part (a) adapted from Ref. [5] with kind permission from John Wiley & Sons, Inc.)

specialized endocrine glands (e.g., the pancreas, adrenal glands, and the gonads) and released into the bloodstream. The endocrine system regulates many processes in the body. These include homeostasis (e.g., blood glucose levels, which are controlled by the hormones insulin and glucagon), the response to external stimuli (e.g., preparation for the fight-or-flight-response, regulated by epinephrine and norepinephrine), and various cyclic and developmental programs. *Paracrine* hormones act only on cells close to the cell that releases them. They are secreted into the immediate extracellular environment and diffuse over relatively short distances. An example is the release of interleukin-12 by macrophages, which stimulates other local immune cells. In some cases, hormones bind to receptors on the cells from which they were released. This is called *autocrine* signaling. Examples of autocrine signaling are the autoactivation of immune cells and the development of tumors through the autocrine production of growth and survival factors.

Alternatively, hormones can be classified by their basic chemical structure (Figure 1.23b). The major groups are peptide hormones (e.g., insulin), steroid hormones (testosterone), and amino acid derivatives (epinephrine and norepinephrine belong to the catecholamine subclass). Only cells with a specific receptor will respond to a given hormone.

1.4.3
Signal Transduction

The message of hormones and other extracellular signal molecules must be transmitted into the cell. With the exception of some nonpolar steroid hormones that can easily pass through the cell membrane and bind to their receptor in the cytoplasm, most signaling molecules are hydrophilic and cannot easily reach the interior of the cell. Instead, they bind to a receptor protein on the cell surface that is specific for that hormone (or other type of ligand). The binding event is transmitted to the cell interior and a series of intracellular signaling events is induced. These responses usually involve the generation of a *second messenger* and protein phosphorylation/dephosphorylation events. Signal intensity is often amplified by a signal cascade.

Signaling by G protein-coupled receptors (GPCRs) resulting in activation of adenylyl cyclase (AC) will be outlined as a paradigm for the various signal transduction pathways (Figure 1.24). A common characteristic of the more than 800 GPCRs encoded in the human genome is that they contain seven transmembrane helices. Binding of a specific ligand results in a conformational change of the GPCR, which transmits the

extracellular signaling event across the plasma membrane to the cytoplasmic face. The intracellular domain of the GPCR is coupled to a heterotrimeric G protein, which consists of a complex of one α-, β-, and γ-subunit each. In the nonactivated state, the α-subunit binds GDP. Activation of the G protein by the GPCR in complex with its ligand induces the exchange of bound GDP for GTP. When GTP is bound, the α-subunit dissociates from the β- and γ-subunits and stimulates adenylyl cyclase. α-Subunit activation is temporary, since it functions as a GTPase and hydrolyzes GTP into GDP + P_i. The inactivated α-subunit carrying GDP then reassembles with the β- and γ-subunits. Inhibition of GTPase activity is the way cholera toxin leads to an overactivation of the signaling cascade. In the intestine, this causes the failure of water resorption and voluminous diarrhea (Section 6.1.1, Box 6.1).

The activated α-subunit carrying GTP stimulates AC to convert ATP into cyclic AMP (cAMP) (Figure 1.25). cAMP acts as a second messenger, that is, as a molecule that transmits the signal intracellularly to the target molecule. cAMP can freely diffuse through the cytoplasm and bind to its target protein kinase A (PKA). In the absence of cAMP, PKA forms an inactive heterotetramer consisting of two catalytic and two regulatory subunits (R_2C_2 in Figure 1.24). Binding of cAMP causes dissociation of the complex so that the catalytic subunit can phosphorylate its substrate proteins to produce a cellular response.

To terminate the signal, the signaling molecule must eventually be eliminated. In the case of cAMP and the related second messenger cGMP, signaling activity is terminated through the action of phosphodiesterases (PDEs) that hydrolyze the cyclic phosphodiesters to AMP and GMP, respectively. In mammals, phosphodiesterases are grouped into 11 families, called PDE1–PDE11. Some PDEs are specific for either cAMP or cGMP, while others can hydrolyze both second messengers. Due to their unique tissue distribution and functional properties, PDEs are attractive drug targets. The drug sidenafil (Viagra, Box 1.1) is an example of a PDE inhibitor. The stimulatory activity of caffeine is (in part) caused by nonselective inhibition of PDEs. In addition, caffeine antagonizes adenosine receptors that act through inhibitory G proteins. Both effects contribute to an increased concentration of stimulatory cAMP in the cell.

An alternative signaling pathway starts with the binding of a ligand, for example, a growth hormone, to a receptor tyrosine kinase (RTK). In contrast to GPCR, RTKs contain only a single-transmembrane segment. Binding of the ligand induces dimerization of the

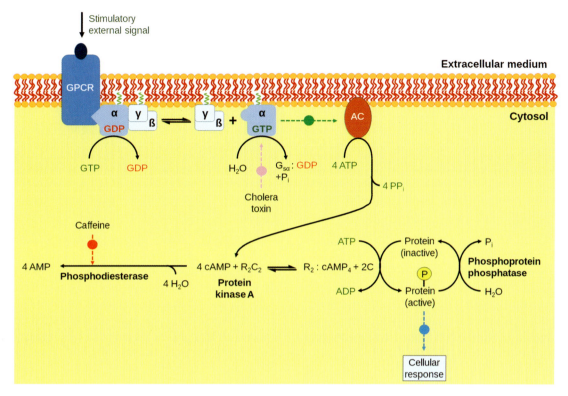

Fig. 1.24   G protein-coupled receptors and the adenylyl cyclase (AC) signaling system. Binding of the ligand to the GPCR causes exchange of GDP for GTP in the α-subunit of the protein, which then dissociates from the β- and γ-subunits and activates adenylyl cyclase. AC generates cAMP from ATP; cAMP activates protein kinase A to phosphorylate its substrate proteins, causing the cellular response. The signaling system depicted here results in activation of AC, but other hormones can bind to inhibitory GPCRs that trigger an almost identical pathway, except that the α-subunits inhibit AC. R_2C_2: the two regulatory and two catalytic subunits of protein kinase A. (Adapted from Ref. [1] with kind permission from John Wiley & Sons, Inc.)

receptor proteins, which brings the intracellular domains close together so that they can phosphorylate each other. This event activates a pathway that is known as the Ras–Raf signaling cascade. A central step in this pathway is the exchange of GDP for GTP by the G protein named Ras. This activates Ras, which then binds to Raf and induces a cascade of several protein phosphorylation events that finally activates various transcription factors. Mutations in Ras are among the most frequent changes

in malignant transformation and can lead to uncontrolled cell growth.

The third signaling pathway, one that mediates the effects of a variety of hormones, is the phosphoinositide (PI) pathway. This pathway is initiated by binding of a ligand to a cell surface receptor, followed by activation of a heterotrimeric G protein, as described above. In the PI pathway, the α-subunit of the G protein in complex with GTP activates phospholipase C

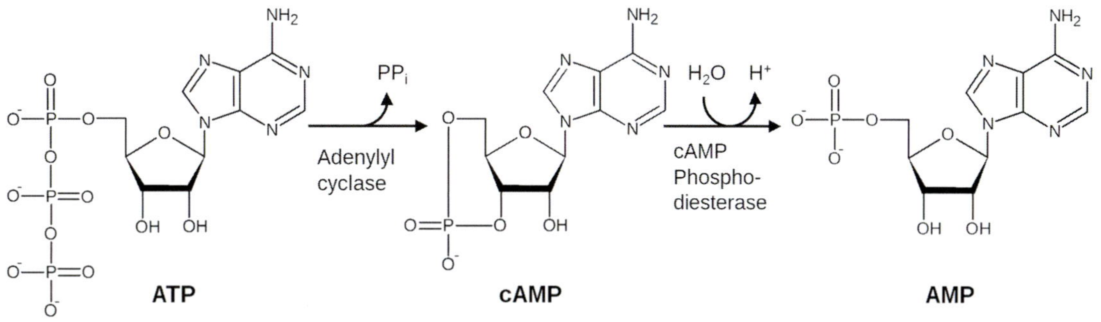

Fig. 1.25   Formation and hydrolysis of cyclic AMP (cAMP). Adenylate cyclase converts ATP into the second messenger cAMP. To terminate the cAMP activity, phosphodiesterases hydrolyze it into AMP.

Box 1.1. The Discovery of Viagra™

The discovery of Viagra as a drug to treat erectile dysfunction was accidental. Its active pharmaceutical ingredient, sidenafil, selectively inhibits the cGMP phosphodiesterase PDE5. Sidenafil was intended as a treatment for *angina pectoris* as it relaxes the arterial wall, which leads to decreased blood pressure. However, the outcome of early clinical trials was disappointing and the trials were terminated. According to the myth, the male participants in the trials refused to return the remaining pills because they claimed they were having firmer and longer lasting erections. In fact, sexual stimulation in males causes penile nerves to release nitric oxide, which activates guanylate cyclase to produce cGMP from GTP. cGMP induces relaxation of the vascular smooth muscle (SM) and blood inflow increases, leading to an erection. Hydrolysis of cGMP by PDE5 interferes with this process; thus, inhibition of PDE5 by sidenafil improves erectile function. Due to this entirely unintended consequence, Viagra has become one of the most frequently prescribed drugs in the world.

(PLC), which then hydrolyzes phosphatidylinositol-4,5-bisphosphate (PIP_2), a phospholipid found on the inner face of the plasma membrane, into the second messengers inositol-1,4,5-trisphosphate (IP_3) and 1,2-diacylglycerol (DAG). While the latter activates protein kinase C, IP_3 stimulates the release of calcium ions from the ER. Ca^{2+} then acts as another second messenger and modulates the activity of a number of proteins to induce a cellular response.

1.5
The Immune System

Animals (as well as virtually all other living organisms) are continually subject to attack by disease-causing pathogens. As a countermeasure, animals have evolved an immune system to protect themselves. The immune system must detect and destroy a wide variety of infectious agents, including viruses, microorganisms, and parasitic worms (Chapters 5 and 6), distinguishing them from the organism's own cells. More recently, it has become evident that the immune system not only fights invading pathogens but also provides an important defense against the development of cancer. It does this by recognizing and destroying transformed cells.

The cells of the immune system are distributed throughout the body and in the bloodstream. Blood cells fall into three different categories: (i) red blood cells (erythrocytes) primarily carry oxygen and carbon dioxide; (ii) platelets (thrombocytes) are involved in the formation of blood clots; and (iii) white blood cells (leukocytes) are the cells of the immune system. The formation of different blood cells, a process called hematopoiesis, is illustrated in Figure 1.26.

Immune cells can be further categorized as being part of the *innate* or *adaptive immune system* (Figure 1.27).

The innate immune system is the first line of defense. It recognizes general characteristics of pathogens and provides a rapid, nonspecific response. In addition, it activates a second layer of response, the adaptive immune system, which provides a specific defense against a given target. Two types of adaptive immunity can be distinguished: cellular and the humoral immunity. This is outlined in Section 1.5.2.

1.5.1
The Innate Immune System

The innate immune system detects macromolecules that represent a danger to the organism and generates the means to destroy them. It provides an immediate defense against infection, but does not confer specific, long-lasting, protective host immunity. The innate immune system does not identify a specific microbe. Rather, it recognizes general patterns that are known as pathogen-associated molecular patterns (PAMPs). The characteristics of invading pathogens that can distinguish them from host cells include components of the outer bacterial membrane (lipopolysaccharides) and viral double-stranded RNA molecules. PAMPs act as agonists for pattern recognition receptors (PRRs) on innate immune cells.

One of the first responses of the innate immune system is the stimulation of inflammation. Acute inflammation is initiated by cells present in the tissue; these cells include macrophages, dendritic cells, and mastocytes. Upon activation, these cells release inflammatory mediators (e.g., histamine, bradykinin, and/or cytokines) that establish a physical barrier against the spread of infection and trigger additional immune system effectors.

Different types of specialized phagocytes destroy harmful particles, such as bacteria, small parasites, fungi, and viruses, by internalization (phagocytosis,

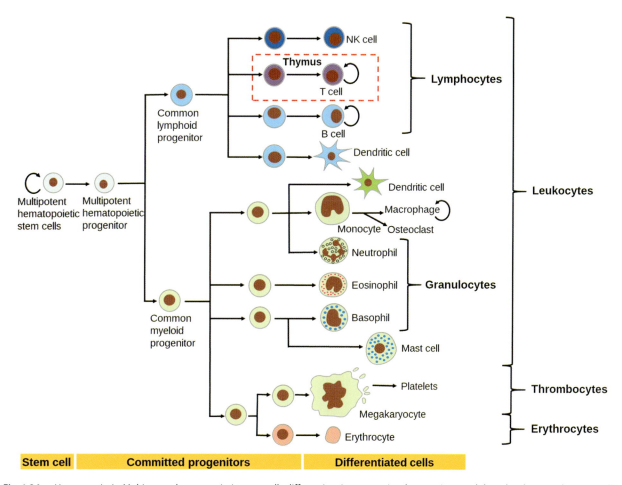

Fig. 1.26 Hematopoiesis. Multipotent hematopoietic stem cells differentiate into committed progenitors and then develop into the terminally differentiated cells. In adult mammals, the cells shown develop primarily in the bone marrow. Exceptions are T lymphocytes, which differentiate in the thymus, and macrophages and osteoclasts that develop from blood monocytes. Dendritic cells may also derive from monocytes. NK cell: natural killer cell. (Reproduced with permission from Ref. [2]. Copyright 2008, Garland Science/Taylor & Francis LLC).

Figure 1.28). Macrophages and neutrophils are the most important phagocytes in the defense against infection. The process of phagocytosis starts with the recognition of the microorganism by PRRs. The invading microorganism is then internalized into the so-called phagosome, a membrane-bound vacuole, which fuses with lysosomes. Lysosomes contain digestive enzymes in an acidic environment that degrade and kill the microorganism.

Antigen presentation is an important process of the adaptive immune system. Some phagocytes (macrophages and dendritic cells) move molecules from engulfed pathogens that have been degraded by the proteasomes (described in Section 1.3.6) back to the surface of their cells. These molecules are then presented to other cells of the immune system. Such cells are known as antigen-presenting cells (APCs). APCs break down foreign proteins to peptides in the

proteasome. The peptides formed are then bound to the so-called major histocompatibility complex (MHC), which carries them to the surface and presents them to lymphocytes. In humans, MHC is also called human leukocyte antigen (HLA). HLA is an important determinant of the compatibility of donors for organ transplants. Cells express different types of MHC molecules: MHC class 1 is expressed on almost all nucleated cells. The expression of MHC class II is restricted to special cells, including thymic epithelial cells, dendritic cells, B cells, and some macrophages. By activating lymphocytes, APCs connect the innate and the adaptive immune systems.

1.5.1.1 The Complement System
The complement system is another part of the immune system that helps to clear pathogens from an infected organism. Since it does not adapt to a specific pathogen

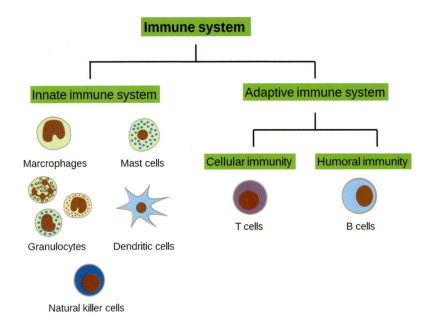

Fig. 1.27 Cells of the immune system. The cells of the innate immune system include macrophages that digest pathogens, mast cells that release histamine and other substances that induce inflammation, granulocytes (neutrophils, eosinophils, and basophils) that release toxic substances and promote inflammation, dendritic cells that present antigens and induce the adaptive immune system, and natural killer cells that destroy infected cells (as well as cancer cells). The key components of the adaptive immune system are T lymphocytes (or T cells) that mediate cellular immunity and B lymphocytes (or B cells) that constitute humoral immunity.

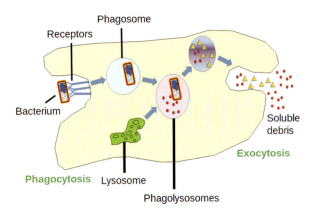

Fig. 1.28 Phagocytosis. Bacteria are recognized by special pattern recognition receptors (PRRs) on the surface of phagocytes. The bacterium is taken up by a process called phagocytosis, and is then trapped in a compartment known as phagosome, which fuses with a lysosome. Several enzymes then destroy the pathogen; the soluble debris is released by exocytosis.

and does not change over the lifetime of an individual, the complement system is considered part of the innate immune system. However, it can be activated by the adaptive immune system. The complement system consists of more than 25 proteins, most of which circulate in the blood as inactive precursors. The major components of the complement cascade are named with the letter "C"

followed by a number. Cleavage products are assigned as "a" and "b," with "b" being the *b*igger fragment.

The complement cascade can be activated by three pathways: the classical, lectin, and alternative pathways (Figure 1.29a). The classical pathway is triggered by activation of the C1 complex, which consists of the C1q protein and the proteases C1r and C1s. C1q binds to antibody–antigen complexes on the surface of pathogens, but can also be activated by direct binding to the surface of bacteria. The C1 complex then splits C2 and C4, and the cleavage products C4b and C2a form the C3 convertase that promotes cleavage of C3 into C3a and C3b.

The lectin pathway functions in an analogous manner, except that the initial step consists of activation of mannose-binding lectin (MBL) by mannose residues on the pathogen surface. This activates MBL-associated serine proteases (MASPs) followed by cleavage of C2 and C4 and formation of the C3 convertase, as in the classical pathway.

The alternative pathway is continuously activated at a low level due to spontaneous hydrolysis of C3. The cleavage product C3b is rapidly inactivated by the two factors H and I. However, binding of C3b to the surface of a pathogen protects the C3b from inactivation and leads to the activation of factors B and D, resulting in the formation of an alternative C3 convertase.

Production of the cleavage product C3b by any of the three pathways mediates the main functions of the complement system. These include coating of microbes and

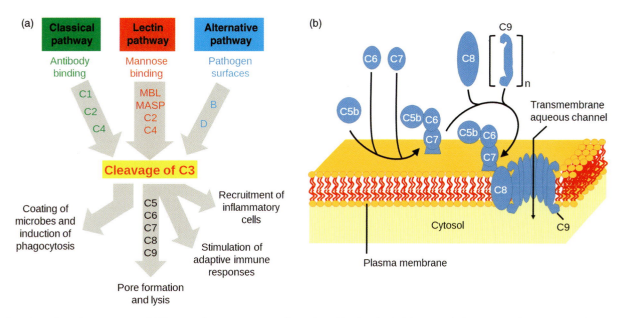

Fig. 1.29 Complement system. (a) The complement system can be activated by the classical, lectin, and alternative pathways. A central step is the activation of C3. Activation of the complement system results in the killing of a foreign cell, stimulation of phagocytosis, triggering of inflammation, and adaptive immune responses. (b) The late components C5b–C9 form a membrane attack complex that leads to osmotic lysis of the microbe or infected host cell. MBL: mannose-binding lectin; MASP: MBL-associated serine protease. (Reproduced with permission from Ref. [2]. Copyright 2008, Garland Science/Taylor & Francis LLC.)

induction of phagocytosis, recruitment of inflammatory cells, and stimulation of the adaptive immune response. In addition, the late components C5–C9 form a pore in the target cell's plasma membrane, the so-called membrane attack complex (MAC, Figure 1.29b). This channel allows free ingress of water and egress of electrolytes, which causes osmotic lysis of the target cell.

1.5.2
The Adaptive Immune System

The adaptive immune response is primarily based on two types of lymphocytes: B cells confer humoral immunity (*humor* is an archaic term for fluid) by producing antibodies (known as immunoglobulins), whereas T cells mediate cellular immunity. The adaptive immune response is triggered by antigens (*anti*body *gen*erators), macromolecules (in most cases proteins or carbohydrates) that are recognized as foreign.

1.5.2.1   Cellular Immunity
The main function of the cellular immune system is to prevent the spread of viral infection by killing virus-infected cells. In addition, the cellular immune system is effective against intracellular bacteria and parasites, where they are protected from attack by antibodies. The cellular immune system is also effective against certain types of cancer. The cells of the cellular immune system

mature in the thymus and are referred to as T cells. The major populations of T cells are helper T cells (T_H), cytotoxic T cells (T_C), memory T cells, and natural killer T cells (not to be confused with the natural killer cells of the innate immune system). A common characteristic of T cells is the presence of a T-cell receptor (TCR) on the cell surface. This receptor is composed of two chains that form the antigen recognition site. The best understood populations are the αβ T cells, in which the TCR is composed of an α and a β chain. TCRs possess unique antigen specificity, which is determined by the structure of the antigen-binding site formed by the two chains. The diversity of TCRs is based mainly on somatic recombination, as described for immunoglobulins in the following section. The T-cell population is further categorized by the additional proteins found on their surface. The *cluster of differentiation* (CD) system is used for immunophenotyping cells according to cell surface markers. Conventional αβ T cells develop into two classes expressing either the CD4 or the CD8 glycoprotein.

TCRs, together with the respective CD proteins, bind to the MHC proteins displaying antigenic fragments on the surface of APCs, as described in the previous section. T cells expressing CD4 (abbreviated as CD4+ T cells) interact with MHC class II molecules, while CD8+ cells interact with MHC class I molecules (Figure 1.30). Binding of a T cell to an APC displaying an antigen–MHC complex causes it to reproduce, a process called

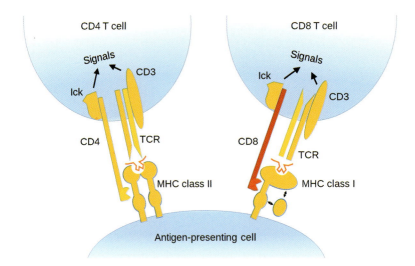

Fig. 1.30 Interaction between T cells and APCs. T-cell receptors (TCRs) and CD4 or CD8 molecules on the surface of T cells interact with proteins of the major histocompatibility complex (MHC) that present a peptide processed from an invading pathogen. CD4 interacts with MHC class II; CD8 binds to MHC class I. CD4 and CD8 are associated with additional factors that mediate signaling to activate the T cell. (Adapted from Ref. [6] with kind permission from Wiley-VCH Verlag GmbH.)

clonal selection. As a consequence, the T cells are produced in large quantity after recognizing an invading pathogen.

CD4 is expressed on T helper cells (T_h cells), which assist other lymphocytes by releasing cytokines. These cytokines induce the maturation of B cells into plasma cells, which release antibodies and activate memory B cells, cytotoxic T cells, and macrophages (Figure 1.31). Upon activation, T_h cells proliferate and differentiate into several subtypes, the two most important of which are known as T_h1 and T_h2 cells. These two types of helper cells are activated by different partners and release different kinds of cytokines that have specific effects. T_h1 cells are important for the immune reaction against intracellular bacteria and protozoa, while T_h2 cells mediate the immune reaction against helminths (parasitic worms). As will be outlined in Section 5.1.1, the human immunodeficiency virus (HIV) uses CD4 as a cellular entry receptor. At a later stage of the infection, massive destruction of $CD4^+$ cells leads to the development of the acquired immunodeficiency syndrome (AIDS).

Cytotoxic T cells (T_C cells) express the CD8 glycoprotein. They destroy cells infected with a virus (or other pathogens) and tumor cells. After recognizing a specific antigen–MHC class I complex, the activated T cell releases cytotoxins, such as perforin, which form pores in the target cell's plasma membrane. This allows water and another toxin called granulysin (a protease) to enter the target cell and kill it.

Some T cells, known as memory T cells, persist for long term after an infection has resolved. They constitute an immune system "memory" of past infections and quickly expand into large numbers of effector T cells upon re-exposure to their cognate antigen. They induce a faster, stronger immune response than the first time the immune system encountered the pathogen.

1.5.2.2 Humoral Immunity

The humoral immune response is largely mediated by B cells, which mature in the bone marrow. The principal function of the humoral immune system is the production of antibodies against soluble antigens. B-cell precursors display immunoglobulins on their surface (Figure 1.32). They can be activated in a T cell-dependent or T cell-independent manner. The latter process is

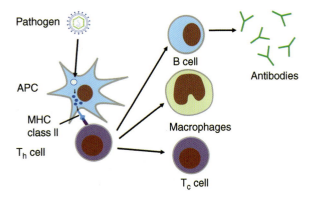

Fig. 1.31 The function of T helper cells (T_h cells). T_h cells are activated by binding to antigen–MHC class II complexes on the surface of APCs. By releasing stimulatory signals (cytokines), they induce maturation of B cells and activation of macrophages and cytotoxic T cells.

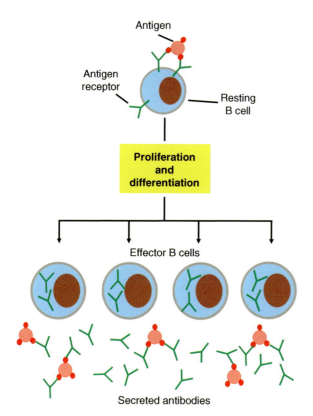

Fig. 1.32 B-cell activation. Activation of a B cell includes recognition of the antigen by the antigen receptor on the surface of the B cell. The activation may be T cell-dependent or T cell-independent. The activated B cell proliferates and differentiates to a plasma cell that secretes antibodies. (Reproduced with permission from Ref. [2]. Copyright 2008, Garland Science/Taylor & Francis LLC.)

triggered when a B cell encounters its matching antigen. The antigen–antibody complex is taken up by the B cell and digested into peptides, which are bound by a MHC class II molecule and then displayed on the cell surface. In doing so, B cells perform the role of APCs. The antigen–MHC complex is recognized by a T cell, which releases cytokines that activate the B cell. B cell activation includes a combination of clonal proliferation and differentiation into effector B cells known as plasma cells. Plasma cells produce and secrete millions of copies of the antibody that recognized the antigen. Although most B-cell progeny are plasma cells, others develop into memory B cells that have a similar function as memory T cells.

Antibodies are large, Y-shaped glycoproteins. In mammals, there are five different classes of antibodies: IgA, IgD, IgE, IgG, and IgM (Ig stands for immunoglobulin). While the monomeric structures of all the isotypes are similar, two classes can form multimers: IgMs are pentamers and IgAs are dimers consisting of linked monomeric units. The five isotypes differ in their biological

properties, functional locations, and ability to deal with different antigens. The antibody isotype of a B cell changes during the differentiation process (*immunoglobulin class switching* or *isotype switching*). This switch occurs by exchanging the constant region of the heavy chain while retaining the variable region that determines antigen specificity. Immature B cells express IgM in a cell surface-bound form. At a later stage, these cells express surface IgD in addition to IgM. When activated to plasma cells, B cells produce antibodies in a secreted rather than membrane-bound form. Secreted IgMs are important for the rapid elimination of pathogens in an early stage of the B-cell response, before there is sufficient IgG. In the final step of isotype switching, cells change from producing IgD or IgM to the production of the IgE, IgA, or IgG classes of immunoglobulin.

IgG is the most abundant class of antibody in the blood, providing most of the antibody-based immunity against invading pathogens. Its general structure is depicted in Figure 1.33. IgGs protect the body from infection by several mechanisms. The antibody may bind to the pathogen and cause its immobilization and binding together via agglutination, preventing its entry into

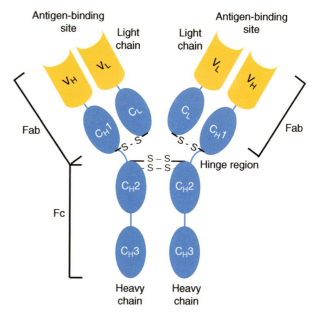

Fig. 1.33 Structure of immunoglobulin G (IgG). An antibody is made up of two identical heavy and two identical light chains connected by disulfide bonds. Each chain is composed of several immunoglobulin domains. The heavy chain has one variable (V_H) and three constant regions (C_H1–C_H3). The light chain is only composed of one variable (V_L) and one constant chain (C_L). The arms of the Y are known as the Fab antigen-binding fragment and are composed of one constant and one variable domain from each heavy and light chain of the antibody. The two variable domains form the site for antigen binding. The base of the Y is called the F_c for crystallizable fragment. It plays an important role in modulating immune cell activity.

host cells. This is also the way IgG neutralizes toxins. An additional defense mechanism mediated by IgG is *opsonization*; here, the IgG coats the surface of the pathogen, permitting its recognition and ingestion by phagocytic immune cells. IgGs also activate the classical pathway of the complement system (Section 1.5.1) and induce antibody-dependent cellular cytotoxicity (ADCC). In this process, the binding of antibodies to surface antigens marks the target cell for lysis by an effector cell of the immune system (e.g., an NK cell).

The human immune system can produce antibodies against virtually any antigen it encounters. Estimates suggest that humans produce more than 10^7 different antibodies (some estimates reach an order of 10^{12}). As the human genome encodes less than 25 000 genes (Chapter 7), each antibody cannot be encoded by a specific gene. Instead, antibody diversity is achieved by two mechanisms: *somatic recombination* and *hypermutation*. Immunoglobulin genes exist as discrete groups of gene segments. Somatic rearrangements are also known as V(D)J recombination. The variable region of each heavy or light chain is encoded by several gene segments. Heavy chains are composed of variable (V), diversity (D), and joining (J) segments; light chains consist of only V and J segments. Multiple copies of each segment in a tandem arrangement exist in the genome. During B-cell differentiation, intrachromosomal recombination occurs, and an immunoglobulin variable region is assembled by randomly combining one V, one D, and one J segment for the heavy chain, or one V and one J segment for the light chain. As the multiple copies of each type of gene segment can be combined in different ways, this process allows the generation of a large number of different antibodies.

An additional layer of diversity is achieved by two types of somatic mutations: (1) During the recombinant joining of the different segments, a terminal deoxynucleotidyl transferase randomly adds up to 30 base pairs that increase variability. (2) The variable regions mutate at rates that are at least a million fold higher than the rates of spontaneous mutation in other genes. This somatic hypermutability is presumably mediated by error-prone DNA polymerases.

The enormous diversity of antibodies has attracted the interest of researchers, who have used them for technical and medical purposes. As will be outlined in detail in Section 10.2.1, monoclonal antibodies are among the most successful classes of modern biomolecular drugs.

This brief introduction does not adequately reflect the enormous complexity of the immune system, which would be beyond the scope of any introduction to molecular medicine. The interested reader is referred to a more in-depth introduction given in the textbook cited in the reference list at the end of this chapter.

References

1. Voet, D., Voet, J.G., and Pratt, C.W. (2013) *Principles of Biochemistry – International Student Version*, 4th edn, John Wiley & Sons, Inc., Hoboken, NJ.
2. Alberts, B., Johnson, A., Walter, P. *et al.* (2007) *Molecular Biology of the Cell*, 5th edn, Taylor & Francis, New York.
3. Igney, F.H. and Krammer, P.H. (2002) Death and anti-death: tumour resistance to apoptosis. *Nat. Rev. Cancer*, **2**, 277–288.
4. Turner, B.M. (2002). Cellular memory and the histone code. *Cell*, **111**, 285–291.
5. Voet, D. and Voet, J.G. (2004). *Biochemistry*, 3rd edn, John Wiley & Sons, Inc., Hoboken, NJ.
6. MacPherson, G. and Austyn, J. (2012). *Exploring Immunology – Concepts and Evidence*, Wiley-VCH Verlag GmbH, Weinheim.

Further Reading

Biochemistry, Molecular, and Cell Biology

Alberts, B., Johnson, A., Walter, P. *et al.* (2007) *Molecular Biology of the Cell*, 5th edn, Taylor & Francis, New York.
Portela, A. and Esteller, M. (2010) Epigenetic modifications and human disease. *Nat. Biotechnol.*, **28**, 1057–1068.
Snustad, D.P. (2011) *Genetics – 6th International Student Edition*, John Wiley & Sons, Inc., Hoboken.
Voet, D. and Voet, J.G. (2011) *Biochemistry*, 4th edn, John Wiley & Sons, Inc., Hoboken, NJ.

Immune System

MacPherson, G. and Austyn, J. (2012) *Exploring Immunology – Concepts and Evidence*, Wiley-VCH Verlag GmbH, Weinheim.

Methods in Molecular Medicine

2

Summary

- Molecular medicine relies on a large variety of experimental procedures. A selection of sophisticated technologies is described in this chapter.
- DNA arrays can be used for genotyping, SNP analysis, and gene expression profiling (as well as numerous other purposes). The arrays consist of up to several hundred thousand short DNA segments that are covalently fixed on a solid support. Fluorescently labeled DNA or RNA samples are added to the array and hybridize with their complementary probes. The fluorescence signal emitted by the labeled sample indicates the presence of a specific genotype and/or may allow the analysis of the expression levels of all cellular genes (the transcriptome).
- Polymerase chain reaction (PCR) is a widely used method to amplify DNA. Real-time experiments allow the online detection of the synthesized fragments. Quantitative PCR (qPCR) has become a standard procedure to measure changes in the expression level of a gene. The amplified DNA is detected either by double-stranded DNA-binding dyes or by sequence-specific probes containing a fluorescent reporter.
- In the past decade, next-generation sequencing (NGS) technologies have had a major impact on research at the genomic level. A common feature of NGS methods is the elimination of a gel electrophoretic step that was rate-limiting step in conventional sequencing. The most recently developed sequencing technologies focus on real-time single-molecule sequencing, which accelerates the sequencing process.
- Wild-type and transgenic animals have become important tools to study biological and pathological processes in living organisms. Knockout technology is widely used to study gene function in mammals. Alternatives to animal models are being developed to reduce the numbers used in biomedical research.
- Fluorescence microscopy can be used to study cellular processes, either of fluorescent proteins or of molecules carrying fluorescent labels. Fluorescence-activated cell sorting (FACS) is widely employed to physically separate cells with different properties and purify populations of interest. Surface plasmon resonance (SPR) allows the label-free real-time characterization of interactions between two biomolecules.

Contents List

DNA Microarrays
Quantitative Polymerase Chain Reaction
Next-Generation Sequencing
Animal Models in Biomedical Research
Additional Methods

- **Fluorescence Microscopy**
- **Flow Cytometry and Fluorescence-Activated Cell Sorting**
- **Surface Plasmon Resonance**

The field of molecular medicine is based on a large number of experimental procedures that are employed in research and clinical application. Since the standard techniques of molecular and cellular biology, such as molecular cloning, protein gel electrophoresis, Western blotting, chromatography, ultracentrifugation, and cell culture, are extensively explained in basic textbooks, they will not be further commented on here. In the following sections, a selection of the advanced techniques that are widely used in molecular medicine will be discussed. Additional technologies used in functional genomics and proteomics will be outlined in Chapter 7.

Molecular Medicine: An Introduction, First Edition. Jens Kurreck and Cy Aaron Stein.
© 2016 Wiley-VCH Verlag GmbH & Co. KGaA. Published 2016 by Wiley-VCH Verlag GmbH & Co. KGaA.

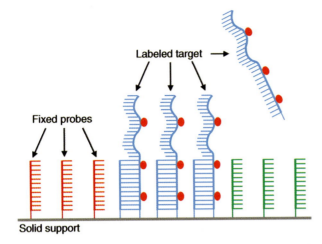

Fig. 2.1 Basic principle of DNA array technology. A large number of DNA sequences (probes) are fixed on a solid support. The different colors indicate different sequences. A labeled target DNA (or RNA) is added to the array and allowed to hybridize with the probes. In the example depicted in the figure, the blue sample hybridizes to its complementary blue probe. The red circles indicate the fluorophore, for example, Cy3.

2.1
DNA Microarrays

DNA microarrays (also known as DNA chips) are widely used to measure the expression levels of a large number of genes or to genotype multiple regions of a genome. They contain up to several hundred thousand different DNA segments in a precise array immobilized on a solid support. Fluorescently labeled DNA (or RNA) samples are then added and allowed to hybridize with the DNA on the microarray (Figure 2.1). After the unhybridized sample is washed away, the light source of the array reader excites the dyes and the resulting fluorescence intensity indicates how much DNA (or RNA) has bound to a particular probe sequence.

DNA arrays can be fabricated by multiple techniques. Most commonly the probes are either spotted onto glass slides (in a manner that resembles inkjet printing) or they are synthesized directly on the array. In spotted microarrays, the probes may be oligonucleotides, cDNAs, or PCR products that have been synthesized prior to deposition on the array and that are then spotted onto the solid support. In an alternative approach used by the company Affymetrix, the oligonucleotides are synthesized on a solid support by photolithography. In the initial step of this process, an anchor with a photosensitive protection group is coupled to a glass surface. Light then removes protecting groups in certain areas of the array as defined by a mask. A single nucleotide is then added and coupled to the deprotected areas. To

synthesize the full oligonucleotide, the process of defined deprotection and addition of one of the four nucleotides is repeated 75–100 times. The final arrayed oligonucleotides typically have a length of 20–25 nucleotides.

Detection of the hybridized sample requires labeling with a fluorescent reporter. A typical procedure employs the incorporation of nucleotides carrying a fluorophore into the cDNA during reverse transcription of the RNA molecules to be analyzed. Alternatively, the RNA can be directly labeled by chemical methods, or the cRNA can be transcribed from the cDNA in the presence of labeled nucleotides. Widely used fluorophores are the cyanines Cy3 (red) and Cy5 (green).

DNA arrays can be employed for a variety of purposes, including the detection of single-nucleotide polymorphisms (SNPs), genotyping, and the measurement of changes in gene expression levels. For the detection of an SNP or the determination of a genotype, probes for different genetic variants are spotted or synthesized on the microarray. With the proper conditions, the fluorescently labeled sample will hybridize with high affinity only to the fully complementary sequence (Figure 2.2a). The SNP or genotype can be directly determined from the spot that emits fluorescence.

Compared to this hybridization approach, a higher signal intensity and improved discrimination between two genetic variants can be achieved by a primer extension reaction (Figure 2.2b). In this assay, fluorescently labeled nucleotides are incorporated by DNA polymerase. Importantly, the enzyme will only extend the primer in the presence of full complementarity, while a mismatch at the 3′ end of the probe prevents the extension reaction. Genotyping is not only used to detect polymorphisms in the human genome but is also valuable in the identification of microorganisms in medicine, food production, or water.

The most common application of a DNA microarray is in gene expression profiling. This approach has in part replaced classical methods such as Northern blotting and RNase protection assays. DNA arrays provide more accurate quantitative results, and in a single experiment can be used to quantify a large number of mRNAs. In fact, modern DNA microarrays contain probes for every human gene and can even differentiate between the different splice variants of an mRNA transcript. Thus, it is possible to analyze the entire transcriptome, that is, the collection of all mRNAs that are rapidly transcribed by the cell.

A typical gene expression profiling experiment by DNA microarrays involves a comparison of two states. These, for example, may be diseased tissue versus healthy tissue (Figure 2.3). Initially, RNA is prepared from two different samples. During reverse transcription, two different

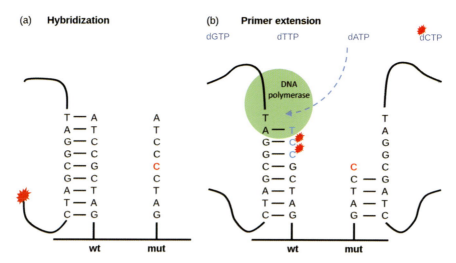

Fig. 2.2 Methods of genotyping by DNA arrays. (a) Direct hybridization. A fluorescently labeled sample is added to the array and hybridizes with high affinity to a fully complementary sequence. In the example shown, the wild-type (wt) sample hybridizes only to the oligonucleotide, which is complementary to the wt sequence. It does not hybridize to the primer for the mutated sequence carrying a mismatch.
(b) Discrimination can be improved by a primer extension reaction. In this assay, DNA polymerase extends only the oligonucleotide to which the sample hybridizes with full complementary. In case of a mismatch on the 3′ end, DNA polymerase does not incorporate labeled nucleotides.

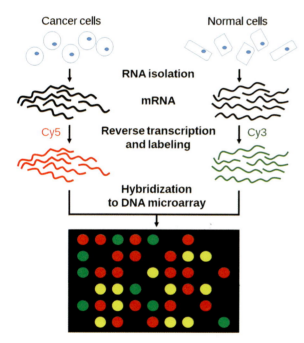

Fig. 2.3 Principle of two-color microarrays. Two-color microarrays can be used to compare the global expression patterns between two different states, for example, between tumor cells and normal cells. After harvesting the RNA, a red fluorescing Cy3 reporter is incorporated into the tumor sample during the reverse transcription procedure. The control sample is labeled with the green fluorescent Cy5 dye. The cDNAs are mixed and hybridized with the DNA probes immobilized on the microarray. A red spot indicates that a gene is transcriptionally activated in the tumor cell, while green spots represent genes that are downregulated. Yellow spots result from a mixture of green and red, and indicate genes whose expression is unaffected. Fluorescence is not detected for genes that are neither expressed in tumor nor control cells.

fluorescent dyes are incorporated into the cDNAs. This strategy is also known as two-color microarray analysis. The labeled cDNAs are mixed and allowed to hybridize with the DNA segments on the microarray. The fluorescence detected in the individual spots indicates up- or downregulation of gene expression.

The results of a microarray analysis are commonly represented as a scatter plot, in which the relative expression of genes in the sample (Y-axis) is plotted against the relative level of the identical transcript in the control sample (X-axis). Figure 2.4 shows a fictitious example in which the expression of genes in a tumor cell is plotted against their expression in a control, normal cell. Each point corresponds to the expression level of an individual transcript. Points on the diagonal represent transcripts that have identical expression levels under both conditions. In contrast, points that lay above the diagonal originate from genes that are overexpressed in the tumor cells, while points below the diagonal represent transcripts that have a lower level of expression in the tumor cells. A more than twofold change in the expression level of a gene is generally considered to be biologically relevant.

A more specific way to visualize results from a DNA microarray analysis is a heat map (Figure 2.5). Upregulated genes are shown in red, while genes with a reduced expression level are shown in green. In the study from which Figure 2.5 was taken, the transcriptomes of 78 sporadic breast tumors were analyzed. Five thousand genes were found to be significantly regulated across the group of samples. This example demonstrates

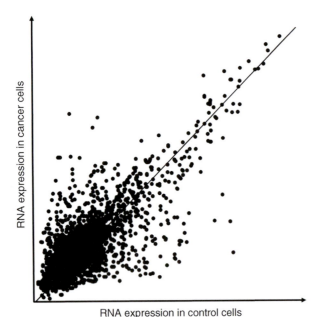

Fig. 2.4 Scatter plot of a DNA microarray experiment. The relative expression levels of genes in tumor cells are plotted as a function of the expression levels in control cells.

the power of bioinformatics to not only handle the enormous amount of data generated by DNA microarray analysis but to also draw biologically meaningful conclusions.

A first step in the evaluation of DNA array experiments is to perform a cluster analysis to group genes

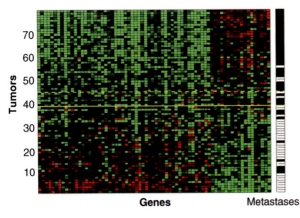

Fig. 2.5 Heat map of a DNA array analysis. Relative expression is shown for 70 prognostic marker genes from tumors of 78 breast cancer patients. Red indicates upregulation of a gene, green indicates reduced expression. Tumors are classified based on prognostic signatures. The metastasis status for each patient is shown in the right panel; white indicates patients who developed distant metastases within 5 years; black indicates patients who remained disease-free for at least 5 years. (Adapted from Ref. [1] with kind permission from Macmillan Publishers Ltd., Copyright, 2002.)

having similar expression patterns. This procedure allows the identification of genes that have similar regulation profiles and are likely to be involved in the same or related pathways. In the example given, patients were divided into a good prognosis group that remained disease-free for at least 5 years, and another group that developed metastases within 5 years. Seventy prognostic marker genes were identified that allowed classification of the patients and correctly predicted the actual outcome of disease for more than 80% of the patients (Figure 2.5). This type of expression profiling may help to optimize clinical diagnosis (Section 8.3.3) and therapeutic strategies.

Confirmation of the bioinformatic analysis of data obtained from a microarray analysis is required to obtain reliable results. This confirmatory analysis should be performed by one or more independent methods. This is because samples may cross-hybridize with unspecific probes; in addition, incorrect fluorescence signals may be recorded. Traditionally, northern blots or RNase protection assays have been carried out to validate the results of microarray analyses. A more advanced method of analysis is to use the quantitative polymerase chain reaction, which will be explained in the next section. More recently, RNA sequencing by next-generation sequencing technologies has become an alternative strategy for expression profiling (Section 2.3).

The overview given here depicts only the most advanced applications of DNA array analysis. While the technology has already become a routine method for profiling noncoding and regulatory RNAs (e.g., microRNAs, Section 13.4), further applications such as splice-variant analysis, epigenetic studies, or large-scale sequencing are at earlier stages of development. For additional information, the reader is referred to the reference given at the end of this chapter.

2.2
Quantitative Polymerase Chain Reaction

Polymerase chain reaction has become one of the most widely used methods in molecular medicine and is routinely applied in virtually every laboratory working in the field. Conventional PCR, as developed by Kary Mullis in 1983 (Box 2.1), permits the exponential amplification of DNA fragments and the endpoint analysis of the final products of the reaction. Quantitative PCR is a further development of the technique that permits monitoring of the amplification of the DNA fragment online during the run. It is also known as real-time PCR. However, the

Box 2.1. Kary B. Mullis

Kary Mullis has been one of the more colorful characters of science over the last few decades. He is famous for the invention of the PCR, which rapidly became a standard technique in modern life sciences and medical diagnostics. In 1983, Mullis was working as a DNA chemist at the Cetus Corporation, a biotechnology company in Emeryville, CA. He described the moment when he had his insight into an article published in *Scientific American* entitled "The unusual origin of the polymerase chain reaction." As Mullis has it, he was driving his car down a moonlit mountain road in northern California as his girlfriend slept next to him. He allowed his mind to wander when all of a sudden he realized that he could amplify a fragment of DNA exponentially by extending primers hybridizing to the target sequence. His sleepy girlfriend did not share his enthusiasm and refused to wake up. After returning to work, his company took Mullis off his usual projects to concentrate on the development of PCR. People were fascinated by the simplicity of the PCR technique and it was difficult to understand, even for Mullis, why nobody thought of it before. In 1993, Mullis received the Nobel Prize in Chemistry.

Soon after the invention of the PCR, Mullis left Cetus and worked for other companies and as a private consultant. He also started a business selling pieces of jewelry and watches containing the amplified DNA of deceased famous people like Elvis Presley, Marilyn Monroe, and Abraham Lincoln. In his autobiography, he wrote about chemically synthesizing and using mind-altering substances, such as LSD and other chemicals. He even stated that he seriously doubts that he would have invented PCR if he had not ever taken LSD. Mullis has also drawn much criticism for promoting pseudoscientific ideas in his later life. For example, he denied that there is a connection between HIV and AIDS, and disagreed with evidence supporting climate change and ozone depletion, although he has never done research in either of these fields.

PCR can amplify small amounts of a DNA fragment by several orders of magnitude, in the process generating millions of copies of a particular DNA sequence. The mix for a PCR reaction contains the target DNA sample, a thermostable DNA polymerase, dNTPs, two primers, and ions (e.g., Mg^{2+}) needed for enzymatic activity. The PCR procedure typically consists of 25–40 cycles of discrete temperature steps (Figure 2.6). In

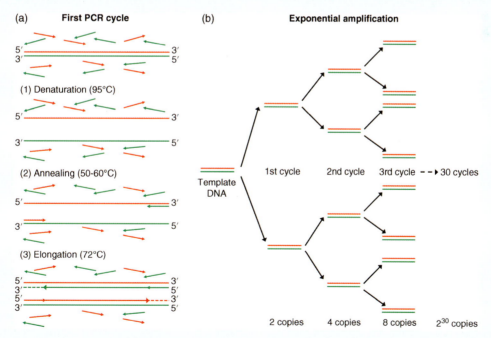

Fig. 2.6 The polymerase chain reaction (PCR). (a) A typical PCR cycle consists of three steps: denaturation, annealing of the primers (the forward primers are indicated as red arrows, the reverse primers are shown as green arrows) and elongation. The number of DNA fragments is doubled in each PCR cycle. (b) Exponential amplification of DNA by the PCR reaction. In the first cycle, the template DNA is doubled. After the second cycle, four copies have been produced, and after the third cycle, eight copies have been produced. After 30 cycles, 2^{30} copies have been theoretically generated.

the first step, the DNA is denatured (usually at 95 °C), that is, the hydrogen bonds between the complementary bases of the two strands are broken, yielding single-stranded DNA molecules. In annealing step, the temperature is lowered to 50–60 °C, allowing the primers to bind to the single-stranded DNA. During the elongation (or extension) step, the DNA polymerase synthesizes a new DNA strand complementary to the DNA template strand. The polymerases used typically have an optimum activity temperature of ~72 °C.

abbreviation RT-PCR, which is sometimes mistakably used for real-time PCR, will here only be used for reverse-transcription PCR. A quantitative RT-PCR experiment is thus abbreviated as qRT-PCR. qRT-PCR (sometimes abbreviated as qPCR) experiments are widely used in medical and biological research to quantify mRNA expression levels. qPCR assays have also gained importance in medical diagnosis for the genotyping of inherited diseases, for viral diagnostics, and for numerous other applications.

In the early PCR experiments, the DNA polymerase had to be added after each round of heating, because heat inactivated the enzyme. It was a major advancement when thermostable DNA polymerases were introduced for PCR applications, since they eliminate the need to add fresh enzyme in each cycle. The most widely used DNA polymerase, Taq polymerase, is derived from the thermophilic bacterium *Thermus aquaticus*. Taq polymerase rapidly synthesizes the new DNA strand; however, due to the lack of proofreading activity of the enzyme, its error rate is comparatively high. If the amplified DNA is to be used in further experiments, the number of mutations must be minimal. In these cases the Pfu polymerase, derived from the thermophilic archeon *Pyrococcus furiosus*, can be used since its error rate is much lower. The drawback of the proofreading step carried out by the Pfu polymerase is its lower reaction velocity. More recently developed DNA polymerases, such as the Phusion DNA polymerase, a *Pyrococcus*-like enzyme fused with a processivity-enhancing domain, combine both desired properties and rapidly amplifies DNA with a low error rate.

Since the amount of DNA is doubled in each cycle, PCR rapidly amplifies DNA in an exponential manner (Figure 2.6b). After 30 cycles, $2^{30} \approx 1$ billion copies of each template molecule initially present in the reaction have been produced. While the PCR reaction can only amplify DNA, its combination with a reverse-transcription reaction (RT-PCR) also permits the detection, amplification, and quantification of RNA molecules. For these applications, the RNA is initially transcribed into complementary DNA (cDNA) by the reverse transcriptase using a reverse primer (Figure 2.7). A DNA polymerase then synthesizes the second strand, and the

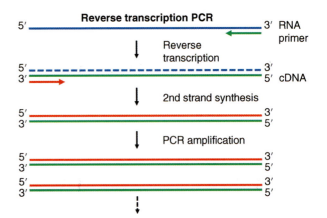

Fig. 2.7   Reverse transcription PCR (RT-PCR). RT-PCR experiments start with the synthesis of cDNA by reverse transcriptase using the reverse primer. The enzyme naturally possesses RNase H activity and degrades the RNA moiety of the hybrid double strand. However, mutant enzymes lacking RNase H activity are sometimes used to produce longer cDNAs with higher yield. After synthesis of the second strand, the cDNA can be amplified by PCR. Technically, the second strand synthesis is usually the first cycle of the PCR reaction.

double-stranded cDNA is then amplified by PCR. RT-PCR is routinely used for two applications: (i) The cDNA synthesized from a cellular mRNA can be cloned into an expression vector so that the encoded protein can be produced from intron-free genetic information. Cloning of a full-length gene from genomic DNA is very demanding, as it is significantly longer due to the presence of introns. In addition, a full-length gene containing introns cannot be expressed in bacteria, which cannot carry out splicing reactions. (ii) Quantitative RT-PCR as a real-time experiment is widely used to quantify the level of a specific RNA in the cell (see below).

The basic principle of qPCR is the same as that of conventional PCR, except for the addition of a fluorescent dye that allows online detection of the amplified DNA. Figure 2.8 shows the typical traces of a qPCR experiment. At the beginning of the experiment, only a small amount of DNA is present and fluorescence is low. After several rounds of DNA amplification, the fluorescence intensity rises above the background level. In the exponential phase, the amount of DNA doubles in each cycle. At the end of the PCR experiment, the

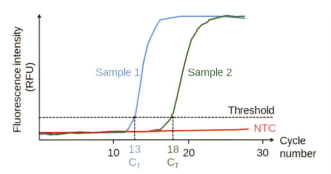

Fig. 2.8 Typical fluorescence signal of a qPCR run. The fluorescence intensity at low cycle number is background noise. At higher cycle number, fluorescence intensity increases above the background level. After additional cycles, a plateau is reached. The threshold is set at the point where the fluorescence intensity exceeds the background level. The threshold cycle numbers (C_T) of the blue and green samples are indicated. The no-template control (NTC) does not give a signal.

enzyme loses activity, and the nucleotides are used up. As a consequence, amplification slows and a plateau phase is reached. An important control confirming the specificity of a qPCR experiment is the no-template control (NTC), which must not give a signal. To evaluate the experiment, a threshold is defined at which the fluorescence intensity exceeds the background level. The cycle number at which the fluorescence intensity of a given sample crosses this threshold is called the threshold cycle (C_T) (sometimes also referred to as the crossing point (C_P) or the quantification cycle (C_q)).

Two types of quantitative real-time PCR experiments can be distinguished: absolute and relative quantification. Both types of assay can be used to quantify the amount of a specific RNA in the cell. The RNA must first be reverse transcribed into cDNA. For the absolute quantification of the amount of (c)DNA in a sample, a standard curve with known amounts of the DNA is usually recorded. From a diagram in which the C_T values are plotted against different amounts of the standard, the absolute amount of DNA in the sample can be calculated after determining its C_T value.

A typical application of relative quantification is the evaluation of the relative changes in gene expression in two samples. For example, this could be a comparison between a disease state (e.g., cancer cells) or a treatment group versus a normal, untreated state (i.e., healthy cells). The traces shown in Figure 2.8 represent a typical example of such an experiment, in which sample 1 may be a disease state and sample 2 the healthy control. After reverse transcription, the C_T values for both samples are determined by qPCR. The C_T value of sample 1 is 13, while the C_T value of sample 2 is 18. A lower C_T value indicates that the gene is expressed at a higher

level. Since the amount of DNA doubles in each PCR cycle in this fictitious experiment, the gene can be calculated to be 32-fold upregulated in the cancer cell. In actual applications, the C_T value of the gene under investigation is given relative to a reference gene to correct small variations in the amount of sample used in the experiment. Typical reference genes are housekeeping genes such as β-actin, glyceraldehyde-3-phosphate dehydrogenase (GAPDH), or ribosomal RNAs (commonly the 18S rRNA). More recently, hypoxanthine phosphoribosyltransferase has become a widely used reference gene. An essential prerequisite is that the expression of these genes does not vary under the experimental conditions. The change in the expression of the target gene is then commonly calculated according to the delta–delta C_T method: Under each experimental condition, the ΔC_T value can be calculated as

$$\Delta C_T(\text{condition X}) = C_T(\text{target}) - C_T(\text{reference}).$$

The ΔC_T values for both experiments (e.g., cancer cell versus normal cell) are then subtracted from each other:

$$\Delta\Delta C_T = \Delta C_T(\text{condition A}) - \Delta C_T(\text{condition B}).$$

The change in the expression level can finally be calculated from the $\Delta\Delta C_T$ value. This calculation assumes optimal amplification efficiency, that is, a doubling of the number of target DNA fragments in each cycle. If the efficiency is lower, a correction factor has to be used for the calculation. This brief description of the evaluation of qPCR experiments demonstrates that numerous factors (amplification efficiency, the nature of reference gene, the quality of the RNA, etc.) must be considered. Therefore, guidelines for the *M*inimum *I*nformation for Publication of *Q*uantitative Real-Time PCR *E*xperiments (MIQE) have been defined (see reference list).

Two detection principles for qPCR experiments can be employed: the addition of double-stranded DNA-binding dyes as reporters and the use of fluorescent reporter probes. Both principles are based on the detection of fluorescence whose intensity is proportional to the amount of DNA amplified.

The most commonly used double-stranded DNA-binding dye is SYBR green. Binding of the dye to double-stranded DNA causes the emission of green fluorescence after excitation with blue light (Figure 2.9). This approach is sensitive, relatively simple, and cost-efficient. Only two PCR primers and the dye are needed. The method can easily be adapted for any target DNA. However, a disadvantage of the use of double-stranded DNA-binding dyes is their lack of specificity. The dye will bind not only to the amplified target DNA but also to the

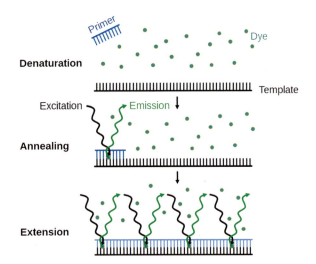

Fig. 2.9 Principle of double-stranded DNA-binding dyes as reporters. The dye binds to double-stranded DNA and emits green fluorescence after excitation. The intensity of the fluorescence is proportional to the amount of double-stranded DNA in the sample.

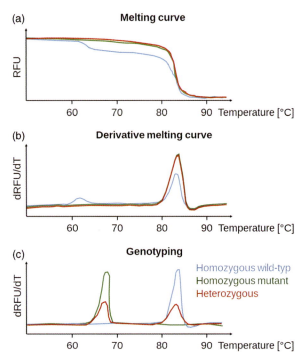

Fig. 2.10 Melting curve analysis of real-time PCR experiments. (a) Melting curves of three samples. (b) First derivative of the melting curves. All samples have a major peak at the melting temperature of 83 °C. However, the blue trace has a second peak at 61 °C indicating the detection of by-products. (c) Genotyping by melting curve analysis. The homozygous wild-type sequence has a melting temperature of approximately 85 °C, while the melting temperature of the homozygous mutant sequence is only 70 °C. The heterozygous genotype has peaks at both melting temperatures.

reaction by-products or primer dimers. This will interfere with the accurate quantification of the target sequence. To verify the absence of by-products in a sample, melting curve analysis is usually carried out: The end product of the PCR experiment is melted by increasing the temperature. When the melting temperature (T_M) of the DNA fragment is reached, the two strands are separated so that the dye can no longer bind and the fluorescence intensity decreases (Figure 2.10a). The inflection point of the first derivative of the melting curve allows the calculation of T_M (Figure 2.10b). In the example shown, the green and red traces have a single transition, confirming the synthesis of a single product. In contrast, the blue trace indicates the existence of a by-product, for example, primer dimers. As a consequence, the quantification of this sample will be inaccurate.

Melting curve analysis of the end product of a real-time PCR run can also be used for the diagnosis of inherited diseases. In the example shown in Figure 2.10c, the homozygous wild-type sequence has a higher melting temperature than the homozygous mutant genotype. The heterozygous genotype has peaks at both melting temperatures. As an example, this type of analysis can be used for the detection of a mutation in human factor V, which causes a hypercoagulability disorder (factor V Leiden).

An alternative to assays using double-stranded DNA-binding dyes are probe-based detection systems with much higher specificity, since the probes specifically hybridize to the amplified target sequence. Sequence-specific probes also allow the analysis of multiple targets in a single reaction (called multiplex PCR). The main

disadvantages of the use of fluorescent reporter probes are high cost and the laborious development of the assay procedure.

Most fluorescent probe systems used in qPCR assays rely on a physical phenomenon known as the *Fluorescence Resonance Energy Transfer* (FRET, sometimes also denoted as *Förster Resonance Energy Transfer*). This form of energy transfer occurs between two chromophores in proximity. In the case of qPCR probes, a quencher prevents emission of fluorescence by a reporter molecule, as the excitation energy of the fluorophore is transferred to the quencher in a nonradiative manner. The efficiency of this energy transfer is extremely sensitive to small distances, as it is inversely proportional to the sixth power of the distance between the donor and the acceptor.

The most widely used probe-based detection method for qPCR assays is based on the 5′ exonuclease activity of DNA polymerase (Figure 2.11a). This type of assay is also known under the trade name of TaqMan assay (the name is derived from the combination of Taq

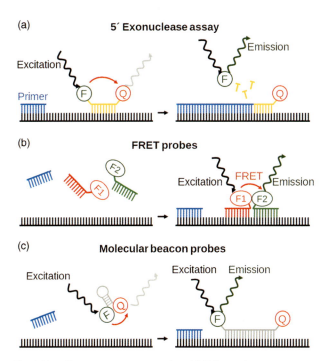

Fig. 2.11 Fluorescent reporter probes. (a) 5′ Exonuclease assay (TaqMan assay). (b) FRET probes. (c) Molecular beacon probes. For further details, see text.

polymerase and the videogame PacMan). The oligonucleotide probe contains a fluorescent reporter at one end and a quencher of fluorescence at the opposite end. Proximity of the fluorescent reporter to the quencher prevents detection of the fluorescence due to the transfer of excitation energy by FRET. As Taq polymerase extends the primer and synthesizes the new DNA strand, the $5′ → 3′$ exonuclease activity of the polymerase degrades the probe that has annealed to the template. Degradation of the probe releases the fluorophore, thus eliminating its proximity to the quencher. As a consequence, the excitation energy is no longer transferred to the quencher and can be detected as fluorescence emitted by the fluorophore.

The utility of FRET probes as reporters (Figure 2.11b) is also based on the nonradiative transfer of excitation energy between two dyes. However, the design of the assay is slightly different from the 5′ exonuclease assay described above. In this case, two probes, each carrying one fluorophore, hybridize to the DNA in proximity. After excitation of fluorophore 1 by the cycler's light source, the energy is transferred by the FRET mechanism to the acceptor fluorophore 2 that is attached to the adjacently hybridized second probe. The energy is then emitted as fluorescence by fluorophore 2, is proportional to the amount of DNA present, and can be detected by the cycler.

Molecular beacons (Figure 2.11c) carry a fluorescent reporter and a quencher like TaqMan probes. Their special feature is that they are somewhat longer than TaqMan probes and form a hairpin structure. The loop of the hairpin harbors the probe sequence complementary to the target DNA. Additional short stretches at both termini are complementary to each other and form a stem in the unbound state. The fluorophore is usually covalently attached to the 5′ end of the oligonucleotide, and the quencher is covalently attached to the 3′ end. As long as the beacon is in the closed-loop shape, the fluorophore and the quencher are in proximity and the fluorescence of the reporter is quenched. When the probe is hybridized to its target sequence, fluorophore and quencher are separated and excitation of the reporter results in fluorescence emission.

In addition to the assay principles described above, additional detection systems such as minor groove binders and so-called "scorpion" probes have been developed for qPCR experiments. All formats differ in ease of probe design, cost, and stability of the probe, in addition to sensitivity and selectivity. Although SYBR green is most widely used as a nonspecific fluorescent reporter and 5′ exonuclease assays are commonly carried out for the sequence-specific detection of amplified DNA, the optimal detection method must be determined for each application.

In addition, a plethora of PCR and qPCR techniques have been developed in the last few decades that cannot be exhaustively covered here. Several applications will be discussed in the relevant sections of the textbook; these include allele-specific PCR for the genetic diagnosis of single-nucleotide polymorphisms (Section 8.3.1), or the use of PCR for the detection of viruses and other pathogens (Chapters 5 and 6).

2.3
Next-Generation Sequencing

Advancements made in DNA sequencing technologies are widely recognized as having produced a major revolution in the life sciences over the past few years. For decades, the chain terminator method developed by Frederick Sanger in 1977 was the dominating sequencing technology. This approach is now considered "first-generation" sequencing. In recent years, various technologies referred to as next-generation sequencing (NGS) or high-throughput sequencing (HTS) have been developed that allow massive parallelization of the sequencing process and fulfill the rising demand for large amounts of sequence data.

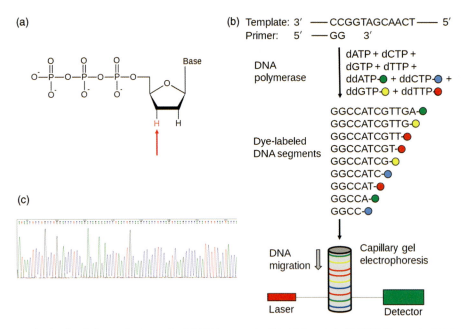

Fig. 2.12 DNA sequencing by the chain-terminator method. (a) Dideoxynucleotide triphosphate (ddNTP). The nucleotide lacks the OH-group at the 3′-position (indicated by the arrow). (b) Automated sequencing by gel electrophoresis in a capillary gel. The ddNTPs are labeled with four different fluorophores. DNA segments are separated by electrophoresis in a capillary gel and the 3′-terminal nucleotide is identified by laser-stimulated fluorescence. (c) Electropherogram of the DNA sequencing. The four colored curves represent each of the four nucleotides. (Part (b) adapted from Ref. [2] with kind permission from John Wiley & Sons, Inc.)

The conventional chain termination method is based on the enzymatic synthesis of a new DNA strand complementary to the template sequence that terminates after specific bases have been added to the growing chain. DNA polymerase uses four deoxynucleoside triphosphates (dNTPs) for the polymerization reaction. The key component of the reaction mixture is a small amount of a dideoxynucleotide triphosphate (ddNTP) lacking the 3′-OH group (Figure 2.12a). When the ddNTP is incorporated into the growing chain, the extension reaction is terminated, because the addition of the next nucleotide requires a free OH-group at the 3′-position. Since only a small amount of the ddNTP is present in the reaction mixture, a series of truncated chains is generated, all of which end with the respective dideoxynucleotide. The fragments are separated according to their size by gel electrophoresis. For many years, the DNA segments were detected by the incorporation of radioactively labeled nucleotides. The introduction of fluorescent labels helped to improve the experimental process.

A major increase in sequencing capacity was achieved by the introduction of automated sequencing devices, which separate the labeled DNA fragments by electrophoresis in a capillary gel (Figure 2.12b). Since each dideoxynucleotide is covalently linked to a different dye, the terminating base of each segment can be determined by detection of laser-stimulated fluorescence. The resulting electropherograms (Figure 2.12c) can easily be analyzed. The complete sequencing process can be automated: Modern devices can carry out 96 or 384 sequencing reactions in parallel in capillary tubes, and up to 1000 bases per reaction can routinely be sequenced.

In the last decade, various NGS technologies have been introduced that dramatically increase sequencing capacity. An important feature of the new methods is the elimination of the rate-limiting electrophoretic step and the massive parallelization of the sequencing process. NGS technologies are sometimes further divided into second- and third-generation sequencing approaches (Table 2.1). Third-generation sequencing approaches have two characteristics in common: A PCR reaction is no longer needed prior to sequencing, which shortens the time for DNA preparation, and the signals are captured in real time during the enzymatic reaction of adding a nucleotide to the complementary strand.

The first NGS devices reached the market in 2005, which was after the completion of the worldwide project to sequence the human genome (Chapter 7). However, this sequencing project had sparked demand to sequence the genomes of other organisms and to sequence individual human genomes. These efforts are reflected in the "$1000 genome" catchphrase, which describes the goal of sequencing the full genome of a human being for

Table 2.1 Next-generation sequencing (NGS) platforms.

Company	Principle	Read length (bases)	Output data per run (Gb)	Run time
Second generation				
Roche/454	Pyrosequencing	700	0.7	24 h
Illumina/Solexa	Reversible terminator	100	600	3–10 days
Life Technologies	Sequencing by oligoligation detection (SOLiD)	50	120	7–14 days
Ion Torrent/Life Technologies	Semiconductor	Up to 400	1	2 h
Third generation				
Helicos BioSciences	Heliscope, single-molecule sequencing	Up to 55	Up to 35	Up to 8 days
Pacific Bioscience	Single-molecule real-time (SMRT)	Up to 6000	Up to 0.09	1 h
Oxford Nanopore Technologies	Nanopore sequencing	N/A	N/A	N/A

Companies marketing the various technologies are given and general features for each of the technologies are given. It should be noted that technical characteristics can change rapidly due to the continuous improvement of these methods; also note that there is no common consensus on the classification of the various methods. In addition, a number of new technologies, including nanoball sequencing, FRET sequencing, and sequencing by scanning tunneling or transmission electron microscopy are in development. The read length describes the number of bases that are continuously sequenced on average in a single reaction; the output data give the total amount of sequence information in gigabases (Gb = 1 billion bases) that is obtained by the device in a single run. N/A, information not available.

$1000. Knowledge of individual genomic information may help to open a new era of predictive and personalized medicine. Several companies currently have the sequencing of a human genome for $1000 or less on offer.

A common initial step in most NGS approaches is the generation of short DNA fragments. To enable this, long DNA segments, for example, chromosomal DNA, are sheared into a library of fragments of a few hundred base pairs in length. These fragments are then sequenced by either of the following methods.

A commercialized first NGS technology was introduced by 454 Life Sciences (acquired by Roche Diagnostics), and was a parallelized version of pyrosequencing. The basic principle of this method is the detection of pyrophosphate (PPi) released during DNA synthesis (Figure 2.13a and b). The four nucleotides are added in a sequential manner to the sequencing reaction, that is, one of the four nucleotides in each round of sequencing. When the particular nucleotide is incorporated into the newly synthesized strand, PPi is released. In a subsequent reaction, ATP sulfurylase uses PPi to convert adenosine-5′-phosphosulfate (APS) into ATP. In the final reaction, ATP provides the energy for luciferase to oxidize luciferin and generate light, which is detected by a very sensitive charge-coupled device (CCD) camera. The signal reflects the incorporation of the specific nucleotide that was added to the reaction in the respective round, and the light intensity is proportional to the

number of incorporated nucleotides. After washing away unincorporated nucleotides, the next nucleotide is added for the subsequent round. With this iterative process, the sequence of the DNA segment can be deduced from the pyrogram (Figure 2.13c).

Since the method is not designed to detect a single fluorescent event, the template must be amplified prior to the sequencing reaction. Amplification of the template is commonly achieved by emulsion PCR (EmPCR) (Figure 2.14). In the first step of this procedure, single-stranded DNA fragments containing an adapter are immobilized by attaching them to the surface of beads. Each bead is linked to a single DNA fragment from the DNA library. The bead–DNA complexes are then compartmentalized into water–oil emulsion droplets. PCR amplification is performed within these droplets so that each bead contains millions of copies of a single DNA template. The beads can then be deposited into individual wells of a picotiter plate, where the sequencing reaction is carried out.

An alternative NGS technology, the Illumina (Solexa) sequencing approach, is based on sequencing by synthesis (SBS) with reversible dye terminators. In the initial process of library generation (see above), oligonucleotide adaptors are ligated to the DNA fragments. These adaptors are necessary to attach the fragments to a solid support by hybridization with the complementary oligonucleotides on its surface. Clusters with a hundred million unique clonal DNA fragments are then generated by

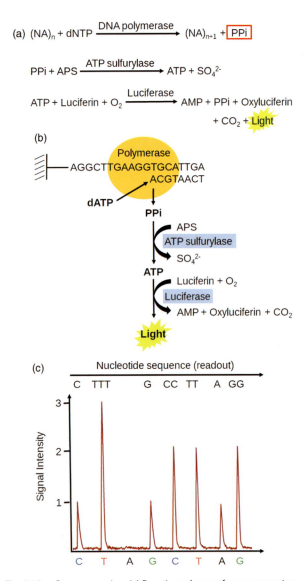

(a) $(NA)_n + dNTP \xrightarrow{\text{DNA polymerase}} (NA)_{n+1} + \boxed{PPi}$

$PPi + APS \xrightarrow{\text{ATP sulfurylase}} ATP + SO_4^{2-}$

$ATP + Luciferin + O_2 \xrightarrow{\text{Luciferase}} AMP + PPi + Oxyluciferin + CO_2 + \text{Light}$

(b)

Polymerase

—AGGCTTGAAGGTGCATTGA
ACGTAACT

dATP

PPi

APS

ATP sulfurylase

SO_4^{2-}

ATP

Luciferin + O_2

Luciferase

$AMP + Oxyluciferin + CO_2$

Light

(c)

Nucleotide sequence (readout)

C TTT   G CC TT   A GG

Signal Intensity

C   T   A   G   C   T   A   G

Fig. 2.13 Pyrosequencing. (a) Reaction scheme of pyrosequencing. The four nucleotides are added sequentially. When the DNA polymerase incorporates a dNTP to extend the nucleic acid chain by one nucleotide, pyrophosphate (PPi) is released. This molecule is subsequently employed by ATP sulfurylase to convert adenosine-5'-phosphosulfate (APS) into ATP, which provides the energy for the luciferase reaction resulting in the emission of light. (b) Schematic representation of the process. (c) Pyrogram. The peak heights are proportional to the number of the respective nucleotides incorporated into the newly synthesized DNA strand. The DNA sequence of the example shown can be read from the pyrogram as CTTTGCCTTAGG.

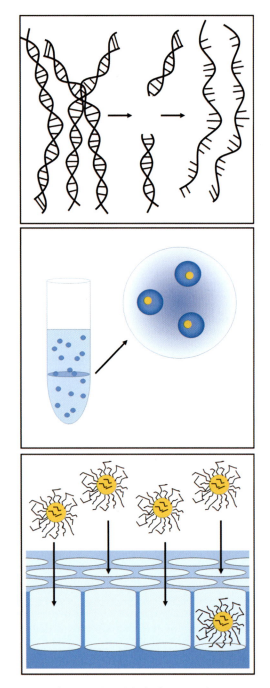

Fig. 2.14 Emulsion PCR (EmPCR). The fragmented genomic DNA is ligated to adapters and separated into single strands. The fragments of the library are linked to beads (only one DNA fragment per bead) and the beads are suspended in water–oil emulsions to perform the PCR reaction. At the end of the reaction, each bead carries copies of a single-DNA template. The beads are deposited into the individual wells of a picotiter plate.

a procedure known as *bridge amplification*. This term reflects that both ends of the DNA are attached to the surface of the chip. For the sequencing reaction, all four nucleotides are added, each containing a specific cleavable fluorescent dye and a removable blocking group (Figure 2.15). DNA polymerase incorporates the base that is complementary to the template nucleotide. Due to the blocking group, DNA synthesis is terminated after the addition of a single nucleotide. Unincorporated nucleotides are washed away, and imaging is performed,

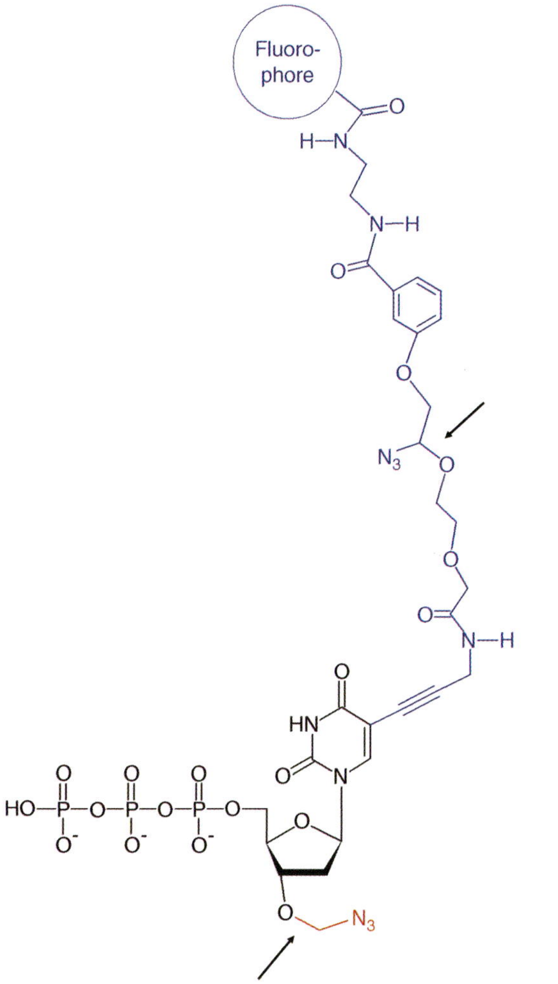

Fig. 2.15 Example of a reversible terminator nucleotide. The nucleotide contains a cleavable dye, which is specific for each of the four nucleotides and a 3′-*O*-azidomethyl group, which terminates the polymerization reaction. After detection of the fluorescence signal, the dye and the terminator are removed (arrows).

producing images of the chip surface to determine the identity of the incorporated nucleotides. The terminating group and the fluorescent dye are then removed, and after an additional washing step, the next base can be incorporated.

Life Technologies' SOLiD system is based on sequencing by ligation rather than by synthesis (Figure 2.16). Because the underlying principle is very sophisticated, only a basic introduction can be given here. Prior to the sequencing procedure, the template DNA is amplified by emulsion PCR as described above, and the bead–DNA complexes are deposited on a glass slide. A library of probes labeled with four different fluorophores is added. The probe oligonucleotides are composed of two fixed starting bases (the interrogation bases AT, AC, AA, and

GA in the example shown in Figure 2.16a) and six degenerate and universal bases, respectively. Each fluorophore represents four different dinucleotide combinations (Figure 2.16c). The probes selectively hybridize to a site complementary to the two interrogation bases (i.e., only the primer with the dinucleotide AT can bind to the TA sequence of the template in the example) and DNA ligase then joins the dye-labeled probe to the primer. Following the imaging step, the last three bases carrying the dye are chemically cleaved and a new set of probes is added. The ligation cycle is repeated several times. Then the universal seq primer is stripped and another round of ligation is performed with a starting primer shifted one nucleotide toward the 5′ end of the template (not shown in the figure). After several rounds with shifting starting primers, the sequence can be deduced from the combined information according to a sophisticated algorithm (Figure 2.16c and d).

Semiconductor sequencing developed by Ion Torrent, Inc. (now owned by Life Technologies) is based on the detection of protons that are released during DNA synthesis, as opposed to most of the other approaches that employ optical methods. In each round of this semiconductor sequencing procedure, a single type of nucleotide is added to a microwell containing the template DNA. Introduction of the nucleotide into the growing DNA strand by DNA polymerase causes the release of a proton that is detected by a very sensitive ion sensor. The sequencer may be thought of as a small sophisticated pH meter. The intensity of the detected signal indicates the number of nucleotide repeats that are incorporated in a single cycle.

A common characteristic of third-generation sequencers is the direct sequencing of a single molecule, eliminating the need for PCR amplification before sequencing. Both true single-molecule sequencing (tSMS) on the HeliScope sequencer (Helicos BioSciences) and the single-molecule real-time (SMRT) sequencing method developed by Pacific Bioscience follow the principle of detecting the incorporation of a nucleotide into a single-molecule template DNA. The detailed procedures, however, differ significantly. For the HeliScope sequencer, the DNA strand is attached to a specific position on the flow cell surface. All four nucleotides are labeled with the same dye and dispensed individually in a predetermined order. A highly sensitive camera images the positions of the fluorescence signals emitted by the strands, which incorporate the specific nucleotide added in each step. After removal of the dye and washing, the next nucleotide is added. These cycles are repeated for reads of up to 55 bases.

The SMRT technology may outpace the tSMS method as it can achieve a longer read length than any of the other next-generation sequencing approaches. This

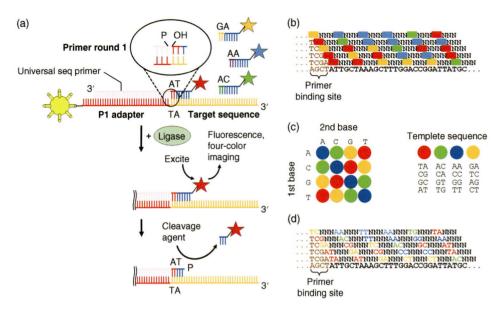

Fig. 2.16 Basic principle of SOLiD sequencing. Primers are added to the target sequence and linked to the sequencing primer by DNA ligase. Following fluorescence imaging, the dye and three nucleotides are removed by chemical cleavage. (b) After several rounds of ligation reactions, a color pattern has been recorded. With the help of the matrix (c) and the known sequence of the primer binding site, the sequence can finally be deciphered (d). (Part (a) adapted from Ref. [3] with kind permission from Macmillan Publishers Ltd., Copyright, 2010.)

platform uses a DNA polymerase anchored to the bottom surface of a very small container named a zero-mode waveguide (ZMW) (Figure 2.17). A sequencing chip contains many ZMWs, each harboring only one enzyme molecule and a single DNA template. The four nucleotides added for the sequencing reaction contain distinct fluorescent dyes that are attached to their phosphate group. When the nucleotide is incorporated by DNA polymerase, the fluorescent dye will be cleaved. A camera captures the signal in a movie format in real time.

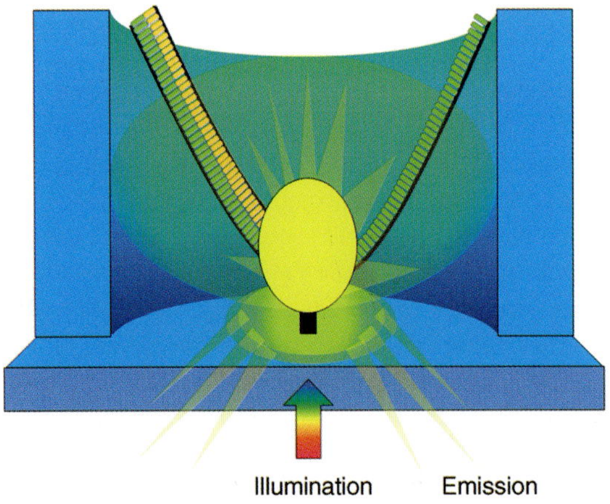

The basic idea underlying nanopore sequencing is the detection of a single molecule of a DNA or RNA strand that passes through a small pore with a nanoscale diameter (Figure 2.18). The *Staphylococcus aureus* toxin α-hemolysin is widely employed to form the pore in a membrane, but other biological as well as nonbiological molecules can also be used. An electric current, due to conduction of ions through the nanopore, is induced by applying a potential difference (voltage) across the membrane. An exonuclease attached to the top of the pore molecule cleaves off the terminal nucleotide of the DNA strand to be sequenced. The single nucleotide is then electrophoretically driven through the nanopore and blocks the channel; this can be detected by electrophysiological techniques. The crucial point is that the current disruption is sensitive to the size and structure of the molecule that passes through the nanopore. Thus, each of the four nucleotides produces a specific signal, and the extent by which the current is reduced permits the identification of the nucleotide passing through the pore. The next nucleotide is then cleaved off by the exonuclease and identified when it passes through the nanopore. These steps are continued until the complete sequence is read. Alternative approaches are in development in which the complete DNA strand is

Fig. 2.17 Single-molecule real-time (SMRT) DNA sequencing. The reaction is carried out in a small well in a zeptoliter (10^{-21} l) volume capturing a single DNA strand at the bottom surface. The four nucleotides are labeled with distinct dyes. Only the fluorescence emitted during incorporation of a nucleotide into the growing chain by DNA polymerase at the bottom of the cell is detected by the camera. The label is detached from the nucleotide upon incorporation into the DNA strand. This procedure allows real-time detection of the DNA synthesis reaction.

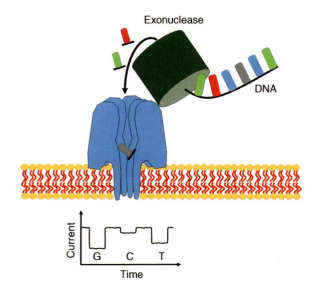

Fig. 2.18 Principle of nanopore sequencing. A voltage is applied across a membrane containing nanopores. The exonuclease cleaves off the terminal nucleotide of the DNA strand, which then moves through the pore. Disruption of the current is measured by electrophysiological techniques. The signal is specific for each nucleotide (diagram) and allows the sequence to be read.

electrophoretically driven through the nanopore without cleaving off the terminal nucleotide by the exonuclease. However, this strategy has been encountering technical difficulties.

The company Oxford Nanopore Technologies has developed a miniaturized nanopore device containing a disposable chip that can be plugged into the USB port of a computer. Other companies are developing different variations of the nanopore technology. The basic principle is attractive as it allows direct reading of the DNA sequence and does not require the introduction of modified nucleotides. Thus, the device can read long nucleic acid (NA) sequences at high speed.

NGS technologies have drastically changed genomic research in modern life sciences as they allow the elucidation of genomes from all organisms. Hopefully, they can to be used routinely for the sequencing of individual human genomes for personalized medical approaches in the not-too-distant future.

However, NGS is associated with a comparatively high error rate. For the *de novo* sequencing of genomes and for the detection of single-nucleotide polymorphisms in well-known genomes such as the human genome, it is important to distinguish between sequencing errors and genomic variations. The accuracy of sequencing results can be increased by sequencing a genome several times. This approach is known as *deep sequencing*. The *coverage* indicates the number of reads performed on a given

nucleotide. For example, a coverage of 30 means that each nucleotide has, on average, been sequenced 30 times. Although there is no strict definition, the term ultradeep sequencing is sometimes used to indicate that the coverage is 100 or higher.

Although initially developed as a tool to sequence DNA, the use of NGS technologies for RNA sequencing (RNA-Seq) has also become an attractive additional feature. RNA-Seq has become an alternative to gene expression profiling by DNA microarrays (Section 2.1) in the transcriptomics field. For mRNA profiling by RNA-Seq, the first step is usually enrichment of coding RNAs with poly(T) oligonucleotides covalently attached to beads. For technical reasons, the RNA is sometimes hydrolyzed into smaller fragments of 200–300 nucleotides. Modified protocols can be used for sequencing of small RNAs without a poly(A) tail such as microRNAs (Section 13.4). Independent of the type of RNA to be sequenced, the next step is conversion of RNA into cDNA, followed by sequencing with a suitable NGS technology. Since the reverse transcription step may introduce a bias, single-molecule-based third-generation approaches offer the advantage of direct sequencing of RNA molecules without their conversion into cDNA. The central postulate of RNA-Seq is that the expression level of a (m)RNA can be deduced from the number of times a sequence is retrieved. In other words, a sequence that is read twice as often in the sequencing run as another sequence is considered to reflect a twofold higher expression level of the respective (m)RNA.

RNA-Seq has several advantages compared to DNA microarrays. One main deficiency of hybridization-based approaches is that they only allow detection of the limited set of sequences that is represented in the set of probe oligonucleotides. Rare allelic variants or new mutations, for example those found in cancer cells, may remain undetected. In contrast, RNA-Seq is independent of any previous knowledge of the expected target molecules. The method goes far beyond expression profiling as it detects new sequences. In fact, numerous new microRNAs (Section 13.4) have been discovered by RNA-Seq. The technology also provides information about alternatively spliced transcripts, gene fusions, new mutations, and so on. RNA-Seq has been a valuable technology in the detection of noncoding transcripts in the ENCODE project (Section 7.2.4).

2.4
Animal Models in Biomedical Research

Animal models have become an indispensable tool in life science research. The sequencing projects of the last two decades (Chapter 6) provided sequence information for more than 20 000 genes of humans and other mammals. However, the functional relevance of many genes and their products still remains elusive. Information can be gained by overexpressing a gene (knockin) or by switching it off (knockout). Genetic manipulation of animals can also be carried out to generate models for human diseases that normally do not occur in animals. Animal experiments are also widely used to test the effects and toxicity of newly developed drugs (and chemical substances in general). To reduce the continuously increasing number of animals needed for biomedical research, alternative methods to *in vivo* testing are being developed (Box 2.2).

Box 2.2. Alternatives to Animal Testing

Estimates of animals used globally for research purposes vary widely and range from tens of millions to 100 million animals per year (compared to approximately 60 billion animals being slaughtered for meat production worldwide every year). The latest overview for the European Union, in 2008, reported the use of 12 million animals for research purposes in member states for experimental and other scientific purposes. Rodents, together with rabbits, represent more than 80% of the total number of animals used (Figure 2.19). The second largest group, at almost 10%, is represented by cold-blooded animals (reptiles, amphibians, and fish) followed by birds. Of the animals used in 2008, 0.08% were nonhuman primates.

The use of animals in medical experiments has provoked much controversy. Opponents argue that research with animals infringes their rights. Data obtained in animal experiments cannot always be applied to humans since species differ significantly and the toxicity of a new drug in humans cannot necessarily be deduced from animal testing. An additional argument against experiments with animals is that it negates the opportunity to use alternative methods that do not harm them. On the other hand, it has been argued that medical advances are entirely dependent on animal testing. According to the proponents of this argument, it will not be possible to replace animal experimentation by alternative methods in the foreseeable future, and it is unjustifiable to initiate clinical trials in humans without prior toxicity evaluation in animals. This view has also been reflected in current regulations for drug development.

Irrespective of these opposing positions, there is common consent that a reduction in the number of animals used for experimental purposes and efforts to reduce their suffering are important goals. The guiding principles for the use of animals in experimental work are based on the "Three R's" first described by Russell and Burch in their book *The Principles of Humane Experimental Technique* published in 1959:

1. *Replacement*: Nonanimal methods should be preferred over animal methods wherever it is possible to achieve identical scientific goals.

2. *Reduction*: Methods should enable researchers to obtain the same level of information from fewer animals or to obtain more information from the same number of animals.

3. *Refinement*: Strategies should be developed to minimize pain and suffering and enhance the welfare of the animals used.

Cultures of cells of animal or human origin are widely employed in life sciences. These cells can be used for the production of biotechnological products such as monoclonal antibodies and other recombinant proteins (Chapter 10), replacing animal-derived products. In addition, much biomedical research is carried out in cell culture. However, two-dimensional cell cultures do not fully reflect the three-dimensional

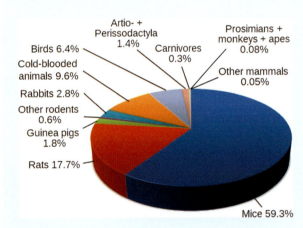

Fig. 2.19 Percentage of animals used for research purposes by classes.[1]

[1] Sixth Report on the Statistics on the Number of Animals Used for Experimental and Other Scientific Purposes in the Member States of the European Union, 2010.

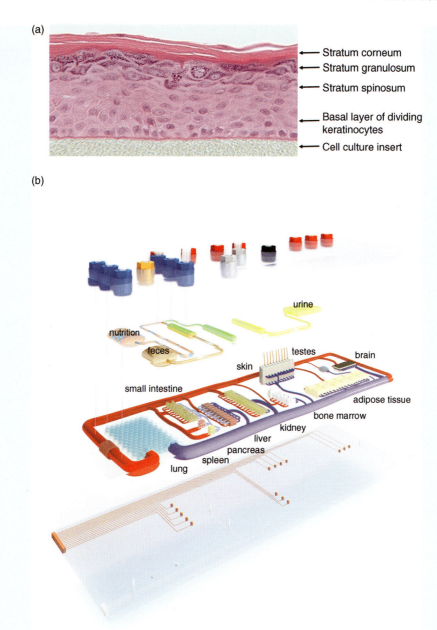

Fig. 2.20 Artificial organs. (a) Three-dimensional human skin model EpiDerm. (b) Human-on-a-chip. The envisioned human-on-a-chip device consists of three layers. One layer carries the equivalents of human organs, each represented by a number of organoids. The upper layer holds the components for the nutrition supply and for the removal of waste products. The lower layer contains sensors for the measurement of various parameters. (Part (a) image courtesy of MatTek Corp. Ashland, MA. (Part (b) reprinted from Ref. [4] with kind permission from the publisher of ATLA journal.)

architecture of organs. In addition, they neglect important biological factors such as cell-to-cell communication and interactions with the cell matrix, among others.

An additional layer of complexity is the use of three-dimensional organ models. Examples of highly developed systems are the engineered human skin constructs EpiDerm™ (MatTek Corp.) (Figure 2.20a) and EpiSkin® (SkinEthics Laboratories). These three-dimensional skin models are produced by human cells and have been approved as replacements for the *in vivo* rabbit test for skin irritation in the investigation of toxic effects resulting from topical exposure of the skin to a test substance.

Another step has been the development of "organ-on-a-chip" systems or even "human-on-a-chip" devices that mimic systemic organ interactions (Figure 2.20b). This approach involves the design of microfluidic systems to provide the microenvironmental architecture that mimics *in vivo* organ function. Cells from tissue biopsies, primary cells, or reprogrammed pluripotent stem (Chapter 12) cells can be grown into miniaturized organs. It has been shown that almost all organs are assembled from multiple identical and functionally self-reliant structural units, called organoids. These sub-structures, which perform the most basic functions of a particular organ, are small in size, ranging from several cell layers up to a few millimeters in diameter. Organoids from different types of organs can be grown on a single chip; they are connected by a microfluidic device that mimics blood flow. Currently, human-on-a-chip devices are primarily envisioned for toxicity evaluation. The formation of organoids composed of human cells will also help to overcome a major problem, which is that due to species differences the results of conventional drug testing on animals is not fully transferable to the human organism. In addition, the microfluidic systems allow unprecedented control over the culture conditions and enable macroscopic-level high-throughput experiments.

A common first step for the establishment of animal models is the stable integration of exogenous genetic information into a genome. This leads to the creation of so-called transgenic animals. One way of creating transgenic animals involves direct microinjection of exogenous DNA, the transgene, into the pronuclei of fertilized eggs (Figure 2.21a). The recombinant DNA is integrated into the genome of the oocyte. The embryo that develops from the injected oocyte is implanted into the uterus of a pseudopregnant female. The developing mouse will then carry the transgene. A major disadvantage of this approach is that the integration of the transgene is random, so that expression occurs in an uncontrolled fashion and important cellular genes may be affected.

To improve the control of genetic modification, the method of gene targeting by homologous recombination has been developed (Figure 2.21b). This approach can be used for knockin approaches, that is, the introduction of a cDNA to be expressed in the transgenic animal, as well as for the knockout of a target gene. In the latter case, a targeting vector is used that encodes a selection marker (e.g., a neomycin resistance gene) surrounded by sequences homologous to the endogenous target gene (insert of Figure 2.21b). After injection into embryonic stem cells (ESCs), homologous recombination between the targeting vector and the endogenous allele takes place. Correctively targeted cells are selected by addition of the antibiotics, expanded and injected into an early mouse embryo. After implantation of the embryo into the uterus of a pseudopregnant female, chimeric mice are born, that is, mice that are composed of different populations of genetically distinct cells. These chimeras can be mated with normal mice to obtain homogeneous knockout mice. Mario R. Capecchi, Martin Evans, and Oliver Smithies first described this technology in the late 1980s and were awarded the 2007 Nobel Prize in Physiology or Medicine for their work. The knockout technology has been widely used to study loss-of-function phenotypes. For the overexpression (knockin) of a gene, the procedure is similar except that the cDNA of interest is inserted in-frame in an exon of an endogenous gene. After homologous recombination, the gene of interest is expressed in place of the endogenous gene.

In conventional knockout approaches, the target gene is permanently inactivated in all cells of the transgenic animal. However, the global disruption of a gene is frequently deleterious to the development of an early embryo (embryonic lethal) and prevents the study of gene function in a specific organ. This shortcoming can be overcome by the conditional regulation of gene expression. The predominant techniques for reversible and irreversible gene inactivation are the tetracycline and Cre/lox systems, respectively.

The Cre/lox system is a site-specific recombinase technology that relies on the Cre recombinase (Cre stands for "causes recombination" or "cyclization recombination"). Cre recombinase was originally derived from the P1 bacteriophage. It recognizes the loxP site, a 34 base pair, asymmetric DNA sequence. Site-specific recombination begins with the binding of one molecule of Cre recombinase to each of two loxP sites in the DNA. The recombinase then joins together the two loxP sites and excises the DNA segment in between. The Cre/lox system requires the generation of double-mutant animals that express the Cre recombinase and contain the target gene surrounded by two loxP sites (the DNA segment is denoted as being "floxed," flanked by loxP sites).

The double transgenic system provides an opportunity that has widely been taken advantage of in recent years:

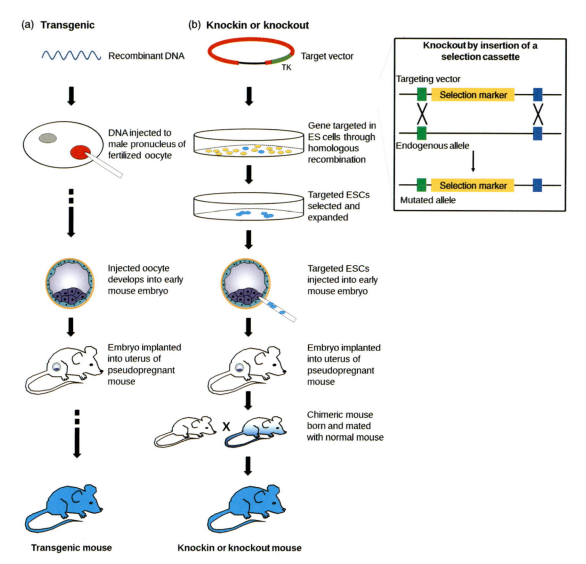

Fig. 2.21   Generation of transgenic mouse models. (a) Generation of transgenic animals by pronuclear microinjection. Recombinant DNA carrying the genetic information to be transferred is injected into the male pronucleus of a fertilized oocyte. The injected oocyte develops into an early embryo, which is implanted onto the uterus of a pseudopregnant mouse. Mice that develop from the modified oocyte carry and express the transgene. They need to be further bred to obtain a homozygote transgenic strain. (b) Generation of knockin or knockout animals by homologous recombination. A targeting vector is introduced into embryonic stem cells by microinjection or electroporation. Cells that carry the transgene after homologous recombination are selected and expanded. Subsequently, the ESCs are injected into an early mouse embryo, which is implanted into a pseudopregnant mouse. Chimeric mice are born and mated with normal mice to obtain a homogeneous mouse carrying the genetic modification. The insert demonstrates the principle of gene knockout by insertion of a selection cassette. The targeting vector contains a selection marker inserted between two sequences homologous to the endogenous allele. After homologous recombination between the targeting vector and the endogenous allele, a mutated allele with the target gene that is disrupted by the selection marker is created. (Adapted from Ref. [5] with kind permission from Macmillan Publishers Ltd., Copyright. 2005.)

A Cre mouse expressing the recombinase under control of a tissue-specific promoter (e.g., a promoter that is active only in the heart muscle) is mated with a floxed mouse that harbors the target gene between two loxP sites (Figure 2.22a). In the resulting mice, the target gene is specifically inactivated in cells that express the Cre recombinase (i.e., in heart muscle cells in the example described above). The advantage of this system is that the Cre mouse expressing the recombinase under control of a specific promoter can be combined with various floxed mice to generate different animals with organ-specific gene disruption.

This system results in inactivation of the target gene from early embryonic development. However, it may be

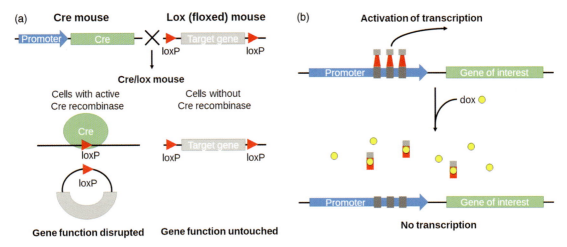

Fig. 2.22 Methods for conditional gene inactivation. (a) Cre/lox system. Cre recombinase excises a DNA segment located between two loxP sites. The recombinase can be expressed under control of an organ-specific promoter. Mating of a Cre mouse with a floxed mouse results in offspring with disrupted gene function in the target organ. (b) Tet-Off system. In the absence of doxycycline (dox), the Tet-transactivator tTA binds to the green tet-operator via the DNA-binding domain depicted in red. The gray activator domain induces transcription. The binding of dox results in a conformational change so that tTA dissociates from the tet-operator and transcription is terminated.

lethal and may prevent the study of the function of the gene in the adult animal. Therefore, methods have been developed to activate Cre recombinase at a later stage of development. One example of this is the fusion of Cre recombinase to a mutated ligand-binding domain of the human estrogen receptor. The resulting fusion construct is active only after addition of the ligand at a chosen time point. An advantage of the Cre/lox-system is that it can not only be used to inactivate genes but also to insert or mutate genes.

However, the Cre/lox-system, as well as the similar but less frequently employed FLP–FRT system, induces irreversible genetic changes. In contrast, the tetracycline (tet)-mediated regulation of gene expression allows the reversible switching on and off of a gene. For this purpose, the bacterial system needed to regulate the expression of a tet resistance gene has been adapted to control transcription in eukaryotic cells. The two most commonly used inducible expression systems are named Tet-Off and Tet-On. The Tet-Off system illustrated in Figure 2.22b makes use of the tetracycline transactivator protein tTA, which is created by fusing the bacterial tetracycline repressor with VP16, the activator domain of the herpes simplex virus protein. In the absence of tet, the tTA protein binds to specific Tet operator (TetO) sequences introduced upstream of a minimal promoter. The combination of a minimal promoter and several TetO sequences is called a *tetracycline response element* (TRE). Binding of tTA to the TRE activates transcription. Addition of tet or the more stable derivative doxycycline (dox) induces a conformational shift in the DNA-binding domain, which leads to release of tTA

from the TRE and termination of transcription. The alternative Tet-On system works in the opposite way. In this case, the transactivator binds to TRE only in the presence of dox, so that addition of the ligand results in activation of transcription. Additional variants of the tetracycline system are available, as well as reversible systems responding to other small molecules.

In addition to the strategies described above, many animal models for biomedical research have been created without genetic modification. Examples include models that restrict blood flow to the brain or heart to induce stroke or myocardial infarction, or that produce neurotoxins that induce symptoms similar to those seen in neurological diseases, or that involve implantation of tumors to study cancer pathways or new anticancer drugs. In addition, many animal experiments involve the toxicological testing of new substances in naïve animals.

2.5
Additional Methods

We will briefly describe three other methods that are widely employed in life science research: fluorescence microscopy, flow cytometry, including fluorescence-activated cell sorting, and surface plasmon resonance.

2.5.1
Fluorescence Microscopy

Optical microscopes have been an important tool in science for centuries. The first microscopes were developed

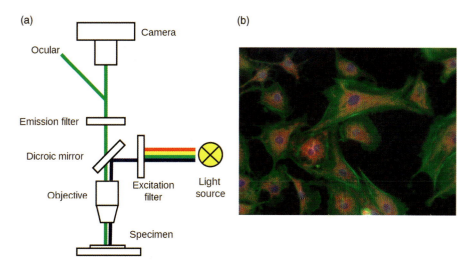

Fig. 2.23 Fluorescence microscopy. (a) Setup of an epifluorescence microscope. The illumination light passes through an excitation filter, so that light of a specific wavelength reaches the specimen. Fluorescence emitted by the fluorophore is separated from the illumination light by a second (emission) filter. Fluorescence microscopes used in cell biology often have an inverted design with the light source on top, while the objectives and turret are below the specimen. This setup is useful for observing living cells (at the bottom of a container such as a tissue culture flask or multiwell plate) under more natural conditions than on a glass slide, as is used for conventional microscopes. (b) Example of a multicolor image. DNA in the nucleus was stained with the dye DAPI. MitoTracker Red® CMXRos was used to stain mitochondria. The cytoskeleton protein F-actin was stained with Alexa Fluor® 488 phalloidin. (Images were taken from the FluoroCells®#1 preparation by Molecular Probes®/Life Technologies.)

in the early seventeenth century and permitted the study of biological structures, which prior to this were invisible to the naked eye. In 1665, Robert Hooke published a book called *Micrographia*, in which he described observations made with microscopes and coined the term "cell." Light microscopes are still used in everyday work in the modern life sciences. Fluorescence microscopes are more advanced versions of optical microscopes that use fluorescence to study cellular processes instead of the simple reflection and absorption of visible light.

In a fluorescence microscope (Figure 2.23a), light from a source such as a xenon lamp or, in more advanced microscopes, a laser or LEDs passes through an excitation filter and illuminates the specimen, to which a fluorescent protein or dye is bound. The excited fluorophore then emits light of longer wavelength, which can be separated from the much stronger illumination light by a spectral emission filter. The fluorescence is viewed through the eyepiece or detected by a sensitive digital camera. The spectral properties of the excitation and emission filters must be adjusted to the characteristics of the fluorescent dye. If more than one type of fluorophore is used to label different components of the specimen, several single-color images are taken and combined to give a multicolor image, as shown in Figure 2.23b.

Standard fluorescence microscopes used in the life sciences are based on the epifluorescence setup illustrated in Figure 2.23a. Despite the magnification of the biological sample, the spatial resolution is low, since all

parts of the specimen in the optical path are excited at the same time; thus, fluorescence from all planes is detected. Confocal microscopes overcome this limitation by using a point illumination so that only fluorescence produced very close to the focal plane is detected. In widely used laser scanning microscopes, a laser beam is focused by an objective lens into a small focal volume within the specimen. The resulting high spatial resolution allows the reconstruction of three-dimensional structures from the images obtained.

Different types of fluorophores are employed to study biological processes by fluorescence microscopy. Some small molecules possessing intrinsic fluorescence properties bind to biological molecules of interest. Examples are nucleic acids stains like DAPI (4′,6-diamidino-2-phenylindole) or Hoechst stains, which bind to the minor groove of DNA. Hoechst stains are widely used to label the nuclei of cells. The fluorophore can also be chemically linked to a molecule that interacts with the target. This strategy is employed for immunofluorescence experiments, which take advantage of the highly specific binding of an antibody to its antigen. The fluorophore can be conjugated directly to the primary antibody. Alternatively, an unlabeled primary antibody can be employed with a secondary antibody carrying the fluorescent reporter added for detection. Immunofluorescence allows the visualization and localization of a specific protein within a cell.

The discovery of the green fluorescent protein (GFP) produced by the marine organism *Aequorea victoria*, for which Roger Tsien, Osamu Shimomura, and Martin Chalfie received the 2008 Nobel Prize in Chemistry, broadened the spectrum of fluorescence microscopy applications. GFP consists of a β-barrel with a central helix that harbors the fluorophore formed by the three amino acids serine, tyrosine, and glycine. No nonproteinaceous moiety is needed to produce fluorescence. Since its original discovery, mutants of GFP have been generated that emit stronger fluorescence, and variants are available that emit fluorescence of different wavelengths. Additional fluorescent proteins have been discovered from other organisms so that a variety of proteins are available, which cover a wide range of the visible spectrum. Fluorescent proteins can be fused to a protein of interest as a reporter to track its intracellular location, even in live cells.

Numerous advanced microscopic techniques are being continuously developed (e.g., multiphoton fluorescence microscopy, confocal high-speed spinning disk systems, stimulated emission depletion microscopy, and total internal reflection fluorescence microscopy). These approaches help to improve the resolution, the spatial decomposition, the speed of image acquisition, and other aspects of imaging experiments. However, a description of these sophisticated methods is outside the scope of this book.

2.5.2
Flow Cytometry and Fluorescence-Activated Cell Sorting

Flow cytometry is a method to characterize heterogeneous cell populations by cell counting or biomarker analysis. A laser-based detection unit is frequently used to differentiate between different types of cells. The technology is routinely employed in clinical diagnosis, for example, of blood cancers, but it is also widely used in basic research. A special variant of flow cytometry is fluorescence-activated cell sorting (FACS, Figure 2.24), used to physically sort particles based on their properties and to purify the cell population of interest.

The basic principle underpinning flow cytometry is that a stream of fluid containing cells flows through a chamber where the individual particles are detected, commonly by optical methods. A laser beam passes through the stream and excites a fluorescent marker incorporated into the cells. The emitted fluorescence is then monitored by the detection unit. Instruments may be equipped with multiple lasers and fluorescence detectors, so that several markers can be analyzed simultaneously. In addition to fluorescence, the scattered light can be detected to provide information about the size and shape of a cell.

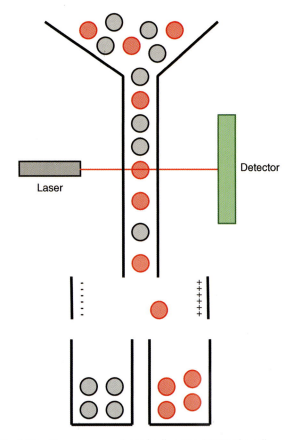

Fig. 2.24 Fluorescence-activated cell sorting (FACS). The cell suspension flows through a narrow chamber. An optical detection unit characterizes the fluorescence of each individual cell. A vibration mechanism causes the stream to break into droplets, each containing a single cell. A charge is placed on the droplets based on the outcome of the fluorescence measurements and the droplets are finally sorted into different containers in an electrical field.

An important prerequisite for most flow cytometry experiments is the introduction of a fluorescent label into the cells to be detected. One way to achieve this is via the expression of fluorescent proteins, such as GFP. However, the use of antibodies linked to a fluorophore that detect certain markers, for example, surface proteins, is a more common approach. Binding of the fluorescent antibody to the marker protein labels the target cell and allows its detection by the system.

For applications that aim not only at counting but also at sorting cells, FACS devices are equipped with a vibrating mechanism just after the flow passes through the fluorescence measuring unit. This mechanism causes the stream of cells to break into small droplets, each containing a single cell. An electrical charging ring is placed at the point where the stream breaks. This ring places a charge on the droplets based on the outcome of the fluorescence measurement. The charged droplets then pass through an electrostatic deflection system that

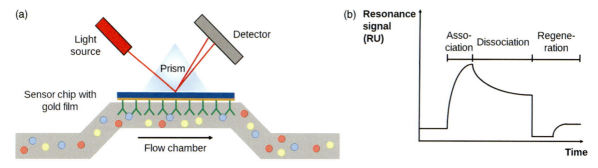

Fig. 2.25 Principle of surface plasmon resonance (SPR) experiments. (a) The Kretschman–Raether configuration utilized for surface plasmon resonance (SPR) measurements. The device consists of a glass surface coated with a thin gold film to which the capture molecule is attached. The analyte is in direct contact with the probe on the chip surface through a continuous flow system. Interactions between the capture molecule and the sample in the flow cell influence the reflection angle of polarized light applied to the optical detection unit. (b) Changes of the reflection angle are recorded. After injection of the analyte, an increase in the resonance signal is observed due to binding of molecules from the sample to the capture molecule. After the desired association time, a solution without the analyte is injected into the flow cell, so that dissociation of the complex can be observed.

distributes them into different containers depending on their charge.

2.5.3
Surface Plasmon Resonance

Surface plasmon resonance (SPR) measurements allow the real-time detection of interactions between biomolecules. The method is sometimes also referred to as a Biacore experiment, after the name of the manufacturer of the most widely used devices (although devices are also available from other manufacturers as well). In the most common configuration employed for SPR experiments, one of the interaction partners is immobilized on a metal surface of a chip, usually a glass surface covered with a thin gold film, which is attached to a prism (Figure 2.25). Different types of biomolecules, such as nucleic acids or proteins, can be linked to the chip by the appropriate surface chemistry. The analyte is injected into a microflow system and can bind to the attached biomolecules. The interaction is monitored by detecting the reflection of polarized light from a laser or other light source directed at the gold surface. The reflection angle is highly sensitive to changes on the surface. These changes, for example, may be due to binding events between the capture molecule and the analyte. The signal intensity, represented in resonance units (RUs), is proportional to the mass bound to the chip surface. SPR measurements permit the measurement of binding affinities (K_d) between the interaction partners, in addition to association–dissociation kinetics. However, these measurements are prone to error.

An important feature of the SPR technology is that it allows the label-free detection of interactions between two partners, in contrast to many optical methods that require the attachment of (fluorescent) reporters to the biomolecules.

SPR measurements have been employed for a wide variety of applications. A classical use has been the characterization of interactions between proteins. However, the technology is not restricted to the study of interactions between two proteins, but can also be used to investigate binding events between other classes of molecules. These interactions, for example, may be between nucleic acids and proteins, or between proteins and carbohydrates. SPR has also been applied to the discovery and development of small-molecular therapeutic agents and to characterize the interactions between these compounds and their targets. SPR measurements can be performed with solutions of pure substances, but they can also be used to identify biomarkers from complex samples such as plasma, serum, or saliva.

References

1. van't Veer, L.J., Dai, H., van de Vijver, M.J. *et al.* (2002) Gene expression profiling predicts clinical outcome of breast cancer. *Nature*, **415**, 530–536.
2. Voet, D., Voet, J.G., and Pratt, C.W. (2013) *Principles of Biochemistry – International Student Version*, 4th edn, John Wiley & Sons, Inc., Hoboken.
3. Metzker, M.L. (2010) Sequencing technologies – the next generation. *Nat. Rev. Genet.*, **11**, 31–46.
4. Marx, U., Walles, H., Hoffmann, S. *et al.* (2012) 'Human-on-a-chip' developments: a translational cutting-edge alternative to systemic safety assessment and efficiency evaluation of substances in laboratory animals and man? *Altern. Lab. Anim.*, **40**, 235–257.
5. Chen, J. and Roop, D.R. (2005) Mouse models in preclinical studies for pachyonychia congenita. *J. Investig. Dermatol. Symp. Proc.*, **10**, 37–46.

Further Reading

DNA Microarrays

Hoheisel, J.D. (2006) Microarray technology: beyond transcript profiling and genotype analysis. *Nat. Rev. Genet.*, **7**, 200–210.

Polymerase Chain Reaction

Bustin, S.A., Benes, V., Garson, J.A. *et al.* (2009) The MIQE guidelines: minimum information for publication of quantitative real-time PCR experiments. *Clin. Chem.*, **55**, 611–622.

Livak, K.J. and Schmittgen, T.D. (2001) Analysis of relative gene expression data using real-time quantitative PCR and the 2(-Delta Delta C(T)) method. *Methods*, **25**, 402–408.

Mullis, K.B. (1990) The unusual origin of the polymerase chain reaction. *Sci. Am.*, **262**, 56–61, 64–65.

Park, D.J. (2011) *PCR Protocols: Methods in Molecular Biology*, 3rd edn, Springer, New York.

Next-Generation Sequencing

Liu, L., Li, Y., Li, S. *et al.* (2012) Comparison of next-generation sequencing systems. *J. Biomed. Biotechnol.*, **2012**, 251364.

McGettigan, P.A. (2013) Transcriptomics in the RNA-seq era. *Curr. Opin. Chem. Biol.*, **17**, 4–11.

Metzker, M.L. (2010) Sequencing technologies – the next generation. *Nat. Rev. Genet.*, **11**, 31–46.

Ozsolak, F. (2012) Third-generation sequencing techniques and applications to drug discovery. *Expert Opin. Drug Discov.*, **7**, 231–243.

Animal Models

Babinet, C. (2000) Transgenic mice: an irreplaceable tool for the study of mammalian development and biology. *J. Am. Soc. Nephrol.*, **11** (Suppl. 16), S88–S94.

Luni, C., Serena, E., and Elvassore, N. (2014) Human-on-chip for therapy and development and fundamental science. *Curr. Opin. Biotechnol.*, **25**, 45–50.

Marx, U., Walles, H., Hoffmann, S. *et al.* (2012) 'Human-on-a-chip' developments: a translational cutting-edge alternative to systemic safety assessment and efficiency evaluation of substances in laboratory animals and man? *Altern. Lab. Anim.*, **40**, 235–257.

Genetic Disorders

3

Summary

- More than 4000 monogenetic disorders that are transmitted by Mendelian inheritance have been described. The most common modes of inheritance are autosomal dominant, autosomal recessive, and X-linked recessive modes. Most monogenic disorders are rare.
- A small number of genetic disorders are not caused by changes in DNA sequence, but rather by changes in DNA methylation patterns. The Prader–Willi syndrome and the Angelman syndrome are two examples of diseases involving genomic imprinting.
- While genetic testing usually permits unambiguous identification of carriers of mutated genes, no established cure exists for most genetic disorders. For example, most children with infantile Tay–Sachs disease inevitably die by the age of three or four. Molecular medicine aims to elucidate both the pathologic mechanisms and to develop new therapeutic approaches for as yet untreatable diseases. Some disorders such as phenylketonuria are routinely diagnosed by newborn screening. Here, immediate treatment with an appropriate diet can prevent the development of disease symptoms.
- Disorders caused by dysfunctional mitochondria due to mutations in the nuclear or mitochondrial DNA are called mitochondriopathies or mitochondrial diseases. The symptoms of mitochondrial diseases frequently are myopathy and neurological problems.
- Many widespread diseases such as cancer, asthma, diabetes mellitus, heart disease, and hypertension are polygenic and are usually influenced by the interplay between different genes and environmental factors. Inheritance patterns are less obvious than in the case of monogenic diseases and genome-wide association studies sometimes produce ambiguous results. Risk prognostication for individuals is often difficult.

Contents List

Single-Gene Disorders

- **Autosomal Dominant Disorders**
- **Autosomal Recessive Disorders**
- **X-Linked Recessive Disorders**
- **Mitochondriopathies**

Polygenic Disorders

- **Asthma**
- **Diabetes Mellitus**

The term *genetic disorder* refers to a disease that is caused by abnormalities in the genome. Traditionally, a genetic disorder implies inheritance as the underlying mutation is found in the germline and transmitted to the offspring. We now know that changes in the genome are also causal for a number of noninherited disorders. In this case, the acquired mutations affect only somatic cells and the abnormality is not heritable. This type of genetic change is particularly important in tumor biology and will be extensively covered in Chapter 4. In a broader sense, chromosomal aberrations such as numerical variations or structural damage may also be considered genetic disorders. These topics will be outlined in Chapter 8. Here, we will focus on genetic disorders in the classical sense as heritable diseases that may be

Molecular Medicine: An Introduction, First Edition. Jens Kurreck and Cy Aaron Stein.
© 2016 Wiley-VCH Verlag GmbH & Co. KGaA. Published 2016 by Wiley-VCH Verlag GmbH & Co. KGaA.

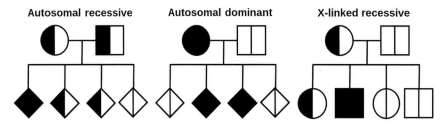

Fig. 3.1 Pedigree depicting Mendelian inheritance. Affected individuals are indicated by filled symbols. Half-filled symbols indicate individuals who do not develop a phenotype, but are genetic carriers of the disease. Males are depicted as squares, females as circles. Diamonds indicate that the gender is irrelevant.

caused by a mutation in a single gene (Section 3.1) or by more complex, polygenetic variations (Section 3.2).

3.1
Single-Gene Disorders

For disorders caused by mutations in a single gene, different modes of inheritance may be grouped into the conventional classes of autosomal dominant, autosomal recessive, and X-linked disorders, as depicted in Figure 3.1. They are transmitted according to the Mendelian rules of inheritance. Another class of inherited disorders caused by dysfunctional mitochondria is referred to as mitochondriopathies.

In autosomal recessive disorders, both alleles must be mutated to cause the disease. Heterozygous carriers of the trait therefore remain unaffected. When both parents are heterozygous carriers, the risk that the offspring are affected is 25%; the other 75% will not develop a phenotype, however, two thirds of this group will again be heterozygous carriers of the mutated allele. In autosomal dominant disorders, one mutated allele is sufficient to cause the disease. Half of the offspring of a heterozygous carrier of the trait will again be heterozygous and develop the disease. Here, the gender is irrelevant. X-linked recessive disorders are normally transmitted from a healthy, heterozygous mother and

50% of the offspring will be heterozygous for the allele. Only sons will develop the disease phenotype. In cases in which both parents are carriers of the mutated allele, daughters can be affected as well. In addition, females may be affected due to disadvantageous imprinting events, as will be outlined in Section 3.1.3.

A few diseases are inherited by an X-linked dominant mode (e.g., Rett syndrome, hypophosphatemia). They cannot be transmitted from male to male, since fathers pass on the Y chromosome to their sons, but females can pass their X chromosome to offspring of both genders and males can pass them on to daughters, so that females are more frequently affected than males. In addition, there are a very small number of Y-linked disorders, that is, disorders that are caused by mutations on the Y chromosome. These conditions can only be transmitted from fathers to their sons. Females can never be affected since they, of course, do not possess a Y chromosome. Y-linked disorders include conditions that cause infertility. Mitochondria are exclusively passed on from the mother to her offspring, so that mitochondrial diseases are transmitted from females to their offspring of either gender. In addition, epigenetic phenomena play an important role in developmental and pathophysiological processes (Section 1.3.4). Sex-specific imprinting determines the characteristics of some diseases such as the Prader–Willi syndrome (PWS) and the Angelman syndrome (AS, Box 3.1).

Box 3.1.   Genomic Imprinting and the Prader–Willi and Angelman Syndromes

The PWS and the AS are the first recognized and best understood human diseases involving genomic imprinting. Both syndromes are distinct neurogenetic disorders in which imprinted genes on the proximal long arm of chromosome 15 are affected. This means that the sequence of the respective chromosomal region is normal, but its methylation (imprinting) pattern results in disease. Each of the two syndromes occurs with a frequency of approximately 1:15 000–1:25 000 newborns. Symptoms of PWS include poor muscle tone, obesity, mental retardation, a characteristic facial appearance, and behavioral problems. AS is characterized by pronounced mental retardation, severe limitations in speech and language, sleep disorder, and seizures. Affected individuals often exhibit an apparently happy demeanor, as is also observed in Down's syndrome in many cases.

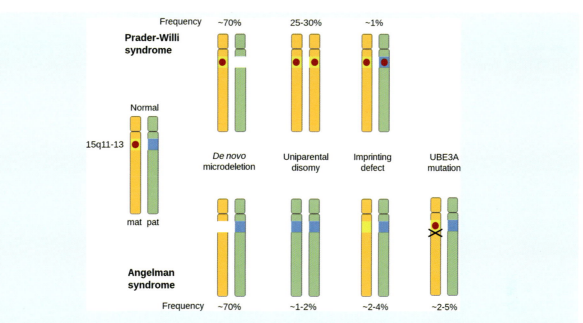

Fig. 3.2 Molecular defects in Prader–Willi syndrome (PWS) and Angelman syndrome (AS). The underlying defects and their frequency of occurrence are given. Inactivation by methylation is indicated by a red dot; deletion of the UBE3A gene is marked by a cross. AS may have additional causes, which are not depicted in the figure. Mat: maternal; pat: paternal.

In individuals with PWS, seven genes on chromosome 15 (q11-13) are deleted or not expressed on the paternal chromosome. The maternal copy of this region of the chromosome is usually imprinted (and thus silenced), while the paternal, nonfunctional copy is active (Figure 3.2). Although several genes are affected, absence of the small nucleolar RNA (snoRNA) gene *SNORD116* seems to produce many of the clinical features of the disease. Furthermore, microdeletions found in a small number of patients with PWS led to the identification of an imprinting center (IC), which acts in *cis* (i.e., on its own chromosome) and regulates methylation and chromatin modification of the entire imprinted domain of the proximal 15q chromosome.

While the paternal chromosome is affected in patients with PWS, AS develops if the affected chromosome is derived from the mother. In this case, loss of function of the *UBE3A* gene, which is expressed only from the maternal chromosome in the brain, was identified as the major determinant of the disease. The gene encodes an E3 ubiquitin ligase, which is involved in the ubiquitination pathway of protein degradation (Section 1.3.6).

Both syndromes can result from different types of chromosomal alteration. In the majority of patients, a 5–7 Mb *de novo* deletion of the proximal region of the chromosome is found. PWS and AS can also be examples of *uniparental disomy* (UPD), which occurs when an offspring receives two copies of a chromosome from one parent and no copy from the other parent. While

UPD often does not result in a phenotypic anomaly, there is a high risk of disease manifestation in the case of uniparental inheritance of an imprinted gene. In the case of PWS, UPD leads to two inactivated chromosomes. This results in the lack of expression of the genes that are normally active on the paternal chromosome. The paternal UPD in individuals with AS results in loss of the active maternal *UBE3A* gene copy. In a few percent of individuals with PWS or AS, the disease is due to an imprinting defect. Maternal imprinting of the paternal chromosome leading to gene silencing of the paternally expressed genes results in PWS, while silencing of the *UBE3A* gene due to a paternal imprinting of the maternal chromosome results in AS. Finally, a fraction of patients with AS harbor deleterious mutations in the *UBE3A* gene.

The risk of siblings of an affected child developing PWS or AS depends on the genetic cause of the disease. It is low (below 1%) if the affected child has a *de novo* mutation or UPD, but will be 50% in case of a familial IC deletion. In AS, the majority of mutations of *UBE3A* are *de novo*, so the recurrence risk is low. However, a mother carrying such a mutation has a 50% recurrence risk for another child with AS. Both syndromes can be diagnosed through genetic testing. In contrast to other genetic disorders, it is not the DNA sequence, but rather its methylation pattern that is analyzed. However, this approach will only confirm a diagnosis. To obtain further information about the underlying defect, fluorescence *in situ* hybridization and/or microsatellite analysis are usually carried out.

Neither of these syndromes can be cured. However, several treatments help lessen the symptoms of PWS, including therapy to improve muscle tone and speech. Obesity due to a chronic feeling of hunger is one of the major problems for individuals with PWS. Recombinant

growth hormones can help to support growth, increase muscle mass, and lessen weight gain. In patients with AS, anticonvulsant medications can help to control epilepsy, and sleep may be improved by treatment with melatonin. Otherwise, the affected individuals are generally happy and interpersonally interactive despite their verbal communication difficulties.

Information on human genes and genetic disorders has been collected since 1966 in a compendium called *Mendelian Inheritance in Man*. The online version of this database is available as the *Online Mendelian Inheritance in Man* (OMIM, www.omim.org/). An OMIM number is assigned for each inherited disorder; hemophilia A, for example, is assigned as OMIM 306 700. In 2015, the database listed approximately 4000 phenotype descriptions for which the molecular basis is known. The number of genetic disorders is therefore frequently cited as being around 4000. However, it should be noted that the database lists a similar number of phenotype descriptions with an unknown basis or only a suspected Mendelian basis. Approximately 92% of the genetic disorders for which the genetic basis is known belong to the class of autosomal disorders; 7% are inherited by an X-linked mode. Only four disorders are Y-linked, and the remaining genetic disorders are mitochondriopathies. Some examples of inherited genetic disorders are given in Table 3.1. Most genetic disorders are rare, affecting only one individual in several thousands or even millions.

It is not possible to provide an exhaustive description of all inherited diseases here. Instead, selected paradigms for autosomal dominant disorders (Section 3.1.1), autosomal recessive disorders (Section 3.1.2), X-linked recessive disorders (Section 3.1.3), and mitochondrial diseases (Section 3.1.4) will be introduced. These examples will demonstrate the great diversity of genetic disorders.

3.1.1
Autosomal Dominant Disorders

In autosomal dominant disorders, only one mutated allele is sufficient for a person to develop the disease, that is, the normal allele cannot compensate for the pathogenic influence of the mutated allele. The mode of transmission can be described as vertical, as the disorder can appear in every generation. Both males and females are affected, and the offspring of an affected individual are at a 50% risk of inheriting the disease. However, in some dominant disorders, a fraction of the individuals who inherit the mutated gene will not develop the disease. This is

reflected by the term *penetrance* of a disorder: If 8 out of 10 carriers of a mutated allele associated with a dominant genetic disease show the clinical phenotype, the penetrance of the disorder is 80%. For some disorders, such as Huntington's disease (HD), penetrance depends on the patient's age. At a young age, carriers of the genetic predisposition for this disease are usually free of the pathological phenotype, but by the age of 60, almost 100% will develop the disease.

The severity of the disease is described as its *expressivity*. This term refers to the variability of the severity for a given mutation. *Pleiotropy* occurs when a genetic variation influences multiple, seemingly unrelated phenotypic traits and involves multiple organ systems. An example is Marfan's syndrome (*vide infra*), which affects connective tissues in the skeletal system, the eye, or the heart. Individuals inheriting the genetic disposition for Marfan's syndrome may have any combination of manifestations with different degrees of severity. The variability may even occur within a family in which the same mutant allele is present. The basis for the variation in expressivity and pleiotropy is not fully understood, but may involve environmental factors, interactions with other genetic variants, or genetic instability.

Dominance of the mutated allele is complete in only a few diseases. In these cases, the phenotype of the heterozygote is completely indistinguishable from that of the dominant homozygote. This is, for example, the case in Huntington's disease, although even in this disorder homozygotes tend to develop the first symptoms earlier in life and the disease progresses more rapidly than in heterozygotes. Most autosomal dominant disorders are characterized by incomplete dominance (or semidominance), that is, the heterozygous genotype is distinct and often milder than the phenotype of the homozygote. Familial hypercholesterolemia (FH) is a semidominant disorder in which homozygous carriers of the mutated allele have a much higher cholesterol level and a more severe disease phenotype than heterozygotes.

In most dominant inherited disorders, the affected individual has one functional copy of the gene, while the other copy is absent. The inability of the functional copy

Table 3.1 Examples of single-gene disorders.

Disease	Mutated gene	Prevalence
Autosomal dominant		
Familial hypercholesterolemia	LDL receptor (ApoB-100)	1:500
Polycystic kidney disease	PKD1, PKD2	1:1000
Neurofibromatosis	Neurofibromin	1:3000
Hereditary spherocytosis	Erythrocyte membrane proteins: ankyrin, band 3, α and β spectrin, protein4.2	1:5000
Marfan's syndrome	Fibrillin-1	1:5000
Huntington's disease	Huntingtin	1:10 000
Autosomal recessive		
Sickle cell anemia (Section 6.2)	Hemoglobin	1:625
Cystic fibrosis	CFTR	1:2500
Tay–Sachs disease	Hexosaminidase A	1:3000 (Ashkenazi Jews) 1:320 000 (general population in the United States)
Phenylketonuria	Phenylalanine hydroxylase	1:10 000
Mucopolysaccharidoses	Lysosomal glycosidases or sulfatases	1:25 000
Glycogen storage disease	Glycogen-degrading enzymes	1:50 000
Galactosemia	Hexose-1-phosphate-uridyl-transferase	1:57 000
Xeroderma pigmentosum	Nucleotide excision repair	1:250 000 (United States)
X-linked recessive		
Red-green color blindness	Opsin	1:10-1:20 (males)
Hemophilia A	Factor VIII	1:10 000 (males)
Hemophilia B (Section 10.2.6)	Factor IX	1:50 000 (males)
Duchenne muscular dystrophy	Dystrophin	1:3500 (males)
Becker muscular dystrophy	Dystrophin	1:20 000 (males)

The disorders are classified as autosomal dominant, autosomal recessive, and X-linked recessive disorders. Disorders marked in bold letters are discussed in the present chapter. Some disorders are outlined in other sections. The prevalence given may vary significantly between different populations. ApoB-100, apolipoprotein B-100; CFTR, cystic fibrosis transmembrane conductance regulator; LDL, low-density lipoprotein; PDK, polycystic kidney disease.

to compensate for the loss of the second copy and restore the wild-type phenotype leads to a disease state. This is also denoted as *haploinsufficiency*. However, in some cases, the mutation results in a structural variant of the protein that inactivates the normal protein encoded by the wild-type allele. This "dominant negative" effect occurs, for example, if the mutated protein is part of a large multisubunit complex or forms insoluble aggregates with the normal version. Here, the consequences of the production of the mutated form can be even more severe than the complete absence of the gene product. Some examples of this phenomenon will be described below. For example, the mutated fibrillin-1 protein in patients with Marfan's syndrome not only has no function, but it also interferes with the generation of extracellular microfibrils that might be formed by the functional protein expressed from the second, wild-type allele.

3.1.1.1 Familial Hypercholesterolemia

Familial hypercholesterolemia (FH) is the most common autosomal dominant disorder, with a prevalence of 1:500. In some populations (Afrikaners, French Canadians, Lebanese Christians, and Finns), the prevalence is higher, presumably because of a genetic phenomenon known as the *founder effect*. The affected populations were founded by a small group of individuals, one or several of whom were carriers of the mutation and who passed the defect responsible for the disorder on to the following generations.

FH is characterized by high cholesterol levels, especially very high levels of low-density lipoproteins (LDLs) in the circulation. Due to their poor aqueous solubility, cholesterol and other lipids are transported as lipoproteins through the bloodstream. LDL particles belong to the class of lipoproteins that transport endogenously produced triacylglycerols and cholesterol from the liver

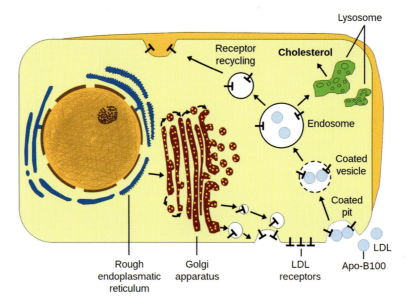

Fig. 3.3 Receptor-mediated endocytosis of low-density lipoproteins (LDL). The LDL receptor is synthesized and inserted into the plasma membrane. The apolipoprotein-B100 (ApoB-100) component of the LDL particle binds to the LDL receptor, inducing receptor-mediated endocytosis and the formation of clathrin-coated vesicles. After depolymerization of the clathrin coats, the vesicles fuse with endosomes. The vesicles separate into membranes harboring the LDL receptors, which recycle them to the plasma membrane. The remaining vesicles fuse with lysosomes. The process eventually leads to the release of cholesterol, amino acids and fatty acids.

to the tissues. LDL is sometimes called "bad cholesterol," because a high LDL level can drive the process of atherosclerosis. In the tissues, LDL particles bind to specific receptors on the cell surface and are internalized by receptor-mediated endocytosis (Figure 3.3). Inside the cell, cholesterol is released from the particles and the receptor is recycled to the plasma membrane.

Most patients with FH have mutations in the LDL receptor. Heterozygotes have two- to threefold increased LDL concentrations in the blood (200–400 mg/dl compared to 75–175 mg/dl in healthy individuals). In rare cases (prevalence 1:250 000), the mutation is homozygous, resulting in even higher blood levels of LDL (more than 450 mg/dl). The high LDL levels result in the deposition of cholesterol in the skin and tendons as yellow nodules called xanthomas. Even greater damage is caused by the development of atherosclerosis. Heterozygotes frequently develop symptoms of cardiovascular disease after the age of 30, while homozygotes may experience heart attacks as early as age 5.

According to the University College London LDL receptor FH database, more than 1000 unique allelic variants are known (http:ucl.ac.uk/fh [1]). These variants include point mutations as well as large deletions in the LDL receptor gene. The mutations can be divided into five groups:

1) Changes in the promoter region, mutations of a splice site, shifts in the reading frame, or large

deletions may result in complete loss of LDL receptor synthesis (null allele).

In all other classes, an LDL receptor protein is synthesized, but it is nonfunctional, because

2) its transport between ER and Golgi apparatus is disturbed.
3) the receptor has lost its ability to bind LDL.
4) the complex of LDL and receptor does not properly cluster in clathrin-coated pits so that it can no longer be internalized by endocytosis.
5) the receptor does not release the LDL particles once inside the cell and is not recycled back to the plasma membrane.

While some of these changes lead to complete loss of function, in other cases the receptor may still have residual activity.

High levels of cholesterol are usually treated with statins, a class of drugs inhibiting 3-hydroxy-3-methyl-glutaryl-CoA reductase (HMG-CoA-reductase), a key enzyme in the endogenous cholesterol synthesis pathway. Normally, cholesterol derived from internalized and degraded LDL particles inhibits HMG-CoA-reductase, thus maintaining a balance between endogenously synthesized cholesterol and cholesterol taken up in the diet. In individuals with a defect in the LDL receptor, this negative feedback loop fails and the rate of endogenous cholesterol synthesis is too high. While statins frequently help to reduce the cholesterol level in heterozygous individuals with FH, it is often less successful in homozygotes. In these cases, LDL may be filtered from the bloodstream by apheresis or the patient may receive a liver with functional LDL receptors by organ transplant. In 2013, the FDA approved mipomersen (trade name Kynamro) for the treatment of FH. Mipomersen is an antisense oligonucleotide that inhibits the expression of apolipoprotein B, as will be described in more detail in Section 13.1.

In addition to defects in the LDL receptor, FH may also be caused by mutations in the apolipoprotein-B100 (ApoB-100), which is a component of LDL particles. ApoB-100 mutations are found in approximately 2–6% of the patients suffering from FH. The most prevalent mutation, exchange of arginine for glutamine at position 3500 in ApoB-100, is located in the region of the protein that normally binds to the LDL receptor; the mutation weakens this binding. Mutations in other genes causing FH are rare.

3.1.1.2 Polycystic Kidney Disease
Polycystic kidney disease (PKD) is characterized by the development of extensive epithelial-lined cysts in the kidneys (Figure 3.4). It is the most prevalent inherited kidney disease and affects more than 12 million people worldwide. In the United States, PKD is the fourth leading cause of end-stage renal disease. About 90% of the cases are inherited as an autosomal dominant disorder, while the autosomal recessive trait is much less frequent.

The pathogenesis of PKD is still largely unclear, but the autosomal dominant form is known to be caused by mutations in the *PKD-1* and *PKD-2* genes, located on

chromosomes 16 and 4, respectively. Mutations in the *PKD-1* gene are responsible for 85% of the cases of autosomal dominant PKD. The encoded protein, polycystin-1, is involved in the regulation of the cell cycle and of intracellular calcium transport in epithelial cells. Malfunction of polycystin-1 leads to abnormal proliferation of tubular epithelial cells and the formation of cysts that accumulate fluid and enlarge and compress the neighboring renal parenchyma, progressively compromising renal function.

Although various therapeutic approaches are in development, a cure for PKD is currently not available. Only its symptoms, such as pain, urinary tract infection, and high blood pressure can be treated. In addition, surgical decompression of the cysts can be performed on progression to end-stage renal disease, which usually happens in the fourth to sixth decades of life, when dialysis or kidney transplantation become necessary.

3.1.1.3 Marfan's Syndrome
Marfan's syndrome is an autosomal dominant disorder of the connective tissue that affects the skeletal system, the eyes, and the cardiovascular system. The prevalence of the disorder is one in 5000 individuals. In approximately one fourth of the patients, the disorder occurs without a family history and is likely to result from spontaneous germ line mutations. Patients with Marfan's syndrome tend to be tall with long limbs and long, thin fingers. The expressivity of the disorder ranges from mild to very severe. Complications can arise from cardiovascular malformations, particularly aortic root dilatation and mitral valve prolapse.

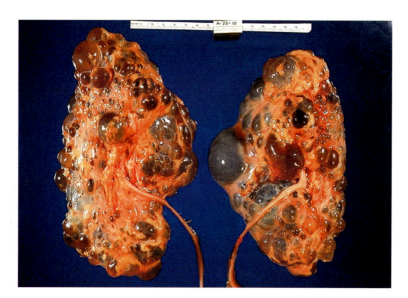

Fig. 3.4 Gross pathology of polycystic kidneys. (Picture taken in 1972 by Dr. Edwin P. Ewing, Jr., Centers for Disease Control and Prevention (CDC).)

Although the molecular basis of Marfan's syndrome has not been fully elucidated, dominant negative effects of mutations in the *fibrillin-1* gene have been identified as the main cause of this disorder. Most known mutations affect a single amino acid of the protein. Fibrillin-1 is a glycoprotein that is a major component of extracellular microfibrils. As the extracellular matrix is critical for the structural integrity of connective tissue, abnormal fibrillin-1 leads to tissue weakness. In addition to its function as part of the structural support for tissues outside the cell, wild-type fibrillin-1 also binds transforming growth factor β (TGF-β). Mutations in fibrillin-1 may disturb this interaction, and the resulting elevated levels of TGF-β can negatively affect the lungs, heart valves, and the aorta.

Although there is no cure for Marfan's syndrome, life expectancy of the affected individuals has increased significantly in the last few decades. For example, patients often undergo prophylactic surgery, such as valve-sparing aortic root replacement. In addition to constantly improving surgical techniques, pharmacological treatment has proven successful. The standard treatment includes beta-blockers, which reduce TGF-β activation, and inhibitors of the angiotensin II receptor.

3.1.1.4 Huntington's Disease

Huntington's disease (HD) is a neurodegenerative genetic disorder that leads to progressive choreic (jerky and disordered) movements, psychological disturbances, cognitive decline, and dementia. It is an autosomal dominant disorder with complete penetrance (at greater age). The severity of the disease does not vary substantially between homozygous and heterozygous carriers of the genetic defect. The onset of the disease is usually between 35 and 45 years of age and life expectancy is generally around 20 years following the first occurrence of symptoms. Complications such as pneumonia and heart disease are the main causes of death. The proportion of patients with HD who develop depression and commit suicide is about 5–10 times that of the general population (5–10%). Approximately 6% of cases start before the age of 21 and are classified as juvenile HD. They usually progress faster than other variants.

In 1993, the genetic basis of the disease was elucidated and the relevant gene was named *huntingtin*. HD belongs to the class of genetic disorders that are caused by expansions of triplet nucleotide repeats. All of these diseases are associated with neurological symptoms. The triplet repeats may be located in the 5′ UTR (fragile X syndrome), the coding region (spinobulbar muscular atrophy and HD), or the 3′ UTR (myotonic dystrophy). In the case of the *huntingtin* gene, the first exon usually contains 6–26 repeats of the cytosine–adenine–guanine

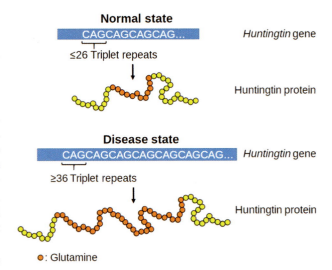

Fig. 3.5   DNA triplet repeats in Huntington's disease (HD). The genetic basis of HD is the extension of the CAG repeats in the *huntingtin* gene. The triplets are translated into glutamine at the protein level. Extended glutamine stretches render the protein nonfunctional.

(CAG) nucleotides. This triplet encodes the amino acid glutamine, thus, the CAG repeats encode the so-called polyglutamine (polyQ) tract. The presence of more than 36 repeats is associated with the development of the disease (Figure 3.5). Genetic testing can be performed to predict the development of the disease (Table 3.2): In the case of 40 or more CAG repeats, there is a 100% probability of HD, and the disorder can be excluded at repeat numbers of 26 and below. The intermediate range is less clear. Between 27 and 39 repeats are considered *premutations*, which will lead to HD in future generations, if they expand. This helps to explain sporadic cases, in which there was no family history, but one of the parents had an intermediate number of triplet repeats. Individuals with 27–35 repeats usually do not

Table 3.2   Correlation of repeat numbers, phenotype, and risk to offspring in Huntington's disease (HD).

Repeat number	Phenotype	Risk to offspring
≤26	Normal	None
27–35	Normal	Elevated, but below 50%
36–39	May develop the disease, but potentially at reduced severity	Elevated (50%)
≥40	Huntington's disease	50%

HD is caused by extension of the CAG triplet repeat in exon 1 on the *huntingtin* gene. The consequences depend on the number of triplet repeats.

develop the disease, but there is a risk that their off-spring will be affected. Carriers of 36–39 repeats may develop the disease, but possibly with reduced severity. In addition, there is a higher risk that any offspring will be affected with HD.

The finding that the repeat number can expand (and in some cases slightly contract) when transmitted to the next generation has helped to explain the phenomenon of genetic *anticipation*, that is, the earlier onset and more severe phenotype as the mutant gene is passed on. In some diseases, the sex of the transmitting parent influences the instability of the triplet repeats. In the case of HD, the instability of the CAG repeat is increased if the transmitting parent is male. This can be explained by greater instability in spermatogenesis than oogenesis, so that maternally inherited alleles are usually of a similar length, while the paternally inherited alleles are more likely to be increased in length.

The molecular basis on which the extended polyQ tract in the huntingtin protein causes disease symptoms is not fully understood. However, the pathology is not believed to be caused by a loss of function of the hun-tingtin protein, but rather by a gain of function with toxic consequences. The onset of neurological symptoms coincides with the deposition of insoluble protein aggre-gates. The polar nature of the polyQ tract of the hun-tingtin protein or its degradation products may lead to the formation of these protein aggregates over time. However, it is not clear whether these aggregates directly cause neuronal death and produce symptoms. In the early phase of disease development, the corpus stria-tum, which is involved in control over movement, mood, and higher cognitive function, is damaged. Later on, other areas of the brain also become affected.

Since currently no cure is available for HD, the goal of treatment is to reduce the severity of the symptoms. In 2008, the FDA approved the antichoreic drug tetraben-zine for the treatment of patients with HD. The drug promotes the rapid metabolic degradation of mono-amines, in particular the neurotransmitter dopamine. In addition, various vitamins with antioxidant functions are used to protect neurons to slow down disease progres-sion. An important part of treatment includes physio-therapy and speech therapy.

HD is an intensively discussed paradigm for the ethical issues associated with genetic testing. A genetic diagnosis can be obtained from a presymptomatic, predictive test long before the disorder develops. Except for the interme-diate range of repeat numbers, it can easily be determined whether or not the individual will be affected. There is an ongoing debate about the age at which an individual is mature enough to decide to perform such a genetic test. In addition, it has been questioned whether employers or insurance companies should be allowed to use genetic information when assessing an individual. While there is broad consensus that the applicant should not be obli-gated to perform a test, it is controversial whether or not an individual will have to disclose information that was already obtained prior to a job or life insurance applica-tion. Acquisition of a substantial life insurance policy without the disclosure of a previously diagnosed extension of the triplet repeats is widely considered insurance fraud. The ethical issues associated with genetic testing are dis-cussed in more detail in Section 15.2.1.

3.1.2
Autosomal Recessive Disorders

Autosomal recessive disorders occur only when both copies of a gene are mutated. The affected individual usually has parents who are heterozygous carriers of the mutated allele, but are phenotypically normal. Mutagen-esis studies have revealed that 90% of mutations result in an allele that is recessive against the wild-type allele. Accordingly, most inherited disorders are autosomal recessive. This mode of inheritance usually gives rise to a horizontal rather than a vertical pattern in the pedi-gree, that is, it is usually not found in every generation, but rather in individuals of a single sibship. Heterozy-gous carriers of a mutated allele of an autosomal reces-sive disorder are clinically normal. There is a 25% chance that the offspring of two heterozygotes are homozygous-normal and a 25% risk of being homozy-gous-affected and of developing the disease. The remain-ing 50% will be heterozygous carriers of the mutated allele without a phenotype. Males and females are affected similarly by autosomal recessive disorders. Most recessive alleles have an incidence of 1:100–1:1000. Thus, the disorder occurs with an incidence of 1:10 000–1:1 000 000. However, the incidence may be higher where there is a high level of consanguinity. Autosomal recessive inheritance is sometimes difficult to detect, since the clinical phenotype may develop only in rare cases and can appear to be sporadic.

One of the best studied and most prevalent autosomal recessive disorders is sickle cell anemia, which is caused by a single amino acid substitution in the β-chain of hemoglobin. Heterozygous carriers of the mutation pres-ent with less severe symptoms when infected with malaria. Thus, this allele confers a selective advantage in regions in which this pathogen is endemic. However, homocygotes of the mutated allele suffer from hemoglo-bin aggregation in the erythrocytes, leading to anemia and the severe pain of sickle cell crisis. Further details about sickle cell anemia will be described in Section 6.2; here, cystic fibrosis (CF), Tay–Sachs disease, phenylketonuria

(PKU), and xeroderma pigmentosum (XP) will be introduced as examples of autosomal recessive disorders.

3.1.2.1 Cystic Fibrosis

Cystic fibrosis (CF) is the most common autosomal recessive disorder in Caucasians, with a prevalence of 1:2500. Approximately 4–5% of the population are heterozygote carriers of a mutated allele. CF most critically affects the lungs, but the pancreas, liver, and intestine are also commonly affected. Other symptoms of the disorder include poor growth and infertility. The disease is characterized by the abnormal transport of chloride ions across the epithelium, leading to thick, viscous secretions. Accumulation of sticky mucus in the lungs causes difficulty breathing and results in frequent lung infections. Bacteria that normally inhabit the thick mucus can cause pneumonia. Cardiorespiratory complications are the most common cause of death in patients with CF.

CF is caused by mutations in the *cystic fibrosis transmembrane conductance regulator* (*CFTR*) gene. The encoded protein transports chloride ions across the cell membrane and out of the cell (Figure 3.6). The protein contains domains for the binding and hydrolysis of ATP, thus generating energy for the transport process. Normal functioning of the protein is required to regulate components of sweat, digestive fluids, and mucus. Over 1500 mutations that produce CF have been described in the *CFTR* gene, but approximately 60% of the affected individuals are homozygous carriers of a deletion of three

nucleotides encoding the amino acid phenylalanine at position 508 of the protein (denoted as ΔF508). Another 35% of European CF patients are *compound heterozygous*, that is, the ΔF508 mutation is combined with a different mutation arising from the second allele. In these cases, the clinical symptoms of CF may be comparatively mild. The high frequency of the ΔF508 mutation in northern Europeans is probably due to a founder effect in Denmark in the Neolithic Era. It has been speculated that heterozygous carriers of the *CFTR* mutation had a selective advantage that enabled them to survive infectious diarrhea as caused by cholera, for example. However, the mutated protein does not fold and function correctly. It may even be degraded by the proteasome before it reaches the cell membrane. As a consequence, chloride ions are not pumped out of the cell, and mucus accumulates.

Different tests can be combined to make a diagnosis of CF. The common sweat test measures the salt content of the sweat. Elevated chloride content is a diagnostic marker for CF. In addition, trypsinogen levels in the blood can be measured, since an increased level in immunoreactive trypsinogen is another indicator of CF. In some countries, this test is part of newborn screening programs. Finally, CF can be diagnosed at the DNA level. Since the *CFTR* gene is large, these tests are usually restricted to the analysis of common mutations, such as the deletion of the triplet encoding phenylalanine at position 508 mentioned above.

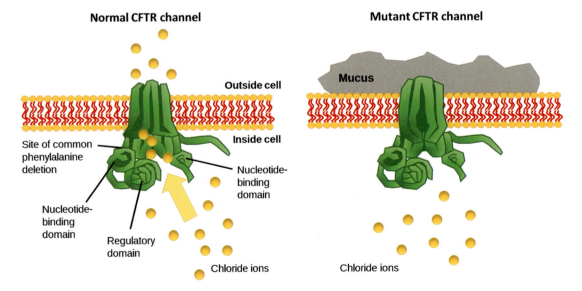

Fig. 3.6 Functional deficiency of the mutant CFTR channel. The normal CFTR channel pumps chloride ions out of the cell. It contains two nucleotide binding domains that hydrolyze ATP, generating the energy required for the transport process. The site of the most common mutation, a deletion of the amino acid phenylalanine, is indicated. The mutated protein is either degraded by the proteasome in the cell or does not pump chloride ions across the membrane, for example in patients carrying a G551D mutation in the CFTR protein. As a consequence, mucus accumulates.

Although to date there is no specific cure available for CF, several treatment methods exist that dramatically improve the prognosis for affected individuals. While infants born with the disease 70 years ago usually died within their first year, the average life expectancy for patients with CF is approximately 40 years today. One of the main aims of CF management is to limit the damage of the lungs caused by the thick mucus. This is accomplished by inhalation therapies, good nutrition, and an active lifestyle. Recombinantly produced deoxyribonuclease (dornase alpha) can be given in aerosolized form to break down DNA in the sputum/mucus and reduce its viscosity in the lungs, promoting improved clearance of secretions. Lung infections are treated with antibiotics and some patients also receive antibiotics when healthy as prophylaxis. Even with these treatments, lung function declines over time and may reach a point where lung transplantation becomes necessary. Problems in the gastrointestinal tract (e.g., pancreatic insufficiency) can be treated by supplying pancreatic enzymes.

New therapeutic approaches for the treatment of CF are continuously being developed. The small molecule drug ivacaftor (trade name Kalydeco) was approved by the FDA in 2012 for the treatment of patients carrying the G551D mutation in the CFTR protein, which accounts for 4–5% of CF cases. The mutated protein is correctly located on the epithelial cell surface, but does not transport chloride ions through the channel. Ivacaftor improves transport by binding to the channel and inducing a nonconventional mode of gating. Additional small molecules aim at overcoming the effects of stop codons introduced by mutations, or at improving the function of the mutant protein with the most prevalent mutation (deletion of phenylalanine 508). In addition, gene therapeutic approaches (Chapter 11) aim to place a functional copy of the *CFTR* gene into affected cells. Adeno-associated virus vectors and adenoviral vectors, which have a natural tropism for lung cells, have been used in these early stage trials.

3.1.2.2 Tay–Sachs Disease

Tay–Sachs disease is a hereditary sphingolipid storage disease that is characterized by progressive destruction of nerve cells. The disease results in deterioration of the mental and physical abilities of affected patients. The disease is classified into three forms: Newborns with infantile Tay–Sachs disease develop normally after birth, but the relentless decline of mental and physical abilities begins at 6 months of age and death usually occurs by the age of 3–4 years. The onset of the juvenile form of the disease is later (between 2 and 10 years). It begins with cognitive and motor skill deterioration and leads to death between the age of 5 and 15 years. Finally, adult-

or late-onset Tay–Sachs disease is rare and usually has its first symptoms during the 30s or 40s. Symptoms include speech and swallowing difficulties, cognitive decline, and psychiatric illness. In contrast to the other forms of the disease, late-onset Tay–Sachs disease is usually not fatal and can stop progressing.

Tay–Sachs disease is caused by a deficiency of hexosaminidase A, a lysosomal enzyme that normally degrades G_{M2} gangliosides. Gangliosides are glycosphingolipids that are located in cell surface membranes. They are found predominantly in the nervous system where they constitute a significant fraction (6%) of total brain lipid. Glycosphingolipids are lysosomally degraded by a series of enzymatically mediated hydrolytic reactions. Hexosaminidase A hydrolyzes the terminal *N*-acetylgalactosamine residue of ganglioside G_{M2} (Figure 3.7a). The enzyme is a hexamer composed of an α- and a β chain. Tay–Sachs disease results from mutations in the *HEXA* gene located on chromosome 15, which encodes the α-subunit. More than 100 mutations in the *HEXA* gene have been identified. A deficiency in hexosaminidase A activity leads to the accumulation of ganglioside G_{M2} in the lysosomes of nerve cells (Figure 3.7b), eventually causing premature cell death.

Infantile Tay–Sachs occurs when a child inherits two alleles encoding the inactive form of hexosaminidase A. In contrast, there is residual enzymatic activity in the juvenile and late-onset forms. The disease's variability may be explained by compound heterozygosity, that is, the affected individuals may have inherited two different variants of the enzyme with diminished activity. Heterozygous carriers with one wild-type allele also have reduced enzyme activity, but the level is still high enough to enable normal neural functioning and to prevent the development of the disease phenotype.

Ashkenazi Jews have a particularly high incidence of Tay–Sachs disease. In the United States, about 1 in 30 Ashkenazi Jews is a recessive carrier of the mutation and the disease incidence is about 1:3000 newborns. In contrast, the incidence in the general population in the United States is approximately 1 in 320 000 newborns. Other populations with an increased incidence of Tay–Sachs disease are Cajuns and French Canadians.

Genetic population studies have identified different mutations arising from small founder populations. In Ashkenazi Jews, a 4 bp insertion in exon 11 of the gene (denoted as 1278insTATC) leads to a shift in the reading frame so that a nonsense protein is synthesized. This mutation leads to the infantile form of the disease. While the same mutation is found in Cajuns, two unrelated mutations predominantly occur in French Canadians living in eastern Quebec.

Tay–Sachs disease can be diagnosed by measuring the activity of hexosaminidase A in serum, fibroblasts, or

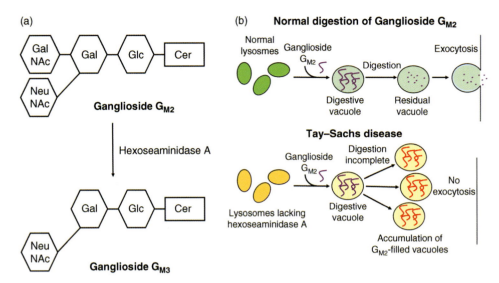

Fig. 3.7   Tay–Sachs disease. (a) The disorder is caused by a deficiency in hexosaminidase A, which normally hydrolyzes the terminal *N*-acetylgalactosamine (Gal NAc) in ganglioside G_{M2}. (b) Normal lysosomes degrade ganglioside G_{M2}. In the absence of hexosaminidase A, the ganglioside accumulates in the lysosomes and interferes with normal biological processes, eventually leading to cell death. Cer: cerebroside; Glc: glucose; Gal: galactose; Neu NAc: *N*-acetylneuraminic acid.

leukocytes. Another sign is a red spot in the retina, originating from gangliosides in the surrounding retinal ganglion cells. A diagnosis based on enzyme activity or ophthalmological examination can be confirmed by genetic analysis. Attempts to reduce the incidence of Tay–Sachs disease include prenatal and preimplantation diagnosis. In addition, some orthodox Jewish organizations offer mate selection programs. These measures have already produced a drop in the incidence in this population.

There is no cure available for Tay–Sachs disease and even with best care, children with the infantile form die by the age of 4. Enzyme replacement therapy is hampered by the large size of the hexosaminidase A protein, so that it is not taken up efficiently by neurons. Alternative approaches involve prevention of ganglioside G_{M2} production by substrate deprivation, or inhibition of enzymes in the biosynthetic pathway. Additional experimental strategies are in development to increase the residual activity of hexosaminidase A and to slow down progression of the late-onset form. However, this approach will not be applicable for juvenile Tay–Sachs disease, since in this case the enzyme is completely absent.

3.1.2.3   Phenylketonuria

PKU is the most prevalent inherited disorder in amino acid metabolism. Untreated PKU can lead to intellectual disability and seizures. However, affected individuals are routinely diagnosed by newborn screening in many countries. Immediate treatment with an appropriate

diet allows patients to have a normal life span and normal mental development.

PKU was the first inherited metabolic disease whose basic biochemical defect had been identified, which occurred in the 1940s. The classical form results from a deficiency in phenylalanine hydroxylase (PAH), the enzyme that metabolizes the amino acid phenylalanine to tyrosine (Figure 3.8). For its catalytic activity, the

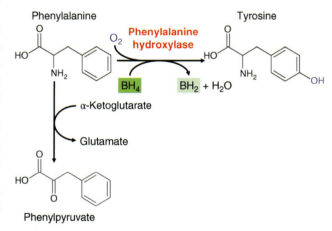

Fig. 3.8   Phenylketonuria (PKU) and the metabolism of phenylalanine. Phenylalanine hydroxylase (PAH) metabolizes the amino acid phenylalanine to tyrosine, a reaction requiring molecular oxygen and the redox cofactor tetrahydrobiopterin (BH_4), which is oxidized to dihydrobiopterin (BH_2). Due to PAH deficiency in individuals with PKU, at elevated levels of phenylalanine, an alternative pathway involving the transamination of the amino acid to phenylpyruvate becomes relevant.

enzyme requires molecular oxygen and oxidizes the redox cofactor tetrahydrobiopterin (BH_4) into dihydrobiopterin (BH_2), which is eventually regenerated to BH_4. The absence of PAH leads to increased blood levels of phenylalanine (hyperphenylalaninemia). Phenylalanine is then transaminated to phenylpyruvate (also denoted as phenylketone) by an otherwise minor pathway. Elevated levels of phenylketone can be detected in the urine, which is the basis for the name of the disease.

Classical PKU is caused by mutations in the *PAH* gene. More than 600 disease-causing mutations have been described. For some mutations, the enzyme retains residual activity, which influences the severity of the disease. Most of the patients are compound heterozygotes and carry two differently mutated alleles. A rarer form of hyperphenylalaninemia occurs when the PAH activity is normal, but there is a defect in the biosynthesis or recycling of the essential cofactor BH_4. The affected individuals have a milder increase in phenylalanine levels, but additional pathways, which require BH_4 as a cofactor, for example, the synthesis of neurotransmitters (e.g., dopamine), are disrupted as well.

The mechanism by which phenylalanine damages the brain is not fully understood. It has been argued that elevated levels of phenylalanine saturate the transport capacity across the blood–brain barrier. The amino acid enters the brain via the large neutral amino acid carrier L-amino acid transporter 1 (LAT1). Tyrosine and tryptophan use the same carrier and their entry may be inhibited by high concentrations of phenylalanine. As a consequence, protein biosynthesis may be compromised. Tyrosine and tryptophan are also precursors of the neurotransmitters dopamine/norepinephrine and serotonin, respectively, which may explain the dysfunction of neurotransmission in patients with PKU. Another symptom of PKU is light hair and skin color, as elevated phenylalanine levels inhibit tyrosine hydroxylation, the first step in the formation of the skin pigment melanin.

As already mentioned, in most countries, newborns are screened immediately after birth for PKU by testing for elevated levels of phenylalanine in the blood (Section 8.1). Screening had commonly been carried out with a bacterial inhibition assay known as the *Guthrie test*. This assay uses the growth of a *Bacillus subtilis* strain as an indicator of high levels of phenylalanine or its metabolites. In the Guthrie test, bacterial growth is inhibited by the compound β-2-thienylalanine. High levels of phenylalanine, phenylpyruvate, or phenyllactate restore growth of the bacteria. More recently, the Guthrie test has been replaced by newer techniques such as tandem mass spectrometry, which can detect a wider variety of metabolic diseases.

PKU can be successfully treated by a phenylalanine-restricted diet. The affected individuals are advised to limit the intake of protein-rich foods such as meat, fish, eggs, milk, and other dairy products. Starchy foods (bread), fruits, and vegetables must also be monitored. Aspartame, a widely used sweetening ingredient in diet soft drinks and other dietetic food products, must be avoided as the methyl ester of the dipeptide aspartate/phenylalanine is broken down into its components in the body. Dietary restrictions may be loosened at the age of 10, but PKU patients are recommended to manage their phenylalanine levels throughout life.

In about 20% of patients, oral administration of BH_4 (marketed under the trade name Kuvan) stimulates PAH activity and can be used as an adjunct medication to increase the amount of natural protein those affected can eat. Therapeutic approaches in development include enzyme substitution therapy with phenylalanine ammonia lyase, a bacteria-derived enzyme that degrades phenylalanine, and restoration of PAH activity by gene therapy.

3.1.2.4 Xeroderma Pigmentosum

Xeroderma pigmentosum (XP) is an autosomal recessive disorder with 100% penetrance in which the cell's ability to repair damage caused by ultraviolet light is deficient. The incidence of this illness varies greatly, between an estimate of 1 in 250 000 in the United States and a higher rate of 1 in 20 000 in Japan. The affected individuals are extremely sensitive to sunlight, resulting in sunburn, pigment changes in the skin, and a greatly elevated incidence of skin cancers. In extreme cases, all exposure to sunlight must be avoided. This is why individuals with the disease are sometime called "children of the night." In the absence of rigorous protection from the sun, skin malignancies, such as multiple basal cell carcinomas, invasive squamous cell carcinomas, and melanomas develop at a young age. The skin cancers occur at a 2000-fold greater rate than normal and are the main cause of death in patients with XP, often at an age of less than 20. Some patients also show progressive neurological degeneration or ocular abnormalities.

XP can result from mutations in any one of eight genes, seven of which are involved in the repair of UV-induced DNA lesions by the nucleotide excision repair (NER) system. UV light leads to the formation of pyrimidine dimers in the DNA. The NER, consisting of at least 16 proteins in eukaryotes, repairs these lesions by excision of approximately 30 nucleotides harboring the damage, synthesizing the missing sequence by the enzyme DNA polymerase and sealing the gap with DNA ligase (Figure 3.9). XP is caused by defective proteins involved in recognition of DNA photoproducts, in opening up the structure of the DNA around the site of the photoproduct, in verifying the correct positioning of the proteins, and in the nucleolytic cleavage of the DNA on

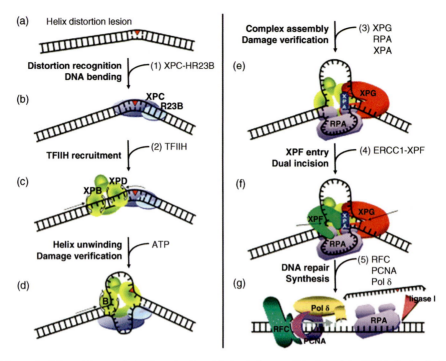

Fig. 3.9 Model of the nucleotide excision repair (NER) mechanism. A lesion induces a distortion of the DNA helix that is detected by XPC-HR23B. This factor recruits the transcription factor TFIIH that unwinds the DNA helix until its XPD helicase subunit encounters a chemically modified base. The second helicase subunit XPB continues unwinding the DNA and creates an open bubble structure. Further factors assemble to form the preincision complex. ERCC1-XPF joins the complex and dual incision occurs. DNA polymerase δ carries out the repair synthesis; ligase I finally seals the nick. (Adapted with permission from Ref. [2]. Copyright 2006, American Chemical Society.)

either side of the damage. The uncorrected lesions in the DNA can cause mutations that may result in cancer if tumor suppressor genes or proto-oncogenes are affected. The eighth gene whose mutation results in XP is not involved in NER, but in the replication of UV-induced DNA damage. In healthy human beings, the specialized DNA polymerase η enzyme replicates DNA lesions caused by UV light. Defects in this polymerase constitute the so-called XP variant group. These patients may display comparatively mild symptoms.

In most cases, the initial diagnosis of XP is made on the basis of extreme sensitivity to UV light or the appearance of brown spots on the face at an unusually young age. The diagnosis can be confirmed by cellular tests. Punch biopsies are taken and fibroblasts cultivated and challenged with UV irradiation. Analysis of the efficiency of DNA repair in cell culture provides the diagnosis of XP. There is currently no cure available for XP, except for the appropriate protection from sunlight.

3.1.3
X-Linked Recessive Disorders

The major difference between autosomal and X-linked recessive disorders is that the latter affects each sex differently. Males have only one X chromosome (i.e., they

are hemizygous for all X-chromosomal variations) and will not compensate a loss of function if they inherit a mutated allele. In contrast, females have two X chromosomes, so that the wild-type chromosome can compensate the loss of function of a mutated allele. In this case, the female will be carrier of a defect but will remain asymptomatic. Only in the rare event that a female inherits two mutated alleles will she develop the disease. As already outlined, the usual mating pattern involves a heterozygous female carrier of a genetic defect and an unaffected male. Daughters will remain phenotypically unaffected, since they will inherit the father's wild-type allele, but they have a 50% chance of becoming an asymptomatic carrier by inheriting the mutated gene from their mother. In contrast, male offspring are at a 50% risk of being affected through inheritance of the mutant maternal allele.

In some cases, the situation becomes more complex in females by the fact that one of the two X chromosomes is inactivated at random during embryonic development. This phenomenon, also known as *Lyonization* (named after Mary Lyon, who discovered the phenomenon of X chromosome inactivation), is necessary as dosage compensation since females have twice as many X-chromosomal gene products as males, who possess only one copy of the X chromosome. Thus, although normal females possess two X chromosomes, only one of them will be active in any given cell. The inactivation occurs at random at the time of gastrulation. It takes place at the cellular level and results in what is known as mosaic expression; that is, some cells have an inactive maternal X chromosome, while others have an inactive paternal chromosome. In females heterozygous for an X-linked disorder, the activity of the normal gene in some cells may be sufficient to compensate for the defect. In addition, it has been hypothesized that the normal allele may have a selective advantage over the mutated one. The process of X chromosome inactivation has not yet been fully understood, but it explains why a female carrier for an X-linked recessive disorder may, to varying degrees, become symptomatic. Typical examples of X-linked recessive disorders are hemophilia A and B (Section 10.2.6), Duchenne and Becker muscular dystrophy, and more common but less serious conditions such as male red-green color blindness.

3.1.3.1 Red-Green Color Blindness

Red-green color blindness (protanopia) is a common X-linked inherited condition caused by the absence of red or green retinal photoreceptors (opsins). Depending on ethnic background, up to 9% of males are affected. Approximately 15% of the females are carriers of the genetic defect, but less than 1% are affected as a consequence of the disadvantageous inactivation of the normal allele. The long arm of the X chromosome normally carries a cluster of 2–9 *opsin* genes. The red photoreceptor gene is the most proximal of the tandem array, followed by one or more green photoreceptor genes. Genes encoding the opsin protein for the absorption of red light (*OPN1LW*, opsin 1 long-wave sensitive) and for the absorption of green light (*OPN1MW*, opsin 1 medium-wave sensitive) are highly homologous, with 98% sequence identity. The minor sequence differences affect the binding of the chromophore retinal, thereby adjusting its absorption maximum. Normal red-green vision requires at least one *opsin* gene that absorbs efficiently in the red and another that absorbs in the green.

The high degree of similarity between the repeat units predisposes them to unequal homologous recombination. Intergenic crossing over may result in the gain or loss of one or more genes (Figure 3.10a). The absence of either of the genes encoding the red- or green-absorbing opsins results in red-green color blindness. In addition, recombination events in the coding regions (intragenic crossing over) create hybrid genes (Figure 3.10b). The encoded proteins will have absorption properties intermediate between the wild-type red and green photoreceptors. This causes difficulty in discriminating between the colors.

3.1.3.2 Duchenne and Becker Muscular Dystrophy

Duchenne muscular dystrophy (DMD) is the most common form of the heterogeneous group of neuromuscular disorders. It is caused by mutations in the *dystrophin* gene. This is the largest protein-coding human gene, being composed of 79 exons and covering approximately 2.6 million base pairs. It is located on the short arm of the X chromosome. The mature transcript is approximately 16 kb long and encodes a 400 kDa protein composed of almost 3700 amino acids. Dystrophin is expressed predominantly in the skeletal and cardiac muscle. The protein is an important structural

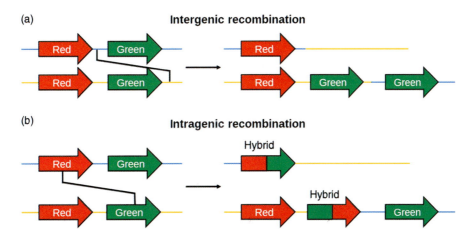

Fig. 3.10 Recombination events in red and green photoreceptor genes. Opsin genes for sensing red and green light are arranged in a tandem array on the long arm of the X chromosome. The red photoreceptor gene is proximal, followed by one or more green photoreceptor genes. (a) Intergenic crossing over results in the gain or loss of one *opsin* gene. (b) Unequal recombination events in the coding regions (intragenic crossing over) create hybrid genes. The encoded hybrid photoreceptors may have intermediate absorption characteristics. (Adapted from Ref. [3] with kind permission from Elsevier.)

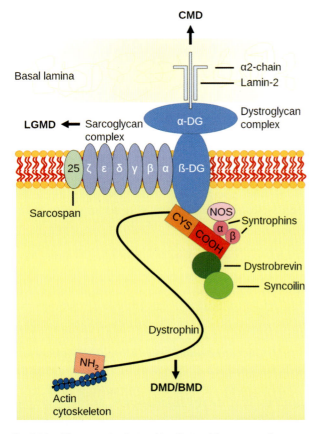

Fig. 3.11 The muscular dystrophies. Dystrophin connects the intracellular actin cytoskeleton with the dystrophin–glycoprotein complex (DGC). This large complex, which includes dystroglycan (DG), the neuronal nitric oxide synthase (nNOS), the syntrophins, dystrobrevin, and syncoilin, provides a link to the extracellular basal lamina. Mutations in the *dystrophin* gene cause Duchenne muscular dystrophy (DMD) or the milder form, Becker muscular dystrophy (BMD). Mutations in other components of the DGC cause different types of muscular dystrophies, such as limb girdle muscular dystrophy (LGMD) or congenital muscular dystrophy (CMD). (Adapted from Ref. [4] with kind permission from Macmillan Publishers Ltd., Copyright 2003.)

component within muscle tissue and links the actin cytoskeleton to the extracellular matrix via the dystrophin–glycoprotein complex (DGC) (Figure 3.11). In addition to its structural role, dystrophin is involved in cell signaling and regulating muscle response to oxidative stress. The absence of the dystrophin protein disrupts the integrity of the cell membrane (sarcolemma). This permits excess calcium to penetrate the cell and alter signaling pathways, eventually resulting in cell death and progressive muscle weakening. The muscle tissue experiences wasting and is eventually replaced by adipose tissue and fibrotic cells. Symptoms usually appear before age 6, and patients are typically wheelchair bound by their early teens. The average life expectancy for individuals afflicted with DMD is around 25 years;

cardiac and respiratory failures are the major causes of death.

A large number of different variations in the *dystrophin* gene have been found to cause DMD. Deletions are the most common mutations, but point mutations, duplications, and translocations have also been reported. In patients with DMD, these mutations usually lead to the complete absence of a functional dystrophin protein due to a shift in the reading frame and/or premature termination of protein synthesis. In contrast, mutations that result in the synthesis of a shortened or mutated protein-retaining residual activity lead to the milder Becker type of muscular dystrophy. In this less common case, symptoms usually first occur after the age of 10 and lead to invalidism in the fourth or fifth decade of life. However, affected individuals may experience a nearly normal lifespan.

Like other genetic disorders, DMD and BMD can by diagnosed by genetic testing (Section 8.2.2). However, since the *dystrophin* gene is large and the disorders may be caused by many different mutations, some of which have not yet been identified, muscle biopsies can be carried out as an alternative test. Muscle tissue is extracted from the patient and analyzed by immuno-histochemistry for the presence or absence of the dystrophin protein.

Although there is no cure for DMD or BMD, various types of treatment have been developed that aim at controlling the onset of symptoms, slowing progression of the disease, and improving the quality of life of the affected individuals. Supportive care involves nonpharmacologic intervention, such as physical and occupational therapy and orthopedic surgery. Pharmacological therapy includes treatment with corticosteroids (e.g., prednisolone and deflazacort) to strengthen the muscles. In addition, the low-molecular weight agent ataluren is believed to improve ribosomal read-through of premature stop codons, thus preventing the synthesis of truncated proteins. However, a phase II trial with the drug produced ambiguous results. Gene therapeutic (Chapter 11) approaches to introduce the functional dystrophin are hampered by the large size of the protein. However, attempts to transfer a modified miniform of *dystrophin* by adeno-associated viral vectors into skeletal and cardiac muscle cells have been promising.

One of the most advanced approaches to treat DMD is the use of antisense oligonucleotides (AS ONs) to modulate splicing. Two experimental drugs (drisapersen and eteplirsen) induce skipping of exon 51, which contains a premature stop codon in approximately 15% of the patients. The AS ONs prevent the synthesis of a truncated protein, and although the product will be shortened, it will still be functional and result in the milder Becker muscular dystrophy. Clinical studies have demonstrated improved muscle function in subjects treated with the AS ONs compared to a control group receiving placebo. For further details about these approaches, see Sections 13.1.

3.1.4
Mitochondriopathies

Disorders caused by dysfunctional mitochondria are called mitochondriopathies or mitochondrial diseases. As outlined in Section 1.2.1, mitochondria contain their own genetic material, the mitochondrial DNA (mtDNA). Mitochondria play a key role in many metabolic activities, particularly in the generation of ATP by oxidative phosphorylation. Defects in mitochondrial function will thus affect the basic energy supply of all cells in the body. The proteins involved in metabolic processes are in part encoded by nuclear DNA and to a smaller extent by mtDNA: The circular mtDNA molecule encodes 13 proteins (in addition to two rRNAs and 22 tRNAs), while the remaining approximately 1000 proteins found in mitochondria are encoded by the nuclear genome.

Mitochondrial diseases may thus be caused by mutations in the nuclear or mitochondrial genome. Mutations in the nuclear DNA are not only restricted to genes encoding proteins of the respiratory chain and other metabolic processes. They may also become important if they affect proteins involved in the assembly of respiratory complexes, the synthesis of phospholipids for organellar membranes or mitochondrial replication, or in the transcription and translation machinery. Mutations in more

than 70 nuclear genes have been identified that interfere with the proper functioning of mitochondria. Fifteen percent of mitochondriopathies are caused by mutations in the mitochondrial genome. More than 200 pathogenic point mutations in the mtDNA and an equal number of large deletions have been reported. Since the developing embryo receives mitochondria almost exclusively from the ovum and not from the spermatozoon, mutations in the mtDNA are usually maternally inherited (Figure 3.12). However, they may also be spontaneously acquired, since new mitochondria are generated during cell division and the repair capacity for mtDNA is lower than for nuclear DNA.

Mitochondrial diseases are complex. Given the fact that each mitochondrion contains 2–10 DNA molecules and eukaryotic cells typically contain 800–2000 mitochondria, there are a large number of mtDNA molecules in a cell. These copies are usually identical (called *homoplasmy*), but may also be a mixture of mtDNA mutant and wild-type versions (called *heteroplasmy*). In case a mutation has particularly severe consequences, it would be lethal if it occurred in all mitochondria. Patients will thus be heteroplasmic as a relatively few wild-type organelles can provide residual mitochondrial activity. For mutations with milder consequences, the homoplasmic mutated state may also be viable.

The consequences of any given mutation may vary between individuals depending on other genomic variations; that is, a mutation that causes liver disease in one person may affect neuronal cells in another. Generally, tissues that require a high energy supply, such as muscle, the cerebrum, or the central nervous system, are usually severely affected by defective mitochondria. Accordingly,

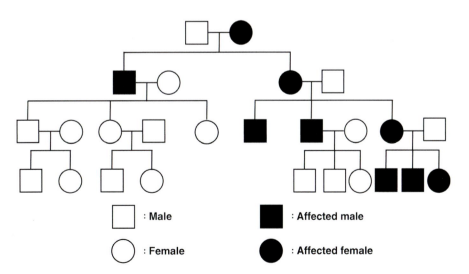

Fig. 3.12 A typical pedigree of an inherited mitochondrial disease. Mutations in the mitochondrial DNA are only inherited from the mother, and can affect male as well as female offspring.

Table 3.3 Some examples of mitochondrial diseases.

Disease	Clinical phenotype	Type of mutation(s)
Kearns–Sayre syndrome	Multisystemic disorder	Large-scale deletions and rearrangements of the mtDNA
Progressive external opthalmoplegia	Myopathy of the extraocular muscles	Large-scale deletions and rearrangements of the mtDNA
Pearson's syndrome	Disorder of the bone marrow and the pancreas	Large-scale deletions and rearrangements of the mtDNA
MELAS syndrome	Encephalo- and myopathy, lactic acidosis	Point mutation in the mitochondrial gene encoding leucine tRNA
MERRF syndrome	Epilepsy, *Ragged Red Fibers*-clumps	Point mutation in the mitochondrial gene encoding lysine tRNA
Leigh syndrome	Fatal neurodegenerative disorder	Mutations in nuclear genes encoding subunits of the respiratory chain
Maternally inherited Leigh syndrome	Fatal neurodegenerative disorder	Mutations in mitochondrial genes encoding a subunit of the ATP synthase
Leber's hereditary optic neuropathy	Blindness	Mutation in a mitochondrial encoded subunit of Complex I of the respiratory chain

MELAS: mitochondrial encephalomyopathy, lactic acidosis, and stroke-like episodes; MERRF: myoclonic epilepsy, with ragged-red fibers.

symptoms of mitochondrial disease include (but are not restricted to) myopathies and neurological problems. Several mitochondriopathies are given in Table 3.3. Approximately 1 in 4000 children born in the United States will develop a mitochondrial disease by 10 years of age.

Mitochondriopathies may also be caused by large-scale rearrangements of mtDNA. These include kilobase-size deletions and partial duplications of the mtDNA. Examples of these types of disorders are the Kearns–Sayre syndrome (KSS), a fatal multisystemic disorder; a less severe progressive external opthalmoplegia; a myopathy leading to paralysis of the extraocular muscles; and Pearson's syndrome, a disorder of the bone marrow and the pancreas. All three disorders are sporadic, that is, they have been acquired through somatic mutation and are not inherited. Dozens of different deletions and rearrangements have been identified that may range from 1 to 10 kilobases in size. The affected individuals harbor a single species of deleted mtDNA that coexists with the wild-type molecule. The deletions usually remove tRNA genes that are required for the translation of mRNAs encoded in the mtDNA. In some cases, mtDNA of a mitochondrion may be completely depleted.

Most of the known pathogenic point mutations in mtDNA are located in genes required for mitochondrial protein translation, the vast majority being located in tRNA genes. These mutations then affect the translation of all 13 mtDNA-encoded proteins, which leads to a general defect in the oxidative phosphorylation process. Mutations of the leucine tRNA are the most commonly maternally inherited point mutations in mtDNA. They often cause the MELAS syndrome (mitochondrial

encephalomyopathy, lactic acidosis, and stroke-like episodes). In contrast, mutations in the gene encoding the lysine tRNA are associated with the MERRF syndrome (myoclonic epilepsy with ragged-red fibers), indicating that the effects on translation may differ depending on mutations in different tRNAs. Other mutations affecting tRNAs or rRNAs are tissue specific and are associated with disorders such as cardiomyopathies or deafness.

Several mutations affect the complexes of the respiratory chain (Figure 3.13). A defect in Complex I is one of the most common causes of the mitochondrial encephalomyopathies. A typical disease associated with dysfunction in Complex I is Leigh syndrome (LS), a devastating neurodegenerative disorder resulting from mutations in the nuclear genome. The symptoms of LS usually appear soon after birth or in early childhood. Progression of the disease leads to severe debilitation as the brain cannot control muscular contraction. Severe forms of the disease caused by the absence of one of the affected proteins cause death at a young age most often due to respiratory failure. In rare cases, the children reach their teenage years. It is not yet fully understood why mutations in the mitochondrial encoded subunits of Complex I result in less severe (although still truly debilitating) phenotypes than those caused by mutations in nuclear genes.

Mutations in the mtDNA-encoded *NADH dehydrogenase* gene encoding a subunit of Complex I are associated with Leber's hereditary optic neuropathy (LHON), a maternally inherited cause of blindness. Males are preferentially affected by LHON for unknown reasons, but which may involve a susceptibility locus on the Y chromosome. Genes on the X chromosome are also thought to influence the development of symptoms.

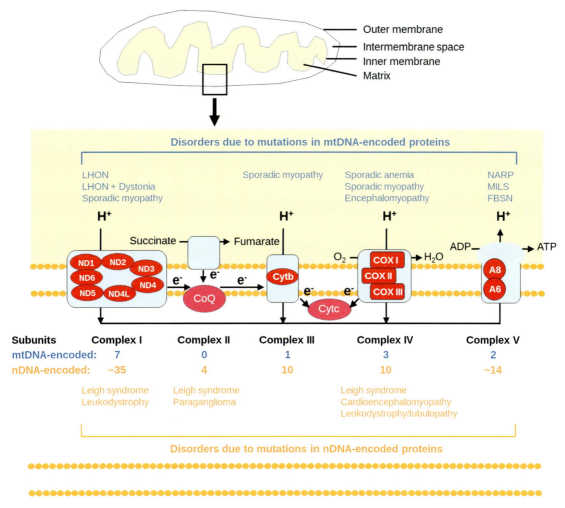

Fig. 3.13 Diseases associated with defects in the respiratory chain. Disorders may either be due to mutations in nuclear- or mitochondrial-encoded proteins. Different disorders may be associated with defects in specific complexes. FBSN: familial bilateral striatal necrosis; LHON: Leber's hereditary optic neuropathy; MILS: maternally inherited Leigh syndrome; NARP: neuropathy, ataxia, and retinitis pigmentosa.

LS may not only be caused by mutations in the nuclear DNA encoding subunits of Complex I but also by alterations of the nuclear genome affecting subunits of Complexes II or IV. In addition, a related disease form, the maternally inherited Leigh syndrome (MILS), is associated with mutations in the mtDNA-encoded subunit A6 of the ATP synthase (or Complex V). In patients with MILS, one of the two prevalent mutations is found in more than 90% of the mtDNA molecules. A lower proportion (70–90%) of mutated mtDNA molecules is associated with a completely different disease, the NARP syndrome (neuropathy, ataxia, and retinitis pigmentosa). Mothers of NARP or MILS patients with less than 70% mutated mtDNA molecules remain asymptomatic, that is, the degree of heteroplasmy and impairment of ATP synthesis determines whether an individual develops fatal MILS, the less severe NARP phenotype, or remains healthy.

Treatment options for mitochondrial diseases are currently still limited, although some progress in research has been made in recent years. As already outlined, mitochondriopathies are extremely diverse, since the causative mutations may be in the nuclear or mitochondrial DNA. In addition, the consequences also depend on additional factors such as gender and other genetic variations. Thus, therapeutic strategies must be tailored for each class of mutation. Current approaches include supplementation with components that enhance the function of the respiratory chain, such as coenzyme Q_{10}, or with antioxidants theorized to prevent mitochondrial dysfunction. Strategies in development involve the enhancement of mitochondrial biogenesis, in addition to a process called heteroplasmic shifting. This approach aims to influence the mode of mtDNA replication, so that the proportion of wild-type mtDNA is increased. As already described, a reduction in the proportion of mutant mtDNA by as little as 10–20% may have dramatic phenotypic consequences. The cure of a mitochondrial disease is envisaged by novel reproductive strategies. One such strategy involves transferring the fertilized nucleus (or the pronuclei) from the oocyte of a woman carrying a mutation in her mtDNA to the enucleated oocyte from a healthy donor (cytoplasmic transfer). The embryo will contain the normal nucleus of the mother (including the genetic nuclear material of the father) without her mutated mtDNA and will also contain the normal mtDNA from the cytoplasmic donor. The feasibility of this approach has already been demonstrated. The procedure is controversial as the child would carry genetic material from one father and two mothers and the germ line modification will be passed on to future generations. Nevertheless, in 2015, the UK government legalized the three-person *in vitro* fertilization (IVF) procedure as a treatment to modulate mitochondrial diseases that are passed on from mother to child.

3.2
Polygenic Disorders

As described in the previous section, the majority of monogenic disorders affect only a small number of individuals. In contrast, most of the more common diseases with large patient numbers are much more complex. They are usually associated with multiple genes and environmental factors often play an additional important role. These diseases often cluster in families, but they do not have a clear Mendelian pattern of inheritance. Their appearance may be highly variable, complicating our understanding of the mode of inheritance. In many cases, it is also difficult to render a prognosis about an individual's risk of being affected by a multifactorial disease. Nevertheless, in many cases it is possible to identify genes that contribute to complex disorders. Examples of polygenic inherited diseases are some types of cancers, as will be explained in Chapter 4, heart disease, hypertension, mood disorders, and obesity. Asthma and diabetes mellitus will be described as two examples of polygenic disorders in more detail in the following sections.

3.2.1
Asthma

Asthma is a common chronic inflammatory disease of the airways, which results in recurrent, reversible bronchial obstruction. Symptoms include wheezing, coughing, chest tightness, and shortness of breath. The disease can start at any age, but in most cases, the first symptoms usually occur in childhood. Several hundred million people globally have been diagnosed with asthma, and it caused almost 350 000 deaths in 2010.

Asthma is a typical example of a disease that is caused by a combination of multiple genetic and environmental factors. Numerous genome-wide association studies have been carried out to identify genes that play a role in the development of this disease. It is clear that there is no single gene for asthma. Rather, dozens of genes putatively contributing to the development of the disease are distributed over many human chromosomes. Many of these candidates are related to the immune system or to the modulation of inflammation. However, most of the genes identified in one association study have not been confirmed by others. An exception is the genetic variability of the chromosome 17q21 locus, which has repeatedly been found to be associated with childhood asthma. Alterations in this locus affect the expression of two genes, *orosomucoid like 3* and *gasdermin B*. Current research focuses on the function of these genes in the pathogenesis of asthma. Analysis of common single-nucleotide polymorphisms (SNPs) also revealed that individually they have a weak influence on asthma risk. More recent studies of the genome-wide sequencing of exons suggested that rare genetic variants might have a larger effect on the heritability of complex diseases, such as asthma, than common variants.

The difficulties in obtaining conclusive data about the genetic contribution to the development of asthma may in part be explained by the fact that the impact of a genetic variation will in many cases depend on both the polymorphisms in other genes and the exposure to specific environmental triggers. A variety of environmental factors have been associated with the development of asthma. Mold, dust mites, various insects (e.g., cockroaches), animal hair, and some viruses have been identified as antigens that can trigger asthma. In addition, exposure to organic chemicals and air pollution are thought to increase the risk of developing the disease. In the last several decades, a marked worldwide increase in asthma prevalence has been observed. This finding was explained by the so-called "hygiene hypothesis," according to which reduced exposure to (harmless) pathogens during early childhood in modern societies unintendedly results in increased susceptibility. In the less sterile environment of the past, exposure to bacterial endotoxins, eukaryotic parasites, and other stimuli in early childhood seemed to prevent the development of asthma. Today, first contact with these triggers usually occurs at an older age and may provoke bronchoconstriction. The hypothesis is supported by the finding that children living on farms have a lower rate of asthma.

Medications to treat asthma can be divided into quick-relief drugs used to treat the acute symptoms, and long-term acting medications designed to prevent exacerbation. The first-line treatment for acute asthma symptoms are short-acting β-adrenoceptor agonists (SABAs), for example, sabutamol, that relieve bronchospasm. Inhaled corticosteroids with anti-inflammatory and antiallergic activity are the standard long-term pharmacological treatment for mild to moderate asthma. They may be given together with long-acting β-adrenoceptor agonists (LABAs) to reduce the need for SABAs. Poor compliance with the required once or twice daily inhalation of corticosteroids remains a major obstacle to therapeutic success. Several novel therapeutic approaches have been developed for the treatment of severe asthma. One example is a humanized monoclonal antibody (mAb) (Section 10.2.1), omalizumab (trade name Xolair), that binds to and neutralizes free immunoglobulin E (IgE). High serum levels of IgE correlate with the occurrence of allergies and treatment with omalizumab decreased asthma exacerbation, especially in children. Additional mAbs directed against cytokines that play an important role in the pathogenesis of asthma, such as interleukins 4, 5, and 13, have also shown some success in clinical investigation.

3.2.2
Diabetes Mellitus

Diabetes mellitus is a metabolic disease in which a person has an elevated blood glucose level. Under normal physiological conditions, uptake of glucose leads to the secretion of insulin from the β cells of the islets of Langerhans in the pancreas. Insulin is transported to peripheral tissue in the bloodstream and activates specific receptors, which belong to the class of receptor tyrosine kinases. This signal results in the uptake of glucose and its catabolism or storage as glycogen. Inability of the pancreas to produce enough insulin or resistance of the target cells to respond to insulin results in an elevated concentration of blood glucose. The high blood sugar level produces the three classical symptoms of polyuria (frequent urination), polydipsia (increased thirst), and polyphagia (increased hunger).

Untreated diabetes causes various complications. The major acute diabetic emergency is ketoacidosis. A less frequent, but equally severe, acute complication is nonketotic hyperosmolar coma. The major long-term complications of diabetes come from damage to blood vessels. These impairments increase the incidence of cardiovascular disease, chronic renal failure, and diabetic retinopathy. Worldwide, almost 300 million people suffer from diabetes and it results in approximately 1.3 million deaths every year, three quarters of which are due to coronary artery disease.

The two major forms of diabetes are designated as type 1 and type 2 diabetes (T1DM and T2DM, respectively). T1DM results from the body's failure to produce insulin. It is usually diagnosed during childhood and requires treatment with insulin. Thus, type 1 diabetes is called juvenile or insulin-dependent diabetes. Type 2 diabetes accounts for 90–95% of all diabetes cases and results from insulin resistance, that is, the target cells fail to response to an insulin stimulus. This lack of response may be combined with insulin deficiency. Type 2 diabetes often occurs at greater age and is also known as maturity-onset or noninsulin-dependent diabetes. Both major types of diabetes are complex diseases involving the interaction between multiple genes and environmental factors (Figure 3.14).

Type 1 diabetes is usually characterized by insulin deficiency due to a loss of the insulin-producing cells in the pancreas. In 70–90% of the cases, the destruction of β cells is due to T-cell-mediated autoimmunity, while in the remaining cases, the specific pathogenesis remains unclear (idiopathic diabetes). In immune-mediated type 1 diabetes, a chronic inflammatory infiltrate affects the pancreatic islets. The autoimmune reaction may be induced by the triggering of the immune system by an antigen that is structurally similar to surface molecules on the islet cells. Cross-reactivity may then result in destruction of the pancreatic cells. Viral infections (for example, with coxsackievirus B4, cytomegalovirus, or herpes virus) have been considered as possible inducers of autoimmunity; however, this hypothesis is debatable. In addition to environmental factors, type 1 diabetes is clearly a polygenetic disorder, with nearly 40 loci identified so far that affect disease susceptibility. The largest contribution to pathogenesis, accounting for almost one-half of the genetic susceptibility, comes from the human leukocyte antigen (HLA) region on chromosome 6, called the *IDDM1* locus, which encodes more than 200 genes for HLA class I and II proteins (Section 1.5.1). Certain HLA class II haplotypes are known to be strongly associated with type 1 diabetes, while others were found to provide disease resistance. In addition, some genotypes for class I molecules seem to influence the risk of developing the disease. HLA proteins are important for the distinction between self (i.e., the body's own proteins) and nonself (i.e., invading pathogens) so that certain genetic predispositions may promote autoimmunity. Some other factors, which are associated with type 1 diabetes, are indicated in Figure 3.14. Most are involved in the immune response, supporting the assumption that certain combinations of haplotypes, in conjunction with environmental factors, induce an aberrant immune reaction. Specific autoimmune antibodies against several islet antigens,

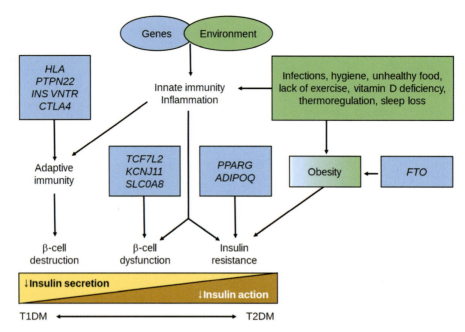

Fig. 3.14 Diabetes mellitus as a complex disease. Interactions between multiple genetic and environmental factors promote inflammatory mechanisms that eventually cause diabetes. Some factors are related to type 1 diabetes mellitus (T1DM), while others are associated with type 2 diabetes mellitus (T2DM). More recently, it has been hypothesized that many cases of diabetes are intermediate forms between the classical type 1 and type 2 classes. ADIPOQ: adiponectin, *C1Q and collagen domain containing* gene; CTLA4: cytotoxic T-lymphocyte-associated protein 4; FTO: *fat mass and obesity associated* gene; HLA: human leukocyte antigen; INS VNTR: insulin gene variable number tandem repeats polymorphism; KCNJ11: potassium inwardly rectifying channel, subfamily J, member 11; PPARG: peroxisome proliferator-activated receptor gamma; PTPN22: protein tyrosine phosphatase, nonreceptor type 22 (lymphoid); SLC3OA8: solute carrier family 30, member 8; TCF7L2: transcription factor 7-like 2 (T-cell-specific, HMG-box). (Adapted from Ref. [5] with kind permission from Macmillan Publishers Ltd., Copyright 2009.)

including insulin (mostly in younger children), can be present in individuals with type 1 diabetes. The detection of autoimmune antibodies not only helps distinguish type 1 diabetes from type 2 diabetes but it can also be used to identify individuals with presymptomatic disease and predict the development of type 1 diabetes.

Type 2 diabetes is caused by insulin resistance, which is likely to involve the insulin receptor on target cells. However, the mechanistic details of the defect are not yet known. Defective responsiveness may be combined with reduced insulin secretion. Type 2 diabetes results from various lifestyle factors, such as obesity, lack of physical activity, poor diet, and stress, in combination with genetic factors. The genetic contribution to type 2 diabetes is even stronger than in the case of type 1 diabetes: The risk of developing disease, if an identical twin is affected, is known as *concordance*. It is ~90% for type 2 diabetes as compared to 40–50% for type 1 diabetes. High concordance is regarded as a strong indication of an important genetic component. An exhaustive survey identified at least 36 diabetes-related genes. However, these explain only about 10% of the heritability of type 2 diabetes. Some of the most relevant factors are indicated in Figure 3.14. It was somewhat unexpected

that most of the genetic factors discovered are linked to β-cell dysfunction rather than insulin resistance. This finding may change our views about the pathogenesis of this disease.

The complexity of the interaction between genotype and lifestyle may best be illustrated for obesity, one of the major environmental factors thought to contribute to the development of type 2 diabetes. However, obesity may not only be strongly inherited, but it may even be caused by diabetes as an effect of changes in metabolism. It is thus not always clear when a particular factor causes a disease as an environmental trigger, or is an inherited predisposition, or is a consequence of the disease itself.

In addition to the two most common forms of diabetes, other types include gestational diabetes in pregnant women without previously diagnosed diabetes, and Maturity onset diabetes of the young (MODY). The latter is a hereditary variant of diabetes caused by mutations in an autosomal dominant gene. It can, therefore, be considered monogenic diabetes, as opposed to type 1 and type 2 diabetes, which involve complex interactions between multiple genes and environmental factors. MODY is usually diagnosed before the age of 20 and accounts for 1–2% of all diabetes cases, although it may cause a higher percentage of presumed type 1 and type 2 diabetes cases. Currently, MODY is grouped into 11 classes, each of which is caused by a specific monogenic defect. Most of the mutations affect transcription factors or metabolic enzymes. All forms of MODY are due to ineffective insulin production or release by the pancreatic β cells. As with any other autosomal-dominant disorder, 50% of the first-degree relatives of an individual carrying a MODY-related variation will inherit the mutation. However, the disease penetrance varies from 40 to 90% between the different types. MODY type 2 and MODY type 3 caused by deficiency in the glucokinase and hepatocyte nuclear factor 1α, respectively, are the most common forms.

Diabetes is a chronic disease for which no cure is available to date. The most common pharmacological treatment of type 1 diabetes consists of multiple daily injections or continuous subcutaneous infusions of insulin. Different forms of recombinantly produced insulin analogs with specific characteristics, for example, rapid-acting or long-acting insulin, are described in Section 10.2.2. Type 2 diabetes requires changes in lifestyle (particularly weight loss). The biguanide metformin is the first-line drug of choice for the treatment of type 2 diabetes. Its major mechanism of action is the suppression of glucose production by inhibition of hepatic gluconeogenesis. In addition, insulin may also be used in type 2 diabetes. Some MODY forms may require specific treatments. Insulin injections, however, are unnecessary in MODY type 2.

References

1. Leigh, S.E., Foster, A.H., Whittall, R.A. *et al.* (2008) Update and analysis of the University College London low density lipoprotein receptor familial hypercholesterolemia database. *Ann. Hum. Genet.*, **72**, 485–498.
2. Gillet, L.C. and Scharer, O.D. (2006) Molecular mechanisms of mammalian global genome nucleotide excision repair. *Chem. Rev.*, **106**, 253–276.
3. Nathans, J. (1999) The evolution and physiology of human color vision: insights from molecular genetic studies of visual pigments. *Neuron*, **24**, 299–312.
4. Khurana, T.S. and Davies, K.E. (2003) Pharmacological strategies for muscular dystrophy. *Nat. Rev. Drug Discov.*, **2**, 379–390.
5. Wentworth, J.M., Fourlanos, S., and Harrison, L.C. (2009) Reappraising the stereotypes of diabetes in the modern diabetogenic environment. *Nat. Rev. Endocrinol.*, **5**, 483–489.

Further Reading

Trent, R.J. (2012) *Molecular Medicine: Genomics to Personalized Healthcare*, Academic Press, Waltham, MA.

Online Mendelian Inheritance in Man

Amberger, J., Bocchini, C., and Hamosh, A. (2011) A new face and new challenges for Online Mendelian Inheritance in Man (OMIM(R)). *Hum. Mutat.*, **32**, 564–567.
Amberger, J., Bocchini, C.A., Scott, A.F. *et al.* (2009) McKusick's Online Mendelian Inheritance in Man (OMIM). *Nucleic Acids Res.*, **37**, D793–D796.

Disorders Involving Imprinting

Buiting, K. (2010) Prader–Willi syndrome and Angelman syndrome. *Am. J. Med. Genet. C Semin. Med. Genet.*, **154C**, 365–376.

Autosomal Dominant Disorders

Canadas, V., Vilacosta, I., Bruna, I. *et al.* (2010) Marfan syndrome. Part 1: pathophysiology and diagnosis. *Nat. Rev. Cardiol.*, **7**, 256–265.
Park, E.Y., Woo, Y.M., and Park, J.H. (2011) Polycystic kidney disease and therapeutic approaches. *BMB Rep.*, **44**, 359–368.
Walker, F.O. (2007) Huntington's disease. *Lancet*, **369**, 218–228.

Autosomal Recessive Disorders

Lehmann, A.R., McGibbon, D., and Stefanini, M. (2011) Xeroderma pigmentosum. *Orphanet. J. Rare Dis.*, **6**, 70.
Prickett, M. and Jain, M. (2013) Gene therapy in cystic fibrosis. *Transl. Res.*, **161**, 255–264.
Blau, N., van Spronsen, F.J., and Levy, H.L. (2010) Phenylketonuria. *Lancet*, **376**, 1417–1427.

X-Linked Recessive Disorders

Leung, D.G. and Wagner, K.R. (2013) Therapeutic advances in muscular dystrophy. *Ann. Neurol.*, **74**, 404–411.

Mitochondrial Disease

Schon, E.A., DiMauro, S., Hirano, M. *et al.* (2010) Therapeutic prospects for mitochondrial disease. *Trends Mol. Med.*, **16**, 268–276.

Polygenetic Disorders

Martinez, FD. and Vercelli, D. (2013) Asthma. *Lancet*, **382**, 1360–1372.
Atkinson, M.A., Eisenbarth, G.S., and Michels, A.W. (2014) Type 1 diabetes. *Lancet*, **383**, 69–82.

Molecular Oncology

4

Summary

- Cancer is a comprehensive description for a family of very diverse diseases. Common characteristics of malignant tumors are abnormal cell growth and the potential to invade or spread to other parts of the body. While conventional cancer therapy is based on surgery, chemotherapy, and radiation therapy, molecular oncology aims at elucidating causative genetic changes and oncogenic pathways to develop novel therapeutic strategies, some of which have already entered clinical practice.
- Breast cancer is a heterogeneous disease. Despite robust basic and clinical research efforts, metastatic breast cancers remain incurable. Personalized treatment is increasingly used to guide breast cancer therapy. Selective clonality develops during breast cancer evolution, which contributes to the development of treatment resistance. Ongoing research efforts are focusing on guided treatment through the identification of molecular targets.
- The two predominant types of lung cancer are non-small cell lung cancer (NSCLC) and small cell lung cancer (SCLC). Approximately 80% of lung cancers are related to carcinogens in tobacco smoke. Targeted therapies modulate pathways that mediate oncogenesis. Tyrosine kinase inhibitors (TKIs) are among the most successful molecular drugs for the treatment of lung cancers.
- Hepatitis is one of the most common risk factors for the development of hepatocellular carcinoma (HCC) and chronic inflammation results from evasion of the immune system. Repeated cellular injury leads to dysplasia and hepatic carcinogenesis. The RAS/RAF/MAPK signaling pathway is important in the development of HCC. Sorafenib, a small-molecule tyrosine kinase inhibitor targeting the RAS/RAF/MAPK signaling pathway, was the first molecularly targeted therapy approved for the treatment of HCC.
- An understanding of its molecular biology has also led to advances in the diagnosis and management of colorectal cancer. Microsatellite instabilities and the epidermal growth factor receptor (EGFR) as well as several other oncogenic pathways are the focus of current investigations. The characterization of common molecular aberrations in colorectal cancer has led to the development of novel screening strategies. These rely on the detection of tumor DNA mutations and epigenetic changes as a strategy to detect high-risk polyps and colorectal cancer in the general population.
- Renal cell carcinoma (RCC) occurs in various subtypes, clear cell RCC being the most common one. Elucidation of the molecular pathways involved in RCC has permitted the development of targeted therapies. Inhibition of angiogenesis by a monoclonal antibody binding vascular endothelial growth factor A (VEGF-A) and small molecules that abrogate signaling through the VEGF–mTOR pathway are currently the most useful treatment options for metastatic RCC.

Contents List

Molecular Medicine: An Introduction, First Edition. Jens Kurreck and Cy Aaron Stein.
© 2016 Wiley-VCH Verlag GmbH & Co. KGaA. Published 2016 by Wiley-VCH Verlag GmbH & Co. KGaA.

- The molecular biology of prostate cancer involves an extremely complex series of events and contributions from genetic, epigenetic, endocrine. and environmental factors. The endocrine component involving signaling through the androgen–androgen receptor pathway is best characterized and has been exploited therapeutically with significant success and clinical benefit for patients. To further improve therapeutic success, modulation of oncogenic pathways will be necessary.
- Hematological malignancies were the first set of cancers for which a clear molecular etiology was demonstrable. Chronic myeloid leukemia (CML) is the paradigmatic disease for how data derived from the genome, transcriptome, and proteome can be used for definitive diagnosis, quantitative evaluation of the response to treatment, highly sensitive detection of minimal residual disease, the use of rationally designed targeted therapy, and the use of mutational analysis to adjust therapy. Tailored, molecularly targeted therapeutic approaches have now been validated for patients with different subtypes of leukemia.

Cancer is one of the world's leading causes of death, with close to 15 million people being diagnosed with cancer and approximately 8 million people dying of the disease every year. Cancer accounts for around 15% of all deaths in the Western world. In 1971, the US President Richard Nixon declared a "war on cancer." Since then, substantial amounts of money have been spent on research into the causes and treatment of cancer. In 2009–2010, the US National Cancer Institute spent $4.7 billion on cancer research, and Europe invested around €1.4 billion. Despite progress in the management of cancer, it remains a terrible societal burden.

Cancer is not a single disease but a group of diverse diseases with similar manifestations. More than 100 different types of human cancers are currently known. Therefore, the concept of a magic bullet, an anticancer drug that would treat all types of malignant tumors, has been abandoned. Rather, specific therapies for each type of tumor or even personalized approaches are necessary.

Many risk factors for the development of cancer have been identified. The leading cause of cancer, accounting for approximately one fifth of cancer deaths, is tobacco smoking. Further important causes are obesity, a poor diet, lack of physical activity, and alcohol abuse. Various environmental factors, including viral infections, exposure to ionizing radiation, and pollutants may also lead to the development of cancer. Although cancer growth is ultimately the result of genetic changes, only about 5–10% of all cancers are due to inherited genetic defects.

By definition, malignant tumors are characterized by abnormal cell growth and the potential to invade or spread to other parts of the body. Some human tumors develop in a multistep progression from normal cells to cells that form a discernible mass. The chain of events, with each step representing a new mutation, is often referred to as the Vogelstein cascade and will be described in more detail in Section 4.4. Many of the genetic alterations affect oncogenes and tumor suppressor genes (Box 4.1). Typically, changes in multiple genes are required to transform a normal cell into a cancer cell. In addition to genetic abnormalities, epigenetic alterations are also frequently found in cancer. These changes may, for example, reduce gene expression of DNA repair genes.

Six hallmarks have been proposed that are both necessary and sufficient to produce cancer (Figure 4.1): sustaining proliferative signaling, evading growth suppressors, resisting cell death, enabling replicative immortality, inducing angiogenesis, and activating invasion and metastasis. Research in the last decade has achieved remarkable progress toward understanding the mechanistic basis of each hallmark. In addition, the reprogramming of energy metabolism and the evasion of immune destruction have more recently been added as additional hallmarks.

When cancer begins, it usually produces no symptoms. However, as the tumor continues to grow, the first signs and symptoms begin to appear. The symptoms depend on the type and location of the cancer and can be very diverse. For example, lung cancer can cause obstruction of a bronchus resulting in cough, while esophagus cancer may make it difficult and painful for a patient to swallow. Tumor ulceration can cause bleeding depending on its location. Weight loss and generalized weakness may be symptoms of late-stage cancer. As a tumor continues to grow, it may interfere with organ function.

Most cancer patients do not die because of the primary tumor but rather from metastases. Cancer can spread from its original site by different routes; these include lymphatic spread to regional or distal lymph nodes, or by hematogenous spread to distant sites. Metastatic tumors may then cause failure of the affected organs.

Conventional cancer treatment has traditionally been based on surgery, chemotherapy, and radiation therapy. Surgery is the primary method used to remove isolated, solid cancers. Removal of the tumor may be followed by treatment with cytotoxic drugs, for example, drugs that kill rapidly dividing cells, one of the main properties of most cancer cells. Radiation therapy damages tumor DNA, which also leads to cell death.

Modern cancer management nicely illustrates the paradigm change in molecular medicine. While the three

Box 4.1. Oncogenes and Tumor Suppressor Genes

Mutations in oncogenes and tumor suppressor genes are frequently found in cancers. An oncogene is a gene that has the potential to cause cancer. It is often mutated or upregulated in tumor cells, causing cells designated for apoptosis to survive and proliferate instead. The first oncogene was discovered in 1970 in a chicken retrovirus and was termed *src*. A few years later, it was demonstrated that most oncogenes are activated forms of cellular proto-oncogenes. These proto-oncogenes typically encode proteins involved in regulation of cell growth and differentiation. Due to mutations or increased expression, the proto-oncogene becomes a tumor-inducing oncogene. Some examples of proto-oncogenes discussed below include *Ras* (*Rat sarcoma*), *Myc* (*myelocytomatosis viral oncogene*), *Erk* (*extracellular signal-regulated kinase*), and the fusion gene *Bcr-Abl* (*breakpoint cluster region–Abelson kinase*).

Tumor suppressor genes are the antagonists of oncogenes as they protect a cell from developing into a malignant tumor cell. The encoded proteins may have multiple functions, including the repression of genes that are essential for the progression of the cell cycle, arrest of the cell cycle when DNA in the cell is damaged, or the induction of apoptosis, that is, the process of programmed cell death that sets limits upon cell growth (Section 1.2.3) if the DNA damage is irreparable. When the tumor suppressor gene loses its function because of mutations that lower its expression level or activity, cells can progress to cancer. For many tumor suppressor genes, both alleles in the genome must be affected before an effect is manifested. This is known as the "two-hit hypothesis." *p53* is one of the best characterized tumor suppressor genes (and an exception to the two-hit hypothesis, as a mutation in one of the alleles is dominant negative). It has been described as "the guardian of the genome," because of its central role in conserving cellular stability by preventing harmful mutations. The p53 protein performs functions typical of a tumor suppressor gene, including the activation of DNA repair when DNA has sustained damage, blockade of cell growth by cell cycle arrest on recognition of DNA damage, and – if necessary – induction of apoptosis. p53 dysfunction is found in two thirds of colon cancers, half of lung cancers, and 30–50% of breast cancers, as well as in many other malignant tumors. Abnormalities of p53 can be inherited, which increases the risk of developing multiple cancers (known as the Li–Fraumeni syndrome (LFS)). Approaches to restore p53 function by gene therapy are described in Section 11.4.2.

classical approaches of tumor therapy, surgery, chemotherapy, and radiation therapy, have been used successfully for decades, impressive progress has been made more recently by investigating the molecular mechanisms underlying tumor development. Oncogenic pathways have

been elucidated that result in abnormal cell growth. As few as 12 pathways have been identified as being the key drivers of cancer processes; usually at least one of these pathways is affected by mutations in malignant cells. Examples of oncogenic pathways that will be discussed in more detail below are the Ras and the JAK-STAT (Janus kinase–signal transducer and activator of transcription) signaling pathways. These findings permitted the development of targeted therapy, that is, a form of therapy that targets specific molecular differences between cancer cells and normal cells. For example, kinase inhibitors specifically inhibit oncogenic signaling cascades.

Molecular oncology has also broadened the spectrum of anticancer drugs used in the clinic. While conventional chemotherapeutic agents belong to the class of small molecules, many cancer therapeutics developed more recently are biologics, for example, monoclonal antibodies (mAbs) (Section 10.2.1). Trastuzumab (trade name Herceptin) is one of the most successful products developed in the era of molecular medicine. This mAb binds to the human epidermal growth factor receptor 2 (Her2), which is overexpressed in very aggressive forms of breast cancer, as discussed in more detail

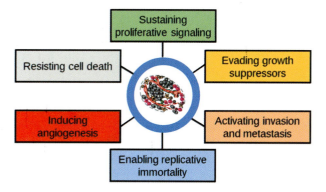

Fig. 4.1 Hallmarks of cancer. Cancers typically have acquired a common set of functional capabilities during their development; they are summarized in the illustration. More recently, the reprogramming of energy metabolism and the evasion of immune destruction have been proposed as additional characteristics of cancer. (Adapted from Ref. [1] with kind permission from Elsevier.)

in Section 4.1.3. Inhibition of the receptor results in cell cycle arrest, reducing cellular proliferation.

Since the early 2000s, cancer stem cells have been a focus of cancer research. Cancer stem cells possess the typical characteristics of normal stem cells, that is, the properties of self-renewal and differentiation into multiple cell types, particularly tumor cells. It has been proposed that cancer stem cells can persist after conventional chemotherapy, when they may cause disease relapse and the development of metastasis, as will be discussed in more detail in Box 12.2 in Section 12.2.

Another important hallmark of cancer growth is the stimulation of blood vessel growth. As tumor size increases, a greater blood supply is required. Cancer cells send signals that induce the growth of new blood vessels, a process known as tumor angiogenesis. The mAb bevacizumab (trade name Avastin) is another successful example of the new generation of targeted molecular therapies. Avastin binds VEGF-A and prevents angiogenesis. Angiogenesis inhibitors were once thought to have the potential to be a single treatment option applicable to all types of (solid) cancers, but this has not been the case in practice. Rather, the combination of angiogenesis inhibitors with other types of cancer therapeutics is a more promising approach to treat malignant disease.

A discussion of all types of cancers is far beyond the scope of this general textbook on molecular medicine. Therefore, the following sections will cover some of the most prevalent tumor types. These examples will illustrate the power of molecular oncology that, investigating the mechanisms of cancer development, enables the development of novel therapeutic approaches. Several molecular anticancer treatments are already being employed clinically. As half of all drugs currently in clinical trials are for a cancer indication, further treatment improvements can be expected to reduce cancer morbidity and mortality in the coming decades.

4.1
Molecular Biology of Breast Cancer and Its Clinical Implications

(Yuan Yuan, City of Hope, Duarte, CA, USA)

Breast cancer is the most common cancer in women worldwide and the second leading cause of cancer-related death in women in the United States. In 2013, an estimated 232 340 new cases of invasive breast cancer were diagnosed in women and approximately 39 620 women died of the disease. During the last two decades, an improved understanding of the molecular biology of breast cancer has led to many breakthroughs in breast

cancer therapeutics, especially in more target-directed approaches. Breast cancer mortality rates have declined by 30% with an improvement in 5 year overall survivals (OSs) to 90%. Our improved understanding of the molecular biology of breast cancer has reshaped the way oncologists treat breast cancer and the way researchers conduct breast cancer research in the laboratory. However, despite improved breast cancer screening and adjuvant therapies, which have contributed to decreased breast cancer mortality, approximately 20–30% of all breast cancer patients develop metastatic disease. Their median survival is only 2–3 years, depending on the molecular subtype.

4.1.1
Intrinsic Subtypes of Breast Cancer

It has been long recognized that breast cancer is a heterogeneous disease that includes distinctive clinical, morphological, and molecular subtypes. Traditionally, breast cancers are classified by pathological parameters such as tumor size, lymph node involvement, histological grade, estrogen receptor (ER) status, progesterone receptor (PgR) status, and HER2 expression. ER-positive and HER2-negative breast cancers account for approximately 60% of all breast cancers, as compared with 20% each for HER2-positive and triple negative breast cancers (TNBCs) (no detectable ER, PgR, or HER2 receptor expression). This clinical–pathological classification has been valuable in tailoring therapy recommendations, but these factors are often insufficient in fine-tuning treatment options and risk stratification. Gene expression microarrays have allowed the simultaneous analysis of thousands of genes in a single experiment to establish the molecular profile of a tumor. Quantitative analysis of multiple genes is clinically employed to provide a more precise biological characterization of the tumor. Breast cancer can be classified into intrinsic subtypes based on gene expression profiles: These subtypes are known as the luminal, basal, normal breast-like and HER2-like subsets. The luminal subgroup can further be divided into two groups: luminal A and luminal B. These classifications have been validated and found to correlate with prognosis regardless of clinical variables. Further analysis of a large cohort of patients utilizing this intrinsic subtype model demonstrated a correlation with disease-free survival and overall survival between the distinctive subgroups.

4.1.1.1   Luminal
The luminal A subtype is usually defined as being ER-positive, HER2-negative, and Ki67 low (<14% cells positive). The luminal B subtype is ER-positive,

HER2-negative, and Ki67 high (≥14% cells positive). Luminal A tumors are low grade, diploid, and have frequent overexpression of *Cyclin D1*. Frequently mutated *PIK3CA*, *GATA3*, and *MAP3K1* combined with whole chromosome arm aberrations are also found. The luminal A subtype is also enriched in ER, along with ER-regulated genes, transcription factors such as GATA3, FOXA1, XBP1, and luminal cytokeratins. Luminal B tumors, on the other hand, are generally higher grade, more proliferative with a higher Ki67 percentage, and often carry mutations of the *PIK3CA* and *TP53* genes in addition to alterations in the RB and MAPK pathways. The HER2-subtype is characterized by enriched *HER2* gene expression and is typically negative for expression of luminal epithelial genes and for genes in the chromosome region adjacent to the *HER2* gene, such as *GRB7*, *MED24*, and *MED1*. Normal-like breast cancer expresses basal and myoepithelial genes. Basal-like breast cancer is characterized by a microarray gene expression signature similar to that of the myoepithelial/basal cells of the breast. They are highly proliferative, aneuploid, and high grade, with common *TP53* mutations and complex genomic rearrangements. Ninety percent of the basal-like breast cancers are TNBC, and they are reported to have transcriptomic characteristics similar to those of breast tumors found in patients with *BRCA1* (*breast cancer1*) germline mutations and high-grade serous ovarian cancers. However, basal-like breast cancer does not totally overlap with TNBC. Potential molecular alterations that have been detected in basal-like breast cancer include those in PTEN, PIK3CA, KRAS, BRAF, EGFR, INPP4B, GFR12, IFR1, KIT, MET, PDGFRA, and the HIF 1-alpha pathways.

4.1.1.2 Subclassification of TNBC

TNBC constitutes 10–20% of all breast cancers. They are unique for the following reasons: They are higher grade, are seen in younger patients, are more prevalent in African-American women, and are biologically more aggressive with a 5 year survival of around 30%. TNBC is a highly heterogeneous group of cancers and further subtyping is required to individualize therapy. The study of 587 TNBC gene expression profiles from 21 breast cancer data sets identified 6 TNBC subtypes displaying unique gene expression profiles. These included two basal-like subtypes (BL1 and BL2), an immunomodulatory (IM) subtype, a mesenchymal subtype (M), a mesenchymal stem-like (MSL) subtype, and a luminal androgen receptor (LAR) subtype. The BL1 and BL2 subtypes were enriched for the expression of cell cycle and DNA-damage response genes; representative cell lines from these two subtypes preferentially respond to cisplatin treatment. The M and MSL subtypes were enriched in gene expression profiles for the epithelial–mesenchymal transition, and cell models were sensitive to PI3K/mTOR and abl/src inhibition. The LAR subtype is characterized by androgen receptor (AR) signaling and LAR cell lines are uniquely sensitive to androgen blockade. These findings may be helpful in biomarker selection, drug discovery, and clinical trial design that will ultimately enable personalized therapy for TNBC patients.

4.1.2
Molecular Profiling of Breast Cancer

Advances in molecular biology in recent years have led to innovative diagnostics and therapeutics. Various molecular prognostic tools have been adapted to routine clinical practice. Among the most popular tools are Oncotype Dx and MammaPrint (Section 8.3.3). These genomic tools assist risk stratification in early-stage ER-positive breast cancers. The Oncotype Dx test analyzes the expression of 21 genes and then calculates a recurrence score (range between 0 and 100); the higher the score, the greater the risk of recurrence. The test is used to estimate a women's risk of recurrence of early-stage ER-positive breast cancer. MammaPrint tests the expression of 70 genes and then calculates a high- or low-risk recurrence score. The test is also used to estimate breast cancer recurrence risk in ER-positive or ER-negative breast cancer. Both tools are currently undergoing validation in clinical trials, where they are utilized to identify breast cancer patients who may benefit from adjuvant chemotherapy. Phase III studies are currently underway to evaluate the predictive value of Oncoytpe Dx in patients with early stage, ER-positive lymph node negative or lymph positive breast cancers. Several other breast cancer molecular profiling tools are also undergoing evaluation. These include the Rotterdam 76 gene signature and the Genomic Grade Index (GGI). Interestingly, there is very little concordance among the candidate genes included in each platform.

4.1.3
Signaling Pathways

Researchers in breast cancer biology have linked breast cancer tumorigenesis and resistance mechanisms to pathways that include HER2, ER, IGF1R, PI3K/AKT, mTOR, and angiogenesis. This improved understanding has led to the development of targeted therapies that are more effective and less toxic compared to conventional chemotherapy agents. Complex biological pathways that involve acquired resistance to therapy include the following: The persistence of proliferative signaling, growth suppression evasion, resistance to cell death, genomic instability, unlimited replicative capacity, metabolic reprogramming, induction of angiogenesis, and invasion and metastasis. Targeting these abnormal pathways represents recent breakthroughs in overcoming acquired resistance.

4.1.3.1 The Role of the Estrogen Pathway in Breast Cancer

Breast cancer cell surface ER and/or PgR are the most important prognostic factors in invasive breast cancer. These receptors are detected in approximately 70% of all breast cancers. The ER signaling pathway plays a critical role in promoting cell proliferation, survival, and invasion in ER-positive breast cancer cells. Estrogen deprivation therapy is the key treatment modality in patients with hormone receptor positive breast cancer. Endocrine therapy options for women with ER-positive breast cancer include selective ER modulators (SERMs), for example, tamoxifen for premenopausal patients, ER antagonists (e.g., Fulvestrant) and aromatase inhibitors (AI). Tamoxifen has been the most commonly used agent for the treatment of ER-positive breast cancer for over 35 years. In postmenopausal women, AIs reduce estrogen synthesis by blocking the conversion of androgens to estrogens in the adrenal glands and in fat tissue (nonovarian tissue), which becomes a primary source of estrogen production in postmenopausal women. Fulvestrant is an estrogen receptor antagonist that competitively binds to the ER with a much greater affinity than tamoxifen. The downregulation of ER protein activity results in reduction of estrogen-sensitive gene transcription.

4.1.3.2 Endocrine Therapy Resistance

In ER-positive breast cancer, antiestrogen therapy produces a risk reduction of approximately 30%. Endocrine therapy resistance develops either through a *de novo* or an acquired mechanism during the course of treatment. The mechanisms of endocrine resistance are complex, and include loss or modification of ER expression and regulation of ER signal transduction pathways. In the so-called canonical pathway, cytoplasmic estrogens bind directly to the nuclear ER and activate gene transcription. In addition, in the "noncanonical" pathway, estrogen binds to plasma membrane receptors and activates the PI3K/mTOR or Ras signaling pathways and other multiple intracellular signaling pathways that crosstalk with the ER pathway: The EGFR, HER2, IGF1, and VEGF pathways also contribute to the development of endocrine therapy resistance.

4.1.3.3 The mTOR/PI3K Pathway and Endocrine Resistance

There have been several breakthroughs in treating ER-positive advanced/metastatic breast cancer for patients progressing through AI. The mammalian target of rapamycin (mTOR) pathway plays a critical role in cell cycle progression and proliferation. Mutations in the PI3K/mTOR pathway are the second most common mutations found in breast cancer. They are the most common activating mutations, being found in as many as 26–45% of breast cancers. Activation of the mTOR/PI3K pathway promotes endocrine resistance. Preclinical models have shown that the AI letrozole and the mTOR inhibitor everolimus act synergistically to augment antitumor activity via inhibition of estrogen-induced breast cancer proliferation. Everolimus inhibits cytokine and growth-factor-dependent cell proliferation through inhibition of the mTOR pathway. Everolimus was found to be synergistic with the nonsteroidal AI exemestane in reversing endocrine resistance. The combination therapy of the nonsteroidal aromatase inhibitor exemestane and everolimus was studied in a phase III randomized trial that enrolled 724 patients with ER-positive breast cancer. The median progression-free survival (PFS) was 10.6 months with everolimus plus exemestane and 4.1 months with exemestane plus placebo. This trial led to the conclusion that everolimus combined with an AI improved PFS in patients with hormone-receptor-positive advanced breast cancer.

4.1.3.4 The CDK 4/6 Pathway

Cyclin-dependent kinases 4 and 6 (CDK4/6) promote the G1/S phase transition by phosphorylating and inactivating the retinoblastoma protein Rb, which releases E2F transcription factors and leads to cell cycle progression. Cyclin D is the upstream activator of CDK4/6; P16-INK4a is the major inhibitor of CDK4/6. Approximately 15–20% of human breast cancer exhibit *CCND1* (the gene encoding cyclin D1) amplification. The Cancer Genome Atlas data revealed the critical role of the cyclin/CDK/Rb pathway in luminal breast cancer. Abnormalities that result in CDK activation are highly enriched in the luminal A and B molecularly defined subgroups, approximately 85% of which were ER-positive/HER2-negative. Cyclin D1 amplifications were observed in 29 and 58% of the luminal A and B subtypes, respectively, and CDK4 amplifications were observed in 14 and 25% of luminal A and B subtypes, respectively. ER/HER2-positive cell lines are preferentially sensitive to CDK4/6 inhibitors. Preclinical data have also demonstrated that despite estrogen deprivation, the ERα protein retains genomic activity and drives CDK4/E2F-dependent transcriptional activity. Thus, CDK4/6 targeted therapy has become an important strategy in the treatment of endocrine-resistant breast cancer. A phase II clinical trial showed significance in PFS for the CDK4/6 inhibitor palbociclib in combination with Letrozole versus Letrozole alone (PFS 20.2 versus 10.2 months). Several phase II/III trials are underway to evaluate the efficacy of CDK4/6 inhibitors in combination with endocrine therapy or chemotherapy.

4.1.3.5 HER2 Pathway and HER2 Targeted Therapy

Historically, breast cancers overexpressing HER2 have aggressive clinical features and belong to one of the subtypes with a poor prognosis. The HER2 receptor belongs to the human epidermal receptor family. HER family member proteins are embedded in the cell membrane. Upon ligand (e.g., EGF) binding, the HER receptor forms homo- or heterodimers that lead to activation of the intracellular tyrosine kinase (TK) domain. Downstream signaling pathways of the HER2 receptors include the Ras/Raf/MAPK, PI3K/AKT/mTOR, and JAK/Stat pathways. These three pathways govern key cellular functions such as cell growth, proliferation, cell survival, and apoptosis, in addition to cell migration and metastases. The introduction of anti-HER therapy has changed the natural history of HER2-positive breast cancer. Trastuzumab is a humanized monoclonal antibody (Section 10.2.1) that binds to an extracellular domain of the HER2 receptor, leading to inhibition of downstream signaling. Adjuvant trastuzumab therapy has reduced the risk of relapse by 50% and improved survival by 33%.

Despite the success with trastuzumab therapy, treatment resistance still develops. Lapatinib is a small-molecule tyrosine kinase inhibitor that reversibly binds to the intracellular tyrosine kinase domain of the EGFR/HER2 receptor. Its binding in the ATP-binding pocket prevents HER2 receptor autophosphorylation and subsequent activation of downstream events. Lapatinib is currently approved as first-line therapy in combination with letrozole for patients with metastatic HER2-positive and ER-positive breast cancers. It is also indicated for the treatment of HER2-positive metastatic breast cancer in combination with capecitabine. Compared with lapatinib, neratinib is a more potent oral small molecule: It inhibits HER1, HER2, and HER4 at the intracellular tyrosine kinase domains through irreversible binding at a targeted cysteine residue in the ATP-binding pocket of the receptor. Both the ability of neratinib to concurrently block signal transduction through the three active tyrosine kinase HER receptors and its irreversible binding to and prolonged inhibition of these growth-promoting pathways provide further clinical improvement in patients with HER2-positive metastatic breast cancer. Early clinical data have demonstrated the activity of neratinib in patients who have failed other small-molecule or antibody-based HER2 targeted therapies. In a first in human, single-agent phase I study of neratinib in solid tumors, a 32% objective response rate (ORR) was observed among patients with trastuzumab refractory HER2-positive disease. A phase II study evaluated the efficacy and safety of neratinib with or without prior trastuzumab treatment. The primary endpoint, the 16-week PFS, was 49 and 78% in prior trastuzumab-treated and trastuzumab-naïve patients, respectively. The ORR was 24 and 56%, respectively, and the median PFS was 22.3 and 39.6 weeks.

Dual HER2 targeting: The combination of the TKIs lapatinib and trastuzumab has set the stage for dual HER2 targeted therapy in treating HER2-amplified breast cancer. This strategy has been proven to improve the pathological complete response (pCR) rate in multiple clinical trials, such as the neo-ALTTO trial and the Neo-SPHERE trial. The Neo-ALTTO trial demonstrated that the combination of lapatinib and trastuzumab with paclitaxel improved pathological complete response rates compared with paclitaxel with either trastuzumab or with lapatinib alone (pCR 51.3 versus 29.5 versus 24.7%, respectively).

Pertuzumab is another humanized mAb; it binds to a different domain of HER2 receptor than trastuzumab and prevents the formation of HER2/HER3 heterodimers. The NeoSPHERE trial studied dual HER2 targeted therapy with trastuzumab and pertuzumab in combination with docetaxel versus docetaxel plus trastuzumab (T) versus docetaxel plus pertuzumab (P) versus the two antibodies alone without chemotherapy in treatment-naïve patients with locally advanced or early-stage HER2-positive breast cancer. The pCR rates were 45.8% in the triple combination arm versus 29% in the trastuzumab plus docetaxel arm versus 24% in the docetaxel plus pertuzumab arm and 16.8% in the trastuzumab plus pertuzumab arm, respectively. This result again confirmed the superior pCR rate with dual HER2 targeted therapy and indicated that in the future, it may be possible to treat some HER2-positive breast cancers with noncytotoxic regimens. The phase III trial called CLEOPATRA demonstrated that adding pertuzumab to trastuzumab plus docetaxel prolonged median PFS by 6.1 months in first-line treatment for HER2-positive metastatic breast cancer. Based on the promising findings of the NeoSPHERE and CLEOPATRA trials, in 2012 the FDA approved pertuzumab use as first-line therapy in metastatic HER2-positive breast cancer, and its use as neoadjuvant therapy for localized disease in 2013.

An additional breakthrough in treating HER2 amplified breast cancer is the novel antibody–drug conjugate (ADC) trastuzumab emtansine (T-DM1), which delivers the emtansine toxin specifically to tumor cells overexpressing HER2. The emtansine is released only after the antibody–drug conjugate has been taken up by a cancer cell via endocytosis, which renders this therapy specific yet reduces its systemic toxicity. In a landmark phase III randomized study, T-DM1 significantly prolonged PFS

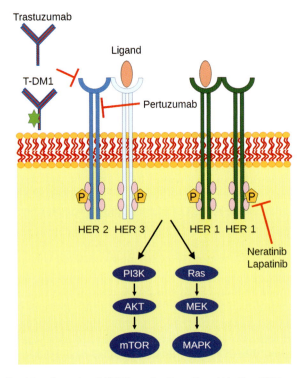

Fig. 4.2 Current anti-HER therapies. Upon ligand binding, HER2 receptors are activated through the receptor tyrosine kinases located in the intracellular domain. Activated HER2 receptors subsequently activate downstream signaling pathways such as the Ras/Raf/MEK/MAPK pathway and PI3K/AKT/mTOR pathways, leading to tumor cell proliferation and survival.

(9.6 versus 6.4 months) and showed a trend toward longer median overall survival time (1 year: T-DM1 84.7% versus capecitabine plus lapatinib (XL) 77.0% in the second-line setting). Figure 4.2 summarizes current anti-HER therapies.

4.1.4
Angiogenesis Pathway

In multiple solid tumors, including lung, colon, renal cell, and breast cancer, inhibition of tumor angiogenesis has been a successful strategy. Currently, there are multiple agents used to target angiogenesis. These include bevacizumab, an mAb targeting vascular endothelial growth factor (VEGF), and TKIs such as sunitinib, sorafenib, and pazopanib. However, multiple studies of bevacizumab in the treatment of metastatic breast cancer have been inconsistent with respect to survival benefit.

4.1.4.1   PARP Inhibitors
Patients with TNBC usually have limited treatment options and have a worse prognosis than those who are ER-positive or HER2-positive. Recent data have shown that poly-ADP ribose polymerase (PARP) inhibitors may

be effective in this subtype of breast cancer and in patients who carry BRCA1 or BRCA2 mutations. PARPs are a family of nuclear enzymes that regulate the repair of DNA single-strand breaks (SSBs) through the base excision repair (BER) pathway. Upon DNA damage, PARP cleaves nicotinamide adenine dinucleotide (NAD) to generate poly-ADP-ribose (PAR) polymers, which are then added onto DNA, histones, and DNA repair proteins. These hetero- and automodification processes mediated by PARP lead to the recruitment of the cellular repair machinery, which facilitates the BER process. Targeting the PARP-mediated DNA repair pathway is a promising therapeutic approach that may potentiate the effects of chemotherapy and radiation therapy and overcome drug resistance. However, the most exciting use of PARP inhibitors may be in utilizing a phenomenon called synthetic lethality. Synthetic lethality is a cellular condition in which simultaneous loss of two nonessential mutations results in cell death, which will not occur if either gene product is present and functional. Tumors with DNA repair defects, such as those arising from patients with BRCA mutations, were found to be more sensitive to PARP inhibition due to synthetic lethality. The BRCA1 and BRCA2 genes encode large proteins that coordinate the homologous recombination repair double-strand breaks (DSBs) pathway. Since BRCA1/BRCA2-mutated tumors cannot utilize homologous recombination to repair DSBs, exposing these cells to a PARP inhibitor, which shuts down the BER rescue pathway, will lead to the accumulation of DNA damage, genomic instability and cell death. However, despite the attractiveness of these mechanisms of action and encouraging early-phase clinical trial data, PARP inhibitors have not been approved by FDA for use outside of a clinical trial setting. A phase II randomized study compared the efficacy and safety of gemcitabine and carboplatin with or without Iniparib, a PARP inhibitor. One hundred and twenty-three patients with metastatic TNBC were included. The addition of Iniparib prolonged the median PFS from 3.6 months to 5.9 months and the median overall survival from 7.7 months to 12.3 months, with no significant difference in toxicity (neutropenia and thrombocytopenia being the most common grade 3 or 4 toxicities) between the two groups. However, a subsequent phase III trial did not identify any overall survival or progression-free survival differences between the two arms.

Despite the disappointing findings from the Iniparib trial, other PARP inhibitors have demonstrated promising clinical activities in BRCA mutation-associated breast cancer. Olaparib is an oral PARP inhibitor that demonstrated a high rate of tumor response in BRCA1 or BRCA2 carriers. A phase II trial in patients with

BRCA1/BRCA2 mutation-associated metastatic breast cancer evaluated olaparib 400 mg daily (cohort 1) versus 100 mg twice daily (cohort 2). Response rates of 41 and 22%, respectively, were observed. Another PARP inhibitor, Veliparib, also showed promising clinical activity in a phase I/II setting either as single agent or in combination with other agents.

4.1.5
Other Biological Therapies/Approaches

Based on various phase I and II studies, other novel targeted agents that inhibit mTOR, PI3K, insulin-like growth factor, fibroblast growth factor receptor (FGFR), heat shock protein 90 (HSP90), in addition to histone deacetylase inhibitors (HDACI) represent promising new leads for the treatment of breast cancer.

Recently, patient-derived xenograft (PDX) models have been generated and utilized to study complex tumor biology that is not sufficiently addressed by using *in vitro* cell lines. PDX models are produced by the transplantation of a patient's primary tumor tissue into immunodeficient mice. Breast cancer PDX models have been successfully established in various laboratories. By this methodology, PDX models made what are known as "preclinical trials" feasible. In a preclinical trial setting, therapies can be tested efficiently, and therapeutic options devised for patients who have developed resistance.

4.2
Lung Cancer

(Karen L. Reckamp, City of Hope, Duarte, CA, USA)

Lung cancer is the most common cause of cancer-related death in men and women in the United States, accounting for more deaths each year than breast, colon, prostate, and pancreatic cancers combined. Over 200 000 new cases of lung cancer were diagnosed in 2012, representing 14% of all new cancer cases. The 5-year survival remains low since most patients are diagnosed with advanced stage disease. Two predominant types of lung cancer are non-small cell lung cancer and small cell lung cancer, which are identified based on clinical, histologic, and molecular characteristics. Approximately 80% of lung cancers are related to carcinogens in tobacco smoke, but our understanding of the molecular basis of "never smoker" lung cancers is improving. Lung cancer is a heterogeneous genomic disease defined by molecular pathways that mediate oncogenesis and that are often driven by genetic alterations. Targeted therapies are available to modulate these pathways and improve patient outcomes.

NSCLC has been at the forefront of an expanding knowledge of genomic alterations that inform therapeutic options and improve cancer patient outcomes. Activating mutations in the *epidermal growth factor receptor* (*EGFR*) gene have been associated with improved progression-free survival when patients are treated with EGFR TKI therapy. The use of clinical characteristics such as smoking status, ethnicity, and tumor histology are associated with *EGFR* mutation, but only genetic data lead to the optimal treatment for patients. NSCLC also harbors translocations in the *ALK* gene; these predict clinical benefit from treatment with crizotinib, a TKI. The establishment of predictive genomic "drivers" has led to improvements in therapy for patients. New genetic alterations have been described that provide for novel therapeutic interventions. Over 60% of patients with the adenocarcinoma subtype of NSCLC have a defined molecular alteration. These growing data must be integrated to determine appropriate treatment.

4.2.1
Genetic Alterations in Non-Small Cell Lung Cancer

4.2.1.1 Epidermal Growth Factor Receptor

The EGFR is a 170-kd glycoprotein with an extracelluar ligand-binding domain (LBD), a transmembrane lipophilic segment, and an intracellular tyrosine kinase domain. It belongs to the HER or ERBB family, which consists of four receptor TKs: EGFR/ERBB1, ERBB2, ERBB3, and ERBB4. Ligand binding causes EGFR to form homo- or heterodimers. This results in the activation of a complex signaling network that promotes tumor proliferation, migration, stromal invasion, neovascularization, and resistance to apoptosis. ERBB signaling can modulate multiple intracellular pathways, including those involving mitogen-activated protein kinase (MAPK), phosphatidylinositol 3-kinase (PI3K), and signal transducer and activator of transcriptions (STATs).

Overexpression of EGFR and its ligands has been demonstrated in multiple tumors, including lung, breast, colorectal, head and neck squamous cell carcinoma (HNSCC), and glioblastoma. Molecular alterations have been demonstrated in a variety of tumors; these can lead to the constitutive activation of EGFR signaling. Mutations in the TK domain of the *EGFR* gene produce increased sensitivity to EGFR TKIs in NSCLC. These mutations stimulate constitutive EGFR signaling and act as "driver" mutations in oncogenesis.

Early studies with EGFR TKI therapy revealed patient characteristics associated with improved response. These include patients with adenocarcinoma, female gender, Asian ethnicity, and never-smoking status. Gefitinib received accelerated FDA approval based on promising phase II data, but a confirmatory phase III trial did not show an improvement in survival. This resulted in limited availability of gefitinib in the United States. A similar reversible EGFR TKI, erlotinib, demonstrated an increase in overall survival (OS) over placebo (6.7 versus 4.7 months) and improved quality of life as a second- or third-line therapy in patients with metastatic NSCLC. Analyses of the patient population suggested that adenocarcinoma histology, never-smoker status, and EGFR expression correlated with the response.

In 2004, activating mutations in the TK domain of *EGFR* were described and associated with improved responses to EGFR TKI therapy. This discovery led to trials focusing on selected groups looking for evidence of significant clinical benefit. To further understand the role of clinicopathologic features versus molecular selection, the phase III Iressa Pan-Asia Study (IPASS) used specific eligibility criteria to identify a group of patients who may derive benefit from gefitinib therapy. The study included chemotherapy-naïve, never-smokers with adenocarcinoma of the lung, most of Asian ethnicity. The number of patients randomized to either gefitinib or carboplatin/paclitaxel was 1217. PFS was superior with gefitinib therapy, and patients with *EGFR* mutations had improved PFS with gefitinib. Patients with wild-type *EGFR* had superior PFS with chemotherapy. This was the first study to definitively identify mutation status as an important predictive marker for EGFR TKI therapy. These results were confirmed in a second trial. Continued evaluation of EGFR TKI therapy in patients with tumors harboring *EGFR* mutations demonstrated increased response rates and PFS with first-line EGFR TKI therapy. However, an analysis of OS has not produced significant differences in survival.

Developing optimal therapies for EGFR-TKI-resistant disease requires a detailed understanding of mechanisms of resistance. Secondary mutations have been described that render initially sensitive tumors that harbor *EGFR* mutations resistant to EGFR therapy. These secondary mutations include mutations at T790M, which has been found in approximately half of tumors that are resistant to erlotinib and gefitinib. Alternatively, bypass mechanisms may lead to resistance, such as amplification of MET and activation of PI3K/AKT signaling. Multiple classes of drugs are in development to overcome these mechanisms of resistance, including second- and third-generation EGFR TKIs that irreversibly bind and inhibit multiple ERBB family members. Combination therapy with EGFR inhibition and MET, HSP90, AKT, SRC, and mTOR inhibitors is under investigation. The most intriguing results come from third-generation EGFR inhibitors that specifically target the T790M mutation and activating mutations, but spare wild-type *EGFR*.

Initial results show 50–60% response rates in patients with refractory, *EGFR* mutation-positive, NSCLC.

4.2.1.2 Anaplastic Lymphoma Kinase

The oncogenic EML4-ALK protein is formed from the fusion of two genes, *echinoderm microtubule-associated protein like-4* (*EML4*) and *anaplastic lymphoma kinase* (*ALK*) (localized on chromosome 2p). Additional chimeric variants and other fusion partners have also been reported. First described in anaplastic large cell lymphoma and neuroblastoma, the fusion gene, *EML4-ALK*, was also identified in NSCLC and occurs in approximately 7% of NSCLC patients. The gene is more prevalent in younger patients with adenocarcinoma who either do not have any history of smoking, or have only a light smoking history.

Crizotinib is an oral tyrosine kinase inhibitor that targets ALK, MET, and ROS1 tyrosine kinases, and is a first-generation ALK inhibitor approved for *ALK* gene rearranged NSCLC. When compared to chemotherapy as second-line therapy for NSCLC, crizotinib resulted in a 65% response rate and a 7.7 month median PFS (compared to 20% and 3 months, respectively, for chemotherapy). Despite exceptional responses, all patients develop resistance and progression on crizotinib is due to second site mutations or activation of bypass pathways. A number of second-generation ALK inhibitors have demonstrated the ability to overcome this resistance. Ceritinib is a second-generation ALK inhibitor that is more selective for ALK, and produced responses in over 50% of patients who had previously progressed on crizotinib therapy. These impressive results led to accelerated FDA approval of this agent.

4.2.1.3 Kirsten Rat Sarcoma (KRAS)

The RAS pathway proteins are GTPases that activate effector proteins essential for cell growth, differentiation, and survival. Stimulation of MAP kinase and phosphatidylinositol-3-kinase (PI3K) signaling and other pathways leads to both positive and negative feedback loops, resulting in complex interactions that mediate proliferation. *RAS* is also the most frequently mutated oncogene in cancer. KRAS is its most common isoform, occurring in approximately 25% of lung adenocarcinomas, where it leads to oncogenesis. *KRAS* mutations differ within various cancers and have distinct transforming capabilities. CDK4 is a novel target in *KRAS*-driven lung cancer, where it exhibits interactions not observed in *KRAS*-mutated cancers of other origins.

GTP has a high affinity for RAS, limiting the potential for competitive inhibition of this kinase. Thus, many studies have focused on blocking downstream pathways. Evaluation of mTOR, PI3K, AKT, and MEK inhibition in KRAS-driven NSCLC resulted in limited clinical benefit and led to the investigation of combination therapies with targeted agents and cytotoxic chemotherapy. This too had modest effects. The MEK inhibitor selumetinib, in combination with the cytotoxic agent docetaxel, compared to docetaxel and placebo in a phase II trial of patients with *KRAS*-mutated NSCLC, demonstrated a significant improvement in PFS, but with increased adverse events and hospitalizations in the combination group. The enthusiasm for the promising activity of MEK inhibition with chemotherapy was dampened by this toxicity, and other combinations are under investigation. The complexity of interactions and pathway redundancies pose challenges for targeting the RAS pathway in lung cancer. However, new insights into the distinct mutations and interactions present opportunities for progress.

4.2.1.4 The Proto-Oncogene ROS1

The ROS1 protein is composed of a glycoprotein-rich extracellular domain, a transmembrane domain, and an intracellular TK, and is located on chromosome 6. A ligand for ROS1 has not been identified. *ROS1* gene rearrangements were initially found in glioblastoma, and now are recognized in a number of malignancies, including NSCLC. In NSCLC, *ROS1* fusions occur in 1–2% of patients and multiple fusion partners have been identified. All involve *ROS1* break points that conserve the TK domain. Potent oncogenic transformation occurs through constitutive kinase activation.

Patients with *ROS1* fusions are similar to those with *ALK* gene rearrangements, which are associated with younger age, a never-smoking history, Asian ethnicity, advanced stage, and adenocarcinoma subtype. However, ROS1 has also been seen in large cell and squamous cell histologies. Both *ROS1* and ALK share significant homology within their tyrosine kinase domains and ALK inhibition has been shown to be beneficial in *ROS1*-fusion NSCLC. Specifically, treatment of patients with *ROS1* fusions resulted in a 50% objective response rate, and additional clinical trials are being conducted in this disease.

4.2.1.5 The Proto-Oncogene BRAF

BRAF mutations, mainly V600E, were initially described as driver mutations in melanoma. Subsequently, mutant *BRAF* was shown to mediate oncogenesis in lung adenocarcinoma. Furthermore, tumor growth was dependent on constitutive *BRAF* oncogene activation, although the frequency of non-V600E mutations is higher in NSCLC than in melanoma (50 versus 20%). Clinical trials evaluating BRAF and MEK inhibitors in *BRAF*-mutated melanoma demonstrated impressive tumor response rates, but also the development of early resistance. Combination therapy with BRAF and MEK inhibitors (dabrafenib and trametinib) led to an improved PFS (9.4 versus 5.8 months compared to monotherapy) and elevated response rates. Preliminary results of another TKI,

dabrafenib, in *BRAF*-mutated NSCLC patients demonstrated significant tumor responses; a combination trial is ongoing in this population.

4.2.1.6 The Human Epidermal Growth Factor Receptor 2 (HER2)

HER2 is activated through homo- or heterodimerization with EGFR or HER3 to interact with downstream signaling pathways. This results in tumorigenesis. *HER2* amplification is infrequently found. Exon 20 mutations in *HER2* are seen in about 2% of NSCLC patients. This mutation has been observed primarily in females, non-smokers, and those with adenocarcinoma histology. Small series and case reports of treatment with HER2 inhibitors and pan-HER inhibitors have produced limited responses, but significant disease control rates. Due to the small population of patients with the *HER2* mutation, few clinical trials have been performed in this group of patients.

4.2.1.7 The RET Proto-Oncogene

RET is a receptor tyrosine kinase (RTK) that mediates neural crest development. Alterations in *RET* have been described in thyroid and lung cancers. *RET* fusions were first described in NSCLC in 2011 and occur in approximately 1–2% of patients. *RET* gene rearrangements are mutually exclusive with other identified mutations in NSCLC. Two multitargeted TKIs that inhibit RET, in addition to several other tyrosine kinases (vandetanib and cabozantinib), have been approved for the treatment of medullary thyroid carcinoma and have demonstrated significant clinical benefit. Limited data are available regarding responses in NSCLC. Ongoing studies will define the utility of these drugs.

4.2.1.8 The MET Proto-Oncogene

MET is a receptor tyrosine kinase that activates the RAS/RAF/MAPK and PI3K/AKT/mTOR pathways following activation by hepatocyte growth factor. Amplification of *MET* has been shown to be an alternative mechanism of resistance to EGFR TKIs in patients with *EGFR* mutation-positive NSCLC. In addition, early studies have shown tumor regression after crizotinib treatment in patients with *MET* amplification.

4.2.1.9 Phosphatidylinositol-3-Kinase (PI3K)

The PI3K pathway is the most frequently altered pathway in human tumors. PI3K signaling is initiated through multiple receptor tyrosine kinases, including EGFR, HER2, insulin growth factor receptor (IGF-1R), vascular endothelial growth factor receptor (VEGFR), platelet-derived growth factor (PDGF), and Src family kinases. NSCLC harbors numerous alterations involved in the PI3K pathway, including *PIK3CA* amplification and mutation, decreased or loss of *tumor suppressor phosphatase and tensin* homolog (*PTEN*), *AKT* mutation, *KRAS* mutation, *LKB1* loss, and *MET*, *EGFR* and *HER2* mutation and amplification. PI3K has also been implicated in the expansion of bronchioalveolar stem cells. PI3K inhibition potentially targets multiple aspects of tumor biology, including angiogenesis, inflammation, B cell survival, the epithelial to mesenchymal transition, and metastasis. Its integral role in immune modulation may make PI3K inhibitors ideal partners for immune checkpoint inhibitors.

4.2.1.10 Immune Checkpoint Inhibition

Although significant progress has been made in the targeting of genomic alterations, it is clear that many patients with lung cancer do not harbor aberrations and cannot benefit from these therapies. The complexity of factors driving lung cancer growth and metastasis highlight the importance of the tumor microenvironment and, beyond mutations, interaction with the immune system. Recent developments in cancer immunotherapy have shown that stimulation of cellular immune responses can promote antitumor effects.

Costimulatory and coinhibitory signals through immune checkpoints modulate the activation of T-cell responses to cancer. Lung cancers produce several molecules that result in inhibition and innate immunosuppression at the tumor site. Several agents in development target inhibition at these checkpoints; these include cytotoxic T-lymphocyte antigen (CTLA)-4 inhibitors, programmed death (PD)-1 inhibitors, and programmed death ligand (PDL)-1 inhibitors. Treatment with PD-1 and PDL-1 inhibitors has demonstrated promising results in melanoma and has produced responses in heavily pretreated patients with NSCLC. Phase III trials have been performed and the results will be reported soon.

The most common oncogenic alterations in NSCLC and the respective therapeutic options are summarized in Table 4.1.

4.3
Hepatocellular Carcinoma

(Vincent Chung, City of Hope, Duarte, CA, USA)

Hepatocellular carcinoma is the most common cancer arising from the liver and is the second most deadly cancer worldwide behind lung cancer. Each year, more than 745 000 people die from this disease. The geographic distribution of hepatocellular carcinoma varies around the world due to environmental exposures or viral pathogens.

Table 4.1 Therapeutic options for oncogenic alterations in non-small cell lung cancer (NSCLC).

Oncogene	Mutation prevalence	Therapy	Predicted response rate
EGFR	Asians 30–40%/ Caucasian 10–20%	EGFR TKIs (most mutations)/pan-HER inhibitors	Erlotinib 60–80% Gefitinib 70% Afatinib 60%
ALK	1–7%	ALK inhibitors/ HSP90 inhibitors	Crizotinib 50–60% Ceritinib 60%
KRAS	Asians 10%/ Caucasian 30%	MEK inhibitors/ CDK4/6 inhibitors	NA
ROS1	1.7%, higher in Asians	ALK inhibitors	NA
HER2	2%	Trastuzumab/pan-HER inhibitors	NA
RET	1.7% (15% in EGFR/ ALK/KRAS negative patients)	RET inhibitors	NA
BRAF	2%	BRAF/MEK inhibitors	NA
MET	7%	Crizotinib	25%

The regions with the highest incidence include Eastern Asia and Africa, with up to 35 cases per 100 000 people. The most common etiology is hepatitis B in Asian countries and hepatitis C in Western countries. In the United States, approximately 27 000 cases are diagnosed each year, with the majority due to hepatitis C. Here, we review the pathogenesis of hepatocellular carcinoma and developments in molecular targeted therapy.

4.3.1
Risk Factors for Hepatocellular Carcinoma

Hepatocellular carcinoma most commonly arises in the setting of chronic inflammation of the hepatocytes. In Asian countries and developing nations, transmission of hepatitis B virus (HBV) from parent to child leads to DNA integration of the virus at an early age. Viral replication as measured by viral load is prognostic for the development of cancer. Due to the early age of HBV infection, hepatocellular carcinoma can develop even without cirrhosis in the setting of minimal liver inflammation, commonly in individuals 40–50 years old. Aflatoxin exposure can play a significant role in carcinogenesis in the setting of chronic hepatitis. This mycotoxin commonly contaminates corn, soybeans, and peanuts, although as a single agent it plays a minor role in the development of cancer. However, for patients with an underlying risk factor such as hepatitis B, it can significantly increase risk. An epidemiology study conducted in Shanghai, China, showed 23% of peanut samples were contaminated with aflatoxin and 65% of peanut butter samples were contaminated. In this study, the relative risk of developing hepatocellular carcinoma in the setting of hepatitis B and aflatoxin exposure was 59 times that in the control population.

Hepatitis C virus (HCV) is transmitted from contaminated blood products. Therefore, the virus is more commonly contracted at a later age. Virtually all patients that develop hepatocellular carcinoma from hepatitis C have some degree of liver fibrosis due to chronic inflammation. The exact mechanism for the development of cancer is not known, but it is believed to be due to chronic inflammation that leads to oxidative DNA damage.

Fatty liver disease is a growing problem in Western industrialized nations due to obesity, diabetes, and hyperlipidemia. Hepatic steatosis, defined as the accumulation of fat in the liver, generally does not cause much fibrosis. However, in the setting of steatohepatitis, which causes inflammation, significant fibrosis can occur. Heavy alcohol use is not only associated with hepatic steatosis but is also directly toxic to the hepatocytes, eventually leading to cirrhosis, a precursor to the development of hepatocellular carcinoma. There are also additional risk factors that can lead to hepatic dysplasia and ultimately cancer (Table 4.2). Here, we will review the sequence of events involved in hepatic carcinogenesis.

4.3.2
Molecular Biology of Hepatocellular Carcinoma

Hepatitis B and C are the most common causes of hepatocellular carcinoma. In the setting of chronic infection, a sequence of events leading to hepatocyte injury and repair results in liver damage. This appears to be immune mediated since the virus can replicate in cell culture without causing cellular damage. By evading the immune system, the hepatitis virus is not completely cleared by the lymphocytes. This causes additional cells, such as nonspecific natural killer cells, to invade the area where they cause cellular damage. Cyclical damage and repair of the hepatocytes eventually leads to dysplasia, a precursor to cancer. But how does the virus evade the immune system?

There are several mechanisms by which the virus can hide from the immune system. Interferons are proteins normally released by host cells in response to pathogens such as bacteria and viruses. Once a pathogen infects a cell, interferon is released, warning neighboring cells and activating a signaling cascade limiting protein synthesis and viral replication. In addition, levels of the tumor suppressor protein p53 are increased, causing apoptosis of the infected cells. Viruses have developed ways of limiting the production of interferon by blocking downstream pathways such as STAT. This protein plays an important role in regulating the growth and survival of cells. Binding of interferon to its cell surface receptor triggers phosphorylation of a JAK protein, which recruits STAT proteins, leading to heterodimerization. Accumulation of JAK-STAT complexes occurs in the nucleus. This activates the transcription of target genes that inhibit viral proliferation. Previous studies have shown that hepatitis C can inactivate the dimerization of STAT (Figure 4.3).

Hepatic carcinogenesis is complex with many affected signaling pathways (Figure 4.4). Discussing each of the pathways is beyond the scope of this section, but one of the most important pathways is the Ras/Raf/MAPK pathway. Receptors on the cell surface bind ligands such as epidermal growth factor (EGF) or platelet-derived growth factor and activate Ras, a family of proteins with GTPase function. In the cell, these proteins control cell proliferation, migration, differentiation, and adhesion. In cancer, this pathway is often dysregulated, resulting in cancer cell invasion and metastasis. Due to its importance in regulating cell proliferation, this pathway represents a potential target for therapy and forms the basis for targeted therapy in the treatment of hepatocellular carcinoma.

Table 4.2 Risk factors for hepatocellular carcinoma.

Hepatitis B
Hepatitis C
Alcoholic liver disease
Nonalcoholic fatty liver disease
Aflatoxin
Diabetes
Hemochromatosis
Alpha-1-antitrypsin deficiency
Acute intermittent porphyria

4.3.3
Development of Sorafenib for the Treatment of Hepatocellular Carcinoma

Sorafenib is a small-molecule tyrosine kinase inhibitor that mainly targets VEGFR, platelet-derived growth factor receptor (PDGFR), and Raf kinase. Based upon the importance of the RAS/RAF/MAPK signaling pathway in hepatic carcinogenesis, the drug was investigated for the treatment of hepatocellular carcinoma.

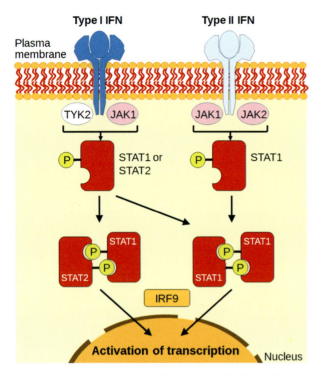

Fig. 4.3 Interferon binding to cellular receptors leading to STAT signaling. Type I interferons (IFNs) bind a common receptor on the cell surface known as the type I IFN receptor. The receptor is composed of two subunits that are associated with the Janus activated kinases (JAKs) tyrosine kinase 2 (TYK2) and JAK1, respectively. IFN-γ, the only type II IFN, binds the distinct type II IFN receptor that is associated with JAK1 and JAK2. Activation of the JAKs results in tyrosine phosphorylation of the signal transducer and activator of transcription (STAT) 1 and 2 proteins and the formation of the STAT1-STAT2-IRF9 (INF-regulatory factor 9) complexes. These complexes translocate to the nucleus and activate transcription of target genes. (Adapted from Ref. [2] with kind permission from Macmillan Publishers Ltd., Copyright 2005.)

In the first-time-in-human phase I clinical trial, one patient with hepatocellular carcinoma responded to treatment. A larger 137 patient phase II trial was conducted to confirm the activity of sorafenib. Survival improved to 9.2 months, which was over 3 months longer than the historical average. Before this drug could be approved by the FDA as standard treatment for hepatocellular carcinoma, an international phase III trial was conducted in which patients randomized to either sorafenib or placebo. The SHARP trial (Sorafenib Hepatocellular Carcinoma Assessment Randomized Protocol) enrolled 602 patients (20% had hepatitis B, 30% hepatitis C, and 25% alcoholic cirrhosis) with advanced hepatocellular carcinoma. The results of the SHARP trial showed a statistically significant improvement in overall survival from 7.9 to 10.7 months in the sorafenib group.

Interestingly, patients with hepatitis C demonstrated a superior outcome in OS of 14 months when treated with sorafenib, compared to 7.9 months with placebo. This may be due to the fact that hepatitis C core protein activates Raf-1 signaling in the cancer. Since sorafenib is a Raf-1 kinase inhibitor, these patients appear to do better by targeting the driving mechanism of cell proliferation. These results illustrate the importance of understanding the signaling pathways that lead to cell proliferation.

4.3.4
Complexity of Cancer

Unfortunately, solid tumors are highly complex and have multiple activated signaling pathways. In the last 10 years, there has been an explosion of targeted therapies developed by pharmaceutical companies designed to block specific pathways in an attempt to slow liver cancer growth. This has led to numerous clinical trials; but in hepatocellular carcinoma, sorafenib was the first molecularly targeted therapy to show an OS benefit. One pathway that is activated in hepatocellular carcinoma involves the fibroblast growth factor receptor. Activated FGFR has been shown to be associated with poor outcomes; inhibition of this receptor seemed to be a logical goal of therapy. With the development of drugs such as brivanib, a small-molecule TKI of VEGF and fibroblast growth factor (FGF), clinical trials were conducted with the hope of improving survival. In the BRISK-FL study, patients with advanced unresectable HCC were randomized to either brivanib or sorafenib. Even though both agents had similar antitumor activity, the primary endpoint of OS was not met. This was considered a negative trial. There were also more side effects with brivanib treatment. The discontinuation rate due to adverse events was 33% for sorafenib and 43% for brivanib.

Hepatocellular carcinoma is a highly vascular tumor and many drugs targeting VEGF have been evaluated with therapeutic intent. Sunitinib, which targets VEGFR1, VEGFR2, PDGFR-alpha/beta, c-kit, FLT3, and RET, was tested in clinical trials and showed promising activity in smaller trials. However, when compared to sorafenib, survival was worse. In addition, the side effect profile did not compare favorably with sorafenib. Both drugs block VEGF, which is important for angiogenesis. Why did sunitinib fail? It is possible that since tissue infected with hepatitis C has an activated Raf-1/MEK and ERK pathway, sorafenib may have the advantage given that it inhibits Raf-1 kinase. This lesson illustrates the complexity we face in understanding the mechanisms that drive cancer progression.

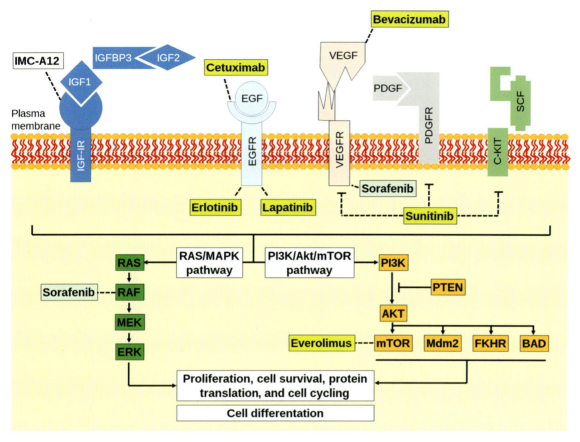

Fig. 4.4 Signaling pathways in hepatocellular carcinoma and targeted therapies. The diagram indicates the targets of monoclonal antibodies (VEGFR: bevacizumab; EGFR: cetuximab), tyrosine kinase inhibitors (VEGFR and PDFGR: sorafenib, sunitinib; *EGFR*: erlotinib, lapatinib), and signal transduction kinase inhibitors (RAF: sorafenib; mTOR: everolimus) in development or clinical use. (Adapted from Ref. [3] with kind permission from John Wiley & Sons, Inc.)

Hopefully, these can be solved in the future with genomic sequencing, in addition to other technologies.

4.4
Molecular Biology of Colorectal Cancer and Its Clinical Implications

(Marwan G. Fakih, City of Hope, Duarte, CA, USA)

Colorectal cancer is the second leading cause of cancer death in the United States. Every year, roughly 140 000 cases of colorectal cancer are diagnosed and approximately 50 000 colorectal cancer patients succumb to their disease. Major progress has been made in the understanding of colorectal cancer since Vogelstein first reported on his carcinogenesis model. The understanding of the colorectal cancer carcinogenesis process and subsequent molecular drivers of growth and resistance has resulted in a major impact on the development of successful screening and prevention strategies, as well as

on the development of novel personalized treatments for advanced colorectal cancers.

4.4.1
Colorectal Cancer Carcinogenesis

Colorectal cancer develops from precancerous intramucosal colonic growths, known as polyps. As colonic polyps progress over time, they transition from the more benign tubular adenomas to tubulovillous and villous adenomas, dysplastic adenomas, and finally to invasive cancers. This knowledge has led to the incorporation of colonoscopies as a backbone of screening for colorectal cancer. It has been recognized that the colonoscopies are one of the most effective modalities of prevention of colorectal cancer, largely through the detection of colonic polyps and their endoscopic removal before they become invasive cancer.

In 1988, Vogelstein first reported on his "Vogelstein Model" for colorectal cancer carcinogenesis. Vogelstein

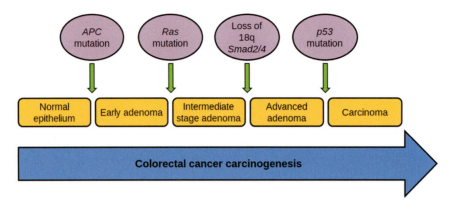

Fig. 4.5 Vogelstein colorectal carcinogenesis model. The sequential accumulation of mutations that silence tumor suppressor genes and activate tumor oncogenes result in a well-defined carcinogenesis process, from early adenoma to invasive cancer.

screened for genetic alterations in polyps from patients with familial adenomatous polyposis (FAP), sporadic polyps without *in situ* carcinoma from patients without FAP, sporadic polyps with *in situ* carcinoma from patients without FAP, and colonic carcinomas. This pivotal study demonstrated that an accumulation of genetic alterations in specific tumor suppressor genes (*APC*, *SMAD4*, and *P53*) and oncogenes (*RAS*) is needed for the progression from the early stages of a polyp to high-risk polyps, and finally to carcinoma (Figure 4.5). It has since been demonstrated that multiple pathways of carcinogenesis exist in colorectal cancer.

4.4.1.1 Chromosomal Instability Pathway
The chromosomal instability pathway is the most frequent pathway of carcinogenesis in colorectal cancer. It is represented by the presence of frequent chromosomal alterations associated with the silencing of tumor suppressor genes and activation of oncogenes, as described by Vogelstein. This pathway is represented in Figure 4.5.

4.4.1.2 Microsatellite Instability Pathway
Microsatellites are sections of DNA that consist of repeats of short sequences of nucleotides. Microsatellites are noncoding and species specific. Microsatellite instability (MSI) occurs in approximately 15% of colorectal cancers. These tumors are characterized by a DNA mismatch repair deficiency leading to widespread genetic mismatch repair abnormalities and a hypermutated tumor profile. A reference panel of five microsatellites has been recommended by the National Cancer Institute (NCI) for MSI testing to reflect the functional status of intratumor mismatch repair. Tumors with two or more aberrant microsatellites are considered MSI-high (MSI-H), while tumors with one aberrant microsatellite are considered MSI-low (MSI-L). Tumors with

five normal microsatellites are considered microsatellite stable (MSS). The significance of MSI-L is unknown, although some of these tumors have similar characteristics as the MSS tumors. MSI-H colorectal tumors are found predominantly in the proximal colon, are associated with mucinous histology, poor grade, and lymphocytic infiltration, and are increasingly seen in the elderly population. MSI-H tumors are associated with a better prognosis than MSS tumors and have a lower tendency to metastasize.

Mismatch repair abnormalities in colorectal cancer are related to mismatch repair protein deficiency (MMRD). These proteins include MLH-1, MSH-2, MSH-6, and PMS-2 and are essential in the process of repairing mismatch errors during DNA-replication. The absence of expression of any of these proteins is a strong indicator of a MSI pathway phenotype. MLH-1 or MSH-2 absence correlates strongly with MSI-H and represents the most common causes of mismatch repair deficiency. Similar to the findings with MSI-H, the absence of MLH1 or MSH2 correlates with a good prognosis and a decreased risk of colorectal cancer recurrence. There are two main drivers for MMRD: Germline mutations and epigenetic changes. Germline mutations involving the mismatch repair genes comprise less than 20% of the MSI-H cases and are related to the inheritance of a defective *MMR* gene (Lynch syndrome, discussed below). Germline mutations in the *MMR* genes are almost never seen in the setting of MSS stable tumor. Therefore, MSI or MMRD screening represents the best initial step in screening for Lynch syndrome. More commonly, MSI-H tumors or MMRD are reflections of the epigenetic silencing of MLH-1 protein due to MLH-1 promoter hypermethylation. This can be distinguished from MLH-1 germline mutation by assaying for MLH-1 promoter hypermethylation, or for the presence of concurrent BRAF mutations.

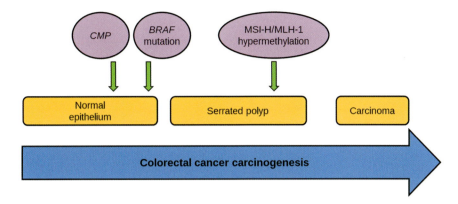

Fig. 4.6 The CPG-island methylator phenotype (CIMP) pathway. The CIMP pathway is characterized by hypermethylation of CpG islands in gene promoters, resulting in the silencing of key tumor suppressor and DNA repair genes.

Neither of these is present in the setting of MLH-1 germ cell line mutations.

4.4.1.3 CpG Island Methylator Phenotype (CIMP) Pathway

This phenotype overlaps with MSI-H tumors and is characterized by hypermethylation of CpG islands. The classification of colorectal tumors as CIMP is typically performed by testing for key gene methylation markers. A considerable overlap occurs between CIMP and MSI-H tumors. Most MLH-1 hypermethylated MSI-H tumors exhibit a hypermethylator profile compatible with CIMP. CIMP colon cancers are associated with *BRAF* mutations, tend to occur in the setting of a serrated adenoma (flat adenoma), involve the right colon, and are associated with rapid carcinogenesis (Figure 4.6).

4.4.2 Hereditary Colorectal Cancers

4.4.2.1 Familial Adenomatous Polyposis

FAP is a dominantly inherited disease characterized by the development of hundreds to thousands of colonic polyps by age 20. FAP affects approximately 1 in 5000–7000 individuals. Without prophylactic colectomy, most individuals with FAP will develop colorectal cancer by age 40, although some patients may develop cancer later in life. In addition to colonic polyposis, FAP patients are prone to develop gastric, duodenal, and small bowel polyps, which in turn can progress to cancer. FAP is also associated with extraintestinal nonmalignant soft tissue tumors such as cutaneous lesions (lipomas, sebaceous and epider-moid cysts) and other benign tumors. The combination of colonic polyposis syndrome and the development of benign soft tissue tumors characterizes Gardner syndrome.

The pathogenesis of FAP is related to an inactivating mutation in the *APC* tumor suppressor gene, located on chromosome 5q21. The APC tumor suppressor protein is critical in the inhibition of Wnt signaling. A complex of APC, GSK3, and CK1 binds and phosphorylates non-cadherin-associated β-catenin, targeting it for destruction by the proteasome. In the setting of *APC* gene mutation, APC protein absence, or in the presence of Wnt activation, β-catenin is spared destruction and accumulates in the nucleus where it associates with TCF to drive the transcription of *MYC* and subsequent cellular proliferation (Figure 4.7).

Most cases of FAP are related to nonsense mutations that are located centrally in the *APC* gene and which result in a truncated APC protein. The location of the *APC* mutation drives the disease phenotype. Less common mutations that cluster in the extreme 3′ and 5′ portions of exon 15 of the *APC* gene are associated with a less aggressive form of FAP, characterized by a less profuse polyposis and a later age of cancer development (attenuated FAP).

4.4.2.2 Management of FAP Patients

Since patients with FAP carry almost a 100% lifetime risk of colorectal cancer, a preventive and intensive screening strategy is indicated. Although a positive impact of nonsteroidal drugs such as celecoxib and sulindac on polyposis progression has been previously demonstrated, the risk of colorectal cancer, despite the use of these agents, remains unacceptable. Therefore, prophylactic colectomy remains the standard approach with patients carrying an *APC* mutation. Patients with *APC* mutations are recommended to have a screening colonoscopy starting in late adolescence. Colectomy is advised as early as polyposis is documented. Colectomy can be in the form of a total colectomy (resection of the colon and rectum with ileostomy formation or with an

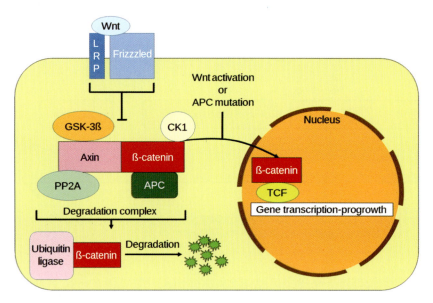

Fig. 4.7 APC mutation and the β-catenin pathway. Mutations in the *APC* gene result in silencing of the APC protein and thus the inability to degrade β-catenin. β-Catenin translocates to the nucleus to form a complex with TCF and activate TCF-regulated genes, including c-MYC.

ileo-anal anastomosis). Alternatively, a subtotal colectomy can be performed.

4.4.2.3 Hereditary Non-Polyposis Colorectal Cancer

Hereditary non-polyposis colorectal cancer (HNPCC), or Lynch syndrome, is linked to approximately 3% of colorectal cancers. HNPCC is inherited in an autosomal dominant fashion and is associated with a lifetime risk of colorectal cancer of approximately 50–80%. In addition to an increased risk of colorectal cancer, carriers of HNPPC-associated germline mutations are prone to endometrial cancer (40–60%), stomach cancer (up to 10%), and, to a lesser extent, ovarian, hepatobiliary, pancreatic, small bowel, brain, skin, and urinary tract cancers.

HNPCC is related to a germline mutation in the mismatch repair genes *MLH1*, *MSH2*, *MSH6*, and *PMS2*. Mutations in *MSH2* and *MLH1* account for approximately 90% of all HNPCC. Mutations in *MSH6* account for approximately 10% of cases, while *PMS2* mutations involve less than 1% of cases. Different mutations have been associated with different risks in HNPCC-associated malignancies. For example, *MSH6* mutations have been associated with a lower risk of colorectal cancer, while the risk of endometrial cancer may be increased in this subgroup of patients. In addition to the germline mutations above, *EPCAM* germline mutations have also been recently associated with approximately 3% of HNPCC cases. Deletions in the *EPCAM* gene result in transcriptional read-through of the mutated *EPCAM*

allele and epigenetic inactivation and silencing of its neighboring gene, *MSH2*. Therefore, HNPCC related to *EPCAM* mutations are characterized by the loss of *MSH2* (or mosaic expression) without evidence of *MSH2* germline mutation.

As detailed above, HNPCC tumors are characterized by microsatellite instability or the lack of the corresponding MMRP as determined by immunohistochemistry (IHC). Therefore, the most effective screening strategy for HNPCC is to conduct MSI testing or IHC testing for MMRP. The absence of MSI or the presence of all MMRPs, as detected by IHC, essentially rules out the possibility of HNPPC and the need for a germline DNA mutation assay.

The clinical diagnosis of HNPCC is based on the satisfaction of the Amsterdam criteria. A family must satisfy all of the following: three affected individuals with an HNPCC-associated cancer (colorectal, endometrial, small bowel, ureter, or renal pelvis), at least one of the affected individuals is a first degree relative of the other two affected members; at least two successive generations are affected; at least one of the affected individuals is less than 50; and there is no evidence of FAP. However, consideration for HNPCC screening should transcend the Amsterdam criteria, as many patients with HNPCC-related germline mutations would be missed. In order to identify a larger number of patients with HNPCC-associated germline mutations, the NCI has recommended less rigid clinical guidelines for the screening of HNPCC (Bethesda guidelines). More recently, some have advocated universal

MSI screening of colorectal cancers and suggest that such a strategy may be the most effective strategy in the identification of HNPCC families.

4.4.2.4 Management of HNPCC-Associated Germline Mutation Carriers

Given the high risks of colorectal cancer in carriers of HNPCC germline mutations, intensive endoscopic screening is recommended in this population. In general, colonoscopies are recommended starting at age 20–25 and are repeated on a 1–2 year basis. In individuals with a HNPCC-associated colon cancer, a total or subtotal colectomy is recommended at the time of surgical intervention. In addition, given the high risk of endometrial and ovarian cancers in females with HNPCC germline mutations, prophylactic hysterectomy and oophorectomy are recommended at age 40 years.

4.4.2.5 MUTYH-Associated Colorectal Cancer

MUTYH-associated polyposis (MAP) is an autosomal recessive polyposis syndrome caused by biallelic mutations in the *MUTYH* gene. The *MYUTH* gene is critical in the BER pathway, which is essential for repairing spontaneous mutations associated with oxidative stress. This condition is typically associated with presence of more than 50 polyps and represents approximately 7% of attenuated polyposis and classical polyposis cases. However, cases of MAP have also been described without the presence of polyposis, often making the diagnosis challenging. MAP is rare, occurring in less than 1% of colorectal cancer cases. The diagnosis should be suspected in cases with more than 10 polyps and without associated FAP. In addition to a lifetime risk of colorectal cancer of 80%, patients with MAP have an increased risk of upper gastrointestinal polyposis and cancer, ovarian cancer, skin cancer, and breast cancer.

4.4.2.6 Management of MAP Patients

Patients carrying a MUYTH biallelic mutation should be screened for colorectal cancer with yearly colonoscopies starting at the age of 20. Those with a significant number of polyps that cannot be removed endoscopically should be considered for prophylactic colectomy. Expanded guidelines to screen for MAP in the general colorectal cancer patient population have recently been proposed. These include patients with colon cancer before age 40 with <10 adenomatous polyps, any patient with more than 10 polyps (adenomas or serrated polyps), polyposis without APC mutation, and colon cancer with a family history consistent with autosomal recessive inheritance.

4.4.3
Clinical Impact of Molecular Markers on the Management of Colorectal Cancer

4.4.3.1 MSI-H Status and Colorectal Cancer

Approximately 15% of colorectal cancers have microsatellite instability. MSI instability (MSI-high) has been associated with a good prognosis and a lower likelihood of distant spread among patients with earlier stages of colorectal cancer. This explains the substantially lower incidence of MSI-H tumors, estimated at 3–4%, in patients with metastatic colorectal cancer. The improved prognosis in this patient population is likely related to the hypermutated profile of the cancer, which leads to a robust host immune response that is reflected in intratumor lymphocytic infiltration.

A differential benefit between MSI-H and MSS tumors after adjuvant therapy with 5-fluorouracil (5-FU) chemotherapy has been demonstrated. Patients with MSI-H colorectal cancer do not derive any benefit from 5-FU after surgery, while a significant improvement in overall survival has been described for MSS tumors. It has been postulated that a deficient mismatch repair mechanism in MSI-H tumors does not allow for the recognition of 5-FU-modified DNA, leading to 5-FU tolerance. Based on these findings, it is currently recommended that the MSI status be evaluated in the event that fluoropyrimidine monotherapy (i.e., 5-FU) is considered in the adjuvant setting. Patients with MSI-H early-stage disease should not be offered fluoropyrimidines and should be followed by observation. Patients with MSI-H tumors and regional disease (lymph node involvement) are more appropriate candidates for oxaliplatin-based therapy and may not derive considerable benefit from 5-FU monotherapy (Figure 4.8).

4.4.3.2 Epidermal Growth Factor Receptor Pathway Targeting and Colorectal Cancer

The EGFR is activated by several ligands. These include EGF, tumor growth factor alpha (TGF-α), epiregulin (EREG), and amphiregulin (AREG). Activation of the EGFR pathway leads to the downstream activation of the RAS-RAF pathway, the PI3K pathway, and the JAK-STAT pathway, leading to cellular growth and antiapoptotic signals (Figure 4.9). Therefore, the inhibition of the EGFR pathway represents an attractive antineoplastic target. Indeed, several anti-EGFR agents have been evaluated in clinical trials in patients with metastatic colorectal cancer, including mAbs to the EGFR receptor and TKIs that target the kinase domain of EGFR. Only the monoclonal antibodies to EGFR have been associated with clinically significant improvement in the disease control rate and survival of patients with metastatic

Stage II colorectal cancer　　　　**Stage III colorectal cancer**

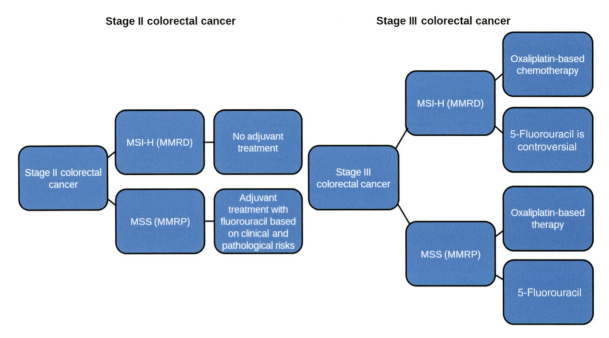

Fig. 4.8 MSI status and adjuvant treatment in colorectal cancer. Patients with stage II colorectal cancer with MSI-H tumors have a good prognosis and do not benefit from adjuvant treatment. MSI-H stage III colorectal cancer carry a better prognosis than MSS stage III tumors but also benefit from oxaliplatin-based adjuvant therapy.

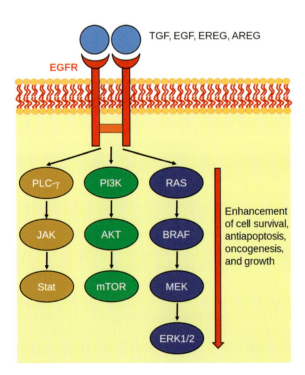

Fig. 4.9 The epidermal growth factor receptor pathway: EGFR activates the RAS-RAF, JAK-STAT3, and PI3K pathways. The RAS-RAF pathway is believed to be the most important driver of cellular proliferation downstream of EGFR. Activating mutations in RAS are associated with resistance to anti-EGFR treatment. Activating mutations in BRAF are associated with a poor prognosis and resistance to anti-EGFR treatment as well as to other cytotoxic therapies.

colorectal cancer. This has led to the FDA approval of cetuximab and panitumumab in the United States.

4.4.3.3 RAS Mutations and Response to Anti-EGFR Therapy

The *RAS* oncogene is mutated in approximately 50% of patients with colorectal cancer. These mutations involve the *KRAS* gene (exons 2, 3, and 4) in approximately 45% of patients and the *NRAS* gene (exons 2, 3, and 4) in approximately 5% of patients. Since these mutations result in constitutive activation of the *RAS* oncogene, it has been postulated that the inhibition of the EGFR receptor upstream of RAS would result in limited antitumor activity in patients with RAS mutations. This has led to the retrospective evaluation of large clinical trials investigating cetuximab or panitumumab in colorectal cancer according to *RAS* status. These studies confirmed that the benefit from EGFR targeting with cetuximab and panitumumab is limited to patients with *KRAS* wild-type tumors. Based on these findings, pharmacogenetic testing can be performed prior to treatment decision (Section 9.3).

4.4.3.4 BRAF Mutations and Colorectal Cancer

The estimated incidence of *BRAF* mutations (V600E) in patients with colorectal cancer is about 12–15%. *BRAF* mutations are mutually exclusive with *RAS* mutations. When mutated, *BRAF* becomes constitutively active, resulting in the activation of the downstream MAPK proliferative pathway. *BRAF* mutations are associated with a more aggressive colorectal cancer phenotype and with chemotherapy resistance. When compared to patients with *BRAF* wild-type tumors, patients with *BRAF* mutant tumors have lower response rates to chemotherapy, shorter progression free survival, and shorter overall survival. *BRAF* mutations may not only act as prognostic markers but may also be predictive of anti-EGFR monoclonal antibody resistance. There has been a significant interest in the development of BRAF inhibitors in this subgroup of colorectal cancer patients. However, the clinical benefits associated with such inhibitors have been modest and limited by the rapid onset of resistance. BRAF inhibition in this subgroup of colorectal cancer patients has been associated with a compensatory EGFR activation, which can be promoted by the concurrent administration of EGFR-targeting monoclonal antibodies. Several studies are currently ongoing to evaluate the combination of BRAF and EGFR inhibitors in BRAF mutant metastatic colorectal cancer, with early signs of clinical activity.

4.5
Molecular Biology of Renal Cell Carcinoma

(Kara DeWalt, Tommy Tong, and Sumanta K. Pal, City of Hope, Duarte, CA, USA)

In the United States, more than 60 000 individuals are diagnosed with RCC every year, and approximately 14 000 die of the disease. For unknown reasons, the disease tends to be more frequent in males than females, occurring in a 3:1 distribution. Risk factors for renal cell carcinoma are poorly defined, although both smoking and obesity have been linked to the disease. Notably, the incidence of RCC has been rising steadily for the past 65 years at the rate of 2% per annum. Although the reasons for this are unclear, many suspect an increase in incidental diagnoses due to more widespread use of radiographic imaging.

The management of RCC varies by stage. Patients with stage I–III disease essentially have disease confined to the kidney and immediately surrounding structures, while patients with stage IV disease have metastases to distant organs. For stage I–III disease, surgical resection of the kidney tumor is the standard of care. At present, there is no role for adjuvant (postoperative) systemic therapy. In stage IV disease, a combination of surgical removal of the kidney tumor with subsequent systemic treatment is typically employed.

The nature of systemic treatment has evolved greatly over the past several decades. Prior to the 1990s, cytotoxic chemotherapy was the mainstay of treatment. Cytotoxic agents act in a nonspecific fashion, interfering with processes including (but not limited to) DNA synthesis, DNA replication, and microtubule spindle formation. Tumor shrinkage and delayed tumor growth occurred infrequently, and thus RCC is commonly referred to as a chemotherapy-resistant disease. In the 1990s, immunotherapeutic agents were increasingly utilized. These agents (specifically, interleukin-2 (IL-2) and interferon-α (IFN-α)) resulted in profound and durable tumor shrinkage in a small subset of patients. However, both drugs produce toxicity. Furthermore, during the so-called "immunotherapy era," estimates of survival in patients with stage IV RCC approached only 1 year.

The most recent transition in the systemic management of stage IV RCC has occurred over the past decade with the advent of "targeted therapies." At present, these targeted therapies are relevant to the predominant biologic subtype of RCC, known as clear cell disease. Clear cell RCC constitutes about 80% of RCC cases, and is thus the principal focus of this section. However, the reader should be aware of other subtypes of RCC such as papillary cancer and chromophobe cancer, constituting about 15 and 5% of cases, respectively. A narrow spectra of rarer RCC subtypes exist, but are beyond the scope of this section. Herein, we discuss the biology of clear cell, papillary, and chromophobe RCC separately. Where relevant, the implications of disease biology to therapeutic strategies are discussed.

4.5.1
Biology of Clear Cell Renal Cell Carcinoma

The microscopic features of clear cell RCC include the presence of eosinophilic cells with abundant cytoplasm (Figure 4.10a).

At a molecular level, the disease is marked by aberrations in the von Hippel–Lindau (VHL) gene. Notably, von Hippel–Lindau syndrome was first described in the late nineteenth century, and includes kidney cancer along with other vascular malformations. In the late 1980s, VHL was mapped to chromosome 3. More recently, it has been suggested that somatic mutations in VHL are found in roughly 50% of sporadic RCC cases. Other modifications of VHL, such as hypermethylation, are found in roughly 10–20% of cases. The VHL protein shuttles from the nucleus to cytoplasm, and appears to play a role in polyubiquitination and subsequent elimination of hypoxia-inducible factor (HIF).

Aberrations in VHL thus lead to increases in HIF. HIF is a transcription factor for VEGF. The VEGF family of ligands includes five moieties (VEGF-A, VEGF-B, VEGF-C, VEGF-D, and VEGF-E). These ligands exhibit distinct and preferential binding to one of three transmembrane receptors: VEGF receptor 1 (VEGFR1), VEGF receptor 2

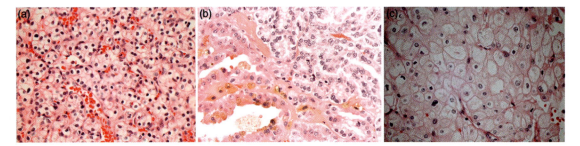

Fig. 4.10 Microscopic view of the predominant subtypes of renal cell carcinoma (RCC). (a) Clear cell RCC. (b) papillary RCC. (c) Chromophobe RCC.

(VEGFR2), or VEGF receptor 3 (VEGFR3). Each of these receptors may serve a distinct biologic function. For instance, activation of VEGFR3 has been associated with lymphangiogenesis, the formation of lymphatic channels. Lymphatic channels drain interstitial fluid and scattered cells to lymph nodes, which are clusters of immune cells that aid in developing a humoral response. VEGFR1 has been associated with formation of the premetastatic niche. The premetastatic niche is a molecular milieu that may serve as a take-off point for future metastasis. VEGFR2 perhaps bears the most relevance to RCC; activation of VEGFR2 is ultimately associated with angiogensis, the formation of blood vessels. Tumors typically exhibit exponential growth curves, often growing at a pace that exceeds their available blood supply. In the setting of RCC, activation of VEGFR2 and the resultant angiogenesis may facilitate tumor growth.

Binding of VEGF ligand to VEGF receptors leads to activation of PI3K in the juxtamembrane region. PI3K in turn activates protein kinase B, also referred to as Akt. Through a series of downstream effectors (including tuberous sclerosis complex 1 (TSC1) and tuberous sclerosis complex 2 (TSC2)), activation of mTOR occurs. mTOR is a serine–threonine protein kinase that forms the core of two key regulatory complexes: mTOR complex 1 (mTORc1) and mTOR complex 2 (mTORc2). Both play key roles in multiple metabolic and oncogenic pathways. As will be discussed later, mTORc2 has been associated with resistance to certain therapies for RCC. mTORc1 and mTORc2 ultimately trigger cell proliferation through several downstream moieties, including S6 kinase (S6K) and 4E-BP1, which increase cellular proliferation.

4.5.2
Approved Drugs for the Treatment of Clear Cell Renal Cell Carcinoma

As noted above, the treatment of clear cell RCC has evolved substantially over the past several decades, owing largely to a better understanding of the biology of

the disease. Several elements of the signaling cascades described in the previous section double as therapeutic targets in RCC. In the extracellular domain, VEGF-A ligand is bound by the humanized IgG2 mAb bevacizumab (Figure 4.11).

Bevacizumab has the microscopic effect of inducing vascular normalization, that is, pruning the erratic blood vessels generated by VEGF signaling.

Treatment with a monoclonal antibody leads to a very specific on-target effect, but a limitation is that intracellular penetration of an antibody is low. Thus, this approach cannot be used for intracellular targets. To target other elements in the VEGF signaling cascade, small molecules have been generated. Small-molecule TKIs inhibit the TK domain of the VEGF receptor.

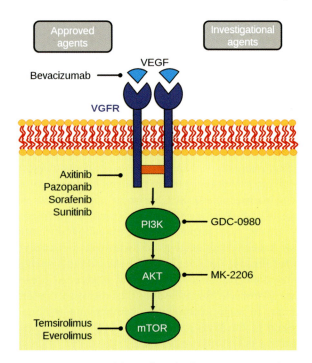

Fig. 4.11 Treatment of clear cell renal cell carcinoma (RCC). The figure summarizes approved agents for metastatic clear cell RCC and selected investigational agents that abrogate signaling along the VEGF–mTOR pathway.

There are a total of four small-molecule TKIs that are approved for metastatic RCC: (1) sunitinib, (2) pazopanib, (3) sorafenib, and (4) axitinib. It is important to note that these small molecules can interact not only with the kinase domain of VEGF receptor but also with the kinase domain of a number of other moieties, including PDGFR, Raf, and others.

A comprehensive discussion of the clinical efficacy of these drugs is beyond the scope of this textbook. In summary, however, clinical trials evaluating VEGF-directed therapies have yielded a survival of nearly two to three times that reported during the immunotherapy era. The mechanism of action of these agents offers a perspective on the toxicities seen with these drugs. Specific blockage of VEGF-A by bevacizumab therapy yields a spectrum of side effects, including blood clots, increased risk of hemorrhage (bleeding), elevated blood pressure, and protein in the urine. From a physiologic standpoint, these clinical effects likely result from decreased angiogenesis as a consequence of decreased activation of the VEGF receptor. In contrast, small-molecule TKIs often have a much broader scope of toxicity as they lack the specificity of bevacizumab. Through "off-target effects" (i.e., by abrogating signaling through moieties outside of the VEGF receptor), these drugs can cause palmar–plantar erythrodysesthesia (sloughing off of the skin on the hands and soles of the feet), decreased blood cell counts, and gastrointestinal toxicity.

The mechanisms of resistance to bevacizumab and the small-molecule VEGF receptor TKIs are not well characterized. Increases in VEGF levels or VEGF receptor expression have been posited. Furthermore, the utilization of alternative pathways (e.g., signaling through the MET receptor) have been suggested. An emerging area of research suggests that the VEGF-directed drugs described in the current section may not act exclusively through decreasing angiogenesis, but may have a complex effect on the immune milieu. For instance, drugs such as sunitinib may potentially reduce the recruitment of myeloid-derived suppressor cells (MDSCs) and regulatory T cells (Tregs) to sites of tumor, and thereby promote an antitumor immune response. When resistance develops, Tregs and MDSCs may reaccumulate. These observations provide a strong theoretical rationale for employing novel immunotherapeutics (discussed in the next section) at the time of resistance to VEGF-directed therapy.

Beyond blocking VEGF and its cognate receptor, a second therapeutic approach that has proved useful in metastatic RCC is to use agents that block mTOR. There are two agents currently approved for clinical use for this indication: temsirolimus and everolimus. Both are small molecules with a common chemical structure and are derived from sirolimus, a drug that has been used as an immunosuppressant in patients who have received an organ transplant. Everolimus and temsirolimus appear to exert most of their anticancer effect through mTORc1. As noted in the previous section, increases in mTORc2 levels may be a means of therapeutic resistance. Thus, efforts are underway to develop more potent dual mTORc1 and mTORc2 inhibitors.

4.5.3
Investigational Approaches for the Treatment of Clear Cell Renal Cell Carcinoma

Although there are a total of seven targeted therapies currently approved for metastatic RCC, the targets of these therapies are redundant. Thus, there is an aggressive effort to develop therapies that target novel points along the VEGF-mTOR signaling axis. As one example, the mAb ramucirumab has affinity for the external domain of VEGFR2. Modest activity has been observed in early studies. Downstream of the VEGF receptor, two unique targets have been exploited: PI3K and Akt. The PI3K inhibitor GDC-0980 (which reportedly also has affinity for both mTORc1 and mTORc2) has been compared directly with everolimus in patients with metastatic RCC. Interestingly, the activity of everolimus appeared to be superior to GDC-0980. Similar results were seen in a trial comparing everolimus with the Akt inhibitor MK-2206. Though the results generated to date challenge the premise of targeting PI3K and Akt, it should be noted that there are a number of agents in the pipeline with unique binding properties that have yet to be explored in RCC.

As noted previously, RCC may potentially be propagated through "bypass pathways," such as the MET signaling axis. MET is a transmembrane receptor that is activated through binding of hepatocyte growth factor (HGF). The activation of MET leads to a multitude of downstream signaling cascades including PI3K-Akt. Cabozantinib, a small-molecule inhibitor of both MET and VEGFR2, has shown profound activity in an early trial in metastatic RCC. On this basis, a large, randomized trial is currently underway to evaluate the agent in patients that have received previous VEGF-directed therapy.

As noted previously, it has been suggested that immune escape may be a means of developing resistance to VEGF-directed agents. To this end, several drugs are currently under development that may promote an antitumor immune response. Understanding the rationale for these drugs requires a fundamental understanding of tumor immunology. As background, the T cell is typically stimulated to elicit a response through binding of the T-cell receptor (TCR) to major histocompatibility complex (MHC) on the surface of the antigen-presenting cell (APC). A number of costimulatory interactions take place as well, such as binding of B7 and CD28 on the APC and T cell, respectively. To moderate the immune response, several inhibitor molecules exist on the T cell, including programmed death-1 (PD-1) and cytotoxic T-lymphocyte-associated protein 4 (CTLA4). PD-1 has two potential ligands on the APC (PD-L1 and PD-L2), while CTLA4 binds to B7 on the APC. These interactions promote T-cell anergy, that is, unresponsiveness. Clinical exploration of the PD-1 inhibitor nivolumab and the PD-L1 inhibitor MDPL3280A is currently underway. Both show modest activity as single agents, and both are being combined with VEGF-directed therapies to identify potential synergy. Notably, nivolumab is also being explored in combination with the CTLA4 inhibitor ipilimumab. Although definitive data from late-stage trials have not emerged, these trials have all accrued patients very rapidly.

4.5.4
Biology and Treatment of Papillary Renal Cell Carcinoma

Papillary RCC can be distinguished from clear cell RCC by the lesser abundance of cytoplasm and nucleolar prominence (Figure 4.10b). From a biological perspective, papillary RCC is driven by distinct molecular aberrations. There are two types of papillary RCC. Type II has been more firmly linked to a familial syndrome known as hereditary leiomyomatosis and renal cell cancer (HLRCC). In addition to RCC, female patients with this inherited disorder may also have uterine tumors. Mutations in the *fumarate hydratase* (*FH*) gene have been noted in families with HLRCC. From a mechanistic perspective, it appears that higher levels of fumarate (as a consequence of defective fumarate hydratase) may lead to stabilization of HIF, and thereby propagate renal tumors. For younger patients with type II papillary RCC, it has been suggested that familial screening for FH mutations may be performed. Extensive gene profiling studies in type II papillary RCC have revealed novel mutations in several genes, including *NRF2*, *SIRT1*, and *CUL3*. The function of their respective gene products has not been completely characterized. In contrast to type II papillary RCC, type I papillary RCC appears to be characterized by aberrations in MET. Recent studies suggest increases in MET gene copy number in nearly 90% of patients with type I disease, as compared to closer to 50% in patients with type II papillary RCC.

Unfortunately, no agents have been specifically approved for the treatment of papillary RCC. Pivotal trials leading to the FDA approval of temsirolimus did include patients with papillary disease, thus leading to

frequent use of this agent in the clinic. Recent trials comparing sunitinib and everolimus have included subsets of patients with papillary RCC. In these subsets, there appears to be a slight advantage to use of sunitinib in this setting. Increasingly, expert opinion suggests that VEGF-directed therapies may be used for patients with papillary RCC.

Several trials exploring agents outside of the VEGF-mTOR signaling axis have been conducted in small cohorts of patients with papillary RCC. Erlotinib, an inhibitor of the EGFR, showed minimal tumor shrinkage but an impressive increase in survival. On the basis of these data, a more recent study explored the putative MET inhibitor ARQ197 either alone or in combination with erlotinib in patients with papillary RCC. Results from this study are still forthcoming.

4.5.5
Biology and Treatment of Chromophobe Renal Cell Carcinoma

Chromophobe RCC appears to be less aggressive than clear cell or papillary RCC (Figure 4.10c). This subset of RCC has been associated with Birt–Hogg–Dube (BHD) syndrome. In addition to chromophobe RCC, patients with BHD syndrome may experience spontaneous pneumothorax (collapse of lung tissue) and benign tissue deposits known as follicular hamartomas. Genetic studies have found that the BHD gene is located on chromosome 17. There is also some suggestion that chromophobe RCC may be associated with aberrations in p53, a tumor suppressor gene that appears to play a role in a wide spectrum of malignancies.

As with papillary RCC, there are no clinical trials that establish a definitive treatment paradigm for chromophobe RCC. As such, the disease is typically approached with the VEGF- and mTOR-directed therapies used for clear cell RCC. Furthermore, although genetic associations have been found in chromophobe RCC (i.e., the BHD gene), there is no gene product for which targeted therapies have been generated.

4.5.6
Further Subtypes of Renal Cell Carcinoma

The biology of RCC is complex and varies by subtype. Although the clear cell, papillary and chromophobe subtypes of RCC are most common, there are several others the reader should be aware of. Collecting duct RCC is infrequent, constituting about 1% of all cases of RCC. Microscopically, the disease appears to be poorly differentiated. The treatment of collecting duct RCC is similar to that for bladder cancer. Xp11.2 translocation-associated

RCC is another rare subtype that occurs at a similar frequency and tends to affect younger patients. Although there is no standard treatment, there may be a role for VEGF-directed agents in patients with Xp11.2 translocation-associated RCC.

A final histologic subtype deserving mention is sarcomatoid RCC. Sarcomatoid RCC is not a distinct subtype, and rather can coexist with clear cell, papillary, or chromophobe RCC. The sarcomatoid component tends to be poorly differentiated, and there are data to support treatment with cytotoxic chemotherapy and with VEGF-directed agents. In fact, trials are ongoing to assess combinations of these modalities given the severity of this disease.

To conclude, RCC represents an array of diseases, each characterized by a distinct biology. In the setting of clear cell RCC, the most common form, a detailed understanding of relevant molecular pathways has led to a treatment paradigm consisting largely of targeted therapies (VEGF- and mTOR-directed agents). In the years to come, research will focus on developing agents that overcome resistance to standard therapies for clear cell RCC. Additional research will focus on developing a rational targeted approach to nonclear cell subtypes of RCC.

4.6
Molecular Biology of Prostate Cancer

(Przemyslaw Twardowski, City of Hope, Duarte, CA, USA)

Prostate cancer is the most common noncutaneous malignancy, accounting for 241 000 cases diagnosed in the United States in 2012. It is usually an indolent disease, but approximately 25% of tumors have an aggressive clinical course, resulting in almost 30 000 deaths annually in the United States. In cases of advanced disease, >80% of patients initially respond to therapy that suppresses androgen hormone production. However, within 18–36 months, the so-called castration-resistant phenotype emerges, which no longer responds to androgen suppression, transforming the disease to a more virulent course with a poor prognosis and a subsequent median survival of approximately 2–3 years. Although the significance of androgen (testosterone) hormone signaling in prostate cancer progression is paramount, the molecular characteristics of prostate cancer development (carcinogenesis) are poorly understood. Environmental and dietary factors have not been conclusively associated with an increased risk of prostate cancer, although the "Western" diet, rich in saturated fats, has been implicated in the increased incidence of this disease. Obesity also appears to be associated with a more aggressive subset of prostate cancer. Men with a first-degree relative with prostate cancer have a two- to- threefold increase in the development of the disease as compared to men with no such history. The risk increases further if more than one first-degree relatives or both first- and second-degree relatives are affected (5–11-fold). All types of inheritance patterns (autosomal dominant, autosomal recessive, and X-linked) have been described. However, highly penetrant susceptibility genes associated with the onset of prostate cancer are yet to be identified.

It appears that multiple genes, each with a small to moderate effect, are involved in prostate cancer carcinogenesis (Figure 4.12). The lack of single "driver" of carcinogenesis in prostate cancer makes preventive and therapeutic efforts extremely complex. The role of androgen signaling remains the most important, best studied, and most therapeutically useful element of prostate cancer biology. Genes associated with the development or progression of prostate cancer can be broadly divided into genes associated with hereditary prostate cancer, sporadic cancer tumor suppressor genes, and sporadic cancer oncogenes.

4.6.1
Genes Associated with Hereditary Prostate Cancer

1) *HPC1 (RNASEL):* This gene is located on chromosome 1q24-q25 and encodes a ribonuclease that degrades viral and cellular RNA and induces apoptosis in virally infected cells. This is one of the most important genes mutated in patients with hereditary prostate cancer.
2) *CAPB (carcinoma prostate brain):* This is located on chromosome 1p36. It was discovered in familial prostate cancer in close relatives suffering from brain tumors. This gene may be important in regulating Sonic Hedgehog pathway signaling, which also has been implicated in prostate cancer carcinogenesis.
3) *CHEK2 (chromosome 22q):* This upstream regulator of p53 is important in the DNA damage signaling pathway and is associated with hereditary prostate cancer.
4) *HOXB13 (chromosome 17q21-22):* The G84E mutation in *HOXB13* gene has been associated with prostate cancer development. However, the mechanism by which it contributes to the pathogenesis of prostate cancer remains unknown. This was the first identified gene to account for a significant fraction of hereditary prostate cancer, particularly of the early-onset variety.
5) *BRCA1 (chromosome 17q21) and BRCA2 (chromosome 13q12.3):* More known for its role in hereditary breast and ovarian cancer, this tumor suppressor gene involved in DNA repair appears to be associated with

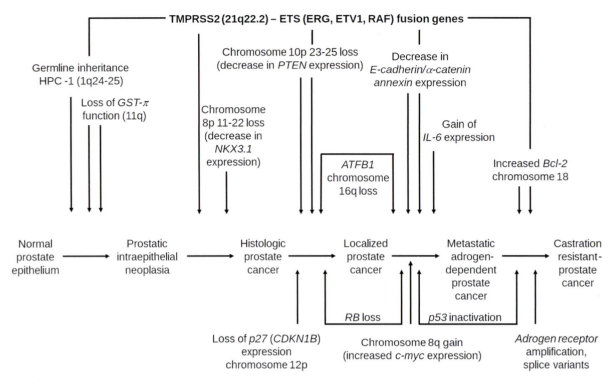

Fig. 4.12   Sequence of selected genetic events involved in prostate cancer carcinogenesis.

increased risk of prostate cancer as well. Prostate cancer occurs among Ashkenazi Jewish carriers of the BRCA2 founder mutation (6174delT). In general, up to one third of *BRCA2* mutation carriers are expected to develop prostate cancer at a relatively younger age. The risk is smaller and its existence disputed in BRCA1 mutation carriers. Some data suggest that BRCA-related prostate cancer has a significantly worse prognosis than prostate cancer that occurs in patients without that mutation. An interesting aspect of the presence of BRCA mutations in cancer cells is their dependence on another DNA repair mechanism, mediated by PARP-1. This enzyme repairs DNA single strand breaks (SSBs) by base exision repair (BER). Inhibition of PARP-mediated DNA repair in BRCA-mutated cells leads to their death. PARP inhibitors are currently being evaluated as therapeutic agents in cancer associated with BRCA mutations including prostate cancer.

4.6.2
Tumor Suppressor Genes in Sporadic Prostate Cancer

1) *NKX3.1*: This gene is located on chromosome 8p21. *NKX3.1* encodes a tumor suppressor protein that decreases cell survival by enhancing mutated *ataxia-telangiectasia* (*ATM*) gene activity after DNA

damage. Loss of chromosome 8p is the most common alteration in the prostate cancer genome and is an early event in prostate cancer carcinogenesis.

2) *p53: (chromosome 17p13.1):* The protein encoded by this gene promotes apoptosis in cells that are disorganized or have damaged DNA. In primary prostate cancer, it is mutated in 10–20% of cases. In advanced metastatic disease, mutations are seen in 40% of patients.

3) *PTEN (chromosome 10q23):* This encodes a phospholipid phosphatase, and acts as a tumor suppressor gene by inhibiting the phosphatidylinositol 3 kinase–protein kinase B (PKB-AKT) signaling pathway that is essential in cell cycle progression and survival. PTEN is involved in the progression of advanced stages of prostate cancer and has no known role in carcinogenesis.

4) *p27 (CDKN1B) (chromosome 12p12.3):* This gene encodes a cyclin-dependent kinase inhibitor. Somatic loss of this gene has been described in 23% of localized prostate cancers and 47% of distant metastases.

5) *MX11:* This gene encodes a protein that acts as a negative regulator of the *c-Myc* proto-oncogene, which plays a significant role in prostate cancer development.

6) The *retinoblastoma* (*Rb*) gene is located on chromosome 13q14.1-q14.2. This protein binds and inhibits transcription factors of the E2F family. Mutations in

this gene have been reported in 34% of localized and 74% of advanced prostate cancers. Loss of this gene has also been associated with the development of castration resistance.

7) *Glutathione S-transferase gene (GSTP1) 11q:* This gene encodes a protein that prevents oxidant and electrophilic DNA damage. *GSTP1* is inactivated by hypermethylation of its promoter region in 96% of primary prostate cancers.

8) *ATFB1:* This gene is located on chromosome 16q22 and encodes a transcription factor. Missense mutations have been noted in 36% of primary prostate tumors.

9) *Annexins:* These proteins regulate the cytoskeleton and cell motility and are typically downregulated in prostate cancer.

10) *CHD1:* This gene is located on chromosome 5q21; deletions in this gene have been identified in 10–17% of prostate tumor samples. *CHD1*-deleted tumors are associated with increased invasiveness, a high frequency of intrachromosomal rearrangements, and a worse prognosis.

4.6.3
Oncogenes

1) *c-Myc:* This gene is located on chromosome 8q24. A translocation involving *c-Myc* is associated with Burkitt's lymphoma (BL). c-Myc acts as a transcription factor that promotes cell proliferation. This oncogene is also overexpressed in 8% of primary prostate tumors and 30% of prostate cancer metastases. It tends to correlate with a high Gleason score (a microscopic measurement of tumor differentiation; the higher the Gleason score, the more undifferentiated the cancer) and poor prognosis. It has been proposed that c-Myc is a ligand-independent AR target gene.

2) *c-ErbB2 (Her2 neu):* This gene is located on chromosome 17q12-21.32. It encodes a transmembrane phosphoprotein and is very important in the pathophysiology of a subset of breast cancers. Overexpression of this gene has been associated with prostate cancer progression and development of castration resistance.

3) *Bcl-2:* This gene is located on chromosome 18. The Bcl-2 protein promotes cell survival through the inhibition of an apoptotic pathway. *Bcl-2* is overexpressed in advanced prostate cancer during progression to the castration-resistant state.

4) *Prostate stem cell antigen (PSCA):* PSCA is overexpressed in 80% of patients with prostate cancer and in a precancerous condition called prostatic intraepithelial neoplasia (PIN). A higher expression of this protein is associated with increasing Gleason score, disease progression, and the development of castration resistance.

5) *Hepsin:* This gene encodes a membrane-bound serum protease. Expression of hepsin correlates inversely with patient prognosis.

6) *Alpha-methyl coenzyme A racemase (AMACR):* This gene is located on chromosome 5. Its protein product is involved in the oxidation of fatty acids. Eighty-eight percent of prostate cancer specimens stain positively for this protein.

7) The SRC family (SRC-1, SRC-2, and SRC-3) of transcription factors mediate nuclear receptor transcription functions. High levels of SRC-3 have been associated with progression of prostate cancer to the castration-resistant phenotype.

8) *Androgens and the androgen receptor:* Signaling via the androgen receptor is the most important component of the pathophysiology of prostate cancer progression and metastasis. This pathway has been exploited therapeutically since the early 1940s after Drs. Huggins and Hodges published their seminal publication describing patients with metastatic prostate cancer who experienced dramatic clinical improvements after surgical castration. The AR is encoded on chromosome Xq11-q12 and belongs to the superfamily of nuclear receptors. Nuclear receptors are ligand-inducible transcription factors that mediate the signals of several fat-soluble hormones. The AR consists of an amino terminal domain, a DNA-binding domain, and a carboxy-terminal ligand-binding domain. The AR in an inactive form is localized in the cytoplasm. It becomes activated by binding to androgen hormones (testosterone, dihydrotestosterone, and others) that induce phosphorylation, homodimerization, and translocation of the AR into the nucleus (Figure 4.13). In the nucleus, the AR then interacts with a variety of coactivators (ARA-70, SRC1-NCOA, and SRC2-NCOA2) and corepressors (NCoR and SMRT) and binds to the androgen response element of the DNA. This triggers transcription of many of the genes responsible for prostate cell and prostate cancer cell proliferation, migration, and prostate-specific antigen (PSA) production. FoxA1 is a transcription factor that is also essential for the expression of AR-dependent proliferative genes.

Androgen hormones (testosterone, dihydrotestosterone, and others) are ligands that bind to and activate the androgen receptor. CYP17 (cytochrome P-450c, 17,20-α-hydroxylase/lyase) is a key enzyme responsible

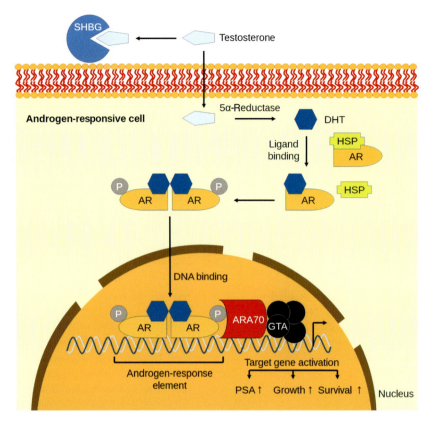

Fig. 4.13 Androgen receptor signaling. Testosterone circulates in the blood bound to albumin (not shown) and sex-hormone-binding globulin (SHBG) and exchanges with free testosterone. Free testosterone enters prostate cells and is converted to dihydrotestosterone (DHT). Binding of DHT to the androgen receptor (AR) induces dissociation from heat-shock proteins (HSPs) and receptor phosphorylation. The AR dimerizes and can bind to androgen-response elements in the promoter regions of target genes. Activation of target genes leads to biological responses, including growth, survival, and the production of prostate-specific antigen (PSA). GTA: GATA transcription factor. (Adapted from Ref. [4] with kind permission from Macmillan Publishers Ltd., Copyright 2001.)

for the synthesis of testosterone and other androgen intermediates (Figure 4.14) in the testis, adrenal glands, and prostate cancer cells.

Therapeutic suppression of androgen signaling can be accomplished by a variety of ways:

1) Surgical castration (orchiectomy)
2) Chemical castration (drugs that interfere with production of androgen hormones). These include the following:
   a) Drugs that interfere with the production of testicular testosterone by interrupting the hypothalamic–pituitary–testicular regulatory axis (e.g., LHRH agonists such as leuprolide and goserelin).
   b) Drugs that interfere with the key enzymes necessary for the synthesis of testosterone (e.g., abiraterone and ketoconazole).
3) Competitive antagonists of androgen receptor (e.g., enzalutamide, bicalutamide, and flutamide).

These hormonal therapeutic strategies, when used sequentially or in combination have provided significant

clinical benefit, including prolongation of life and palliation in patients with advanced prostate cancer. However, blockade of the androgen pathway eventually loses its effectiveness due to the development of resistance, when the disease is known as castration-resistant prostate cancer (CRPC). This later stage of prostate cancer is mediated by several mechanisms (Figure 4.15), which include the following:

a) Overexpression of the AR. This is observed in approximately 60% of specimens of metastatic CRPC.
b) Mutations in the AR are rare in primary prostate tumors, but can develop in metastatic CRPC. These mutations allow activation of AR by ligands that are normally not involved in AR stimulation (e.g., glucocorticosteroids, spironolactone, eplerenone, and anti-androgens such as flutamide and bicalutamide). This is sometimes referred to as the "promiscuous pathway" of AR activation. Constitutively active AR splice variants are also capable of enhancing the expression

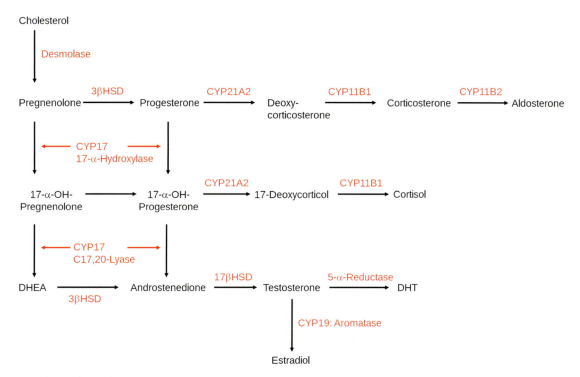

Fig. 4.14   Pathway of steroidogenesis.

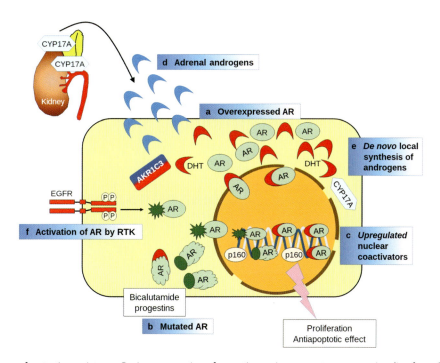

Fig. 4.15   Mechanisms of castration resistance. During progression of castration-resistant prostate cancer, signaling from the androgen receptor (AR) is maintained or upregulated by (a) increased quantity of AR (e.g., overexpression of AR), (b) changes in the structure of AR (e.g., mutations of AR), (c) alterations in coregulators of AR, and (d) continuous production of androgens by the adrenal glands and/or the prostate cancer itself. EGFR: epidermal growth factor receptor; RTK: receptor tyrosine kinase; CYP17A: cytochrome P17A; DHT: dihydrotestosterone; AKR: aldoketoreductase; p160: SRC family of AR coactivators. (Adapted from Ref. [5] with kind permission from Macmillan Publishers Ltd., Copyright 2010.)

of AR-dependent genes in a ligand-independent fashion. Recently, it has been reported that metastatic tumors expressing a truncated AR variant (ARV7) lacking the ligand binding site are resistant to therapy with androgen receptor antagonist enzalutamide.

c) Upregulation of AR coactivators (p160-SRC family).
d) Synthesis of androgens and testosterone in organs other than the testicles (e.g., adrenal glands).
e) *De novo* synthesis of testosterone and androgens intratumorally. Upregulation of CYP17 (C17-20-α-hydroxylase/lyase) enzyme activity plays a critical role in this mechanism (Figure 4.14). Mutations and upregulation of other enzymes involved in the synthesis of androgens (e.g., 3βHSD1) have also been proposed.
f) Activation of AR by pathways able to phosphorylate AR independent of the ligand-binding site. This includes receptor tyrosine kinases, and the PI3K-AKT, MAPK, RAS-RAF-MEK, SRC-kinases, Her2 or c-Met pathways. This is sometimes referred to as the "outlaw pathways."

9. *ETS (E26 transformation-specific) gene family: ERG* (21q22.3) and *ETV1* (7p21.2) gene rearrangements typically involve the fusion of the promoter of an androgen-responsive gene, such as TMPRSS2 (21q22.2), to a gene encoding the ETS transcription factor.

These rearrangements appear to be present in approximately 50–60% of all prostate cancers, including metastases, making them the most common genetic translocations associated with any solid malignancy. These fusions typically result in the androgen-driven overexpression of ETS transcription factors, such as ERG or ETV1. The predominant ETS fusion, TMPRSS2:ERG, results in a 20–120-fold overexpression of ERG protein. Genetically engineered mice expressing ERG or ETV1 under androgen regulation exhibit PIN-like lesions that signify early stages of prostate cancer carcinogenesis.

In preclinical models, prostate cancer cells are "addicted" to these specific alterations, meaning that knockdown of the specific oncogene results in abrogation of the malignant phenotype. Conversely, overexpression of the oncogene reconstitutes the malignant features of the cell.

ETS proteins are active transcription factors that drive cellular invasion through the induction of a transcriptional program highly enriched for invasion-associated genes.

Clinically, ERG-positive high-grade PIN is frequently found in proximity to ERG-rearranged cancers. These data suggest that *ETS* gene fusions are an early driving event in prostate carcinogenesis. Additionally, high levels of ERG and ETV1 gene expression continue to be maintained in advanced or metastatic disease. One study has suggested that metastases may be relatively enriched for ERG positivity; the authors found that in a cohort of prostate cancer patients with multifocal disease containing both ERG-positive and ERG-negative foci, the positive foci have a greater predilection for metastases. Together, these data suggest that ETS fusions are addicting oncogenes that contribute to both carcinogenesis and progression and, therefore, represent a promising target for oncologic therapy in prostate cancer.

Recent evidence indicates that AR signaling leads to site-specific DNA double-stranded breaks at the androgen response elements (AREs), which, in the presence of DNA repair enzyme PARP-1, may lead to rearrangements responsible for the generation of fusion genes. Increased PARP-1 enzymatic activity has also been associated with AR transcriptional function and the maintenance of the castration-resistant phenotype.

A comparatively novel approach for the treatment of prostate cancer is a cell-based immunotherapy. Sipuleucel-T (trade name Provenge) involves triggering of the patient's immune system against tumor cells by presenting a tumor antigen on dendritic cells (Box 5.2 in Section 5.2)

4.7
Molecular Biology of Hematological Malignancies

(Joseph C. Alvarnas, City of Hope, Duarte, CA, USA)

The hematological malignancies are a diverse set of cancers that originate in cells derived from the blood, bone marrow, or lymph nodes. The English physician Thomas Hodgkin was the first to describe a hematological malignancy, when in 1832 he recognized a proliferative disorder of the lymph nodes, liver, and spleen that came to be known as Hodgkin lymphoma. This was followed in 1845 by Rudolf Virchow's initial description of leukemia. The hematological malignancies are relatively rare and account for only 9.4% of all new cancer diagnoses in the United States. More than 150 000 people are annually diagnosed with a blood-related cancer. In its classification system for hematological malignancies, the World Health Organization (WHO) differentiates more than 120 distinct disorders. Table 4.3 details the projected 2014 US incidence for the most common hematological malignancies.

Hematological malignancies are challenging to treat because many of them are disseminated at the time of diagnosis and the associated tumor burden is massive:

Table 4.3 Projected 2014 US incidence of the major hematological malignancies.

Hematological malignancy	Number of new diagnoses
Non-Hodgkin lymphoma	70 800
Hodgkin lymphoma	9190
Total leukemia diagnoses	52 380
Acute lymphoblastic leukemia	6020
Acute myelogenous leukemia	18 860
Chronic myelogenous leukemia	5980
Chronic lymphocytic leukemia	15 720
Multiple myeloma	24 050

Data taken from Ref. [6].

Patients with acute leukemia may have 1 trillion cancer cells in their bodies at the time of diagnosis. Over the past 50 years, the prognosis for patients diagnosed with blood-related cancers has improved profoundly. For example, Surveillance, Epidemiology and End Results Program (SEER) data indicate that between 1975 and 2011, the 5-year survival for acute lymphoblastic leukemia (ALL) has risen from 40.6 to 70.0%.

The hematological malignancies are a bellwether for the importance of molecular medicine in effective diagnosis, risk stratification, pharmacogenomic testing, detection of minimal residual disease, development of targeted therapeutics, and the use of risk-adapted therapies in the care of cancer patients. These advances in molecular biology have helped to upend the idea that the key to curing blood cancers always requires the "bigger hammer" of more intensive therapies. Instead, patients may receive therapy whose intensity is based upon a rational risk assessment.

4.7.1
The Importance of Cytogenetics in Diagnosis and Treatment Decision-Making

Chronic myelogenous leukemia (CML) was the first malignancy for which a clonal chromosomal abnormality was described. CML is a lethal disorder in which patients develop a high white blood cell (WBC) count, and where the majority of circulating WBCs are normal-appearing neutrophils. Patients may also develop increased numbers of circulating basophils and marked enlargement of the spleen. The median survival for untreated patients is 3 years. Eventually, patients progress to an accelerated phase and ultimately toward what is known as blast crisis, in which their disease resembles an aggressive, refractory form of acute myelogenous leukemia (AML).

Prior to 1959, most cancer-associated chromosomal abnormalities were thought to be incidental. In 1959,

Peter Nowell and David Hungerford described a consistent chromosomal abnormality in patients with CML. They named their finding after the city where they made their discovery: the Philadelphia chromosome. They originally postulated that the Philadelphia chromosome involved a partial deletion of chromosome 22.

By the early 1970s, staining of chromosome bands allowed for a greater understanding of the nature of the Philadelphia chromosome. Janet Rowley was able to utilize chromosomal stains to more successfully identify structural changes in CML. She was ultimately able to identify that the Philadelphia chromosome involved a balanced, reciprocal translocation between chromosomes 9 and 22 (Figure 4.16a). This translocation is designated t(9;22)(q34;q11.2). As patients progress to the accelerated phase or blast crisis stages of their disease, they may acquire additional clonal karyotypic abnormalities (Figure 4.16b).

In subsequent years, Dr. Rowley's laboratory described the involvement of the Philadelphia chromosome in ALL, the presence of the t(8;21)(q22;q22) translocation in selected patients with AML, and the t(15;17)(q22;q12) translocation in patients with acute promyelocytic leukemia (APL). More than a hundred chromosomal and subchromosomal abnormalities have now been described in blood-related cancers. Rigorous analyses of these translocations have helped to advance the idea that the hematological malignancies are not clinically identical, monomorphic disorders. What we perceive as AML, for instance, may represent dozens of biologically distinct disorders, rather than one.

In a large evaluation of 1612 patients with AML performed by the Medical Research Council of the United Kingdom (MRC), the investigators found that cytogenetic studies performed at the time of diagnosis were the most robust predictor of patient outcomes after treatment. The estimated overall 5-year survival for the entire group of patients was 44%. The investigators found that patients could be divided into three prognostic groups based upon karyotype. Patients with favorable risk disease had a 5-year estimated OS that ranged from 61 to 69%, those with intermediate risk ranged from 23 to 60%, and those with unfavorable risk ranged only from 4 to 21%. Table 4.4 details the karyotypic abnormalities that differentiate low, intermediate, and unfavorable risk patients with AML.

Cytogenetic information now forms an essential part of risk stratification for many hematological malignancies. In ALL, patients may also be segregated in prognostic groups based upon karyotype at the time of diagnosis. Patients with hyperdiploidy and the t(12:21) (p13;q22) translocation have a favorable prognosis, while those with hypodiploidy, t(9;22)(q34;q11.2), t(4;11)(q21;

(a)   Karyotype 46,XY,t(9;22)(q34.1;q11.2)   (b)   Karyotype 54,XY,+Y,+5,+6,+8,+8,t(9;22)
(q34.1;q11.2),+15,+19,+der(22)t(9;22)

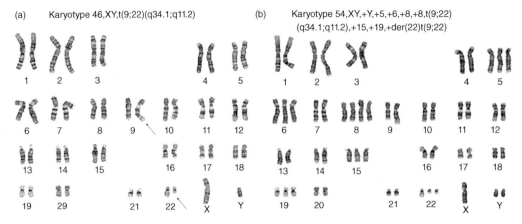

Fig. 4.16   Karyotypes of patients with chronic myelogenous leukemia (CML). (a) Karyotypes for a patient with newly diagnosed CML in chronic phase. This demonstrates the Philadelphia chromosome as only a cytogenetic abnormality. (b) Karyotypes for patient with CML that has progressed to the blast crisis stage of his disease. The patient now demonstrates hyperdiploidy with multiple additional acquired karyotypic abnormalities. (Images courtesy of Dr. Joyce Murata-Collins, City of Hope.)

q23) translocations and complex karyotypic abnormalities (more than four) have a very poor prognosis. These findings have helped, in part, to explain some of the age-related differences in ALL survival outcomes. The estimated 5-year OS for patients less than 45 years of age is 77.8%, while it is only 12.2% for patients 65 years of age or older. Favorable cytogenetic abnormalities are found principally in younger patients, while the more unfavorable cytogenetic abnormalities are found principally in older patients.

Karyotypic data may also provide important prognostic information for patients with non-Hodgkin lymphoma (NHL). Mantle cell lymphoma (MCL) is a very aggressive

Table 4.4   Prognostic karyotypic abnormalities in acute myelogenous leukemia.

Risk group	Clonal karyotypic findings
Favorable risk	t(15;17)
	t(8;21)
	inv(16)
Intermediate risk	Normal cytogenetics
	t(9;11)
	Trisomy 8
	Other clonal abnormalities that are not favorable or poor risk
Poor risk	Complex karyotypic changes with three or more clonal abnormalities
	Monosomy 5 or 7
	5q- or 7q-
	11q23 associated translocations
	t(9;22)
	t(6;9)

Modified after Ref. [7].

form of NHL that is often characterized by extranodal involvement by malignant cells in sites that may include the blood, bone marrow, central nervous system, spleen, and gastrointestinal tract. MCL is often refractory to standard chemotherapeutic regimens used to treat patients with aggressive forms of NHL. The t(11;14)(q13;q32) translocation is almost invariably present.

Burkitt's lymphoma is characterized by its aggressive clinical behavior, which includes the rapid growth of malignant cells, frequent extranodal involvement, and bulky tumor masses. Patients with BL have characteristic karyotypic abnormalities that include 8q14 as a partner in translocations involving 14q32, 22q11, or 2p12, respectively.

Chronic lymphocytic leukemia (CLL) is a form of leukemia characterized by elevated numbers of well-differentiated small lymphocytes. Patients with CLL have a median survival greater than a decade; many patients do not require treatment for many years. Deletion of the short arm of chromosome 17 (−17p) and deletion of the long arm of chromosome 11 (−11q) are associated with an inferior CLL prognosis. Patients with −17p may have a median survival of less than 3 years, and are often refractory to fludarabine-based chemotherapy (the typical initial treatment for this disease). Those with the −11q abnormality have disease that is more likely to exhibit extensive lymph node involvement at the time of diagnosis; these patients also have a lower response rate to chemotherapy. Patients with an isolated deletion of 13q14, however, have a more favorable prognosis.

In the past, the diagnosis and classification of hematological malignancies was based principally upon morphological studies. The availability of cytogenetic data has upended this convention and helped to achieve an unprecedented degree of specificity in diagnosis and prognostic prediction. The WHO classification system

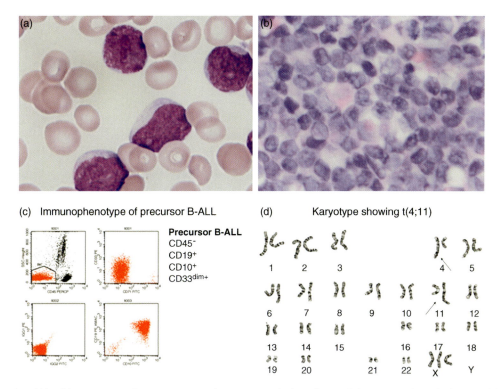

Fig. 4.17 Peripheral blood, bone marrow, flow cytometry, and cytogenetics findings for an adult patient with newly diagnosed, high-risk acute lymphoblastic leukemia (ALL). (a) Peripheral blood smear from a patient with newly diagnosed leukemia. The majority of the white blood cells are blasts that demonstrate a high nuclear: cytoplasmic ratio, an absence of cytoplasmic granules, and rare nucleoli in the blasts. (b) Bone marrow aspirate from this patient. This demonstrates complete replacement of the bone marrow by undifferentiated blasts. By morphology alone it is impossible to distinguish whether this patient has AML or ALL. (c) Flow cytometry analysis of blast surface proteins. The blasts express CD10, CD19, and show dim expression of CD33. This is consistent with the diagnosis of pre-B-cell ALL. (d) Karyotype for this patient that demonstrates t(4;11)(q21;q23). This is an adverse risk cytogenetic abnormality. The patient's final diagnosis is high-risk, pre-B-cell ALL. This patient will likely require allogeneic bone marrow or blood stem cell transplantation for management of his leukemia. (Figure 4.17a and b courtesy of Dr. Karl Gaal, City of Hope. Figure 4.17c and d courtesy of Dr. Joyce Murata-Collins, City of Hope.)

for hematological malignancies includes cytogenetic information as part of the diagnostic criteria for a number of diseases. Figure 4.17 illustrates the peripheral blood, bone marrow, flow cytometry, and cytogenetics findings for an adult patient with newly diagnosed, high-risk ALL.

4.7.2
Recognition of a Genetic Basis for the Hematological Malignancies

The karyotype of the malignant cells provides only a crude indication of what is happening on a molecular level. Each chromosomal band consists of approximately 5 million base pairs. Further understanding of the molecular basis of hematological malignancies requires an understanding of altered gene expression that occurs following nonrandom translocations and deletions.

Following the identification of the Philadelphia chromosome in CML, further investigation by Janet Rowley's group and others led to the finding that this abnormality was not an incidental marker, but was directly related to the underlying molecular biology of the disorder. Rowley's group first identified that the Philadelphia chromosome represented a balanced translocation between chromosomes 9 and 22. Further experiments demonstrated that this translocation leads to the rearrangement of the Abelson proto-oncogene (*abl1*) from chromosome 9 to what is called the breakpoint cluster region (*bcr*) on chromosome 22. The *bcr* cluster on chromosome 22 consists of a 5.8 kb region that includes multiple exons. Most of the gene rearrangements in CML occur within what is called the major breakpoint cluster region (M-BCR); this consists of exons 12 through 16. Rearrangement of the *bcr-abl1* gene at this region results in a novel 8 kb mRNA transcript that encodes a 210 kD fusion protein referred to as p210. In some variants of CML, the rearrangement may occur at exons 17 through 20; this results in a longer fusion gene product that produces the p230 fusion protein. In Philadelphia chromosome-positive ALL, the *bcr-abl1* rearrangement most often occurs in exons 1 or 2, leading

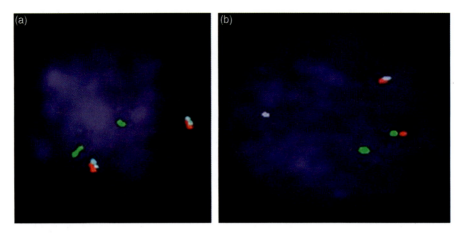

Fig. 4.18   Fluorescence *in situ* hybridization (FISH) for a normal and chronic myelogenous leukemia (CML) cell. (a) Assay of an interphase nucleus from a normal cell. The red-marked FISH probe marks the position of *ABL1*, which is found on chromosome 9q34.1. The blue-marked FISH probe demonstrates the position of *ASS1*, which is normally found adjacent to *ABL1* on chromosome 9q34. The green-marked probe demonstrates the position of the BCR region, which is found on chromosome 22q11.2. In this normal patient, the red and green probes remain separate and the red probe remains adjacent to the blue probe. (b) FISH assay for an interphase nucleus from a patient with CML. The red and green probes are now adjacent to each other, demonstrating the presence of t(9;22)(q34.1;q11.2) in this cell. (Images courtesy of Dr. Joyce Murata-Collins, City of Hope.)

to the expression of a relatively truncated 190 kD fusion protein, *p190*. This molecular rearrangement resulting in the *bcr-abl1* fusion gene can be seen in fluorescence *in situ* hybridization (FISH) studies. These studies demonstrate the abnormal clustering of the *bcr* and *abl* genes on chromosome 22 (Figure 4.18 and Section 8.2.2). The fusion gene may also be detected through Southern blot analysis and by reverse-transcriptase polymerase chain reaction (RT-PCR).

The protein encoded by the *bcr-abl1* fusion gene is a potent TK. Expression of this protein in mice that have been genetically engineered to express *bcr-abl1* results in the mice developing lethal leukemias of myeloid, T-cell, and B-cell origin. The bcr-abl1 fusion protein phosphorylates multiple intracellular proteins; this results in increased cellular proliferation and inhibition of apoptosis. Thus, the molecular changes produced by the appearance of the Philadelphia chromosome are the mechanism of the malignant transformation in CML. This recognition has been heralded by many as the "Rosetta Stone" for understanding the molecular basis for blood-derived cancers.

CML is only one of a number of hematological malignancies for which a clonal karyotypic abnormality has translated into an understanding of the molecular basis of the resulting disease. Acute promyelocytic leukemia is characterized by the t(15;17)(q22;q12) translocation. Patients with APL often present with a very high WBC count characterized by the presence of hypergranular immature cells, or blasts. They may also develop severe disorders of coagulation that can result in catastrophic episodes of hemorrhage following treatment with

standard chemotherapy. On a molecular basis, the t(15;17)(q22;q12) translocation results in the fusion of the *PML* gene, derived from chromosome 15q22, with the retinoic acid receptor alpha (*RARA*) gene expressed at 17q12. The resulting *PML/RARA* fusion gene leads to the expression of an aberrant retinoic acid receptor alpha. Retinoic acid is involved in many tissue differentiation pathways; the retinoic acid receptor alpha is expressed differentially in tissues of hematopoietic origin. The dysfunction in retinoic acid signaling ultimately results in a functional maturational arrest of myeloid cells at the level of the promyelocyte.

Mantle cell lymphoma is an aggressive variant of NHL that is characterized by the t(11;14)(q13;q32) translocation. This translocation results in apposition of the immunoglobulin heavy chain gene found at 14q32 and the *Cyclin D1* gene found at 11q13. The translocation promotes overexpression of the *Cyclin D1* gene. The resulting kinase causes increased phosphorylation of proliferation-related intracellular proteins and stimulates entry of the cell into the S-phase (synthesis) portion of the cell cycle. Additional genetic changes acquired in *MCL* may also lead to inhibition of apoptotic cell death.

The *mixed lineage leukemia (MLL)* gene is found on chromosome 11q23 and is a partner in several chromosome translocations that have been identified in patients with poor risk AML and ALL (Table 4.4). The *MLL* gene encodes a histone methyltransferase that enhances gene expression; this likely plays a role in the epigenetic control of cell growth and development.

In some leukemia subtypes that are associated with rearrangements involving *MLL*, there is also an increase in *FLT3* expression. *FLT3*, or the FMS-like tyrosine kinase 3, is a cell surface receptor tyrosine kinase that plays a role in cell proliferation and survival. AML associated with the overexpression of *FLT3* (including those with internal tandem duplications of this gene but absent 11q23-related translocations) have a significantly worse prognosis.

For patients whose malignancies, such as those with adverse-risk CLL, are associated with the loss of the short arm of chromosome 17 (−17p), the loss of endogenous tumor suppressor activity is an essential part of the molecular biology of the disease. The *p53* gene is a potent tumor suppressor that is found at 17p13.1. This gene plays a significant role in DNA repair and in apoptosis. Loss of *p53* expression is associated with a markedly increased risk of developing cancer. A number of hematological malignancies, such as CLL and MCL, may be associated with loss of *p53* gene expression. The resulting malignancies are also associated with a poor prognosis.

Hematological malignancy-related chromosomal translocations and deletions may be associated with expression of novel tyrosine kinases, aberrant cell surface receptors, abnormal expression of transcription regulators, and loss of tumor suppressor gene activity. An understanding of these molecular mechanisms for blood-derived cancers has helped direct efforts toward the development of rational molecular therapeutics.

4.7.3
Targeted Therapeutics for Hematological Malignancies

CML was the first disease for which a molecularly targeted therapeutic agent became commercially available. As noted above, the protein encoded by the *bcr-abl1* fusion gene is a TK that phosphorylates multiple intracellular proteins. This leads to abnormal cellular proliferation and inhibition of apoptosis. Imatinib mesylate (marketed by Novartis as Gleevec in Canada, South Africa, and the United States or Glivec in Australia, Europe, and Latin America) is a TKI that was designed to bind to the ATP binding pocket of the fusion protein. This prevents ATP binding, which subsequently results in a blockage of phosphorylation activity. This results in cessation of the downstream proliferative and antiapoptotic effects that are ordinarily induced by the fusion protein.

The initial clinical experience with imatinib was both astonishing and unprecedented. In a landmark trial, 1106 patients with newly diagnosed CML were randomized to receive either imatinib (553 patients) or the then-prevailing standard therapy that consisted of IFN-α combined with cytarabine (553 patients). The patients were evaluated for both response to therapy and progression

of the disease to accelerated phase or blast crisis. Patients who responded were assessed for hematological response rates (normalization of blood counts) and cytogenetic response rates. Patients in the standard treatment group were allowed to cross over to the imatinib group if they failed to respond or had progression of their disease. The patients were followed for a median of 60 months. By 60 months, 87% of patients treated with imatinib achieved a complete cytogenetic response, and the overall survival for patients who were treated with imatinib was 89%. Patients who had complete cytogenetic responses and had a three-log suppression of their *bcr-abl1* mRNA demonstrated superior survival.

Prior to this trial, the standard of care for management of patients with CML was to identify those patients for whom an allogeneic bone marrow or blood stem cell transplant (in which a sibling or unrelated donor was the stem cell source) was feasible, and proceed to transplant as rapidly as possible. While allogeneic transplant could cure a majority of these patients, treatment-related mortality ranged from 10 to 20% and a significant portion of patients were affected by long-term immunological complications of their treatment. Under this new treatment paradigm, transplant could be reasonably deferred in those responding to imatinib.

Through the use of quantitative RT-PCR, it is possible to assess a patient's tumor burden throughout the course of treatment. The standard of care for managing patients with CML now includes assessment of treatment response every 3 months using quantitative RT-PCR. The goal of therapy is to achieve *bcr-abl1* transcript suppression that is >3 logs below baseline. Patients who fail to respond to quantitative benchmarks for the suppression of *bcr-abl1* transcripts require reassessment and consideration of switching to an alternative TKI.

Therapeutic switching between TKI agents may be guided by a molecular analysis for the presence of resistance mutations. These resistance mutations are frequently point mutations with the *bcr-abl1* fusion gene that result in altered binding of the TKI. There are now five commercially available TKIs for patients with CML.

In APL, all-trans retinoic acid (ATRA) has proven to be an effective targeted agent. ATRA binds to the aberrant *PML/RARA* receptor and induces differentiation of the leukemic cells. While APL was once one of the most dreaded and lethal forms of acute leukemia, ATRA when combined with chemotherapy may produce remission rates of 90% or greater in patients with APL. Patients treated for APL also benefit from molecular monitoring of their response. Patient responses can be assessed by FISH for t(15;17) and quantitative RT-PCR for the *PML/RARA* fusion transcripts. Following consolidation therapy (additional chemotherapy administered after a patient achieves a remission) with ATRA, chemotherapy, and/or arsenic trioxide, patients with APL may achieve cure rates of nearly 80%.

Ibrutinib (trade name Imbruvica) was recently approved for the treatment of B-cell malignancies, including MCL and CLL. Ibrutinib is an inhibitor of the Bruton tyrosine kinase (BTK). It prevents phosphorylation of multiple intracellular targets of BTK, which results in decreased cell proliferation and the recovery of apoptotic function.

Scientists continue to evaluate hematological malignancies for molecular targets that can be leveraged in the development of novel therapeutics. This represents the antithesis of the one-size-fits-all approach to cancer care. The growing body of targeted molecular agents are the harbingers of a future of tailored therapeutics that are specifically suited toward a vast array of molecular subtypes of hematological malignancies. While slow and laborious, this approach has largely upended our long-standing paradigm for the management of hematological malignancies.

4.7.4
Risk-Adapted Therapies

For many newly diagnosed patients who are suffering from hematological malignancies, data gleaned from cytogenetics and molecular diagnostic studies may play an essential role in risk stratification and the application of risk-adapted therapeutic approaches. Patients with favorable risk ALL and AML may be appropriate candidates for conventional dose chemotherapy followed by appropriate monitoring. Patients with CML and APL may be treated with targeted therapeutic agents and followed for the effectiveness of their treatment with molecular monitoring of their malignancy-related markers. Patients with adverse-risk disease may proceed with more intensive therapeutic approaches, including blood or marrow transplantation, without unnecessary delay. By applying risk-adapted strategies to the care of patients with blood-related cancers, we have seen marked improvements in overall survival over the past four decades.

4.7.5
Epigenetics and Hematological Malignancies

In this digital age, it is tempting to conceive of the molecular biology of hematological malignancies as a binary phenomenon, that is, the presence or absence of a particular oncogene translates into the presence or absence of a cancer. While a vast array of 0s and 1s (offs and ons) underpin much of the digital computer science innovation of the past 50 years, human biology does not follow this model. The cell is best thought of as a symphony orchestra with multiple musicians acting in intricate coordination. Each is an independent agent whose aggregate activity produces a complex result. Epigenetic control of gene expression (Section 1.3.4) may allow some genes or networks of genes to be silenced, underexpressed, or overexpressed. This may involve methylation, resulting in the silencing of some genes, or the induced overexpression of others. This may be especially true for some patients with AML or one of the myelodysplastic syndromes (MDS) (syndromes of ineffective blood cell maturation that may devolve into acute leukemia). Epigenetic control of cancer is complex and its impact extends over a network of genes, rather than reflecting the hyper- or hypoactivity of a single gene. Mutations in genes that are responsible for DNA methylation (*DNMT3A*) result in gene silencing. Mutation in genes responsible for the removal of methyl groups (*TET1-2*) may result in enhanced gene expression. Mutations in genes responsible for the direct or indirect modification of histone proteins may allow them to block transcription (*IDH1* and *IDH2*). The aggregate effect of these alterations in gene expression may result in the development of AML or MDS. Epigenetic chemotherapeutic agents, such as azacitadine and decitabine, may mitigate the aggregate effects of these disparate molecular mechanisms and result in enhanced differentiation and/or apoptosis of abnormal clonal cells in patients with AML and MDS. Epigenetic control of hematological malignancies is an area of current intense investigation.

4.7.6
The Unknown Unknowns – The Future of Molecular Oncology

A recent American Secretary of Defense opined that the nature of knowledge and knowing falls into three categories: the known knowns (the things that we truly know), the known unknowns (the things that we know that we do not know), and the unknown unknowns (those things that we think that we know, but truly do not understand). As much as our understanding of hematological malignancies has advanced over the past 20 years, the known unknowns and unknown unknowns far outweigh our accumulated knowledge.

Data obtained from the sequencing of the human genome and the recent characterization of the human proteome suggest that we have only begun to scratch the surface of what is knowable about the molecular biology of the hematological malignancies. The human genome consists of approximately 20 000–25 000 protein-coding genes. Given the added complexities of epigenetic control of gene expression, we finally have, for the first time, a sense of the scope of the question – which genes or networks of genes result in malignant

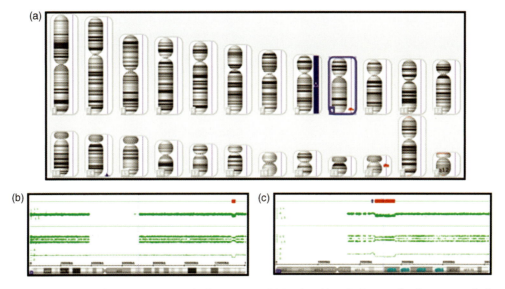

Fig. 4.19 Pictorial representation of SNP microarray results from genomic DNA isolated from leukemic cells of a patient with chronic myelogenous leukemia (CML). (a) The karyoview displays a genome-wide summary of the aberrations present. The dark blue bar adjacent to chromosome 8 indicates an increase in copy number for the entire chromosome – in this case demonstrating the presence of Trisomy 8. Chromosome 9 is enclosed in a blue box, identifying it as a chromosome of interest. A small red bar adjacent to chromosome 9 indicates loss of a chromosomal region on chromosome 9, corresponding to concurrent deletion of the ASS1 gene locus in 9q34, a larger red bar adjacent to chromosome 22 indicates a larger region of loss (deletion) in 22q11.2. (b) A detailed view of chromosome 9 identifies the breakpoints of the 9q34 deletion that includes the ASS1 gene locus, commonly deleted in a subset of CML patients. (c) A detailed view of chromosome 22 identifies a larger deletion of 22q11.2, adjacent to the BCR region involved in the BCR/ABL1 fusion. (Images courtesy of Dr. Joyce Murata-Collins, City of Hope.)

cellular transformation; what proteins are relevant to that process; and how does modulation of gene expression impact the end result.

These questions have provoked a broader examination of the genome and transcriptome for further clues regarding diagnosis and treatment of the hematological malignancies. Single-nucleotide polymorphisms (SNPs) are single nucleotide changes in the DNA of a single gene. They may help to identify genes or gene networks that differ between patients with similar cancers, and help to refine therapeutic decision-making. A comparison of SNPs evaluated from genomic DNA in patients with ALL demonstrate which genes that affect the metabolism of drugs such as methotrexate can effect therapy-related toxicity and the likelihood of achieving a cure. In time, such data may be used to more effectively tailor the dose of selected chemotherapeutic agents in patients (Chapter 9). A number of investigators are exploring the use of SNPs as a means of more effective risk stratification in hematological malignancies.

Microarray analysis (Section 2.1) of cellular DNA or mRNA from leukemia or lymphoma cells can also provide a robust indication of the presence and activity of selected genes and networks of genes relevant to the molecular biology of a hematological malignancy. Gene microarrays may contain more than 30 000 genes. Microarray analysis of DNA derived from leukemia and lymphoma cells may be able to identify gene rearrangements, gene duplications, and chromosomal abnormalities that may not be detectable through the use of conventional cytogenetic studies (Figure 4.19).

In addition to acting as a supplement to standard cytogenetic studies, microarray-based analysis of tumor DNA or mRNA can help identify networks of genes that play an integral role in the molecular biology of a number of hematological malignancies. Identification of these gene networks can help to develop a genetic "fingerprint" that can help to achieve more effective diagnosis and direct appropriate therapy for patients with seemingly similar cancers.

References

1. Hanahan, D. and Weinberg, R.A. (2011) Hallmarks of cancer: the next generation. *Cell*, **144**, 646–674.
2. Platanias, L.C. (2005) Mechanisms of type-I- and type-II-interferon-mediated signalling. *Nat. Rev. Immunol.*, **5**, 375–386.
3. Llovet, J.M. and Bruix, J. (2008) Molecular targeted therapies in hepatocellular carcinoma. *Hepatology*, **48**, 1312–1327.
4. Feldman, B.J. and Feldman, D. (2001) The development of androgen-independent prostate cancer. *Nat. Rev. Cancer*, **1**, 34–45.
5. Seruga, B., Ocana, A., and Tannock, I.F. (2011) Drug resistance in metastatic castration-resistant prostate cancer. *Nat. Rev. Clin. Oncol.*, **8**, 12–23.

6. Leukemia and Lymphoma Society (2014) Facts Spring 2014 (www.LLS.org).
7. Grimwade, D., Walker, H., Oliver, F. *et al.* (1998) The importance of diagnostic cytogenetics on outcome in AML: analysis of 1,612 patients entered into the MRC AML 10 trial. The Medical Research Council Adult and Children's Leukaemia Working Parties. *Blood*, **92**, 2322–2333.

Further Reading

Hanahan, D. and Weinberg, R.A. (2011) Hallmarks of cancer: the next generation. *Cell*, **144**, 646–674.
James, N. (2011) *Cancer – A Very Short Introduction*, Oxford University Press, Oxford.

Breast Cancer

Clark, O., Botrel, T.E., Paladini, L. *et al.* (2014) Targeted therapy in triple-negative metastatic breast cancer: a systematic review and meta-analysis. *Core Evid.*, **9**, 1–11.
Kumler, I., Christiansen, O.G., and Nielsen, D.L. (2014) A systematic review of bevacizumab efficacy in breast cancer. *Cancer Treat. Rev.*, **40**, 960–973.
Kumler, I., Tuxen, M.K., and Nielsen, D.L. (2014) A systematic review of dual targeting in HER2-positive breast cancer. *Cancer Treat. Rev.*, **40**, 259–270.

Lung Cancer

Govindan, R., Ding, L., Griffith, M. *et al.* (2012) Genomic landscape of non-small cell lung cancer in smokers and never-smokers. *Cell*, **150**, 1121–1134.
Herbst, R.S., Heymach, J.V., and Lippman, S.M. (2008) Lung cancer. *N. Engl. J. Med.*, **359**, 1367–1380.
Lovly, C.M. and Shaw, A.T. (2014) Molecular pathways: resistance to kinase inhibitors and implications for therapeutic strategies. *Clin. Cancer Res.*, **20**, 2249–2256.

Hepatocellular Carcinoma

Galuppo, R., Ramaiah, D., Ponte, O.M. *et al.* (2014) Molecular therapies in hepatocellular carcinoma: what can we target? *Dig. Dis. Sci.*, **59**, 1688–1697.
Page, A.J., Cosgrove, D.C., Philosophe, B. *et al.* (2014) Hepatocellular carcinoma: diagnosis, management, and prognosis. *Surg. Oncol. Clin. N. Am.*, **23**, 289–311.
Villanueva, A., Hernandez-Gea, V., and Llovet, J.M. (2013) Medical therapies for hepatocellular carcinoma: a critical view of the evidence. *Nat. Rev. Gastroenterol. Hepatol.*, **10**, 34–42.

Colorectal Cancer

Fearon, E.R. (2011) Molecular genetics of colorectal cancer. *Annu. Rev. Pathol.*, **6**, 479–507.
Ogino, S., Chan, A.T., Fuchs, C.S. *et al.* (2011) Molecular pathological epidemiology of colorectal neoplasia: an emerging transdisciplinary and interdisciplinary field. *Gut*, **60**, 397–411.

Renal Cell Carcinoma

Pal, S.K., Hu, A., and Figlin, R.A. (2013) A new age for vaccine therapy in renal cell carcinoma. *Cancer J.*, **19**, 365–370.
Rini, B.I., Campbell, S.C., and Escudier, B. (2009) Renal cell carcinoma. *Lancet*, **373**, 1119–1132.

Prostate Cancer

Attard, G. and de Bono, J.S. (2011) Translating scientific advancement into clinical benefit for castration-resistant prostate cancer patients. *Clin. Cancer Res.*, **17**, 3867–3875.
Lorente, D. and De Bono, J.S. (2014) Molecular alterations and emerging targets in castration resistant prostate cancer. *Eur. J. Cancer*, **50**, 753–764.
Mazaris, E. and Tsiotras, A. (2013) Molecular pathways in prostate cancer. *Nephrourol. Mon.*, **5**, 792–800.

Hematological Malignancies

Alvarnas, J.C., Brown, P.A., Aoun, P. *et al.* (2012) Acute lymphoblastic leukemia. *J. Natl. Compr. Canc. Netw.*, **10**, 858–914.
Cortes, J. and Kantarjian, H. (2012) How I treat newly diagnosed chronic phase CML. *Blood*, **120**, 1390–1397.
Guzman, M.L. and Allan, J.N. (2014) Concise review: leukemia stem cells in personalized medicine. *Stem Cells*, **32**, 844–851.

Molecular Virology

5

Summary

- Viruses are a highly diverse class of pathogens. Their genetic material may consist of DNA or RNA, single- or double-stranded. Retroviruses carry an RNA genome that is converted into DNA in the infected cell. While some viruses consist only of genetic material and a protein capsid, others are surrounded by an additional membrane. The human immunodeficiency virus (HIV), the cause of acquired immunodeficiency syndrome (AIDS), the hepatitis B and C viruses, and influenza viruses are major killers worldwide.
- Vaccines can be used to confer immunity against viral infection. Different classes of vaccines include the use of attenuated live vaccines, inactivated non-living vaccines, subunit vaccines (recombinant proteins), DNA vaccines, and recombinant live vaccines.
- Several methods for the diagnosis of viral infection have been developed. These include microscopic techniques, pathogen cultivation, and immunological assays. While these assays are still useful in medical diagnosis, molecular techniques, such as quantitative PCR and modern sequencing technologies have gained importance in this field.
- In recent years, the repertory of antiviral drugs increased. Nucleoside analogs that inhibit viral polymerases are the dominant class of drugs used to treat viral infection. Additional antiviral agents can interfere with viruses at virtually any stage of their life cycle.
- More recently, misfolded infectious proteins have been discovered that cause and transmit disease (e.g., bovine spongiform encephalopathy (BSE), Creutzfeldt–Jakob disease (CJD)) in the absence of nucleic acids. These so-called prion proteins are a novel class of pathogen distinct from viruses, bacteria, or eukaryotic pathogens.

Contents List

The Basics of Virology

- Human Immunodeficiency Virus
- Hepatitis B Virus
- Influenza Virus

Vaccination

- Live Vaccines
- Recombinant Virus Vaccines
- Inactivated Virus Vaccines
- Subunit Vaccines
- DNA Vaccines
- HIV Vaccines

Detection of Viruses

- Cytopathic Effects
- Electron Microscopy
- Hemagglutination Assay
- Enzyme-Linked Immunosorbent Assay (ELISA)
- Indirect ELISA
- Polymerase Chain Reaction (PCR)
- Antiviral Susceptibility Testing

Antiviral Therapy

- Human Immunodeficiency Virus (HIV)
- Hepatitis C Virus
- Influenza Virus
- Other Viruses

Prions

Viruses are small infectious agents that can replicate only inside the cells of other organisms. They depend exclusively on their host's metabolism for their own metabolism. Viruses are sometimes considered a life form, since they carry genetic material, reproduce, and evolve through natural selection. However, several key

Molecular Medicine: An Introduction, First Edition. Jens Kurreck and Cy Aaron Stein.
© 2016 Wiley-VCH Verlag GmbH & Co. KGaA. Published 2016 by Wiley-VCH Verlag GmbH & Co. KGaA.

characteristics that define life are lacking, such as a cell structure or any kind of metabolic processes. The term "virus" is derived from the cognate Latin word meaning poison or noxious substance. Viruses are very diverse, with an estimated 100 million different types existing. However, this number includes phages that infect and replicate within bacteria. Fortunately, the number of viruses known to infect humans is much smaller, numbering about 200 species that belong to approximately two dozen virus families. The first human virus to be discovered was the yellow fever virus in 1901, but several new species are still being found every year.

In some cases, newly emerging viruses such as the severe acute respiratory syndrome (SARS) coronavirus or variants of well-known viruses (e.g., avian influenza viruses) have rapidly disseminated over large parts of the world and have raised concerns about possible future pandemics. The close proximity of humans and livestock in some regions of the world facilitates transmission of viruses from animals to humans, and extensive global traffic systems enable their rapid worldwide spread. The HIV pandemic, beginning in the early 1980s, has infected approximately 70 million people, half of whom have died of AIDS. A similar number of individuals are currently living with HIV. These facts demonstrate the destructive potential of previously unknown viruses.

In addition to HIV, which claims approximately 1.5 million lives every year, the hepatitis B and C viruses and the influenza viruses are the virus classes causing the highest numbers of annual deaths (Table 5.1). The discovery of the principle of vaccination in the late eighteenth century offered a powerful strategy for long-term protection from viral infections. Potent vaccines are available for many of the most dangerous viruses (with important exceptions such as HIV), as will be outlined in Section 5.2. Modern techniques of molecular biology, particularly real-time PCR, have permitted substantial progress in the specific diagnosis of viral infection (Section 5.3). However, the development of efficient and nontoxic antiviral agents still remains a challenge. Since viruses do not have a metabolism of their own, it is difficult to identify substances that specifically inhibit a virus without interfering with normal processes in the host. Nevertheless, the antiviral repertory has increased steadily in the last several decades (Section 5.4). Understanding the molecular mechanisms of a viral life cycle aids in the development of new vaccines and antiviral agents.

More recently, a new and extremely simple type of infectious agent was discovered, which is now considered a separate class of pathogen: Prions are infectious proteins that cause transmittable diseases in the absence

Table 5.1 Selected viruses with high clinical importance.

Virus		Annual number of deaths in 2010
Human immunodeficiency virus		1 465 400
Hepatitis B virus	Acute infection	132 200
	Liver cancer	341 400
	Cirrhosis	312 400
Influenza virus		507 900
Hepatitis C virus	Acute infection	16 000
	Liver cancer	195 700
	Cirrhosis	287 400
Respiratory syncytial virus		253 500
Rotavirus		250 900
Human papillomavirus	Cervical cancer	225 400
Measles virus		125 400
Hepatitis A virus		102 800
Hepatitis E virus		56 600
Rabies virus		26 400
Dengue virus		14 700
Varicella-zoster virus		6800

Deaths numbers based on Ref. [1].

of nucleic acids. The best known prion diseases are BSE and CJD, the human form of a prion disease. Prion diseases will be briefly introduced in Section 5.5.

5.1
The Basics of Virology

Viruses are small pathogens with sizes ranging from below 20 nm to approximately 450 nm. A virus particle (also known as virion) consists of genetic material (DNA or RNA) surrounded by a protein coat, the capsid (Figure 5.1). The unit composed of the viral genome and the capsid is also referred to as the nucleocapsid. In addition, some viruses are surrounded by a lipid bilayer, the envelope, which is derived from the plasma membrane of the host cell and contains various membrane proteins. Morphologically, viruses are very diverse (Figure 5.2). For example, poxviruses are brick shaped, while rabies viruses are bullet shaped. A large number of animal and human viruses (e.g., herpes-, adeno-, papilloma-, parvo-, rota, and poliovirus) are icosahedral, a shape that approximates a 20-sided "sphere."

Viral nucleic acids can be DNA or RNA. The viral genome can be circular, as in polyomaviruses, or linear, as in adenoviruses. It may be either single-stranded or double-stranded. A plus-stranded RNA genome indicates

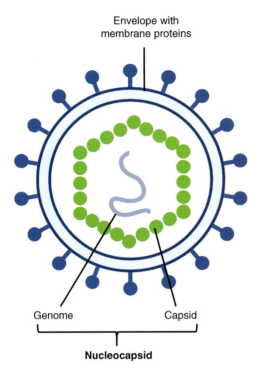

Envelope with
membrane proteins

Genome

Capsid

Nucleocapsid

Fig. 5.1 Basic structure of a virus. The genome of the virus is shielded by a capsid composed of proteins. Some viruses are enveloped by a lipid bilayer containing membrane proteins.

that the single-stranded genomic RNA has the same sense as the mRNA and can be immediately translated. In contrast, a minus-strand RNA is complementary to the mRNA and must first be converted to the positive-sense RNA before translation. The viral genome may be divided into separate parts, in which case it is termed a segmented genome. Table 5.2 summarizes the genomic organization and structural characteristics of the most important virus families.

Although viruses have many different life cycles, these cycles have several steps in common (Figure 5.3). Infection of a host cell starts with viral attachment to a cellular receptor. This interaction is a major determinant of viral host range and tissue tropism. For example, the glycoprotein gp120 of HIV interacts with the CD4 molecule present on the surface of T cells. Frequently, a coreceptor is required for virus entry into the host cell. Once inside the cell, the virus releases its nucleic acid, a process named uncoating. The genetic material must then be replicated. The details of this process depend on the organization of the viral genome. Although there are some exceptions, most DNA viruses replicate within the nucleus, while the majority of RNA viruses replicate within the cytoplasm. Retroviruses initially carry out a reverse transcription step to convert their RNA genome

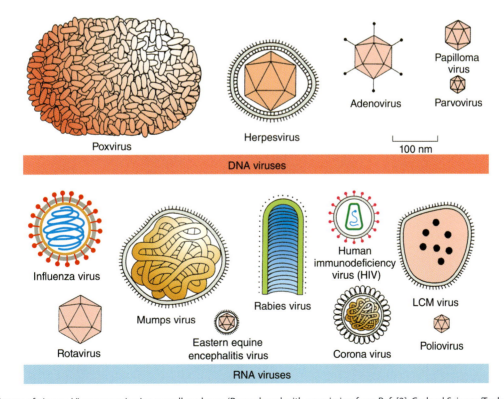

Fig. 5.2 Shapes of viruses. Viruses vary in size as well as shape. (Reproduced with permission from Ref. [2], Garland Science/Taylor & Francis LLC, Copyright 2008.)

Table 5.2 Genomic organization and structural characteristics of virus families.

Genetic material	Genome	Envelope	Families
DNA	Single-strand	Nonenveloped	Circoviridae, Parvoviridae
	Double-strand	Nonenveloped	Papovaviridae, Adenoviridae
		Enveloped	Herpesviridae, Poxviridae, Iridoviridae
	Single-/double-strand	Enveloped	Hepadnaviridae
RNA	Single-strand (plus)	Nonenveloped	Picornaviridae, Astroviridae, Caliciviridae
		Enveloped	Togaviridae, Flaviviridae, Coronaviridae, Retroviridae
	Single-strand (minus)	Enveloped	Paramyxoviridae, Rhabdoviridae, Filoviridae, Orthomyxoviridae
	Single-strand (ambisense)	Enveloped	Bunyaviridae, Arenaviridae
	Double-strand	Nonenveloped	Reoviridae, Birnaviridae

into DNA. While the RNA of plus-stranded viruses can be translated directly, other viruses must first generate mRNA. In addition to the replication of the genome, new virus proteins must also be synthesized. The newly synthesized virus genomes, along with the viral proteins, then assemble into new virus particles. Finally, the virus is released from the cell. While most nonenveloped viruses are liberated by cell lysis, enveloped viruses leave the cell by budding, and are surrounded by a cell membrane.

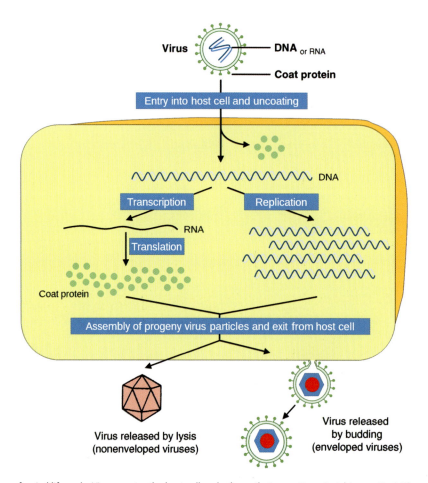

Fig. 5.3 General steps of a viral life cycle. Viruses enter the host cell and release their genetic material (uncoating). The genome is then replicated and viral proteins are synthesized by the cellular machinery. After the assembly of new virus particles, they exit the cell either by lysis or by budding. (Reproduced with permission from Ref. [2], Garland Science/Taylor & Francis LLC, Copyright 2008.)

Viral transmission can occur via several routes. These include oral–oral, fecal–oral, by blood, by a vector (e.g., mosquitoes), by animal bites, by skin-to-skin contact, or sexually. Some viruses remain locally at the site at which they first enter the body. Examples are the common cold and other respiratory viruses that remain in the lungs. Norovirus remains in the intestine and produces severe gastroenteritis. Other viruses spread from the initial site of infection into different regions of the body. These systemic infections may cause disease in distant organs. Examples include the hepatitis viruses, HIV, measles, and smallpox. The pathogenicity of a virus may strongly depend on the route of infection. For example, the rabies virus is not harmful if swallowed, but it inevitably causes severe disease if it reaches the tissues of an unvaccinated human.

An important feature of a virus is its *tropism*. The host tropism refers to the ability of a virus to infect a certain species. Here, we will exclusively deal with human viruses, but animal viruses such as foot-and-mouth disease causes profound problems in animal husbandry. The cell tropism of a virus denotes its ability to infect a certain cell type or cells of a certain organ. For example, the hepatitis B virus (HBV) specifically infects the liver.

Viruses may cause disease in different ways. They may be pathogenic by themselves and directly harm the host by destroying cells or interfering with their function. For example, polioviruses kill motor neurons in the spinal cord, and HIV kills CD4-positive T cells. However, for many viruses, the major symptoms of infection are caused by the induction of inflammation that damages the host's own cells, tissues, and organs. While most infections of the respiratory tract (e.g., the common cold and influenza) and systemic infections (e.g., measles) cause acute inflammation, infection with hepatitis B or C virus (HBV and HCV, respectively) may cause chronic inflammation. Viral liver infection stimulates a strong T-cell-mediated response that kills infected cells. In the acute phase, liver failure may result if too many liver cells are infected, and die as a result. If the virus is not successfully eliminated in the acute phase, the organismal immune response may continue at a low level over a long period of time. The inflammatory response may be accompanied by a healing process that leads to liver fibrosis. This, in turn, may cause cirrhosis, which can also lead to liver failure. In other patients, carcinoma of the liver may develop.

Tumor viruses were initially discovered in animals. In humans, papillomaviruses are the only viruses to date that have been proven to be tumorigenic, causing cancer of the cervix. The risk of developing cervical cancer can be greatly reduced by vaccination (Section 5.2.4). In addition, a close relationship between tumor formation and viral infection has been observed for the Epstein–Barr virus and Burkitt's lymphoma, hepatitis viruses and liver cancer as mentioned above, and human herpesvirus in Kaposi's sarcoma in patients with AIDS.

Viruses have developed many strategies for infection and replication and have numerous mechanisms of pathogenicity. Since it is beyond the scope of this textbook to discuss all types of viruses, three viruses with high clinical relevance will be described in more detail: HIV as a representative of the retrovirus family, HBV as a DNA virus, and influenza virus as an RNA virus.

5.1.1
Human Immunodeficiency Virus

AIDS was first clinically observed in the United States in 1981. Shortly thereafter, the HIV was independently recognized by the research groups of Robert Gallo and Luc Montagnier as a novel retrovirus that caused AIDS (although the virus was differently named at that time). Contradictory hypotheses exist about the origin of HIV, but the most plausible theory states that it originated in nonhuman primates in West-Central Africa, presumably through the evolution of the related simian immunodeficiency virus (SIV).

Two types of HIV have been characterized, denoted as HIV-1 and HIV-2. HIV-1 is more virulent and the cause of the majority of HIV infections globally. HIV-1 is divided into the M (main), N (non-M, non-O), and O (outlier) groups. In 2009, the putative group P, pending the identification of further human cases, was identified. HIV-1 group M is responsible for more than 90% of all infections. In is further subdivided into eight subgroups (or clades) that are geographically distinct. More than 35 million people have died from AIDS since the early 1980s, and the same number of people are currently living globally with HIV. The number of new infections continues to fall: According to UNIAIDS, the number of people acquiring HIV infection in 2013 was 2.1 million, ~38% lower than in 2001. The number of AIDS deaths is also decreasing with approximately 1.5 million deaths annually, down from the peak of 2.3 million in 2005. HIV is transmitted by three main routes: sexual contact, exposure to infected body fluids (e.g., blood), and from mother to child.

An infection with HIV usually proceeds through three stages; the acute infection, the clinical latency phase, and the development of clinical AIDS (Figure 5.4). During the initial acute phase, most individuals may have no significant symptoms, or may develop an influenza-like illness. HIV infection is, therefore, not often recognized at this stage, although it is at this time that HIV levels

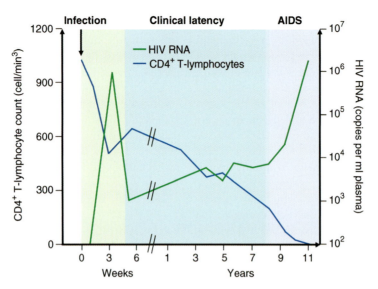

Fig. 5.4   Time course of HIV infection. The time course shows the viral load (as HIV RNA) and the CD4+ T-cell count [3].

peak in the bloodstream. The viral load then diminishes and the second stage, clinical latency, follows. This may last from a few years to over 20 years. During this time, most individuals remain asymptomatic, at least initially. HIV levels remain very low to nondetectable, but HIV virions are constantly produced and the CD4+ cell count begins to decline. There are various mechanisms by which HIV infection leads to low levels of CD4+ T cells, including the direct killing of infected cells, the killing of infected CD4+ cells by CD8 cytotoxic lymphocytes that recognize infected cells, and the apoptosis of uninfected bystander cells. CD4+ cell counts below 200 cells/µl in conjunction with the occurrence of a specific disease in association with HIV infection define the clinical manifestation of AIDS. Without treatment, the HIV load will increase while the number of CD4+ cells will continue to decline. At this stage, the patient's immune system is no longer capable of proper functioning and the patient acquires secondary infections, cancers (e.g., Kaposi's sarcoma), or other manifestations of HIV infection (e.g., neuropathy).

HIV-1 virions carry two copies of a positive single-stranded RNA, which are bound to p7, a nucleocapsid protein. The RNA encodes enzymes required by the virus, such as reverse transcriptase, a protease, and an integrase (Figure 5.5a). The conical capsid is composed of ~2000 copies of the viral p24 protein. The capsid is surrounded by a membrane that contains the envelope protein (env). This consists of a cap composed of three molecules of glycoprotein 120 (gp120) and three gp41 molecules that anchor env into the viral envelope. A layer of the p17 matrix protein is attached to the inside of the membrane.

The HIV-1 genome (Figure 5.5b) encodes nine genes that can be classified into three functional groups. These are the structural (*gag, env, pol*), regulatory (*tat, rev*), and accessory genes (*vpu, vpr, vif, nef*). The group-specific antigen (*gag*) codes for the p17 matrix protein, the p24 capsid protein, and the p7 nucleocapsid protein; *pol* encodes the reverse transcriptase, the protease, and the integrase; and *env* codes for the envelope protein that is synthesized as a single polyprotein in the cellular endoplasmic reticulum, glycosylated in the Golgi apparatus, and cleaved into two glycoproteins, gp120 and gp41.

The life cycle of HIV-1 begins with the adsorption of the glycoproteins on its surface to the CD4 receptor on macrophages and CD4+ T cells (Figure 5.6). The complex undergoes structural changes exposing the chemokine-binding domains of gp120, permitting them to interact with a coreceptor, in most cases the C-C chemokine receptor type 5 (CCR5). Some individuals carry a deletion in the CCR5 gene, known as CCR5Δ32, which protects them against infection with some viral strains (but not against strains that use the alternative coreceptor C-X-C chemokine receptor type 4 (CXCR4), Section 5.4.1). The viral and the cellular membranes then fuse and the RNA genome is released by uncoating.

The viral polymerase is a reverse transcriptase that converts the single-stranded RNA genome into a complementary DNA (cDNA) molecule. This process is extremely error-prone: With an average of around 1 error per 1000 nucleotides, almost every copy will contain multiple errors. The set of viral genomes in an individual at any given time is, therefore, like a swarm around a particular mean, rather than a fixed individual genome as is found in most other organisms. While

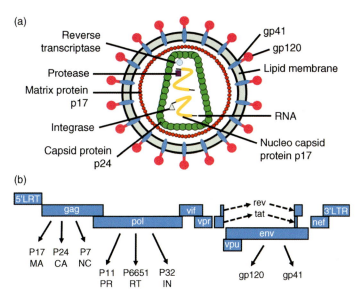

Fig. 5.5 Structure and genome of HIV-1. (a) Genomic RNA is surrounded by a conical capsid, which is enveloped by a membrane containing the transmembrane glycoproteins gp120 and gp41. (b) The HIV-1 genome is approximately 10 kb in length and encodes the structural, regulatory, and accessory genes embedded between the long terminal repeats (LTRs) [3]. Further details are given in the text.

many of the mutated viral progeny will be defective, numerous viral mutants also emerge that can evade the host's immune system and may be resistant to antiviral treatment. The reverse transcriptase not only copies RNA to DNA but also has RNase H activity that degrades the viral RNA during the synthesis of the cDNA. It also has DNA-dependent DNA polymerase activity, which generates the second DNA strand. The double-stranded cDNA is then transported into the nucleus and integrated into the host genome by the viral

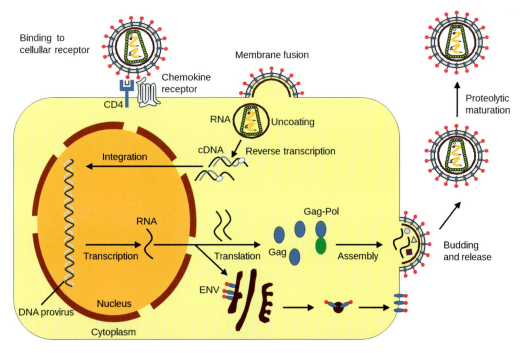

Fig. 5.6 The HIV life cycle. After attachment and entry into the cell, the viral RNA is released and converted into cDNA. The proviral DNA is integrated into the host genome. Viral genes are expressed and assembled into new virus particles with genomic RNA molecules. The virus leaves the cell by budding and finally undergoes proteolytic maturation steps.

integrase. The integrated DNA, called the provirus, is transcribed into mRNA that is spliced, exported from the nucleus into the cytoplasm, and translated.

Full-length RNA can also serve as a new virus genome. The packaging sequence, φ, binds to the Gag protein and is packaged into new virus particles that leave the cell by budding. The final steps of proteolytic maturation take place in the packaged virion outside the cell. Antiviral agents have been developed that inhibit HIV at virtually every stage of its replication cycle (Section 5.4.1). Although there is currently no cure available, antiretroviral drugs can slow the course of the disease, leading to a near-normal life expectancy with a reasonable quality of life, despite the side effects of treatment. However, the development of an HIV vaccine has proven difficult due to the high mutation rate of the virus (Section 5.2.6).

5.1.2
Hepatitis B Virus

The hepatitis B virus (HBV) causes inflammation of the liver. Approximately one third of the world's population has been infected with HBV at one point in their lives, and 350 million people are estimated to be chronic carriers. The virus is transmitted by blood, blood products, and sexual contact. Upon infection, HBV causes an acute infection, which is manifested by the typical symptoms of infection but which then progresses to jaundice. Some people develop fulminant hepatic failure, which may be lethal. In 95% of otherwise healthy adults, the infection is self-limiting and is cleared within weeks to months. In some cases, however, the virus is not fully cleared and may persist in the body, causing chronic inflammation of the liver. Over a period of several years, cirrhosis of the liver may develop; this dramatically increases the incidence of hepatocellular carcinoma (Table 5.1). A subunit vaccine is available to protect against HBV infection (Section 5.2.4). Treatment options include interferon-α as an immune stimulator, in addition to other molecules that are inhibitors of reverse transcriptase activity (Table 5.4).

HBV is an enveloped virus with a circular DNA genome. This is unusual, because the DNA is not fully double-stranded (Figure 5.7). To distinguish HBV from hepatitis viruses containing an RNA genome, it has been classified as a hepadnavirus (for *hepa*titis-*DNA*-virus). Interestingly, the life cycle of HBV resembles that of retroviruses despite carrying a DNA genome. It is, therefore, also denoted as a pararetrovirus. After entering the target cell, the viral DNA is transferred to the nucleus by proteins of the host. The partially double-stranded viral DNA is then made fully double-stranded and transformed into covalently closed circular DNA (cccDNA).

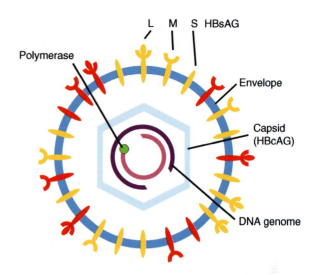

Fig. 5.7 Structure of the hepatitis B virus. The genome of HBV is comprised of circular DNA, but the DNA is not fully double-stranded. The viral polymerase is linked to one end of the DNA. The genome is surrounded by an icosahedral capsid comprised of the hepatitis B core antigen (HBcAg). The nucleocapsid is enveloped by a lipid bilayer that contains the large, medium, and small surface antigens (hepatitis B surface antigen, HBsAg). Some of these antigens (indicated in red) carry carbohydrate moieties.

The cccDNA serves as a template for transcription of the HBV mRNAs. The viral polymerase uses the longest mRNA to synthesize new genomic DNA via its reverse transcriptase activity. After assembly, the new virions are released from the cell. The virus does not cause direct cytopathic effects (CPEs); damage to the hepatocytes is induced primarily by the immune response to the viral infection.

5.1.3
Influenza Virus

Influenza is a severe infectious disease caused by Orthomyxoviridae. It should not be confused with influenza-like illnesses, such as the common cold. Typical symptoms of influenza include, among others, chills, fever, severe headache, and coughing. Influenza can lead to pneumonia, either directly or as a secondary bacterial infection, which can cause severe, even lethal complications. Influenza viruses are typically transmitted through the air via droplet secretion.

The Orthomyxovirus family is divided into several genera. Influenza A viruses cause all the flu pandemics. Influenza A viruses are further classified according to their surface proteins hemagglutinin (HA or H) and neuraminidase (NA or N). Sixteen H subtypes (H1-16) and nine N subtypes (N1-9) have been identified, differing in their tropism. For example, the subtypes H1, H2, and H3 (and less frequently H5, H7, and H9) infect humans, while all 16 subtypes infect aquatic birds. Influenza spreads around the world in seasonal epidemics with 3–5 million cases of severe illness and up to 500 000 deaths reported every year. In the last 120 years, influenza A caused six major pandemics. The most devastating pandemic was the "Spanish flu," produced by the H1N1 virus in 1918/19. Approximately one third of the world's population was infected and exhibited clinical illness. The total number of deaths was estimated at ∼50 million and may have been as high as 100 million. An unusual feature of this pandemic was that it mostly killed young adults (influenza is normally most deadly for very young or very old), in addition to the physically frail.

New influenza strains frequently appear when an existing virus spreads from animals to humans or a human-specific strain picks up genetic material from a virus that usually infects birds or pigs. An example is the H5N1 avian influenza that emerged in Asia in the 1990s after transmission of avian influenza to humans. In 2009, the so-called H1N1 swine flu was caused by a strain that combined genes from human, pig, and bird flu. Although it spread around the world rapidly, it caused comparatively mild forms of disease and the percentage of cases requiring treatment in an intensive care unit, as well as the mortality rate, were low.

Influenza viruses contain a segmented single-stranded RNA genome in the negative orientation. In the case of influenza A, the genome is composed of eight segments and encodes 11 proteins. The RNA is wrapped by a nucleoprotein and each segment is associated with a polymerase complex consisting of the PB1, PB2, and PA proteins (Figure 5.8). The virus is enveloped with a lipid bilayer that contains the surface proteins NA, HA, and the M2 ion channel. A layer of the M1 matrix protein is attached to the inside of the membrane.

Influenza viruses bind to their target cells via interaction in the HA with sialic acid sugars on the surfaces of epithelial cells in the lung. The plasma membrane encloses the virion and imports it by endocytosis. The acidic pH of the endosome induces a conformational change of the HA protein that is followed by fusion of the viral envelope with the endosomal membrane. A pore is formed through which the viral RNA (vRNA) molecules, accessory proteins, and the RNA-dependent RNA polymerase (RdRP) are released into the cytoplasm. The proteins and the vRNA form a complex that is transported into the nucleus, where the RdRP transcribes the negative-oriented RNA into a positive-oriented complementary RNA (cRNA). The latter is exported to the cytoplasm where it is translated into new viral proteins. The membrane-associated proteins

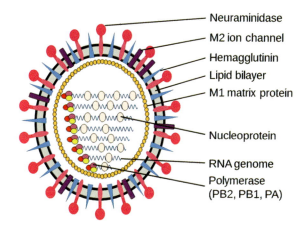

Fig. 5.8 The structure of the influenza A virus. The influenza A virus has a segmented single-stranded RNA genome in the negative orientation. The envelope membrane contains the surface proteins neuraminidase, hemagglutinin, and the M2 protein. The M1 protein forms a protein layer on the inside of the membrane.

(HA, NA, and M2) are synthesized, trafficked into the ER, and transported to the cell surface via the Golgi apparatus. The other proteins have signals for transport back into the nucleus, where they further assist the replication process. Newly synthesized negatively oriented vRNA molecules associate with nucleoproteins and proteins of the polymerase complex. The resulting viral ribonucleoprotein complexes then bind the matrix protein M1 and exit the nucleus. New viruses are assembled that leave the cell by budding, but the HA remains bound to the sialic acid receptor until NA cleaves the sialic acid residues from the host cell and detaches the virus (see also Section 5.4.3, Figure 5.19).

New influenza viruses constantly evolve via two mechanisms, known as *antigenic drift* and *antigenic shift*. Antigenic drift is caused by the high error rate of the viral polymerase. Mutations emerge in each replication cycle, some of which will modify the NA and HA surface proteins. These variants may evade the immune system and infect people who are immune to preexisting strains. In some cases, the mutations can even change the viral tropism so that an avian influenza virus can infect mammals. Still, new strains generated by antigenic drift will be similar to the older strains.

In viruses with a segmented genome, such as the influenza virus, reassortments of the segments can produce strains with completely new properties. This process is called antigenic shift (Figure 5.9). For example, a host cell may be infected with a highly pathogenic avian strain and a human strain of low virulence. Reassortment of the genomic segments may then produce a new strain that is highly pathogenic and infects humans.

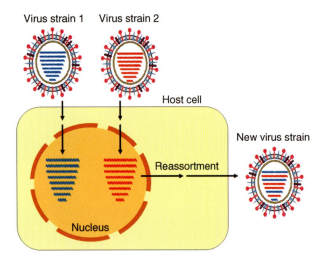

Fig. 5.9 Antigenic shift. Antigenic shift represents a major change in a viral genome and results from gene reassortment in viruses with a segmented genome. In the example shown, the host is infected with two virus strains. The progeny virus contains segments from both parental strains and may have completely new properties (virulence, tropism).

The options for the prevention and treatment of influenza infection are different from those described for HIV-1. An inactivated virus vaccine is available that confers protection against infection with the included strains (Section 5.2.3). However, due to the high genetic variability of influenza viruses, vaccination has to be repeated annually against the seasonal strains. Antiviral agents for the treatment of influenza inhibit the process of viral uncoating and the enzymatic activity of NA, but their efficiency is limited and many strains have become resistant (Section 5.4.3).

5.2 Vaccination

The object of vaccination is to induce immunity to a specific pathogen. The term "vaccine" originated in Edward Jenner's use of cowpox (Lat. vacca, cow) to inoculate humans and provide protection against smallpox (Box 5.1). Typically, a vaccine contains an agent that resembles the disease-causing microorganism. It is often produced from weakened or killed forms of the microbe, from its toxins, or from one of its surface proteins. A vaccine stimulates the body's immune system to recognize the agent as foreign and to remember it so that it can more easily recognize and destroy the microorganism when it encounters it again.

Vaccines are usually given in a prophylactic manner to prevent or ameliorate a future infection with a pathogen. A more recent concept involves using vaccination in a therapeutic manner, to treat, for example, cancer (Box 5.2).

Basically, two types of vaccination can be distinguished: active and passive immunization. The latter is achieved by the transfer of antibodies against a pathogen. Passive immunization is effective only for a comparatively short period, since the antibodies are rapidly degraded and the body does not develop an immunological memory. The procedure is, therefore, employed only under certain conditions, for example, to protect an individual directly after exposure to an infectious pathogen (postexposure prophylaxis). Health personnel are often treated with antibodies against HBV after contact with blood from infected patients and nonimmunized individuals can receive passive immunization after being bitten by animals that may have been infected with the

Box 5.1. Edward Anthony Jenner

The British physician Edward Jenner was a pioneer of smallpox vaccination and is sometimes called the father of immunology. He observed that infection of milkmaids with cowpox conferred immunity to the more dangerous smallpox virus. In 1796, he inoculated 8-year-old James Phipps, the son of his gardener, with the pus from cowpox blisters on the hands of a milkmaid who had caught cowpox. Phipps had fever and some uneasiness, but no signs of severe disease. Later, when the boy recovered, Jenner challenged him with material from a blister of a real smallpox infection, but Phipps remained healthy. Jenner's approach demonstrated that immunity to smallpox could be safely induced by what was called variolation.

At this time, Jenner did not know that smallpox is caused by a virus. The smallpox and cowpox viruses are two members of the poxvirus family that share antigens, thus enabling cross-protection. Although the vaccination and subsequent challenge with smallpox material was unethical by today's standards, Jenner's work probably saved more lives than the work of any other single individual. It also provided the basis for the eradication of smallpox by the WHO in the twentieth century (Box 5.3).

Box 5.2. Cell-Based Immunotherapy

A comparatively new therapeutic approach is the use of vaccines for therapeutic rather than prophylactic purposes. To this end, the patient's immune system is triggered to recognize and destroy cancer cells. As outlined in Section 1.5.1, dendritic cells belong to the class of antigen-presenting cells (APCs) that display foreign antigens complexed with major histocompatibility complexes (MHCs) on their surfaces. The displayed complexes prime lymphocytes to kill cells that also present the antigen. Several strategies have been developed to employ this natural function of the immune system to destroy tumors. One method involves vaccination with short peptides derived from antigens on cancer cells. To stimulate a strong immune response against the tumor, the peptides may be given with highly immunogenic substances known as adjuvants (see below). A second approach to induce an antitumor response by the immune system is to make tumor cells express signal molecules such as the granulocyte macrophage colony-stimulating factor (GM-CSF). This can be achieved *ex vivo* by removing cancer cells from the patient and transferring the GM-CSF cDNA by a viral vector before reinjecting the genetically modified cells into the patient. Alternatively, the gene can be transferred *in vivo* by treating the tumor with an oncolytic virus (Box 11.1 in Section 11.2.2) encoding the GM-CSF cDNA.

The most advanced cell-based immunotherapy is sipuleucel-T (trade name Provenge), which was approved by the FDA in 2010 for the treatment of prostate cancer. The patient-specific preparation of this immunotherapy consists of three basic steps: Initially, dendritic cells are extracted from the patient's blood. The cells are then incubated with the fusion protein PA2024, which consists of two parts: prostatic acid phosphatase (PAP), which is an antigen present in 95% of prostate cancer cells, and GM-CSF, which helps the dendritic cells to mature. The activated cells are reinfused into the patient and cause an immune response against cancer cells carrying the PAP antigen.

rabies virus. In these cases, the antibodies must be given immediately after the exposure to neutralize the virus and prevent its spread through the body. In addition, passive immunization is provided prior to traveling to high-risk countries, where the time is too short for active immunization.

The aim of active vaccination is to acquire long-term protection from a pathogen. The immune response is induced by exposure to an antigen. Ideally, protection is based on a combination of neutralizing antibodies and cytotoxic T cells.

Different types of vaccines can be distinguished (Figure 5.10 and Table 5.3). Most of the early vaccines (including cowpox) as well as some more recently developed recombinant virus vaccines belong to the class of live vaccines. The majority of the nonliving vaccines are inactivated (or killed) viruses. Another class of nonliving vaccines are the so-called subunit vaccines, which consist of a viral envelope protein (or peptide) rather than the whole pathogen. In addition, DNA vaccines can be used to produce the antigen in the immunized host. Toxoids are inactivated toxins whose immunogenicity has been maintained. Since toxoid vaccines are mostly used for immunization against bacteria, they will be discussed in Section 6.1.2.

5.2.1
Live Vaccines

Live vaccines are replication-competent pathogens. They infect cells of a vaccinated individual and replicate but do not induce a clinically severe disease. They elicit a full immune response that includes cytotoxic T cells and specific neutralizing antibodies that bind to viral surface proteins. Since live vaccines induce a humoral as well as a cellular immune response, they provide efficient and long-term protection against a pathogen. One problem associated with live vaccines is the risk of reversion to the virulent virus or of uncontrolled spread, which can lead to disease in immunocompromised individuals. Some thermolabile vaccine viruses require the maintenance of special conditions, such as low temperature, during transport and storage.

Vaccination against smallpox, which is caused by the variola virus, was not only the first vaccination ever to be carried out (Box 5.1), but also led to the eradication of the disease in the 1970s, one of history's major public health achievements (Box 5.3). Interestingly, the origin of the vaccinia virus used for centuries for vaccination against smallpox remains obscure. It appears to be distinct from both the variola virus and the cowpox virus

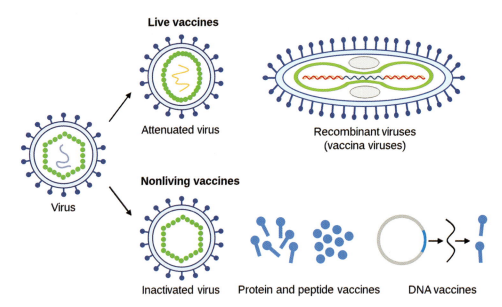

Fig. 5.10   Types of vaccines. Vaccines can be grouped into live vaccines (attenuated and recombinant viruses) and nonliving vaccines (inactivated viruses, protein and peptide vaccines, and DNA vaccines). Further details are given in the text.

that was initially employed by Edward Jenner. Infection with the vaccinia virus is typically asymptomatic or causes a mild rash and fever, but it confers immunity against the lethal variola virus. Occasionally, complications arise after vaccination with the vaccinia virus, particularly in immunocompromised individuals. Therefore, the vaccine is currently only administered to health care workers and military personnel. The threat of variola

virus use as a bioterror weapon has increased interest in developing safer new vaccines.

In the case of the variola virus, closely related viruses (cowpox and vaccinia virus) exist naturally that allowed cross-protection without inducing severe disease. However, such low-virulence relatives of serious pathogens are rare. In cases in which a safe live vaccine is not available from natural sources, it is sometimes possible to generate strains that retain immunogenicity, but are much less virulent than the parental virus. These so-called attenuated viruses can be obtained by passage of the virus through a tissue culture or embryonated eggs. The virus adapts to the culture conditions and loses its virulence. The molecular basis of virus attenuation lies in the accumulation of mutations in the genome of the wild-type virus that do not affect its proliferation in culture, but decrease its viability in hosts.

One of the two widely used vaccines against poliovirus (Figure 5.11), the OPV, is a prominent example of an attenuated live vaccine. It was produced by Albert Sabin in the late 1950s by passage of the virus through non-human cells at subphysiological temperature. This procedure led to the accumulation of spontaneous mutations in the viral genome. The mutations are now known to be located in the internal ribosome entry site (IRES) and alter its stem-loop structure, reducing the virus' ability to translate its RNA into protein. The attenuated poliovirus only replicates in the gut, the primary site of infection, but does not replicate efficiently within the nervous system. It mimics the natural infection and induces long-lived

Table 5.3   Selection of licensed vaccines.

Virus	Disease	Type of vaccine
Hepatitis A virus	Hepatitis A	Inactivated virus
Hepatitis B virus	Hepatitis B	Subunit vaccine
Human papillomavirus	Cervical cancer, genital warts	Subunit vaccine
Influenza virus	Influenza	Inactivated virus
Measles virus	Measles	Attenuated virus
Mumps virus	Mumps	Attenuated virus
Rubella virus	Rubella	Attenuated virus
Varicella-zoster virus	Chicken pox, herpes zoster	Attenuated virus
Poliovirus	Poliomyelitis	Attenuated virus/ inactivated virus
Rabies virus	Rabies	Inactivated virus
Variola virus	Smallpox	Live vaccine
Yellow fever virus	Yellow fever	Attenuated virus

intestinal antipolio IgA secretion. Three different Sabin strains are usually combined for oral vaccination. OPV can be easily produced and it largely replaced the inactivated Salk polio vaccine (IPV, see below) in the 1960s. In 1988, the WHO launched an initiative aimed at the global eradication of poliovirus, which is largely based on the use of OPV (Box 5.3). However, a major concern about OPV is its ability to revert to a virus form (vaccine-derived poliovirus, VDPV) that can achieve neurological infection and in rare cases cause paralysis (vaccine-associated paralytic poliomyelitis, VAPP). Since reversion is not possible with inactivated virus, many industrialized nations returned from OPV to IPV.

Immunization against the *measles*, *mumps*, and *rubella viruses* is usually provided by a combined vaccine that contains a mixture of attenuated viruses of the three diseases (MMR vaccine). The component viral strains were obtained by propagation of the viruses in animal and human cells. More recently, a combination vaccine, to which an attenuated *varicella virus* is added as a fourth component (known as the MMRV vaccine), has become available

The history of mumps vaccination is a good example of what is referred to as herd immunity. The incidence of mumps drastically declined long before vaccine coverage was extensive. The chain of infection can be

Box 5.3.   The Eradication of Infectious Diseases

The eradication of an infectious disease means that its prevalence in the global population has been reduced to zero. The variola virus, which causes smallpox, was the first human pathogen that has successfully been eradicated. The fact that humans are the only reservoir for smallpox infection was an important prerequisite for its successful eradication. The overall mortality rate of the variola major virus was 30% and it was responsible for several hundred million deaths in the twentieth century – in addition to countless survivors that suffered lifelong morbidity from the infection. Since the 1960s, intensive efforts have been made to eradicate the variola virus by worldwide vaccination. In 1975, the last person infected with naturally occurring variola virus was living in Bangladesh. In 1980, smallpox was declared eradicated. Since then, only two stocks of variola virus have been officially maintained, one in the Centers for Disease Control and Prevention in the United States and one in the State Research Center of Virology and Biotechnology in Russia. The WHO has repeatedly recommended destroying the stocks, but these requests have to date been denied, the argument being given that the stocks are necessary for smallpox researchers to develop new vaccines, antiviral drugs, and diagnostic tests.

Poliovirus was the second virus chosen for global eradication, since it does not have a nonprimate reservoir in nature. Further, survival of the virus in the environment is possible only for a short period of time. Interruption of person-to-person transmission by vaccination should, therefore, allow eradication of the virus. In 1988, the WHO launched a program to eradicate polio worldwide by the year 2000. Although the number of polio cases was reduced by more than 99% from an estimated 350 000 paralytic polio cases

occurring in 1988 to roughly 1200 cases in 2004, the original aim was not achieved. New cases have continuously been reported from Nigeria, Pakistan, and Afghanistan. The worldwide eradication campaign is based mainly on the attenuated oral polio vaccine (OPV, see below). However, fear of infectious virus revertants is one of the reasons for resistance to vaccination with OPV in some developing countries. It was estimated that the frequency of vaccine-associated paralytic poliomyelitis (VAPP) is approximately 1 per 1 100 000 cases. Due to opposition to the vaccination program in some countries, the WHO has had to constantly renew its strategy for the elimination of the last reservoirs of wild poliovirus circulation.

The posteradication strategy is also controversial. Indefinite vaccination carries risks, such as the occurrence of VAPP from the use of OPV. Also, problems exist with correct handling of inactivated polio vaccine (IPV) in poor countries. Substantial economic resources would be required for immunization against a disease that had been certified as eradicated. However, cessation of vaccination would carry the risk that an unimmunized global population would be highly susceptible to an outbreak of a poliovirus infection from an unknown source, like a bioterrorism attack. A study from 2002, in which Eckard Wimmer's group from the State University of New York at Stony Brook succeeded in synthesizing poliovirus from its nucleotide code, received much attention. The synthetic virus was able to infect mice, replicate, and cause paralysis or death. This experiment demonstrated that bioterrorists could produce a virus from publically available sources, which might be particularly dangerous to a world population no longer vaccinated.

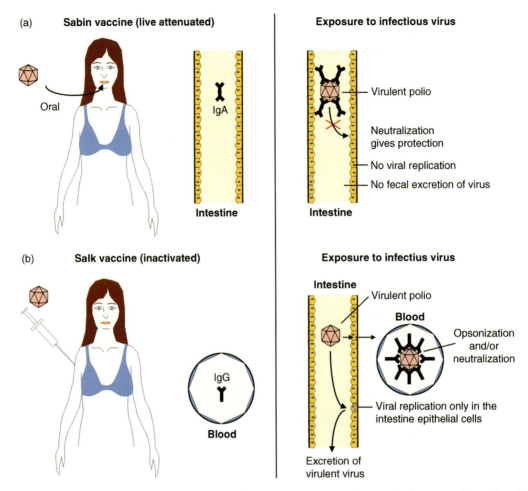

(a) **Sabin vaccine (live attenuated)**

Oral

IgA

Intestine

Exposure to infectious virus

Virulent polio

Neutralization gives protection

No viral replication

No fecal excretion of virus

Intestine

(b) **Salk vaccine (inactivated)**

IgG

Blood

Exposure to infectius virus

Intestine

Virulent polio

Blood

Opsonization and/or neutralization

Viral replication only in the intestine epithelial cells

Excretion of virulent virus

Fig. 5.11 Polio vaccines. (a) The attenuated live (Sabin) vaccine is given orally and mimics the natural infection. It induces a long-lived local IgA response. As a consequence, the natural virus is neutralized and cannot attach to epithelial cells. (b) In contrast, the inactivated (Salk) vaccine is given by injection and mainly induces an IgG antibody response. This does not prevent infection of the intestinal epithelial cells, but prevents spread beyond the intestine by opsonization and neutralization. (Adapted from Ref. [4] with kind permission from Wiley-VCH Verlag GmbH.)

disrupted even by partial coverage of the population, thus indirectly protecting unvaccinated individuals. The critical portion of the population that must be immunized for extensive protection of a community varies with the virulence of the pathogen.

5.2.2
Recombinant Virus Vaccines

A novel, promising strategy is the use of live, recombinant virus vaccines. The basis of this approach is to use a well-known virus with little pathogenicity (e.g., adenovirus), or a virus that has successfully been used as a live vaccine, usually the vaccinia virus. Genes encoding surface proteins of the pathogenic virus against which the vaccine will be developed are introduced into the genome of the vaccination virus. After inoculation, the

intrinsic virus genes as well as the foreign gene will be expressed. The advantage of the recombinant virus vaccine approach over an inactivated virus (see below) is that because it is a live vaccine, it will induce a cellular immune response in addition to inducing neutralizing antibodies. Recombinant virus vaccines have been employed to protect wild foxes against infection with the rabies virus.

5.2.3
Inactivated Virus Vaccines

Nonliving virus vaccines are available for a variety of viruses, including the influenza, hepatitis A, and rabies viruses. These vaccines are prepared from viruses grown in cell cultures or eggs. Subsequently, the virus is inactivated with chemicals such as aldehydes (mostly formalin) or alcohols. These agents remove the genetic material without destroying the protein epitopes. The great advantage of nonliving virus vaccines is the low risk of infection. Only in rare cases are such vaccines contaminated with residual infectious virus that has resisted the inactivation procedure. However, a great disadvantage of nonliving vaccines is that they usually do not induce a cellular immune response, which requires (a) active protein synthesis, (b) processing of the antigens through the proteasome, and (c) presentation of peptides by MHC molecules on the cell surface. Thus, only neutralizing antibodies are generated. In many cases, several doses of inactivated vaccine are given to boost the immune response.

To improve the efficiency of a (nonliving) vaccine, immunologic adjuvants may be added. These adjuvants do not confer immunity themselves, but maximize the immune response to the antigen. Typical adjuvants are oil/water emulsions and aluminum salts. The most commonly used adjuvant in human vaccination is alum, which is, for example, included in human papillomavirus (HPV) and hepatitis vaccines. More recently, a new generation of adjuvants has been developed that includes complex combinations of salts, lipids, liposomes, organic macromolecules, emulsions, and oligonucleotides (Section 14.4). However, only very few of these novel adjuvants have yet been approved for human use so far.

Adjuvants can act via several mechanisms. One is to provide a depot of the antigen at the site of injection and increase the amount that is available to immune cells, such as dendritic cells. A second mode of action is the activation of the innate immune response. Adjuvants can trigger dendritic cells to release cytokines that stimulate B- and T-lymphocytes. Interestingly, adjuvants may be employed not only to enhance the immunostimulatory effect of an antigen with limited immunogenicity but they may also help to save vaccine in cases in which availability is limited. For example, one third of the normal dose of an influenza vaccine was sufficient to induce a protective immune response when an adjuvant was added. The adjuvants may thus be helpful in case of a pandemic, in which a large number of individuals need rapid vaccination.

As described above, although attenuated poliovirus has widely been used as a vaccine due to the ease of administration, the first *polio vaccine* that was developed by Jonas Salk in the early 1950s was an inactivated virus. The IPV combines the three pathogenic poliovirus serotypes. Those viruses are grown in mammalian cells and are subsequently inactivated by treatment with formalin. IPV needs to be administered by subcutaneous or intramuscular injection and induces mainly an IgG rather than an IgA response (Figure 5.11).

Since *influenza viruses* are continuously changing antigenicity, annual vaccination is necessary for protection against emerging virus variants. The seasonal composition of a virus vaccine is defined by the WHO approximately half a year prior to the influenza season. The vaccine is intended to confer optimal resistance against the most common virus strains; it, therefore, contains two subtypes for influenza A (H1N1 and H3N2) and up to two influenza B strains. The viruses are separately grown in fertilized chicken eggs and are usually inactivated by chemical disruption, which results in a so-called split-virus vaccine. Influenza vaccines may be administered as an injection, known as the flu shot, but are also available as a nasal spray. More recently, an influenza vaccine was approved that is produced in a mammalian cell culture (trade names Flucelvax in the United States and Optaflu in the European Union) rather than in chicken eggs. This should allow upscaling and rapid production in addition to reducing allergic reactions against the vaccine.

Additional inactivated vaccines are available against the *rabies* and *hepatitis A viruses*.

5.2.4
Subunit Vaccines

In some cases, it is not necessary to use the entire pathogen for vaccination. If the antigenic protein against which the immune response is induced is well known, it is possible to produce this protein by heterologous expression and employ it for immunization (Section 10.2.8). This class of vaccine is known as a subunit vaccine. Proteins that self-assemble into virus-like particles (VLP) are especially well suited for this approach, because their particulate structure induces a strong T- and B-cell immune responses. Since VPLs only resemble viruses but do not contain any viral genetic material, they cannot cause a vaccine-induced infection.

Since the 1980s, a subunit vaccine has been available for the prevention of *hepatitis B virus* infection. The hepatitis B surface antigen (HBsAg) is produced in yeast cells. HBsAg self-assembles into VLPs and usually induces a protective immune response even in the absence of an adjuvant. A course of two to three intramuscular vaccine injections is given to achieve sufficient protection against HBV. However, approximately 10–15% of adults either respond poorly or fail to respond to the HBV vaccination at all. In these cases, the immunogenicity of the HBsAg can be enhanced by the addition of an adjuvant, for example, an aluminum salt or a CpG-rich oligonucleotide (Section 14.4).

A pediatric dose of recombinant hepatitis B vaccine is also part of the (Pediarix) pentavalent vaccine, which also contains a combination of vaccines for polio, diphtheria, tetanus, and acellular pertussis.

The development of a subunit vaccine to prevent infection with the *human papillomavirus* (HPV) has been one of the major breakthroughs in the prevention of disease in recent years. The virus has a specific tropism for squamous epithelial cells and is the most common sexually transmitted infection in adults worldwide. More than 100 types of HPV have been identified, of which at least 15 types have been implicated in cervical cancer. HPV DNA is detectable in more than 95% of cervical cancer cases. Worldwide, there are almost 530 000 new cases and 275 000 deaths per year, making cervical cancer the third most common cancer by incidence among women. HPV types 16 and 18 account for 70% of newly diagnosed cervical cancer cases and are implicated in anal, vaginal, and vulvar cancers as well. HPV 6 and 11 cause 90% of the cases of genital warts, but these viral types do not promote the growth of life-threatening cancers.

The two major HPV vaccines available since 2006, Cervarix and Gardasil, contain the L1 protein, the major capsid protein of the virus (Figure 5.12). The protein is produced by recombinant expression in insect and yeast cells, respectively (Section 10.2.8), and spontaneously forms VLPs such as the HBsAg. The VLPs trigger an immune response that confers protection against HPV. The vaccination can almost completely prevent cervical intraepithelial neoplasia, the precancerous change occurring in cervical cells caused by the HPV types included in the vaccine. Vaccination is recommended for young females prior to sexual activity. Cervarix contains the L1 proteins of HPV types 16 and 18, which, as mentioned above, account for the majority of newly diagnosed cervical cancer cases. Gardasil also contains the L1 proteins of the HPV types 6 and 11, and thus protects not only against cervical cancer but also against genital warts.

Subunit vaccines have also been approved or are being developed for other viruses, including influenza and HIV. In addition, the concept of using synthetic peptides as vaccines has also been evaluated. In this case, only the epitopes of those viral proteins that induce an immune response are chemically synthesized. This procedure facilitates the production process and increases the safety of the vaccine. However, the peptide vaccines tested to date did not achieve sufficient protective immunogenicity in humans.

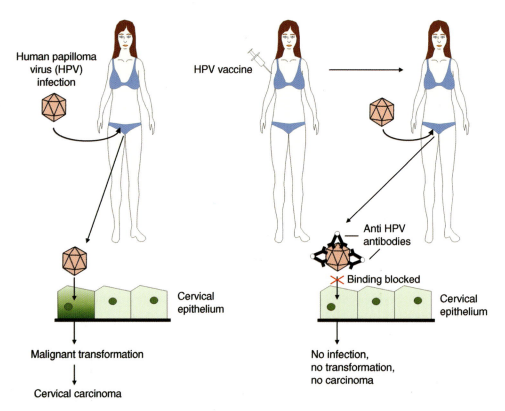

Fig. 5.12 HPV vaccine. Sexually transmitted HPV can lead to malignant transformation of cervical epithelium cells. An anti-HPV subunit vaccine induces an antibody-based immune response that prevents the virus from infecting the epithelial cells and protects against subsequent tumor development. (Adapted from Ref. [4] with kind permission from Wiley-VCH Verlag GmbH.)

5.2.5
DNA Vaccines

A more recent approach is the development of DNA vaccines, which represent a special form of gene therapy. For DNA vaccination, a plasmid is engineered to produce an antigen of the pathogen. The DNA vaccine is injected into the cells either by a syringe or a gene gun (Section 11.2.4). The cellular machinery then produces the viral proteins that are displayed on the cell surface and triggers an immune response. Like the other non-living vaccines, DNA vaccines do not produce vaccine-induced infections. In addition, DNA is a very stable molecule that can easily be handled and distributed. A major advantage of DNA vaccines is that they can be adapted to new viruses or new variants (mutants) of a known virus by simply altering the DNA sequence of the expression plasmid. However, the ability of this experimental approach to evoke an efficient immune response has been limited to date.

5.2.6
HIV Vaccines

A special challenge is the development of an HIV vaccine. Immediately after the discovery of HIV-1, the search for a vaccine that protects from infection with the virus was initiated. However, the outcome of most clinical trials was disappointing. In some cases, the number of newly infected individuals was even greater in the group of vaccinated test persons than in the control group. An exception was a large-scale clinical trial known as RV144, which demonstrated that an HIV vaccine could modestly reduce the incidence of HIV infection. In this study, two HIV vaccines were combined that had failed on their own: ALVAC-HIV-1 is a live recombinant viral vector that produces three HIV genes (env, gag, and pol). The immune response was then boosted with AIDSVAX B/E, a protein-based vaccine containing genetically engineered gp120. The major challenge for the development of an HIV vaccine is the genetic diversity and mutability of HIV-1. In addition, structural features, that is, the presence of carbohydrate moieties that shield potential epitopes from antibodies, complicate the induction of neutralizing antibodies. A very recent and innovative approach is the direct delivery of neutralizing antibodies against the virus by gene transfer rather than by coaxing the immune system to produce them. In early phase clinical testing, an adeno-associated virus vector (Section 11.2.3) encoding the HIV neutralizing antibody PG9 was given by intramuscular injection. However, it still remains to be seen whether this strategy produces enough antibodies to neutralize the virus in humans.

5.3
Detection of Viruses

For many years, the diagnosis of viral infection has lagged far behind that of bacterial disease. Originally, viruses were identified as infectious agents that passed through filters with a pore size small enough to trap bacteria. The invention of the electron microscope in the 1930s allowed viruses to be visualized. A broad variety of methods for the laboratory diagnosis of viral infection is currently available. The virus can be diagnosed by directly detecting the pathogen itself, its proteins, or its nucleic acids. However, in some cases, the virus is only detectable prior to the symptomatic phase, requiring indirect diagnosis by detection of virus-specific antibodies in the plasma.

In recent years, the diagnosis of viral infection has been revolutionized by the application of PCR techniques for sensitive and specific detection. Real-time PCR (Section 2.2) permits the quantitative detection of a virus and can be combined with sequencing technologies to further characterize the detected virus. For any method of virus detection, the diagnostic gap (or window period), which describes the timeframe between infection and the time point when a test can reliably detect that infection, is critically important. For example, approximately 12 weeks must pass after infection before a typical ELISA test for the detection of an HIV infection (see below) will produce a reliable result.

Due to the fact that many diagnostic assays can be expected to be replaced by PCR-based assays in the near future, only a few major tests will be described below.

5.3.1
Cytopathic Effects

Many viruses can be grown in cultured cells where they induce morphologic changes in the cells they infect; this is referred to as a cytopathic effect (CPE). For example, a CPE is induced by herpes simplex virus grown in rabbit kidney cells or by adenoviruses grown in human embryonic kidney cells. The changes in cell morphology can be used to diagnose a viral infection. An advantage of this method is that only functional viruses are detected. However, either some viruses do not induce a CPE or no assays for viral propagation in cell culture have been established. Another disadvantage of this assay is that it usually takes several days and is not necessarily specific.

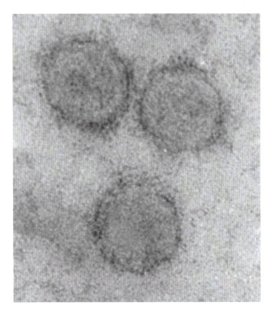

Fig. 5.13 Electron micrograph of poliovirus 1 particles. (Image courtesy of Heinz Zeichhardt, Charité Berlin, Germany.)

5.3.2
Electron Microscopy

Electron microscopes use an electron beam instead of visible light to illuminate a specimen and produce a magnified image. Therefore, the resolving power of an electron microscope is much greater than that of a light microscope. This allows the visualization of very small objects such as viruses (Figure 5.13). A skilled operator can distinguish various types of viruses. Advantages of electron microscopy include speed of measurement and the lack of requirement for viral viability. However, the technique often does not allow the exact classification of a virus species. Specificity and sensitivity can be improved by preincubation of the sample with immune serum (immune electron microscopy).

5.3.3
Hemagglutination Assay

Several virus families (and many bacteria) have envelope or surface proteins that bind to *N*-acetylneuraminic acid and agglutinate red blood cells. Since each agglutinating molecule attaches to multiple others, a lattice structure will form. The hemagglutination assay is carried out with erythrocytes from a suitable species (e.g., sheep). It can be employed for paramyxo- and orthomyxoviruses as well as flavi-, toga-, corona-, and parvoviruses. The assay is very rapid and simple, but does not provide specific information about the virus. The result indicates

only whether or not the infection is caused by a virus that belongs to the class of hemagglutinating viruses.

A common variation of the assay is the hemagglutination inhibition assay, in which antibodies prevent hemagglutination by interfering with viral attachment to the red blood cells. This procedure can either help to determine the virus type more specifically or to measure plasma antibody levels (for example, against influenza or rubella virus).

5.3.4
Enzyme-Linked Immunosorbent Assay (ELISA)

ELISA assays can be used for the direct or indirect diagnosis of a viral infection. An antigen ELISA, also known as a sandwich ELISA, allows the direct detection of viral protein (Figure 5.14a). The test is based on the specific interaction of an antibody and a viral antigen. A second antibody, which also interacts with the viral protein, is covalently coupled to an enzyme that converts a substrate into a colored product. The assay allows the quantitative determination of the amount of viral antigen that is present in the sample. Examples of antigen ELISA tests are the detection of HIV from blood samples or of influenza, parainfluenza, adeno-, rota-, and noroviruses in the sputum or feces. Results from an antigen ELISA test are sometimes confirmed by Western blotting.

5.3.5
Indirect ELISA

In contrast to the antigen ELISA already described, the indirect ELISA does not detect virus proteins but rather antibodies produced against the virus by the patient (Figure 5.14b). In this assay, the virus protein, which is usually obtained by heterologous expression, is fixed on the surface of the microtiter plate. Antibodies in the specimen then bind to the antigen. They are detected by a second antibody that is coupled to the enzyme, which converts the substrate into a colored product. In addition to the level of the antibody, the prevalent type of antibody in the patient's plasma provides valuable information. IgM antibodies appear early after the infection and indicate an acute infection. In contrast, IgG antibodies are a sign of an infection that occurred sometime in the past.

5.3.6
Polymerase Chain Reaction (PCR)

The application of PCR to diagnostic virology was a major advancement for this field. This technique allows the rapid, specific, and sensitive detection of a virus in a

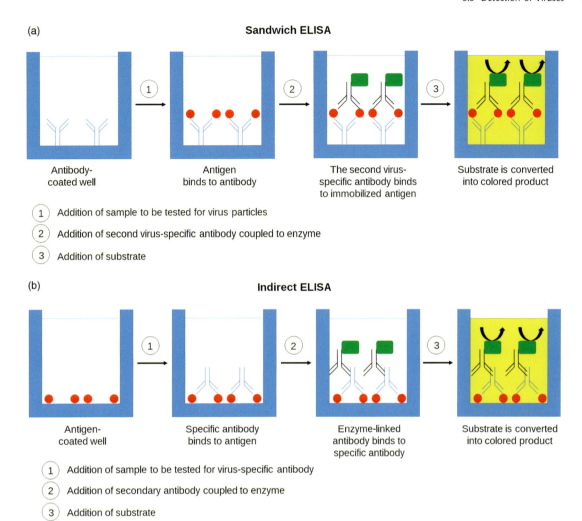

(a) **Sandwich ELISA**

Antibody-coated well	Antigen binds to antibody	The second virus-specific antibody binds to immobilized antigen	Substrate is converted into colored product

1 Addition of sample to be tested for virus particles

2 Addition of second virus-specific antibody coupled to enzyme

3 Addition of substrate

(b) **Indirect ELISA**

Antigen-coated well	Specific antibody binds to antigen	Enzyme-linked antibody binds to specific antibody	Substrate is converted into colored product

1 Addition of sample to be tested for virus-specific antibody

2 Addition of secondary antibody coupled to enzyme

3 Addition of substrate

Fig. 5.14   Enzyme-linked immunosorbent assay (ELISA). (a) The antigen or sandwich ELISA can be employed for the direct detection of a viral protein. The assay is usually carried out in microtiter plates. The first antibody, immobilized on a solid support, is called the capture antibody and is directed against the protein of interest. The solution to be assayed is then applied to the well. While unspecific proteins are washed away, the protein of interest binds to the antibody. Subsequently, a second antibody against the antigen is added, which is covalently linked to an enzyme. Finally, a substrate is added to the well, which is converted into a colored product by the enzyme. The intensity of the color is proportional to the amount of antigen. (b) An indirect ELISA is used to detect antibodies rather than virus proteins in the specimen. Viral antigens, which can be produced by heterologous expression in bacteria, are fixed on the surface of a microtiter plate. Antibodies in the patient's plasma bind to the antigen. A second antibody is directed against the human antibody. This antibody is coupled to an enzyme that converts the substrate into a colored product that can be quantified.

sample. For RNA viruses, the genomic RNA must be converted into DNA by reverse transcription prior to amplification by PCR. Several methods have been developed to increase the sensitivity of the test. One example is *nested PCR*, which involves the application of two successive runs with two different primer sets. In the first run, the material is amplified, but unwanted sequences may be amplified as well. The second run amplifies the first run product with a set of primers whose binding sites are different from those of the primers used in the first reaction. Since it is very unlikely that any unwanted product from the first run will also contain binding sites for the two new primers, the second PCR product will be free from contamination.

Real-time PCR technology permits the online detection of the PCR product during the run (Section 2.2) and allows fast and convenient analysis of patient samples for the presence of a virus. The major advantages of real-time PCR are its higher sensitivity compared to conventional PCR and its ability to accurately quantify the viral load in clinical specimens. The determination of the viral load in plasma allows a distinction to be made between latent and active infections and is also useful for monitoring the efficacy of antiviral therapy.

Amplification of a DNA fragment by PCR requires exact knowledge of the nucleotide sequence for the design of the primers. Primers binding to a conserved region may be used to amplify several subtypes of a species of virus. However, it may be desirable to test for any possible virus that could be the cause of an infection of, for example, the respiratory tract. Therefore, *mutiplex PCR* assays are widely used. This involves the combination of different primer pairs. By this approach, a range of pathogens in a specimen or sequence variations of a pathogenic strain can be identified in a single reaction. Advances in sequencing technologies (Section 2.3) allow the rapid and easy analysis of a viral sequence in a PCR product, helping to determine serotypes and mutations and providing valuable information for potential therapeutic strategies (see below).

5.3.7
Antiviral Susceptibility Testing

As will be discussed in the following section, the development of an increasing number of antiviral drugs has been accompanied by the appearance of resistant mutants, which pose a major challenge to successful antiviral treatment. Here, susceptibility testing can guide the choice of effective drugs.

Antiviral susceptibility testing can be divided into both phenotypic and genotypic methods. In phenotypic tests, the effect of a drug on the growth of a virus is evaluated. The assays are attractive because they are a direct measurement of the efficiency of a drug. However, the method is labor-intensive and slow, since the virus needs to be cultivated. Genotypic tests are much faster. They analyze the genetic basis for resistant phenotypes by sequencing the virus genome. A limitation of this approach is that it is useful only in cases in which the genetic basis for resistance is known. For example, six different point mutations in the HIV reverse transcriptase gene are known to confer resistance against the antiviral drug azidothymidine (see below). If these mutations are detected, another antiretroviral drug must be used to treat an infected patient.

5.4
Antiviral Therapy

While antibiotics for the treatment of bacterial infections were introduced in the mid-twentieth century, no drugs to treat viral infections were available at that time. In contrast to bacteria, viruses are obligate, intracellular

parasites without a metabolism of their own and were, therefore, believed not to be susceptible to therapeutic intervention. It took until the 1960s for this dogma to be reversed and for scientists to develop the first drugs against herpes infection. At the beginning of the second decade of the twenty-first century, approximately 50 antiviral drugs have been approved (Table 5.4), about half of which are intended for the treatment of HIV infection. While multiple drug combinations are used for HIV and HCV therapy, for other viruses, such as herpes-simplex virus (HSV), cytomegalovirus (CMV), varicella-zoster virus (VZV), poxvirus, influenza, HBV, and filoviruses, current therapies are based on the use of single agents only. Antiviral drugs share a common feature, in that they primarily inhibit the replication of the pathogen, rather than destroying it as antibiotics do.

Since viruses do not have their own metabolism, the antiviral drug must target specific viral functions without affecting cellular processes. The chemotherapeutic index, defined as the ratio of the concentration of a substance that is required for virus inhibition and the concentration at which cytotoxic effects are observed, should be high. Many viruses use their own polymerases, which differ from the cellular polymerases. Since these enzymes are essential for viral replication, they are often targeted by antiviral agents. In fact, idoxuridine, the first virus inhibitor discovered, is a modified form of deoxyuridine that inhibits replication of the herpes simplex virus (Figure 5.15). The nucleoside analog is incorporated during viral replication, but the iodine atom added to the uracil prevents further DNA synthesis. However, first-generation antiviral nucleosides were not selective enough for virus-infected cells and were too toxic for systemic administration. Idoxuridine is, therefore, only used topically to treat herpes simplex keratitis.

Second-generation nucleoside analogs are highly selective for the targeted virus and less systemically toxic. An example is aciclovir (Figure 5.15; chemical name acycloguanosine, marketed under various trade names such as Acyclovir, Herpex, and Acivir), a guanosine analog that is widely used to treat herpesvirus (HSV-1, HSV-2, and VZV) infection. The key to the high virus-specificity of aciclovir is the selective phosphorylation of the acyclic guanosine nucleoside by a herpesvirus-encoded thymidine kinase (Figure 5.16). Aciclovir is a prodrug that is initially administered in an inactive form and is then metabolized into the active species in the body. Following selective first phosphorylation by the viral thymidine kinase, cellular kinases further phosphorylate the monophosphate to give the active triphosphate form of the drug. The acyclo-GTP is incorporated into the newly synthesized viral DNA by the herpesvirus polymerase, but since it does not have a 3′ OH group, no further

Table 5.4 Selection of approved antiviral drugs.

Virus	Antiviral substance	Substance	Mode of action
HSV, VZV	Aciclovir	Guanosine analog	Nucleoside analog inhibitor of viral polymerase
	Valacyclovir	Guanosine analog esterified to valine	Nucleoside analog inhibitor of viral polymerase
CMV	Foscarnet (also indicated for HSV)	Phosphonic acid	Nonnucleoside inhibitor viral polymerase
	Ganciclovir	Guanosine analog	Inhibitor of viral polymerase
	Valganciclovir	Guanosine analog esterified to valine	Inhibitor of viral polymerase
	Cidofovir	Cytosine analog	Inhibitor of viral polymerase
HIV	Azidothymidine	Thymidine analog	Nucleoside analog reverse-transcriptase inhibitor
	Lamivudine	Cytosine analog	Nucleoside analog reverse-transcriptase inhibitor
	Abacavir	Guanosine analog	Nucleoside analog reverse-transcriptase inhibitor
	Nevirapine	Organic substance	Nonnucleoside analog reverse-transcriptase inhibitor
	Rilpivirine	Organic substance	Nonnucleoside analog reverse-transcriptase inhibitor
	Enfuvirtide	Peptide	Fusion inhibitor
	Raltegravir	Organic substance	Integrase inhibitor
	Maraviroc	Organic substance	Inhibitor of CCR5 coreceptor
	Saquinavir	Peptide mimetic	Protease inhibitor
	Ritonavir	Peptide mimetic	Protease inhibitor
	Tipranavir	Nonpeptidic molecule	Protease inhibitor
	Raltegravir	Organic substance	Integrase inhibitor
HCV	PEG-IFN-α	Protein	Stimulates immune response
	Ribavirin	Guanosine analog	Nucleoside analog inhibitor of viral polymerase
	Sofosbuvir	Uracil analog	Nucleoside analog inhibitor of viral polymerase
	Telaprevir	Peptide mimetic	Protease inhibitor
	Boceprevir	Peptide mimetic	Protease inhibitor
HBV	PEG-IFN-α	Protein	Stimulates immune response
	Lamivudine	Cytosine analog	Nucleoside analog reverse-transcriptase inhibitor
	Adefovir	Adenosine analog	Nucleotide analog reverse-transcriptase inhibitor
	Entecavir	Guanosine analog	Nucleoside analog reverse-transcriptase inhibitor
	Telbivudine	Thymidine analog (L-isomer)	Nucleoside analog reverse-transcriptase inhibitor
	Tenofovir	Adenosine analog	Nucleotide analog reverse-transcriptase inhibitor
Influenza	Amantadine	Organic substance	Uncoating inhibitor
	Zanamivir	Sialic acid analog	Neuraminidase inhibitor
	Oseltamivir	Sialic acid analog	Neuraminidase inhibitor

HSV: herpes simplex virus; VZV: varicella-zoster virus; HIV: human immunodeficiency virus; HCV: hepatitis C virus; HBV: hepatitis B virus.

nucleotides can be added to the growing strand. The incorporation of acyclo-GTP, therefore, results in premature chain termination. Since the viral DNA polymerase, which lacks the proofreading function of cellular polymerases, cannot remove the region of DNA containing the acyclo-nucleoside, further elongation is inhibited.

However, acyclic nucleosides such as aciclovir are poorly soluble in aqueous solution and have low bioavailability following oral administration; hence, intravenous administration is necessary if high plasma concentrations are required. To overcome these drawbacks, analogs with higher oral bioavailability have been developed. One example is valacyclovir, in which a valine is joined to aciclovir via an ester linkage. This modification increases the oral bioavailability of the drug from 10–20% to about 55%. Valacyclovir is converted by esterases to aciclovir in the liver.

Nucleoside analogs designed to inhibit viral polymerases have been the dominant class of molecules in the field of antiviral therapy for decades. Nonnucleoside inhibitors of viral polymerases inhibit the same enzyme as the nucleoside analogs, but their structures are fundamentally different. While nucleoside analogs mimic natural nucleosides and bind as a competitor to the active center of the enzyme, nonnucleoside inhibitors have diverse structures (Figure 5.15) and bind to sites other than the catalytic core. Foscarnet is a phosphonic acid derivative that selectively inhibits the pyrophosphate binding site of the CMV and HSV polymerases. Another example is nevirapine, a nonnucleoside inhibitor of the HIV reverse transcriptase (nonnucleoside reverse transcriptase inhibitor, NNRTI, see below).

In addition to the viral polymerase, several other targets relevant to various stages of the viral life cycle have been proposed as antiviral drugs. For HIV, inhibitors of cellular entry, integration into the host genome, and protein maturation by proteolytic cleavage have been developed, as outlined in more detail below (Figure 5.17). Other approaches have been used for viruses with different life cycles; for example, the development of influenza can be inhibited by inhibitors of uncoating following cellular uptake or by blocking the release of the intact virus from host cells. In addition,

viral or cellular messenger RNAs can be blocked by antisense approaches.

While small-molecule drugs have dominated the class of antiviral drugs for decades, biologics have gained importance more recently. As can be seen in the overview in Figure 5.15, most of the approved antivirals are nucleoside analogs or other organic molecules that are produced by chemical synthesis. While this market is dominated by the pharmaceutical industry, smaller biotech companies focus on the development of innovative new therapeutic strategies. Some biological therapeutics, such as interferon-α, have already become part of standard therapy (e.g., for HCV in combination with the low-molecular weight inhibitor ribavirin). The synthetic peptide enfuvirtide prevents fusion of HIV-1 with CD4⁺ cells and monoclonal antibodies (mAbs, see Section 10.2.1) can be used to bind to the virus and prevent its interaction with cellular receptors. Palivizumab is an example of an mAb that can be used for the prevention

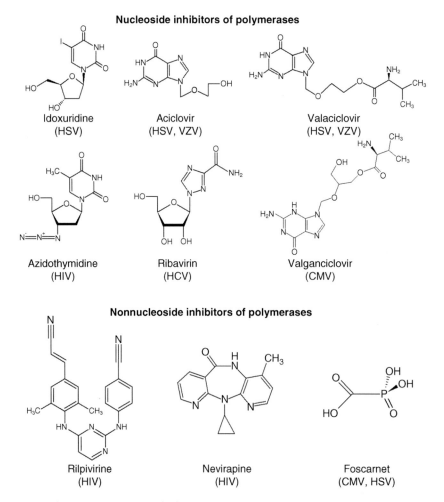

Fig. 5.15   Chemical structure of some important antiviral substances.

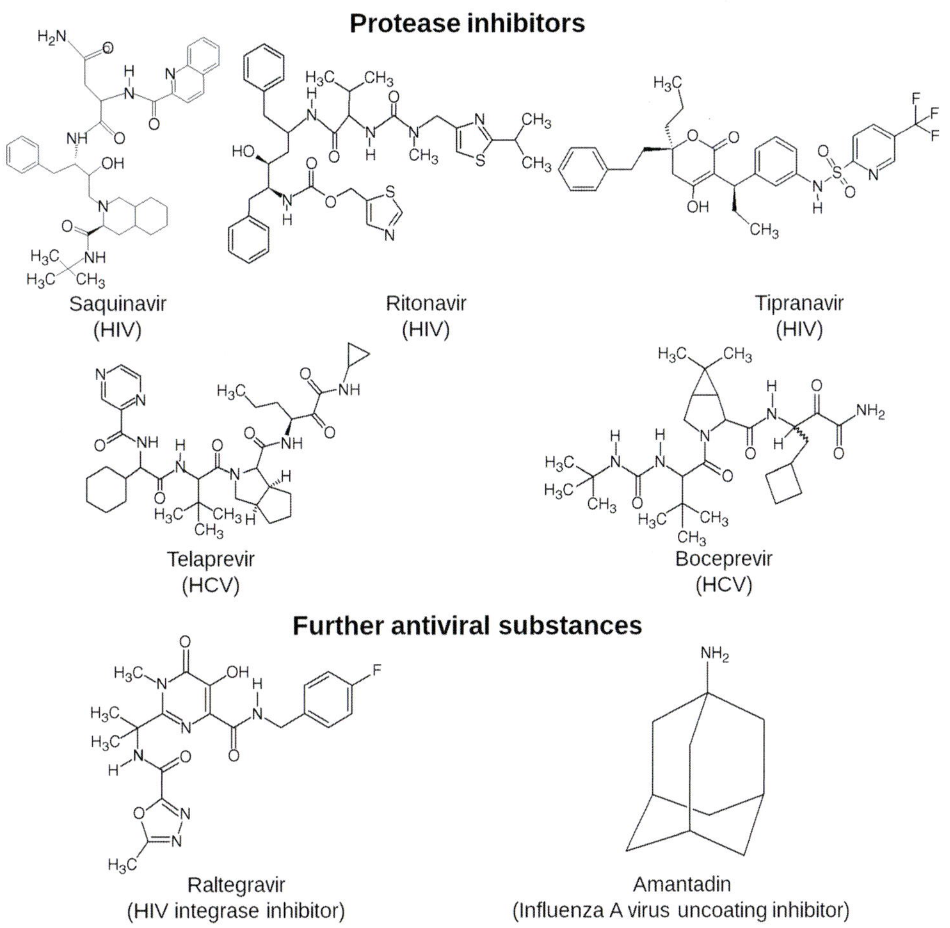

Protease inhibitors

Saquinavir
(HIV)

Ritonavir
(HIV)

Tipranavir
(HIV)

Telaprevir
(HCV)

Boceprevir
(HCV)

Further antiviral substances

Raltegravir
(HIV integrase inhibitor)

Amantadin
(Influenza A virus uncoating inhibitor)

Fig. 5.15   (*Continued*)

of RSV infections. Antisense strategies based on the use of antisense oligonucleotides, ribozymes, and small interfering or short hairpin RNAs (Chapter 13) are also attractive, since they can be directed against virtually any viral sequence without any deeper knowledge of the features of a given virus. The general principles of the treatment of some major viruses will be described in the following section.

5.4.1
Human Immunodeficiency Virus (HIV)

As already mentioned, by far the greatest number of antiviral substances have been approved for the treatment of HIV. A major breakthrough came when azidothymidine (AZT, also called Zidovudine, Figure 5.15) was approved as the first drug for the treatment of HIV in 1987. AZT is a nucleoside analog reverse-transcriptase inhibitor (NRTI). After crossing the cell membrane, cellular enzymes convert AZT to the effective

triphosphate form, which is incorporated into newly synthesized DNA strands by the HIV reverse transcriptase. Since the 3′ OH-group of the ribose is replaced by an azide group, the chain cannot be extended, resulting in premature termination of reverse transcription. Upon prolonged treatment with AZT, HIV develops resistance to the drug due to mutations in its reverse transcriptase. Six point mutations in the reverse transcriptase have been identified. These occur in a sequential manner and confer treatment resistance. Therefore, additional nucleoside and nucleotide inhibitors of the HIV reverse transcriptase have been developed.

Nonnucleoside reverse-transcriptase inhibitors (NNRTIs) are another widely used class of drugs for the treatment of HIV infection. In contrast to the NRTIs, they are noncompetitive inhibitors of the enzyme, binding to a pocket distinct from the active center. They block DNA synthesis by RT by inducing allosteric conformational changes. The advantages of NNRTIs are their high specificity and low toxicity. However, the NNRTI-binding pocket tolerates

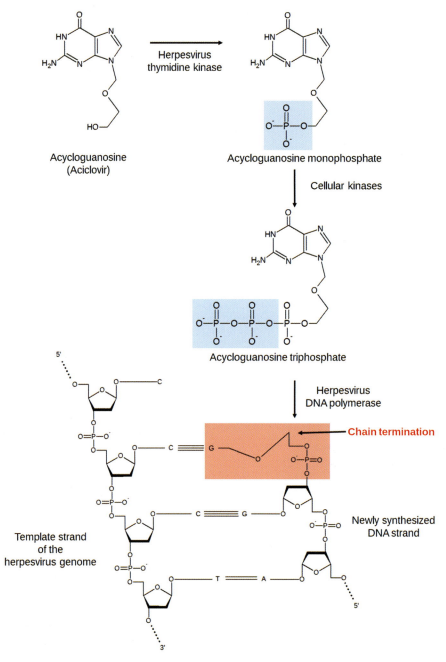

Fig. 5.16   Mode of action of Aciclovir. Aciclovir (acycloguanosine) is administered as an inactive prodrug. The herpesvirus thymidine kinase phosphorylates the nucleoside analog to give the monophosphate (acyclo-GMP). Cellular kinases add two further phosphate groups. Acyclo-GTP is then incorporated into the growing DNA strand by the viral DNA polymerase. Since the analog does not have a 3′ OH-group, the synthesis of the new strand is terminated.

mutations that do not affect the enzymatic activity of the catalytic site, where the NRTIs bind. Thus, HIV rapidly develops resistance against NNRTI treatment. For first-generation NNRTIs, such as nevirapine, resistance mutations are frequent even after a single treatment. In consequence, second-generation NNRTIs (e.g., rilpivirine) have been designed with a more flexible structure that still

permits binding to the RT despite the accumulation of mutations.

As described above, the protease is an essential enzyme of HIV, cleaving the newly synthesized polyprotein to create mature protein components. Without this cleavage step, HIV virions are produced but remain uninfectious. The HIV protease is, therefore, an

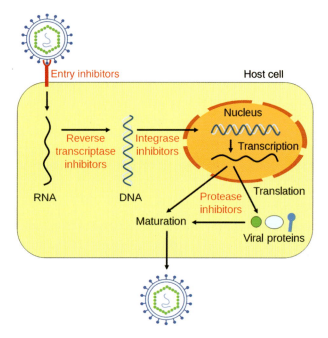

Fig. 5.17 Inhibition of HIV. Antiviral drugs intervene at various points of the HIV life cycle, including cellular entry, reverse transcription, integration into the host genome, and proteolytic maturation of the viral proteins.

attractive target for antiretroviral approaches. Since the mid-1990s, several inhibitors of the enzyme have been approved. Most of the currently available protease inhibitors mimic the structure of the peptide substrates that are recognized by the enzyme, thus blocking the catalytic core (Figure 5.15). The development of these inhibitors was facilitated by the X-ray crystallographic structure of the HIV protease, which allowed computer-based drug design of peptide mimetics specific for the viral enzyme, but which do not inhibit cellular proteases. However, the early protease inhibitors (e.g., saquinavir) had poor bioavailability. The development of ritonavir, which not only blocks the HIV protease but also inhibits cytochrome P450 3A4 in the liver, was a step forward. Cytochrome P450 enzymes metabolize xenobiotics including toxic chemicals and drugs (Section 9.2.1). Inhibition of cytochrome P450 3A4 prevents the breakdown of anti-HIV protease inhibitors and allows lower drug doses to be employed, thus reducing toxic side effects.

Protease inhibitors usually induce the rapid development of resistant viral mutants and are used to treat patients as one component of combination therapy (see below). Since the representatives of this class of drug are structurally similar and have a common mode of action, viruses that are resistant against one protease inhibitor are often also cross-resistant against others. More recently, protease inhibitors such as tipranavir

have been developed that do not mimic the peptide structure. This permits their use as salvage therapy against virus variants that have become resistant to peptidic protease inhibitors. A disadvantage of tipranavir is that it causes more severe side effects, including hepatoxicity and intracranial hemorrhage.

Raltegravir was the first integrase inhibitor to be approved in 2007. It was demonstrated to be safe and effective even in patients failing therapy with multidrug-resistant virus mutants. The drug is believed to affect latent viral reservoirs.

In 2003, a completely different molecular entity was approved for the treatment of HIV infection. Enfuvirtide is a synthetic peptide consisting of 36 amino acids that prevents entry of the virus into host cells (Figure 5.18).

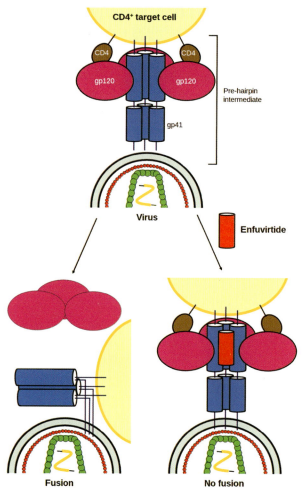

Fig. 5.18 Enfuvirtide. The viral glycoproteins gp120 and gp41 interact with the CD4 receptor of the target cell. Conformational changes are necessary to bring the viral and cellular membranes in proximity for membrane fusion. The peptide enfuvirtide blocks these conformational changes, preventing virus entry into the cell. (Adapted from Ref. [5] with kind permission from Macmillan Publishers Ltd., Copyright 2003.)

The envelope glycoprotein of HIV-1 is composed of two subunits, gp120 and gp41. In the first step, gp120 interacts with the CD4 receptor of the target cells. Conformational changes of gp120 and gp41 bring the viral and cellular membranes in close proximity for membrane fusion. Enfuvirtide binds to gp41, thereby preventing the conformational changes. As a consequence, the membranes cannot fuse, preventing viral entry into the cell.

A second viral entry inhibitor is maraviroc. This compound binds to the CCR5 coreceptor and prevents its interaction with gp120, protecting cells against HIV-1 infection. CCR5 is a particularly attractive target for antiviral drugs in the treatment of HIV infection. Between 5 and 14% of the Caucasian population carry a deletion of 32 bp in this molecule, named CCR5-Δ32, which renders the protein inactive. Carriers of the mutation are protected against infection with the dominant HIV-1 strains that use CCR5 as a coreceptor. In fact, it was shown that replacement of the bone marrow of a HIV-infected patient with stem cells from a donor with the CCR5-Δ32 mutation can cure the affected individual from HIV (Box 5.4). Importantly, the deletion mutation in CCR5 does not seem to have any negative impact on the carrier's health, although CCR5-Δ32 individuals may be more susceptible to infection with the West Nile virus. Nevertheless, CCR5 is an important target for HIV therapy, not only with low-molecular weight drugs, but also with ribozymes, RNA interference, and genome

engineering approaches (Chapter 13 and Box 11.3, Section 11.4.3).

Soon after the introduction of the first drugs for the treatment of HIV infection, the rapid emergence of resistant mutant strains was noticed. This occurs mainly because HIV RT lacks proofreading activity. The high error rate of the polymerase, in combination with the rapid replication cycle of the virus, results in high genetic variability. As a consequence, viral resistance against antiretroviral drugs has emerged, followed as a countermeasure, since the 1990s, of combining multiple antiretroviral drugs. This strategy is widely known as HAART (highly active antiretroviral therapy), recently and more correctly called cART (combination antiretroviral therapy). Frequently used cART regimens combine two NRTIs with an additional NNRTI, a protease inhibitor or an integrase inhibitor. Other combinations consist of four drugs. Another aim of therapeutics is to combine the complex drug regimens into a single pill. These so-called fixed-dose combinations increase the ease of maintaining correct therapy and patient compliance. However, improper use of these drugs is a major reason for the development of resistant virus strains. cART currently allows long-term suppression of the virus in infected individuals. However, the treatment may have severe adverse effects, including liver toxicity, renal insufficiency, neuropathy, or lipodystrophy.

Box 5.4. The Berlin Patient

The Berlin Patient, Timothy Ray Brown, is the first man in the world considered to be cured from an HIV infection. This US citizen, who was living in Berlin, tested positive for HIV in 1995 and received antiretroviral therapy. In 2006, he developed acute myelogenous leukemia (AML). Since chemotherapy was not successful at disease eradication, he received a stem cell transplant (SCT). The German hematologist Gero Hütter found a bone marrow donor with a homozygous CCR5-Δ32/Δ32 mutation. Originally, the recipient himself was heterozygous for CCR5-Δ32, but after the transplantation, the patient's CD4+ T-cells circulating in the blood became homozygous for CCR5-Δ32. No HIV could be detected in the blood and bone marrow and biopsies from various organs were negative for HIV as well. Brown's antiretroviral drug treatment was discontinued. In a

follow-up 45 months after SCT, the patient still had no evidence of HIV. The authors of the study therefore concluded that the patient was cured of his infection. Unfortunately, the mortality risk associated with bone marrow transplants contraindicates its use in HIV-positive individuals without leukemia or lymphoma. In addition, there is an ongoing debate whether it is premature to consider the patient cured of HIV, since the virus might still be present in a latent form in organs that are not easily biopsied. Furthermore, virus strains that use CXCR4 instead of CCR5 as a coreceptor may become dominant and resist this type of treatment. The case of the Berlin patient has contributed to a recent shift in the research paradigm; that is, whether to use lifelong antiretroviral therapy to suppress the virus, or to attempt to cure individuals infected with HIV.

5.4.2
Hepatitis C Virus

For many years, the standard of care for HCV infection has been combination therapy with PEGylated interferon-α (PEG-IFN-α) and ribavirin. IFN-α production results from the innate immune response to virus infection and enhances the defense against viruses by stimulating T-lymphocytes. PEGylation of the cytokine extends its duration of action (Chapter 10). Ribavirin (Figure 5.15) is a nucleoside analog that inhibits the viral polymerase. However, depending on the genotype, combination treatment achieves disease cure in only 40–80% of patients, and is associated with severe side effects such as fever, fatigue, and anemia. Therefore, new drugs have been developed for the treatment of HCV infection. Two examples are the protease inhibitors telaprevir and boceprevir, which are given in combination with PEG-IFN-α and ribavirin. These protease inhibitors, which bind to a serine residue in the catalytic core of the protease, increase the cure rates in genotype 1 patients from approximately 40% to about 80%.

A breakthrough in the treatment of HCV infections was the 2013 approval of sofosbuvir (trade name Sovaldi) by the FDA. The produg is metabolized to the active antiviral agent inside hepatic cells and inhibits the viral RNA polymerase. In combination with other drugs, sofosbuvir can effectively cure hepatitis in 90% of patients. It is administered in combination with ribavirin for therapy of HCV genotypes 2 and 3. This is the first regimen that is all-oral and interferon-free, which reduces side effects. Although for genotypes 1 and 4 pegylated interferon must be injected along with oral administration of sofosbuvir and ribavirin, the approval of sofosbuvir has ushered in the postinterferon era in HCV treatment; the development of new drugs and regimens can be expected in the near future. Another innovative approach to inhibiting HCV, by blocking liver-specific microRNA 121, an essential host factor during virus replication, is described in Section 13.4.2.

5.4.3
Influenza Virus

In the 1960s, two adamantane derivatives, amantadine (Figure 5.15) and rimantadine, were described as the first inhibitors of influenza A virus (but not type B or C). Both are uncoating inhibitors, but it took decades to elucidate their mechanism of action. Amantadine binds to the viral M_2 protein (Figure 5.8), which forms a tetrameric proton channel in the virion envelope. Protons that enter the virion through this channel induce a conformational change that allows uncoating and release of the viral genome from the endosome. This process is blocked by amantadine and rimantadine. Since influenza strains occurring today are virtually completely resistant to these drugs, they are no longer recommended for the treatment of influenza.

Current anti-influenza treatment is based on the inhibition of neuraminidase, which cleaves sialic acid from membrane glycoproteins and releases new virions from infected cells (Figure 5.19a). Based on the three-dimensional structure of neuraminidase, the analog zanamivir (Figure 5.19b) was developed. This drug binds to the active site of the enzyme and blocks its cleavage activity. Since zanamivir is not orally bioavailable and must be administered as an aerosol, an alternative neuraminidase inhibitor with improved pharmacokinetic properties was developed. Oseltamivir (trade name Tamiflu) has high oral bioavailability as it is administered as an ester prodrug. After cleavage, the prodrug gives rise to the active carboxylate molecule (Figure 5.19b). However, the efficacy of oseltamivir has been seriously questioned. In addition, widespread resistance to oseltamivir in influenza viruses from recent flu samples has been observed. A single point mutation in the neuraminidase enzyme is sufficient to confer resistance to oseltamivir.

5.4.4
Other Viruses

Due to the large number of infected individuals, *HBV* is a major health problem worldwide (see above). HBV utilizes an RNA-dependent DNA polymerase that is similar to the reverse transcriptase of HIV. In addition to interferon-α, which stimulates the immune system, a number of nucleoside and nucleotide analogs that inhibit the polymerase have been approved for the treatment of HBV infection. Some of these compounds are also in use for the treatment of HIV infection. These inhibitors are usually given as prodrugs that are converted to their active form *in vivo*.

Aciclovir is one of the most commonly used antiviral drugs. This guanosine analog is primarily indicated for the treatment of *HSV* and *VZV* infections. The prodrug is phosphorylated by the viral thymidine kinase to give the monophosphate, which is further phosphorylated into the active triphosphate by cellular kinases. Incorporation of the compound into the growing DNA strand results in premature chain termination (Figure 5.16). As outlined above, aciclovir has poor oral bioavailability and more recently developed molecular variants such as valacyclovir and famciclovir demonstrate improved oral absorption.

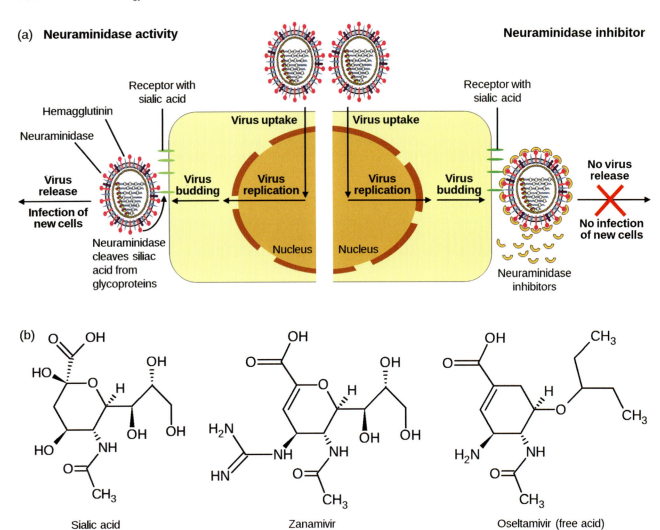

Fig. 5.19 Neuraminidase inhibitors. (a) Mode of action of neuraminidase inhibitors. Neuraminidase cleaves sialic acid from membrane glycoproteins. Inhibition of the enzyme prevents virions from being released from the surface of infected cells. (b) Structures of sialic acid and the neuraminidase inhibitors zanamivir and oseltamivir.

CMV infections are especially threatening for immunosuppressed individuals, such as AIDS patients. Treatment is based on foscarnet, a nonnucleoside inhibitor of the viral polymerase, in addition to ganciclovir and cidofovir, which are nucleoside analogs. However, these compounds may have severe toxic side effects (neutropenia and nephrotoxicity). They are also not orally bioavailable. This problem has been circumvented by the development of valganciclovir, a valine ester of ganciclovir. This drug has improved oral bioavailability and is now the favored treatment for CMV infection.

Although smallpox was declared eradicated by the WHO, the identification of antiviral agents that inhibit the *variola virus* has gained new interest as

the virus may be employed as a biological weapon. Several compounds are in development to inhibit the orthopoxviruses.

Hemorrhagic fever viruses (e.g., dengue, yellow fever, Ebola, Marburg, and Lassa) cause severe disease characterized by fever and coagulopathy that in many cases progresses to high fever and death. In 2014, an Ebola outbreak spread outside of Africa for the first time. Although frequent outbreaks of illness have been recorded and these viruses are also considered to be potential biological weapons, antiviral agents are in a comparatively early stage of development. In addition to small molecular weight substances, small interfering RNAs and antisense morpholino oligomers have also been evaluated (Chapter 13).

5.5
Prions

For decades it was thought that all pathogens employed nucleic acids to direct their replication. This dogma was revised when confronted by the "protein-only hypothesis," which states that the infectious agents of prion diseases are proteins in a misfolded state. The term prion is derived from the expression "*Pr*oteinaceous *in*fectious *on*ly," indicating protein pathogenicity occurring in the absence of nucleic acids. Prions were discovered by Stanley B. Prusiner, who faced much opposition before his idea was accepted. Prusiner was awarded the Nobel Prize in Physiology or Medicine in 1997 for his achievements.

Prion disease in cattle, also known as BSE or mad cow disease, drew much attention in the late 1980s when the disease reached epidemic proportions among cattle in the United Kingdom. Several hundred cases have been reported in which BSE was transmitted to humans by eating meat from BSE-infected cattle. The human form of the disease was called (new) variant CJD, since it closely resembled sporadic CJD, a rare human disease that arises spontaneously. Sheep develop a similar disease known as scrapie. All of these prion diseases share a common feature: Neurons develop large vacuoles that give brain tissue a sponge-like appearance. The diseases are thus collectively known as transmissible spongiform encephalopathies (TSEs).

Prion diseases are transmitted by a hydrophobic protein named "prion protein" (PrP). The protein is a normal constituent of the cell; however, its normal biological function still remains elusive. The prion protein exists in two conformations, the normal cellular conformation known as PrP^C and the pathogenic scrapie conformation known as PrP^{Sc}. Both conformations have the same amino acid sequence, but PrP^C is rich in α-helices (Figure 5.20a), whereas PrP^{Sc} has higher β-sheet content (Figure 5.20b). If a prion (PrP^{Sc}) enters a healthy organism, it induces existing, properly folded proteins (PrP^C) to convert into the misfolded prion (PrP^{Sc}) form. The prion acts as a template to guide the misfolding of more proteins into the prion form, thus mediating the spread of the infection. The exact mechanism by which this conversion happens is still debated.

The disease initially develops slowly; the incubation period for prion diseases can take decades. However, once the first symptoms arise, the disease progresses rapidly and leads to death within a few months. To date, there is no reliable method to detect PrP^{Sc} except for a postmortem neuropathological examination of the brain. Intensive efforts aimed at detecting PrP^{Sc} in the plasma

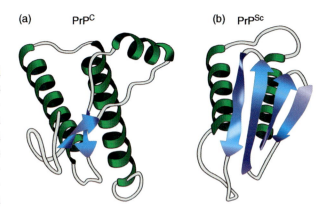

Fig. 5.20 Prion protein conformations. (a) The cellular conformation of the prion protein (PrP^C) is rich in α-helices (green) and has a low content of β-sheets (blue). (b) The pathological conformation (PrP^{Sc}) has a higher overall content of β-sheets, since the N-terminal region, which is flexible and disordered in the normal cellular conformation, adopts a β-sheet structure.

by immunological methods or on the basis of its high resistance to proteases, are ongoing. In addition, non-invasive imaging approaches are being developed. However, the usefulness of prion diagnosis has been questioned, as no treatment is available. Therapeutic developments include the use of antibodies that bind to the cellular prion protein and prevent the conformational transition into PrP^{Sc}. However, the availability of an approved therapy for CJD is not expected in the near future.

References

1. Lozano, R., Naghavi, M., Foreman, K. *et al.* (2012) Global and regional mortality from 235 causes of death for 20 age groups in 1990 and 2010: a systematic analysis for the Global Burden of Disease Study 2010. *Lancet*, **380**, 2095–2128.
2. Alberts *et al* (2008) *Molecular Biology of the Cell*, 5th edn., Garland Science/Taylor & Francis LLC.
3. Costin, J.M. (2007) Cytopathic mechanisms of HIV-1. *Virol. J.*, **4**, 100.
4. MacPherson, G. and Austyn, J. (2012) *Exploring Immunology: Concepts and Evidence*, Wiley-VCH Verlag GmbH, Weinheim, Germany.
5. LaBonte, J., Lebbos, J., and Kirkpatrick, P. (2003) Enfuvirtide. *Nat. Rev. Drug Discov.*, **2**, 345–346.

Further Reading

Virology

Barre-Sinoussi, F., Ross, A.L., and Delfraissy, J.F. (2013) Past, present and future: 30 years of HIV research. *Nat. Rev. Microbiol.*, **11**, 877–883.

Cann, A.J. (2011) *Principles of Molecular Virology*, 5th edn, Academic Press, Waltham, MA.

Carter, J. and Saunders, V. (2013) *Virology: Principles and Applications*, 2nd edn, John Wiley & Sons, Inc., Chichester, UK.

Costin, J.M. (2007) Cytopathic mechanisms of HIV-1. *Virol. J.*, **4**, 100.

Vaccines

Couch, R.B. (2008) Seasonal inactivated influenza virus vaccines. *Vaccine*, **26** (Suppl. 4), D5–D9.

Garcon, N. and Goldman, M. (2009) Boosting vaccine power. *Sci. Am.*, **301**, 72–79.

Kew, O.M., Sutter, R.W., de Gourville, E.M. *et al.* (2005) Vaccine-derived polioviruses and the endgame strategy for global polio eradication. *Annu. Rev. Microbiol.*, **59**, 587–635.

Kwong, P.D., Mascola, J.R., and Nabel, G.J. (2012) The changing face of HIV vaccine research. *J. Int. AIDS Soc.*, **15**, 17407.

Markowitz, L.E., Tsu, V., Deeks, S.L. *et al.* (2012) Human papillomavirus vaccine introduction: the first five years. *Vaccine*, **30** (Suppl. 5), F139–F148.

Virus Diagnosis

Ratcliff, R.M., Chang, G., Kok, T. *et al.* (2007) Molecular diagnosis of medical viruses. *Curr. Issues Mol. Biol.*, **9**, 87–102.

Antivirals

Das, K. and Arnold, E. (2013) HIV-1 reverse transcriptase and antiviral drug resistance: Part 1. *Curr. Opin. Virol.*, **3**, 111–18.

De Clercq, E. (2013) Antivirals: past, present and future. *Biochem. Pharmacol.*, **85**, 727–44.

Fox, J.L. (2007) Antivirals become a broader enterprise. *Nat. Biotechnol.*, **25**, 1395–1402.

Prions

Roettger, Y., Du, Y., Bacher, M. *et al.* (2013) Immunotherapy in prion disease. *Nat. Rev. Neurol.*, **9**, 98–105.

Bacteria and Eukaryotic Pathogens

6

Summary

- Humans face attack by a large variety of pathogenic microorganisms, including bacteria, eukaryotic fungi, and parasites that may be single-celled or multicellular. Major pathogen killers worldwide include *Mycobacterium tuberculosis*, the cause of tuberculosis, and the single-celled eukaryotic parasite *Plasmodium falciparum*, the cause of falciparum malaria. Each claims approximately 1.2 million lives every year.
- Vaccines can be employed to confer immunity against pathogens. In addition to vaccines, which are also used to protect individuals from viruses, toxoid vaccines (inactivated bacterial toxins) and conjugate vaccines, which consist of a carbohydrate linked to a protein carrier, are also available to protect against bacterial infection.
- Bacteria can be diagnosed by cultivation and microscopic examination, or by more modern methods such as real-time PCR or genome sequencing.
- For almost 80 years, antibiotics have successfully been used to treat bacterial infections. Most of the antibacterial drugs target specific bacterial metabolic pathways (e.g., cell wall synthesis). However, the increasing prevalence of multiresistant bacteria, which are not susceptible to any known antibiotic, has become a serious medical problem in recent years.
- Fungal infections are particularly severe in immunocompromised individuals. Single-celled eukaryotic parasites often have at least two hosts. For example, protozoans of the genus *Plasmodium* are transmitted to humans from *Anopheles* mosquitoes. Infections with multicellular parasites often have a comparatively low mortality rate, but may have devastating consequences as chronic illnesses.

Contents List

Bacteria

- Pathogenic Bacteria
- Bacterial Vaccines, Diagnostic, and Antibiotics

Eukaryotic Pathogens

While the majority of microorganisms are harmless to humans, some cause disease. Pathogens can be grouped into different classes according to their complexity (Figure 6.1). Viruses are the smallest classical pathogens that contain their own genetic material. They have extensively been covered in Chapter 5. Bacteria are approximately 100–500 times larger than viruses. The bacterial cell is surrounded by a cell membrane and contains a double-stranded DNA genome, mostly as a single circular chromosome. As bacteria are prokaryotes, they do not have membrane-bound organelles (e.g., a nucleus) within the cell. Bacteria have evolved diverse metabolic strategies. Bacteria can live either outside of cells or within a host's cells. Eukaryotic pathogens can be divided into several classes: The large group of yeasts and fungi can be either single-celled or multicellular. Parasites comprise another class of eukaryotic pathogen. Protozoal (single-celled) parasites usually have at least two hosts, which they switch between during their reproductive cycle. Metazoal (multicellular) pathogens, mostly worms, can grow up to several meters in length. They are obviously too large to live inside cells, and live in body cavities, such as the intestines.

Three major worldwide killers are evenly distributed between these different classes of pathogen: The human immunodeficiency virus (HIV), which was introduced in Section 5.1.1, claims approximately 1.5 million lives every year, while the *M. tuberculosis* and the protozoal parasite *P. falciparum*, the cause of falciparum malaria, each kill approximately 1.2 million people annually.

Molecular Medicine: An Introduction, First Edition. Jens Kurreck and Cy Aaron Stein.
© 2016 Wiley-VCH Verlag GmbH & Co. KGaA. Published 2016 by Wiley-VCH Verlag GmbH & Co. KGaA.

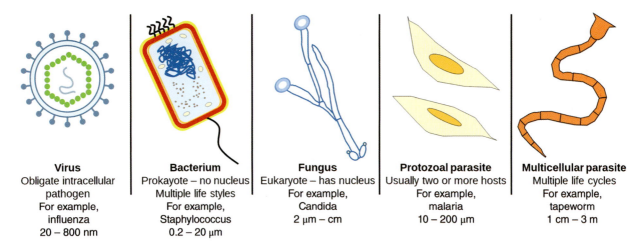

Virus	Bacterium	Fungus	Protozoal parasite	Multicellular parasite
Obligate intracellular pathogen	Prokayote – no nucleus	Eukaryote – has nucleus	Usually two or more hosts	Multiple life cycles
For example,	Multiple life styles	For example,	For example,	For example,
influenza	For example,	Candida	malaria	tapeworm
20 – 800 nm	Staphylococcus	2 µm – cm	10 – 200 µm	1 cm – 3 m
	0.2 – 20 µm			

Fig. 6.1 Major classes of pathogens. Viruses are the smallest pathogens containing their own genetic material. Bacteria are a large group of prokaryotes lacking a nucleus. Eukaryotic pathogens may be single-celled or multicellular yeast and fungi. Parasites are also eukaryotic pathogens and can be single-celled (protozoal) or multicellular (metazoal) parasites. (Adapted from Ref. [1] with kind permission from Wiley-VCH Verlag GmbH.)

Many potentially infectious agents are normally non-pathogenic to healthy individuals. *Opportunistic pathogens* usually do not cause disease in a healthy host. They take advantage of certain situations (opportunities) to become pathogenic. A compromised immune system is one of the most frequent reasons for the development of an *opportunistic infection*. Patients infected with HIV serve as an example. When infected with HIV, these patients become susceptible to the fungal organism *Pneumocystis jirovecii*. This organism is normally not harmful to healthy individuals, but can cause *Pneumocystis pneumonia* in immunocompromised hosts.

As described previously for viruses, the pathogenicity of bacteria may depend on the route and site of infection. For example, the bacterium *Escherichia coli* forms a part of the normal gut flora, where it produces vitamin K and prevents the establishment of pathogenic bacteria. However, if *E. coli* penetrates the urinary tract, these bacteria can cause inflammation of the bladder and urosepsis. In addition, highly pathogenic strains of *E. coli* can cause severe, sometimes lethal, diarrhea.

Vaccines are available to protect individuals from many infections. The basic principles of vaccination were extensively described in Section 5.2. Only some specific aspects of bacterial vaccination, for example, the use of toxoid or polysaccharide vaccines, will be outlined in the present chapter. Advancements made through the application of real-time PCR and sequencing technologies have similar applications in the diagnosis of both viral and bacterial infection. However, treatment options differ substantially. The difficulties in developing antiviral agents were discussed in Section 5.4. Many antibiotics have been successfully employed for almost 80 years to treat bacterial infections. However, the prevalence of multiresistant bacterial strains has increased in recent years. Intensive efforts are ongoing to develop novel antibiotics based on our increased molecular understanding of critical biochemical processes in prokaryotes. There is also an urgent and unmet medical need for effective treatment of parasitic infections. However, these diseases are not the main focus of most pharmaceutical companies as the majority of individuals infected by parasites live in underdeveloped countries.

Like bacteria, microorganisms may cause disease in two general ways: (i) They may directly harm the host, for example, by secreting toxins. (ii) Alternatively, the induction of inflammation subsequent to a host immune response may be responsible for the cell damage that causes disease symptoms.

Because of the dazzling diversity of pathogens, it is not possible to discuss all of them in detail. A few paradigmatic cases will be described in the following sections. The interested reader is referred to the books and articles listed at the end of the section for additional in-depth information.

6.1
Bacteria

Bacteria constitute a large kingdom of prokaryotic microorganisms and are the most abundant form of life on earth. The number of bacterial cells has been estimated to be on the order of $4–6 \times 10^{30}$. Bacteria can be classified by their ability to take up Gram stain. This procedure was developed by Christian Gram in 1884

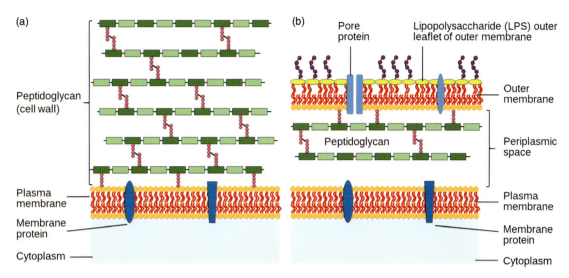

Fig. 6.2 Gram-positive and Gram-negative bacteria. (a) Gram-positive bacteria are characterized by a thick cell wall covering the plasma membrane. (b) In contrast, Gram-negative bacteria have a thin cell wall located in the periplasmic space between two membranes.

and involves the treatment of heat-fixed cells with a crystal violet/iodine dye followed by destaining by ethanol or acetone. Gram-positive bacteria (e.g., *Streptococcus* and *Staphylococcus*) have a single plasma membrane covered by a thick cell wall made of peptidoglycan (Figure 6.2), which is stained purple by crystal violet. In contrast, Gram-negative bacteria such as *E. coli* and *Salmonella* have two membranes. The cell wall is much thinner and is located in the periplasmic space between the membranes. The outer membrane excludes toxic substances (including the Gram stain). Gram-negative bacteria are often, but not always, more resistant to antibiotics than are Gram-positive bacteria.

Alternatively, bacteria can be classified by their morphology. Bacteria typically have one of three basic shapes: spheroidal (cocci), rod-like (bacilli), and helically coiled (spirilla) (Figure 6.3). Modern bacterial classification emphasizes molecular systematics based, for example, on gene sequences that have not undergone extensive lateral gene transfer (e.g., the rRNA gene).

In contrast to viruses, bacteria are self-contained organisms that are generally capable of reproducing independent of their host. Their metabolic pathways are the common targets of antibacterials. Many bacteria are found as symbionts in humans. For example, more than 1000 bacterial species are present in the normal human gut flora of the intestine and perform important functions. However, a comparatively small fraction of bacteria are pathogenic, as outlined in more detail in the following section.

6.1.1
Pathogenic Bacteria

Pathogenic bacteria are a major cause of human disease and death and claim millions of lives every year. Some examples of bacteria with significant clinical relevance are summarized in Table 6.1. As already mentioned, one of the bacterial diseases with the highest burden is tuberculosis, which kills 1.2–1.5 million people every year, mostly in low- and middle-income countries. Several hundred thousand deaths occur in individuals coinfected with HIV. Other globally important diseases caused by bacteria are pneumonia (e.g., by *Streptococcus*), meningitis (e.g., by *Neisseria meningitides*), and foodborne illnesses (e.g., *Shigella*, *Campylobacter*, and *Salmonella*). Some bacterial pathogens have caused devastating pandemics in the past, but can now be treated with the help of antibiotics (Box 6.1).

Coccus   **Bacillus**   **Spirillum**

Fig. 6.3 Typical shapes of bacteria. Bacteria have a spheroidal (coccus), rod-like (bacillus) or helically coiled (spirillum) shape. (Reproduced with permission from Ref. [2], Garland Science/Taylor & Francis LLC, Copyright 2008.)

Table 6.1 List of selected pathogenic bacteria with high numbers of fatalities worldwide.

Species	Gram stain	Disease	Annual number of deaths (2010)
Mycobacterium tuberculosis	Positive	Tuberculosis	1 196 000
Streptococcus pneumoniae	Positive	Pneumococcal pneumonia	827 300
		Pneumococcal meningitis	118 400
Haemophilus influenzae	Negative	H. influenzae type B pneumonia	379 900
		H. influenzae type B meningitis	83 000
Escherichia coli	Negative	Enteropathogenic E. coli	88 700
		Enterotoxogenic E. coli	120 800
Salmonella typhi	Negative	Typhoid and paratyphoid fever	190 200
Shigella sonnei	Negative	Shigellosis	122 800
Treponema pallidum	Negative	Syphilis	113 300
Campylobacter jejuni	Negative	Campylobacter enteritis	109 700
Bordetella pertussis	Negative	Whooping cough	81 400
Salmonella	Negative	Salmonellosis	81 300
Neisseria meningitidis	Negative	Meningitis	75 000
Clostridium tetani	Positive	Tetanus	61 300
Vibrio cholerae	Negative	Cholera	58 100
Corynebacterium diphtheriae	Positive	Diphtheria	2900

Source: Deaths numbers based on Ref. [3].

Box 6.1. The Plague, Cholera, and Tuberculosis

Today, pandemics are usually associated with viral infection by, for example, HIV, Corona-SARS, or swine flu. However, bacteria were the cause of devastating pandemics in the past (and still cause more deaths than viruses, since many deaths associated with a viral infection are actually caused by a secondary bacterial infection). Plague and cholera are two examples of diseases that have killed millions of people in the past, but are controllable today. In contrast, tuberculosis is still a leading cause of death worldwide.

Plague is caused by the bacterium *Yersinia pestis*, which is spread to humans via fleas that are carried by rodents (mostly rats). Plague caused three major pandemics. The second, and most devastating pandemic, known as the "Black Death," is thought to have originated in China in 1334, and then traveled along the great trade routes to Constantinople, and finally to Europe. The pandemic claimed an estimated 40–60% of the European population in the mid-fourteenth century.

When a flea bites a human, *Y. pestis* bacteria are passed into the tissue and enter the lymphatic system. The bacterium causes the lymph nodes to swell, producing the characteristic bubo associated with the disease. When bacteria enter the bloodstream, they can cause septicemic plague. The bacterial endotoxins cause disseminated intravascular coagulation. This results in the depletion of the body's clotting factors, and cutaneous and internal bleeding. People often died on the day the first symptoms appear, as septicemic plague could not be treated. Pneumonic plague infects the lungs and can be transmitted from one person to another through respiratory droplets. Untreated, the mortality is almost 100%. Protective clothing was developed as a countermeasure to prevent infection (Figure 6.4).

The pathogenicity of *Y. pestis* is caused by endotoxins and exotoxins. With the advent of antibiotics and improved sanitation, plague has almost disappeared as a pandemic disease; only a few cases have been reported in recent years. If diagnosed early, plague is highly responsive to antibiotics such as streptomycin, chloramphenicol, tetracycline, gentamicin, and doxycycline (Section 6.1.2). An attenuated live vaccine has been developed, but it is recommended only in special cases because of limited disease prevalence.

As a highly contagious disease with an extremely high mortality rate, *Y. pestis* has a long history as a biological weapon. It had been used as such in ancient China and against Genoese in the Crimea by the

Fig. 6.4 The plague doctor. The copper engraving shows Doctor Schnabel, a plague doctor in seventeenth-century Rome. He wears a hat, a mask that looks like a bird beak, goggles, and a long gown. His clothing identifies him as a plague doctor and is intended as protection. The beak of the mask was often filled with spices, thought to prevent the wearer from becoming infected.

Mongols in the 1340's, and was being developed in research programs after World War II. Appalling strategies considered included contamination of drinking water, catapulting plague victims over city walls (like the Mongols did), and – more recently – dropping infected fleas from airplanes or spreading aerosols.

The bacterium *Vibrio cholerae* has been responsible for seven cholera pandemics in the last 200 years. The seventh pandemic began in Indonesia in 1961 and is currently ongoing. The disease spread through Asia to Africa, Europe, and Latin America. However, because of improved case management, the number of deaths has decreased from an estimated 3 million per year in the 1980s to 100 000 or less more recently.

The primary symptoms of cholera are watery diarrhea and vomiting, which can lead to dehydration and electrolyte imbalance, and death in some cases. The cholera bacterium is transmitted primarily by fecal contamination of food and water.

During infection, *V. cholerae* secretes cholera toxin, an oligomeric protein complex that crosses the cell membrane and inhibits the GTPase activity of a G protein. As a consequence, the activity of adenylate cyclase increases so that the level of cAMP increases more than 100-fold above normal. This leads to an overactivation of protein kinase A, which phosphorylates and activates a chloride channel. The imbalance of ion homeostasis results in diminished entry of water into enterocytes and rapid fluid loss from the intestine.

In addition to continued eating, drinking, and rehydration, the therapy of cholera includes antibiotic treatment. Typically, doxycycline is used as the first-line antibiotic, but some strains of *V. cholerae* have become resistant. Several other antibiotics have proven to be effective in treating cholera. Antibiotic therapy of *V. cholerae* infection can decrease mortality from more than 50% to less than 0.2%. Two oral vaccines are available; they contain killed bacterial cells supplemented with a recombinant cholera toxin subunit. Although they are safe and effective, their use is limited for logistic and financial reasons.

While plague has lost much of its terrifying potential and major advancements have been made in the management of cholera, tuberculosis (TB) remains a major medical problem. Approximately one third of the world's population has been infected with *M. tuberculosis* and 1.2–1.5 million people die of TB every year, making it second to HIV/AIDS as the greatest infectious killer worldwide. Double infections with HIV and *M. tuberculosis* represent a particular problem, as TB causes approximately one quarter of all deaths of people living with HIV.

The mycobacterium is usually transmitted as an aerosol and most commonly affects the lungs. The infection remains asymptomatic and latent in 90% of those infected. The bacterium is phagocytized by alveolar macrophages, but survives and replicates within cellular endosomes. Lymphocytes (and other cells) surround the infected macrophages and collagen is then laid down, giving rise to structures termed granulomas. These nodules, also known as tubercles, prevent dissemination of the pathogen, but mycobacteria inside the granuloma can become dormant, resulting in latent infection. Necrosis may develop in the center of the tubercles. If the immune system is compromised, as in patients with AIDS, the infection may be reactivated. When mycobacteria replicate in the endosomes and eventually kill the macrophages, molecules are released that induce local inflammation. An important part of the tissue damage in TB is not caused by the bacteria, but rather is a result of the host's own immune response. In active TB, bacteria may not only disseminate throughout the lungs, but may even gain entry to the bloodstream and spread throughout the body.

Bacillus Calmette–Guérin (BCG) (Section 6.1.2), an attenuated live vaccine, can provide protection against infection with *M. tuberculosis* for up to 15 years. However, the efficacy of the vaccine is considerably lower in countries where the bacterium is still epidemic.

Streptomycin, one of the first antibiotics, has been used for the treatment of TB. However, effective treatment is complicated due to the structure and composition of the mycobacterial cell wall, which hinders the entry of drugs. Currently, the first-line agents for the treatment of TB infection are isoniazid, rifampicin, ethambutol, and pyrazinamide. Treatment requires daily administration of antibiotics for 6 months. A major issue in TB treatment is the recent increase in resistant mycobacterium strains. Multidrug-resistant TB is defined as resistance to isoniazid and rifampicin, the two most effective first-line drugs. In these cases, treatment requires more expensive and more toxic drugs and an even longer treatment regimen.

The WHO declared TB a global public health emergency in 1993 and has initiated programs to control the disease. Subsidized diagnostic tests and drugs have been made available in low- and middle-income countries. New vaccination strategies are being developed, either by improving the BCG vaccine or by creating novel live vaccines. In addition, new drugs have been evaluated for the treatment of resistant TB. In 2012, bedaquiline (trade name Situro), which inhibits the mycobacterial ATP synthetase, was approved by the FDA to treat multidrug-resistant TB. It was the first new tuberculosis drug in 40 years. However, the efficacy and safety of the drug has been disputed since a published study indicated that five times as many patients given bedaquiline died compared to the control group.

Bacteria can be transmitted in multiple ways: from droplets in the air, via contaminated food and drinking water, through sexual contact, and by cuts and abrasion, or via insect bites. Severe disease can be caused by bacterial infection of the lungs (bacterial pneumonia) or the meninges, that is, the protective membranes covering the brain and spinal cord (bacterial meningitis). Infection of the gastrointestinal and urinary tract as well as skin and vaginal infections are common. In some cases, pathogenic bacteria belong to the same group as harmless bacteria. An example is *E. coli*, which is an important constituent of the normal gut flora, but may cause severe, sometimes fatal, gastroenteritis.

Pathogenic bacteria have developed strategies to invade a host, to establish themselves at an optimal site for nutrition, and to be released from the host. To facilitate this, they produce molecules known as *virulence factors*, which may be responsible for causing disease. In many cases, virulence factors are encoded on extrachromosomal genetic elements (e.g., on bacteriophages or plasmids) and can easily spread through horizontal gene transfer. There are a variety of different types of virulence factors, including adhesins, aggressins, impedins, invasins, and toxins. Adhesins are molecules that help the bacteria to bind at a specific optimal host site. Virulence factors of the aggressin class are usually enzymes that have a localized effect. They produce a barrier between the bacteria and the host defense or help to destroy phagocytes. Impedins are virulence factors that act against the host defense system. An example is the polysaccharide capsule of *S. pneumonia*, which protects the bacteria and makes them resistant to phagocytosis. Invasins confer the ability to move within the host. For example, pathogenic strains of *E. coli* differ from commensal strains of *E. coli* in their ability to invade the host by migrating out of the gastrointestinal tract and into the bloodstream.

A large and important group of virulence factors, many causing severe symptoms, are bacterial toxins. These toxins are classified as endotoxins and exotoxins. While exotoxins are secreted by bacteria, endotoxins are normally part of the bacteria and are released only after destruction of the bacterial cell wall (Figure 6.5). Enterotoxins are a subclass of exotoxins (and should not be confused with endotoxins) that are released by bacteria and target the intestine to facilitate bacterial spread. For example, *Staphylococcus aureus* releases an enterotoxin that elicits vomiting, thus spreading the bacteria into the environment. In a similar manner, diarrhea-causing bacteria produce toxins that are released back to into the environment.

Exotoxins have several modes of action: While some cause damage to the host by destroying cells, others interfere with normal cellular metabolism. An example of the latter is the diphtheria toxin: After entering the cell, it inactivates elongation factor-2, thus inhibiting translation. Although the diphtheria toxin is extraordinarily lethal, it is also in clinical evaluation as anticancer drug. By coupling the toxin to ligands of cancer cell-specific receptors, it can be directed to tumor cells and destroy them.

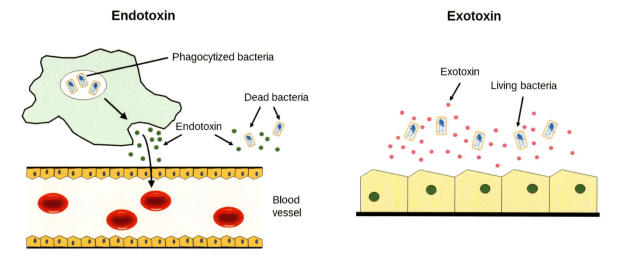

Fig. 6.5 Endotoxins and exotoxins. Endotoxins are part of the bacteria and are only released after destruction of the bacterial cell wall, when the bacteria are phagocytosed or dead. Exotoxins are released by live bacteria into the environment.

The major cytotoxic agent released by the bacterium *S. aureus* is its α-toxin, which destroys the cell membrane by forming a pore. The pore allows the upregulated exchange of ions and eventually results in cell death. The tetanus toxin is one of the most deadly known exotoxins. *Clostridium tetani* produces this neurotoxin when it enters damaged tissue. The toxin remains inactive inside the living bacterium and is released and activated by proteases when the bacterium dies. The toxin amplifies signals from nerves to muscles, which causes a generalized contraction of the musculature, termed a tetanic spasm. As will be outlined in more detail in the next section, vaccination against tetanus is performed with an inactivated form of the toxin (called a toxoid), which induces the generation of antibodies that can neutralize the deadly toxin, preventing its binding to nerves.

The most important endotoxins are lipopolysaccharides. These substances are found in the outer membrane of Gram-negative bacteria (Figures 6.2b and 6.6). Large amounts of endotoxins are mobilized when Gram-negative bacteria are destroyed. The endotoxins then bind to toll-like receptor-4 (TLR-4) on immune cells such as monocytes, macrophages, and dendritic cells and induce the release of a broad spectrum of immune mediators. Septic shock can be the result of this "cytokine storm."

However, endotoxins are not only bacterial pathogenic factors but are also frequent contaminants of biotech products such as recombinant proteins (Chapter 10) or plasmid DNA (Chapter 11). Since endotoxins are highly toxic clinically, they must be removed during the production and purification process. A very sensitive assay for the detection of residual endotoxins is the *Limulus amebocyte*

lysate assay. The assay is based on a lysate from horseshoe crabs. If a sample is contaminated with endotoxins, a protease cascade resulting in coagulation will be activated. To avoid killing large numbers of horseshoe crabs, alternative assays have been developed more recently.

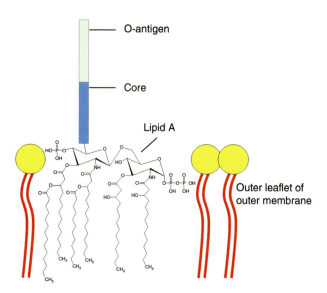

Fig. 6.6 Lipopolysaccharides. Lipopolysaccharides (LPS) are found in the outer membrane of Gram-negative bacteria. They consist of a lipid covalently joined to a polysaccharide. The membrane anchor, known as lipid A, is composed of two glucosamine sugars, to which phosphate groups and six fatty acids are linked. A long, branched polysaccharide denoted as the core is attached to lipid A. The outer O-antigen consists of repetitive oligosaccharide units and is the main determinant of the serotype of a bacterium.

6.1.2
Bacterial Vaccines, Diagnostic, and Antibiotics

6.1.2.1 Vaccines

The basic principles of antiviral vaccination were described in Section 5.2. Only selected vaccines that protect against infection with bacteria will be discussed here, since general strategies, such as the use of attenuated live vaccines, are similar for viral and bacterial vaccines. One example is a vaccine against tuberculosis (Box 6.1), which is named Bacillus Calmette–Guérin after the scientists who developed it (Albert Calmette and Camille Guérin). The BCG vaccine is prepared from a strain of the weakened bovine tuberculosis bacillus, *Mycobacterium bovis*. Its efficacy is limited but produces approximately a 50% reduction in the transmission of tuberculosis. However, this varies depending on the vaccine batch and on the presence of other strains of mycobacteria in the environment of the vaccinated person. The WHO recommends use of the BCG vaccine for children born in countries in which tuberculosis is endemic.

As outlined earlier, the pathogenicity of many bacteria is mediated by toxins rather than by the bacterial organism itself. One vaccination strategy achieves protection by the induction of antibodies against the toxin. Since it is impossible to administer the active toxin for the purpose of immunization, an inactivated form called a toxoid is used. Examples are vaccines for protection against tetanus and diphtheria. Production of toxoids starts with the cultivation of pathogenic bacteria. The released exotoxins are harvested and inactivated by formalin treatment. It is important that the bacterial protein retains its immunogenicity but loses its toxicity. Toxoids are usually administered together with an adjuvant to enhance the immune response. The toxoid vaccines against tetanus and diphtheria are commonly given as a triple combination (DPT) together with killed cells of *Bordetella pertussis*, the cause of whooping cough.

Another class of nonliving bacterial vaccines is directed against the carbohydrate capsule that surrounds many pathogenic bacteria. For example, the vaccine against *S. pneumoniae* consists of purified polysaccharides from 23 serotypes of bacteria. However, polysaccharides induce only a weak immune response in B cells, which release IgM without the assistance of T cells. Thus, the vaccine is ineffective in young children and some adults and immunity is not lifelong. To overcome this problem, conjugate vaccines have been developed in which the carbohydrate is linked to a protein carrier. In a conjugated pneumococcal vaccine, an inactivated and nontoxic form of the diphtheria toxin containing a single amino acid substitution (CRM197) is used. Due to the protein component present, the conjugate is processed and presented to T cells. B cells specific for epitopes on the carbohydrates receive signals from the T cells to produce high levels of IgGs. In addition, long-term immunological memory is induced. Conjugate vaccines for protection against pneumococcus are available as hepta-, deca-, and tridecavalent vaccines that contain polysaccharides from, respectively, 7, 10, and 13 serotypes of *S. pneumonia*. Additional conjugate vaccines are available for meningococcus (*Neisseria meningitidis*) and *Haemophilus influenzae*.

6.1.2.2 Diagnostic

Traditional methods of bacterial identification rely on the phenotypic identification of the causative organism performed, for example, by microscopic examination or culturing. However, microscopic identification of bacteria is unspecific, and culturing may be slow and difficult. As outlined for viruses previously (Section 5.3), molecular techniques have proven helpful in overcoming the limitations of earlier methods. Some of these approaches are based on an analysis of nucleic acids such as the 16S or 23S rRNA. Conserved regions of ribosomal RNA can be amplified by PCR and analyzed by microarray technology or sequencing. Comparing the results to sequences in the databases will identify the pathogenic microorganism.

6.1.2.3 Antibiotics

In contrast to viruses, bacteria have their own metabolism. Some metabolic pathways (e.g., cell wall synthesis) are specific for prokaryotes and are absent in human cells. In other cases, such as the synthesis of proteins by ribosomes, basic cellular processes are similar among prokaryotes and eukaryotes. However, the molecular machinery differs sufficiently so that specific inhibitors can be identified. Over the past century, these specific prokaryote characteristics have allowed the development of a broad variety of antibacterial drugs. Going forward, the terms antibacterials and antibiotics will be used synonymously, although the term antibiotics was originally restricted to substances produced by a microorganism that are antagonistic to the growth of other microorganisms, even at high dilution.

Although mixtures with antimicrobial properties have been used for over 2000 years, the discovery and isolation of penicillin by Alexander Fleming in 1928 (Box 6.2) initiated the modern antibiotic era. Shortly after the isolation of penicillin from a microorganism, Gerhard Domagk at Bayer AG tested chemically synthesized dyes for their antibacterial activity. He discovered that a red dye, later called Prontosil, treated streptococcal infection in mice. French researchers discovered that Prontosil is a prodrug that is metabolized to the active

Box 6.2.   Alexander Fleming

Sir Alexander Fleming was a Scottish biologist who is best known for his discovery of the antibiotic penicillin. On September 3, 1928, he returned to his laboratory from a holiday. He began to sort through Petri dishes containing cultures of S. aureus, which he had grown and left behind on a bench. He noticed that one culture was contaminated with a fungus and that colonies of the staphylococci surrounding it had been destroyed. In contrast, colonies farther away remained unaffected. Fleming realized that the mold had secreted something that inhibited bacterial growth. Fleming grew the mold in pure culture and identified it as Penicillium notatum. He surmised that a low molecular weight substance released by the fungus must have been able to travel through the relatively viscous agar and named this compound penicillin. In subsequent experiments, he observed the antibacterial effect of penicillin on other Gram-positive bacteria, including meningococcus and the diphtheria bacillus, but not on Gram-negative bacteria. Fleming also discovered that bacteria can develop resistance to penicillin when treated with insufficient concentrations of penicillin or for short treatment periods.

However, Fleming did not succeed in purifying sufficient amounts of the compound for further experiments. The substance also proved to be unstable. In the late 1930s, Ernst Chain and Howard Florey, chemists at the Oxford University, began research on penicillin and finally succeeded in producing and purifying larger amounts in an effective and stable form. Clinical trials were conducted achieving cures deemed miraculous. Laboratories in the United States then produced large quantities; many soldiers were successfully treated with penicillin in the last years of World War II. Fleming, Chain, and Florey received the Nobel Prize in Physiology or Medicine in 1945. Although penicillin was only effective against Gram-positive bacteria, its discovery ensured that many microbiologists began to search for new antibiotic-producing organisms.

compound sulphanilamide, a sulfonamide. However, like penicillin, sulfanilamide was only effective against Gram-positive bacteria. In the early 1940s, Selman Waksman and colleagues looked for microorganisms from the soil that release substances with antibacterial properties. By screening a large number of samples they found a promising candidate produced by Streptomyces bacteria, later named streptomycin. The substance was active not only against Gram-positive but also against Gram-negative bacteria, including M. tuberculosis. Many antibacterials from the Streptomyces genus were discovered soon thereafter. In fact, these bacteria are ultimately the source of two thirds of all our antibiotics. Domagk and Waksman received the Nobel Prize in Physiology or Medicine in 1939 and 1952, respectively.

Antibiotics are often called "secondary metabolites" of fungi and bacteria, since they are not required for the growth of the organism, though they provide it with a survival advantage under certain conditions. Since many bacteria and fungi live in the soil, it is no surprise that numerous antibiotics have been isolated from soil microorganisms which compete for a finite supply of environmental nutrients and inhibit the growth of competing bacteria and fungi.

Although a great number of compounds with antibacterial properties have been isolated, many of them were shown to be too toxic for use in humans. In addition,

the pharmacological properties of these antibacterials were inadequate in some cases. For example, penicillin does not survive the acidic conditions of the stomach and must therefore be administered by injection. Chemical modification of the natural product drastically improved the properties of the compound. The addition of an amino group (Figure 6.7) prevented the molecule from being destroyed by stomach acid, permitting oral administration. Furthermore, chemical modification produced antibiotics that were effective not only against Gram-negative bacteria but also against Gram-positive bacteria. One of these penicillin derivatives is known as ampicillin, the most widely used antibiotic in the world. Other types of natural antibiotics (e.g., cephalosporins) have also been optimized by chemical modification.

The production of these semisynthetic substances relied on the production of the natural precursor molecule by, for example, fermentation. Chemists then developed methods for the total synthesis of antibiotics from small and cheap building blocks. These efforts included the synthesis of known compounds such as penicillin, but also included new antibiotics. George Hitchings succeeded in synthesizing a novel compound that inhibited bacteria by blocking the enzyme dehydrofolate reductase. This substance, called trimethoprim, has been used to treat bacterial infections since the 1960s. For his discovery of several important principles of drug treatment, Hitchings

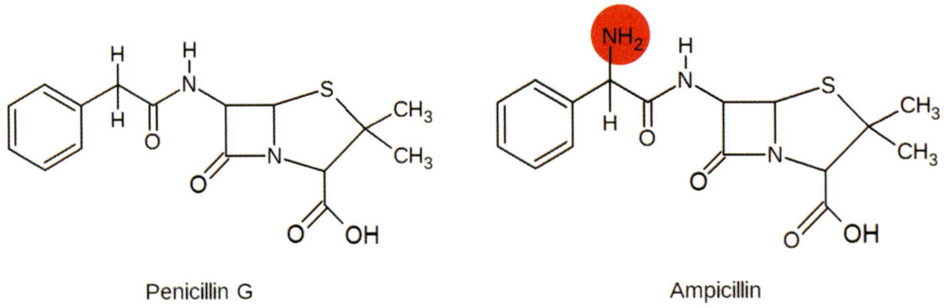

Penicillin G

Ampicillin

Fig. 6.7   Penicillin G and ampicillin. Penicillin G and ampicillin differ only in a single amino group, but ampicillin has substantial advantages: It can be administered orally and is active against Gram-negative as well as Gram-positive bacteria.

received the Nobel Prize in Physiology or Medicine in 1988 (together with Sir James Black and Gertrude Elion).

Antibiotics can be either bactericidal or bacteriostatic. The latter class of antibiotics prevents bacteria from reproducing, but does not kill them. These antibiotics usually interfere with bacterial protein production, DNA replication, or other aspects of bacterial metabolism. Tetracyclines, sulfonamides, and spectinomycin are some examples of this class of antibiotics. Bactericides are substances that kill bacteria, such as the β-lactam antibiotics that inhibit cell wall synthesis. While some antibiotics have a narrow spectrum of activity, that is, only against certain types of bacteria, broad-spectrum antibiotics affect a wide range of bacteria. Antibiotics have many targets in the bacterial cell (Figure 6.8). The main chemical classes of antibiotics and their modes of action are described next.

β-Lactam antibiotics are the most widely used group of antibiotics (Figure 6.9a). They are characterized by the presence of β-lactam ring in the chemical structure. The class of a β-lactam antibiotics includes penams (penicillin derivatives), cephems (cephalosporins), monobactams, and carbapenems. Most β-lactam antibiotics kill bacteria by interfering with cell wall biosynthesis. They bind to and inactivate enzymes that cross-link the peptidoglycan strands. During cell growth, bacteria must open their rigid cell walls to insert new cell wall material. Inhibition of subsequent cross-linking results in cell lysis.

Glycopeptide antibiotics are composed of glycosylated cyclic or polycyclic nonribosomal peptides. A representative of this class of antibiotics is vancomycin. Glycopeptide antibiotics also affect the biosynthesis of the bacterial cell wall, but their mode of action differs from that of β-lactam antibiotics. Glycopeptide antibiotics

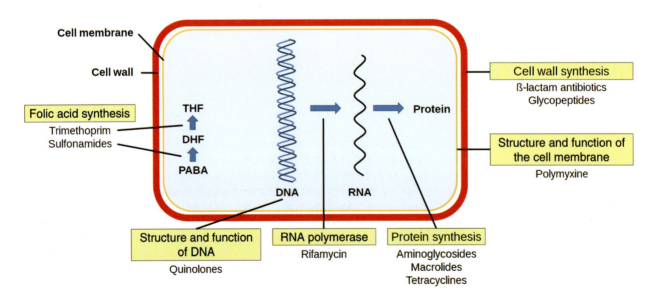

Fig. 6.8   Cellular targets of antibiotics. The major targets of antibiotics are shown. For further details, see the text. PABA: *para*-aminobenzoic acid; DHF: dihydrofolate; THF: tetrahydrofolate.

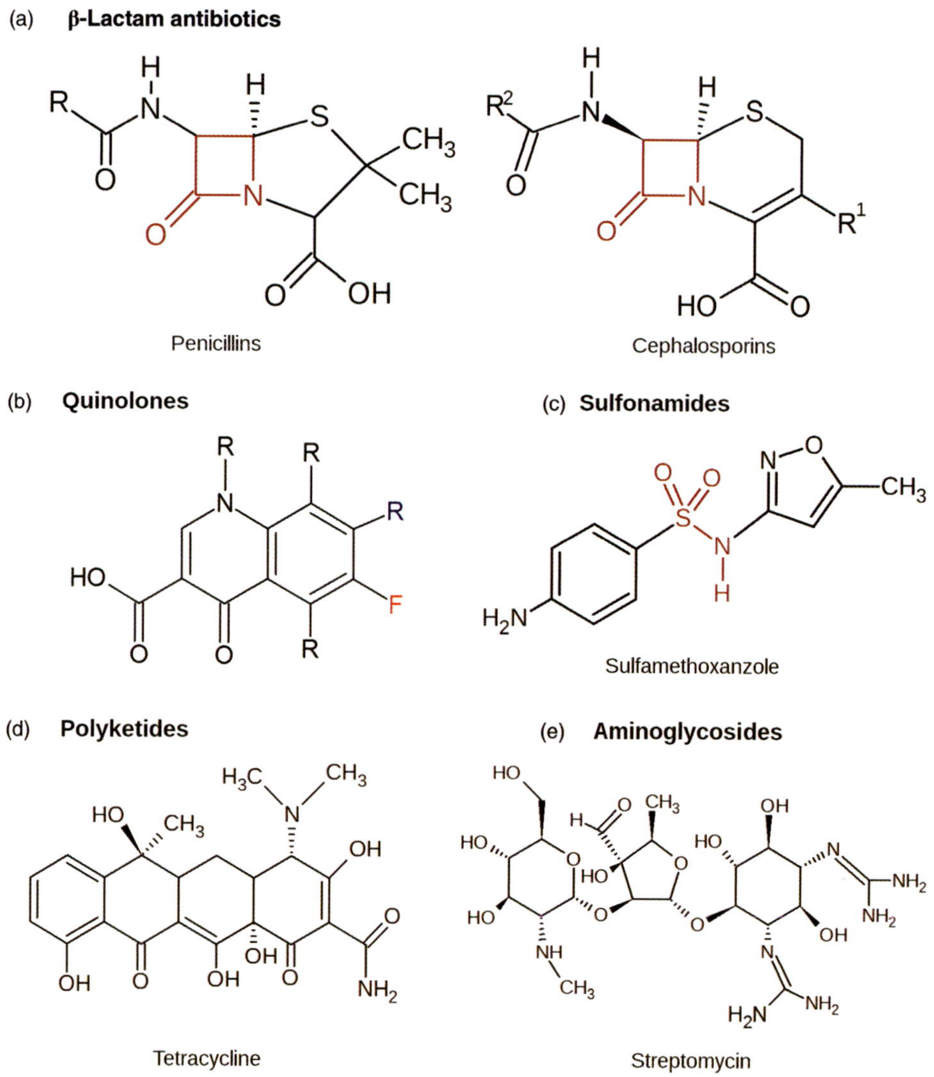

(a) β-Lactam antibiotics

Penicillins

Cephalosporins

(b) Quinolones

(c) Sulfonamides

Sulfamethoxanzole

(d) Polyketides

Tetracycline

(e) Aminoglycosides

Streptomycin

Fig. 6.9 Examples of antibiotics. (a) β-Lactam antibiotics are characterized by the β-lactam ring, indicated in red. (b) Basic structure of quinolones. The R substituent colored blue is usually piperazine. If the compound contains fluorine, drawn in red, it is a fluoroquinolone. (c) The functional group of a sulfonamide antibiotic is shown in red. (d) Tetracycline is a member of the polyketide class. (e) Aminoglycosides are amino-modified sugars.

bind to terminal D-alanyl-D-alanyl sequences in a peptidoglycan and prevent the addition of new units to the peptidoglycan. This leads to the perforation of the bacterial cell wall and subsequent cell lysis.

Several antibiotics act at the level of the bacterial DNA. *Quinolones* (Figure 6.9b) interfere with DNA unwinding and inhibit replication. For many Gram-negative bacteria, DNA gyrase is the target of quinolones, while topoisomerase is the target for many Gram-positive bacteria.

Sulfonamides (Figure 6.9c) interfere with the synthesis of folic acid, an important metabolite in DNA synthesis. They act as competitive inhibitors of the enzyme

dihydropteroate synthetase. Trimethoprim, as above, inhibits the same metabolic pathway, but at a later stage. It binds to dihydrofolate reductase and prevents the reduction of dihydrofolic acid (DHF) to tetrahydrofolic acid (THF). Sulfonamides are usually bacteriostatic.

Rifamycin and its derivatives (e.g., rifampicin) are a class of antibiotics that specifically inhibit prokaryotic RNA polymerases. They are effective against mycobacteria and can be used to treat tuberculosis and leprosy.

Many antibiotics block translation. Determination of the X-ray structure of the ribosome allowed the identification of the exact binding sites of different antibiotics (Figure 6.10). Several inhibitors of protein biosynthesis

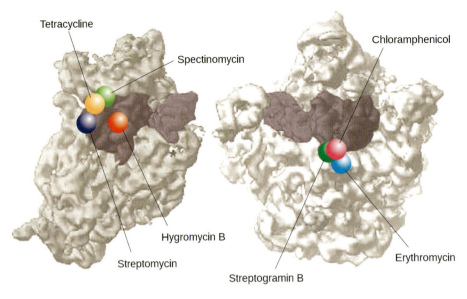

Small ribosomal subunit   **Large ribosomal subunit**

Fig. 6.10 Antibiotics bind to the prokaryotic ribosome. Structural elucidation of the ribosome permitted the exact determination of the antibiotic binding sites. (Adapted from Ref. [4] with kind permission by Macmillan Publishers Ltd., Copyright 2005.)

belong to the group of polyketides, a heterogeneous class of molecules that are usually synthesized through decarboxylative condensation of malonyl-CoA to produce polyketide chains, which are then further derivatized. An example is tetracycline (Figure 6.9d), which binds to the small subunit of the prokaryotic ribosome and prevents the entry of aminoacyl-tRNAs into the A site (Section 1.3.5). Macrolides (e.g., erythromycin) are polyketides that bind to the large ribosomal subunit. They block the tunnel through which the nascent polypeptide chain leaves the ribosome. As a consequence, protein biosynthesis is terminated. Aminoglycosides, for example streptomycin (Figure 6.9e), are composed of amino-modified sugars. They bind to the 30S ribosomal subunit and induce it to misread the mRNA. Nonsense proteins are synthesized that cannot be used by the bacteria. At higher concentrations, translational initiation is prevented, which can cause host cell death.

Polypeptide antibiotics contain nonprotein polypeptide chains. They interfere with the transport mechanisms of the cell membrane. Many polypeptide antibiotics are toxic and cannot be used for systemic treatment, and are usually administered topically. An example is polymyxin B, which is primarily used for infections with resistant Gram-negative bacteria.

Ever since antibiotics have been used to treat infections in humans, bacterial resistance has been a serious problem. Misuse of antibiotics has been identified as the main reason for the rapid emergence of resistant bacteria in recent years. For example, antibiotics are often inappropriately prescribed for people with a viral infection (e.g., the common cold). In addition, many patients take less than the required antibiotic dosage because their symptoms have been relieved. Over the years, the development of multidrug resistant (MDR) pathogens has become a major challenge. This is particularly true for nosocomial infections. Hospital-acquired infections affect millions of patients every year and account for tens of thousands of deaths worldwide. Nosocomial infections are frequently difficult to treat as the pathogens have developed resistance against all available antibiotics. *S. aureus* is one of the major nosocomial pathogens and is often the cause of postsurgical wound infections. The estimated numbers of infections with resistant *S. aureus* strains vary dramatically, since it is difficult to distinguish between causative and commensal infections (*S. aureus* is therefore not listed in Table 6.1). However, approximately 19 000 deaths caused by resistant *S. aureus* have been estimated for the United States in 2005.

Antibacterial treatment usually begins with a first-line antibiotic that is characterized by an acceptable safety profile and low cost. Should resistance develop, a second-line antibiotic is usually broader in spectrum, but has a less favorable risk–benefit profile and is more expensive. Third-line antibiotics are used in cases in which treatment with second-line antibiotics failed. However, some bacteria, for example *S. aureus*, have developed resistance even against the drugs of last resort.

In many cases, resistance is caused by the accumulation of mutations. If, for example, the concentration of an antibiotic is too low during treatment, bacteria will continue to replicate and mutate. Some of these mutations will confer resistance to the antibiotic; the mutant strains will survive and grow. Interestingly, some bacteria become resistant against several antibiotics at the same time. Here, resistance is usually caused by horizontal gene transfer. Several resistance genes may be imported en bloc by a plasmid or transposon.

Various other strategies have been identified that help bacteria to become resistant to antibiotic treatment (Figure 6.11): The most straightforward mechanism is the alteration of the target site, so that the antibiotic can no longer bind to and inactivate the target protein (Figure 6.11b). An example is the development of resistance to treatment with penicillins by alteration of the active site of penicillin-binding proteins (PBP). As already outlined, β-lactam antibiotics inactivate these enzymes and thus prevent synthesis of the cell wall. Widely recognized methicillin-resistant *S. aureus* (MRSA), which is a particular problem in hospitals, has become resistant against all β-lactam antibiotics, including penicillins (e.g., methicillin) and cephalosporins. They have accomplished this by expressing a variant of the enzyme that catalyzes the transpeptidation required for peptidoglycan cross-linking during cell wall synthesis.

The most widespread mechanism of developing resistance against antibiotics is the expression of a resistance gene that inactivates the drug by degradation or covalent modification (Figure 6.11c). An example is β-lactamase, which cleaves the amide bond of the β-lactam ring of antibiotics such as penicillin. Many bacteria encode β-lactamase genes on their chromosome, but in many clinically relevant bacteria, the enzyme has been acquired by plasmid transfer. To overcome resistance, β-lactam antibiotics can be given in combination with β-lactamase inhibitors (e.g., sulbactam) to maintain their antibacterial activity.

A third strategy of acquiring resistance to antibiotics is to decrease the active concentration of the drug inside the bacterial cell. This can be achieved either by the expression of an efflux transporter that pumps the toxic substance out of the cell (Figure 6.11d) or by the reduction of permeability of the cell for the drug. An example of this is tetracycline resistance, which is often acquired by horizontal transfer of a gene that encodes an efflux pump, which transports the antibiotic from the cell.

The resistance mechanisms described above usually derive from bacteria that produce antibiotics to destroy other microorganisms in their immediate environment.

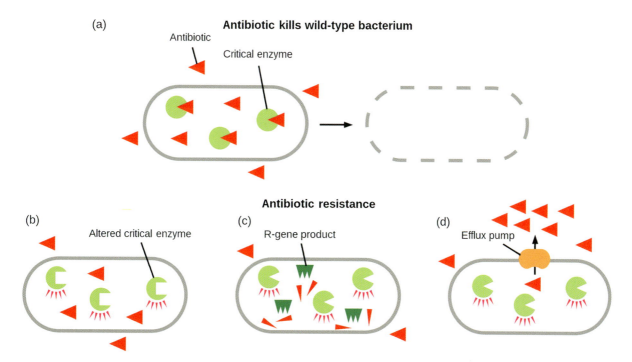

Fig. 6.11 Mechanisms of resistance against antibiotics. (a) Initially, antibiotics will kill wild-type bacteria. However, bacteria have also developed several strategies to resist treatment. (b) The target enzyme may be modified by mutations so that the antibiotic can no longer bind to and inactivate it. (c) Bacteria may express a resistance gene that inactivates or degrades the drug. (d) Some bacteria express an efflux transporter that pumps the antibacterial agent out of the cytoplasm. (Reproduced with permission from Ref. [2], Garland Science/Taylor & Francis LLC, Copyright 2008.)

To avoid killing themselves, these bacteria needed to evolve a defense system. However, that system spread and conferred resistance to other bacteria as well. This method of the acquisition of resistance was not expected to occur for synthetic antibiotics that do not have a counterpart in nature, such as the sulfonamides and trimethoprim. However, resistance against these antibiotics also emerged within just a few years after their introduction, though by a different mechanism. Here, plasmids encoded additional enzymes (dihydropteroate synthase in the case of sulfonamides and dihydrofolate reductase in the case of trimethoprim) that were far less susceptible to the drug and bypassed antibiotic inhibition of the chromosomal enzyme. The origin of these enzymes is still not clear.

There has been a so-called antibiotic innovation gap for the four decades between the early 1960s and the year 2000, a time when no new classes of antibiotics were approved. However, the rapid emergence of MDR strains and the inability to treat several types of bacterial infection increased efforts to develop new antibiotics. Two strategies were followed; the modification of known molecules and the discovery of new compounds. An example of the former is the development of tigecycline. This drug is a chemically modified tetracycline that cannot be pumped out of the cell by efflux transporters and can be used to treat tetracycline-resistant bacterial strains.

New classes of antibiotics, such as oxazolidinones, lipopeptides, pleuromutilins, and lipiarmycins, have different targets in bacterial cells. *Oxazolidinones* are heterocyclic organic compounds that contain a nitrogen molecule and an oxygen atom in a five-membered ring. They block bacterial protein synthesis by interfering with the binding of the *N*-formylmethionyl-tRNA to the bacterial ribosome. Oxazolidinones are the drugs of last resort to treat Gram-positive pathogens such as MRSA. Daptomycin is a *lipopeptide* antibiotic that disrupts the cell membrane and leads to its depolarization. Retapamulin belongs to the class of *pleuromutilin* antibiotics that inhibit bacterial protein synthesis by binding to the peptidyl transferase center of the large subunit of the ribosome. *Lipiarmycin* (Fidaxomicin) is a macrocyclic antibiotic that inhibits the bacterial RNA polymerase.

Although these new drugs represent advances, they are all specific for the treatment of infections with Gram-positive bacteria. In contrast, no new antibiotic classes have been discovered in the last 45 years to treat Gram-negative bacteria, although pan-drug resistant strains (PDR, that is, resistant against all approved antibiotics) have been described. Newer molecular techniques promote novel strategies for the discovery of substances with antibacterial activity. High-throughput sequencing approaches (Section 2.3) are also valuable, as they allow the rapid analysis of bacterial genomes.

6.2
Eukaryotic Pathogens

Many eukaryotic parasites and fungi are pathogenic for humans. Their cellular metabolite pathways are closely related to those of humans, which complicates the development of therapeutic agents. Drugs against fungi and parasites are often less effective and more toxic than antibacterials. The development of therapeutic agents for this class of pathogen is further complicated by the specific life cycles of these organisms, which often vary between different stages and hosts. In many cases, a drug is usually effective only against a certain stage of the life cycle.

Fungi may be unicellular yeasts or multicellular molds. Yeast and fungal infections are very common, but usually induce comparatively mild diseases in otherwise healthy individuals. A common cause of superficial infection is *Candida albicans*, which can normally be treated relatively easily. However, in immunocompromized individuals, for example, AIDS patients, systemic infections with *C. albicans* are frequently associated with morbidity and mortality. Several antifungal drugs, such as amphotericin B, caspofungin, and fluconazole are available for the treatment of systemic infection.

Many eukaryotic parasites, ranging from single-celled protozoa to very large metazoan worms, infect humans. Protozoan parasites claiming large numbers of lives are summarized in Table 6.2. Malaria is the most common disease caused by a parasitic protozoon and will be discussed as a disease paradigm in more detail below.

Approximately 250 million new cases of malaria, resulting in more than 1 million deaths, have been estimated to occur yearly. Malaria is a mosquito-borne infectious disease caused by protozoans of the genus *Plasmodium*. It is endemic in a broad band around the

Table 6.2 List of selected protozoan parasites with high numbers of fatalities worldwide.

Species	Disease	Annual number of deaths (2010)
Plasmodium falciparum	Malaria	1 169 500
Cryptosporidium parvum	Cryptosporidiosis	100 000
Entamoeba histolytica	Amebiasis	55 500
Leishmania	Leishmaniasis	51 600
Trypanosoma cruzi	Chagas disease	10 300
Trypanosoma brucei	Sleeping sickness (African trypanosomiasis)	9100

Source: Deaths numbers based on Ref. [3].

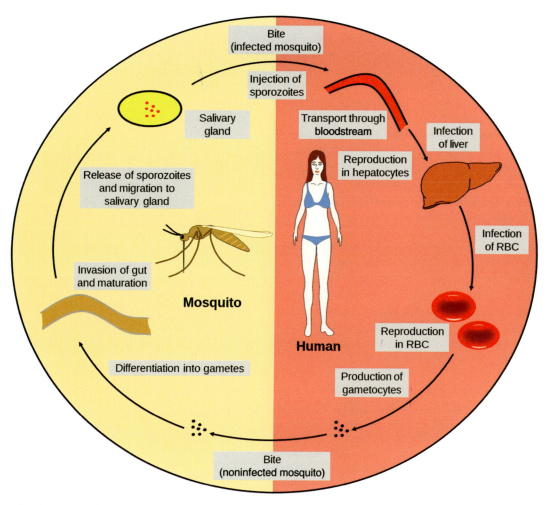

Fig. 6.12 The Plasmodium life cycle. *P. falciparum*, the most dangerous malaria pathogen, switches between humans and mosquitoes as hosts during its life cycle. Details are given in the text. RBC, red blood cells.

equator, with 85–90% of the malaria fatalities occurring in Sub-Saharan Africa. Of the five species of *Plasmodium* that can infect humans, *P. falciparum* is the deadliest. The parasite has a complex life cycle (Figure 6.12). Initially, a female *Anopheles* mosquito transmits the infective form (sporozoite) to a human host. The sporozoite travels in the blood stream to the liver, where it reproduces asexually, generating thousands of merozoites. These infect red blood cells and initiate a series of asexual multiplication cycles to produce additional infective merozoites. Eventually, the red blood cells burst and release a large number of parasites that invade new red blood cells. Other merozoites develop into immature gametes, called gametocytes. Gametocytes are taken up with infected blood when a mosquito bites an infected person; they then differentiate into gametes. Male and female gametes fuse and form zygotes, which develop into new sporozoites. Finally, the sporozoites migrate

from the mosquito's gut to the salivary glands and can infect a new vertebrate host.

The initial signs of malaria are typically flu-like symptoms and include headache, fever, and shivering. The symptoms are usually cyclical, with sudden chills followed by rigors, fever, and sweating. Falciparum malaria may lead to respiratory distress and neurological symptoms, which can progress to coma or death.

The great impact of malaria can be seen in the evolutionary selection pressure on populations living in endemic regions. An example is sickle-cell anemia, which is characterized by rigid and sickled erythrocytes that cannot easily pass through blood capillaries. Sickling is caused by a mutation in the hemoglobin gene, which produces sickle-cell hemoglobin (hemoglobin S). The mutated hemoglobin protein contains valine rather than glutamate at position six in the β-chain. The hydrophobic amino acid forms intermolecular interactions that lead to the aggregation of hemoglobin fibers in the erythrocyte. Individuals who are heterozygous for the hemoglobin gene usually lead a normal life, but homozygotes suffer from sickle-cell anemia, a painful, debilitating, and sometimes fatal disease.

Approximately one fourth of the sub-Saharan African population carries a single copy of the mutated hemoglobin. The prevalence of the sickle-cell gene coincides with regions where malaria is endemic. Heterozygotes have less severe symptoms when infected with malaria. The mechanism by which mutated hemoglobin confers resistance to malaria is not fully understood, but seems to involve enhanced clearance of infected erythrocytes by the spleen.

Several medications for travelers are available to prevent malaria when visiting endemic countries (prophylaxis). In addition, a variety of antimalarial drugs have been developed. The currently recommended treatment for malaria is artemisinin combination therapy (ACT). Artemisinin was originally discovered by screening Chinese antipyretic herbal medicines. Artemisinin derivatives have improved bioavailability compared to the original drug. The derivatives are prodrugs that are active during the stage when the parasite is located inside red blood cells. The exact mechanism of action of artemisinin is still controversial, but seems to include a drug interaction with the mitochondrial electron transport chain. This interaction generates reactive oxygen species from the peroxide moiety of the drug, which damage the parasite. Since monotherapy with artemisinin rapidly leads to drug-resistant malarial parasites, it is recommended that artemisinin be combined with another antimalarial drug. Nevertheless, evidence is accumulating that resistance to artemisinin derivatives is also developing, requiring the development of a new generation of antimalarial drugs.

Due to the great genomic plasticity of the malaria parasite, the development of resistance against any antimalarial drug is inevitable. Simultaneously, its high antigenic variability makes the development of an efficient malaria vaccine, one that protects against all strains of the pathogen, quite difficult. After decades of setbacks, a vaccine named RTS,S or Mosquirix, has become the most promising candidate. This vaccine is composed of the C-terminal portion of the major coat protein of *P. falciparum* chemically linked to the hepatitis B virus surface antigen (HBsAg, see Section 5.2.4). This conjugate is formulated with an excess of HBsAg to form virus-like particles and is administered together with an adjuvant. The vaccine induces immunity against *P. falciparum* in addition to HBV. Although RTS,S is considered an advance in protection against malaria infection, its efficacy is only 30–50%.

The second most prevalent disease in Africa after malaria is schistosomiasis (or bilharzias). This parasitic disease is caused by worms belonging to the genus *Schistosoma*. It is estimated that 200 million people are infected in over 70 countries in tropical and subtropical regions. Although the mortality rate is low, schistosomiasis often is a chronic illness that can damage internal organs and impair growth and cognitive development in children. Furthermore, the disease has devastating socioeconomical consequences. Vaccine candidates have been unsuccessful and only a single drug, praziquantel, which has been used for more than 35 years, is available.

References

1. MacPherson, G. and Austyn, J. (2012) *Exploring Immunology: Concepts and Evidence*, Wiley-VCH Verlag GmbH, Weinheim, Germany.
2. Alberts *et al.* (2008) *Molecular Biology of the Cell*, 5th edn, Garland Science/Taylor & Francis LLC.
3. Lozano, R., Naghavi, M., Foreman, K. *et al.* (2012) Global and regional mortality from 235 causes of death for 20 age groups in 1990 and 2010: a systematic analysis for the Global Burden of Disease Study 2010. *Lancet*, **380**, 2095–2128.
4. Poehlsgaard, J. and Douthwaite, S. (2005) The bacterial ribosome as a target for antibiotics. *Nat. Rev. Microbiol.*, **3**, 870–881.

Further Reading

Bacteria

Amyes, S.G.B. (2013) *Bacteria: A Very Short Introduction*, Oxford University Press, Oxford, UK.

Antibiotics

Dantas, G. and Sommer, M.O.A. (2014) How to fight back against antibiotic resistance. *Sci. Am.*, **102**, 42–51.
Shlaes, D.M. and Spellberg, B. (2012) Overcoming the challenges to developing new antibiotics. *Curr. Opin. Pharmacol.*, **12**, 522–526.
Walsh, C.T. and Fischbach, M.A. (2009) New ways to squash superbugs. *Sci. Am.*, **301**, 44–51.

Eukaryotic Pathogens

Miller, L.H., Ackerman, H.C., Su, X.Z. *et al.* (2013) Malaria biology and disease pathogenesis: insights for new treatments. *Nat. Med.*, **19**, 156–167.
Riley, E.M. and Stewart, V.A. (2013) Immune mechanisms in malaria: new insights in vaccine development. *Nat. Med.*, **19**, 168–178.

Genomics and Proteomics

7

Summary

- The Human Genome Project (HGP) was launched in 1990 to elucidate the sequence of the more than 3 billion bases of the haploid human genome. The publicly funded project used an ordered shotgun approach. Celera, a competing company, applied the whole genome shotgun strategy to human genome sequencing. Both projects simultaneously reported the complete sequence of the human genome, which has become an indispensable part of biomedical research and development. One of the surprising results of the sequence analysis was that the number of proteins encoded in the human genome is between 20 000 and 25 000, much lower than previously thought.

- Following the HGP, various large-scale, worldwide projects have been initiated to further understand the human genome. Several of these projects investigate variations in the human genome and their relevance to human health and disease. Thousands of genomes have been sequenced. The genomes of two unrelated human individuals differ by approximately 0.1%. Approximately 40 million single-nucleotide polymorphisms (SNPs) have been discovered, epigenetic patterns have been elucidated, and genomes of various tumor types have been sequenced.

- The Encyclopedia of DNA Elements (ENCODE) project aims to identify all the functional elements in the human genome. It revealed that at least 80% of the human genome is presumed to have biological function, although most of the information contained in the DNA is not translated into proteins. A large fraction of the genome is transcribed into noncoding RNAs (ncRNAs) with regulatory functions. Many functional elements at the DNA level play an important role in the modulation of gene expression. These findings reverse the traditional view that more than 98% of the human genome does not encode proteins and can thus be regarded as "junk."

- While initial sequencing of the human genome took 13 years and cost approximately $3 billion, next-generation sequencing (NGS) technologies now permit the elucidation of a genome within a few days for less than $1000. Sequencing of the genome, or the relevant parts of the genome of a (cancer) patient, may, therefore, become a routine approach for personalized decision-making for therapeutic intervention.

- Proteomics is the large-scale investigation of all of the proteins expressed by an organism, a tissue, or a cell type. Proteomics usually involves the use of mass spectrometric techniques for the identification

Contents List

of proteins. Isotope labeling of proteins is used for quantitative analysis. Modern shotgun proteomics permit the analysis of protein abundance, protein modifications, and their interactions in complex mixtures, for example, from whole cells. Clinical proteomics aims to identify new targets for therapeutic intervention and new biomarkers of disease or of pharmacologic treatment.

- Structural genomics aims at solving three-dimensional protein structures via a high-throughput approach. This is achieved by automation of many steps of the pipeline, such as protein expression and purification, the optimization of crystallization conditions, data acquisition, and structure determination.

The discovery of the double helix structure of DNA is widely acknowledged as the greatest achievement in molecular biology in the twentieth century. The elucidation of the sequence of the human genome at the beginning of the twenty-first century is considered to be another major breakthrough, one which will revolutionize our current understanding of human biology and pathophysiology. This chapter will describe the general principles behind the sequencing of large genomes (Section 7.1) and the lengthy path that led to the deciphering of the human genome, a path driven mainly by the publicly funded HGP and the company Celera (Section 7.2). Several follow-up projects aim at investigating genetic variations (International HapMap Project, 1000 Genomes Project and Personal Genome Project) and at identifying all the functional elements in the human genome (ENCODE). A major challenge of current research is to elucidate the function of the gene products (usually proteins) encoded in the human genome. These approaches will be described in Section 7.3, which discusses functional genomics and proteomics.

7.1
Whole Genome Sequencing

Whole genome sequencing is used to determine the complete genomic DNA sequence of an organism. Currently, however, most completely deciphered genomes are microbial. In contrast, for the larger genomes of eukaryotes, analysis of more than 95% of the genome is usually considered to be whole or full genome sequencing.

The first free-living organism to have its entire genome sequenced was the bacterium *Haemophilus influenzae*, in 1995. This genome consists of a single circular DNA molecule of approximately 1.8 million base pairs. One year later, the 12 million base pairs of

the baker's yeast *Saccharomyces cerevisiae* was the first eukaryotic genome sequence to be released. In 1998, the first sequence of a multicellular eukaryote, *Caenorhabditis elegans*, which contains a genome of 97 million base pairs, was completed. Sequencing of the human genome, consisting of more than 3 billion base pairs, was completed in 2003 and will be described in Section 7.2.

7.1.1
Cloning of a Genome

Conventional sequencing of complete genomes by the chain-termination method using dideoxynucleotides (Section 2.3) is usually preceded by cloning of the genetic material into a vector (Table 7.1). The standard strategy for cloning of DNA is the insertion of the DNA fragment of interest into a *plasmid* (Figure 7.1). Commonly, a plasmid is a small circular, double-stranded DNA molecule that is separate from the chromosomal DNA within a cell and can replicate independently. The plasmid is cut open by restriction enzymes and a DNA fragment with compatible ends can be ligated into it. The plasmid with the insert is then transformed into bacteria and amplified.

A shortcoming of standard plasmids is their size limitation. Plasmids with inserts of more than 10 kb are unstable and difficult to amplify in bacteria. These limitations led to the development of vectors that can harbor larger DNA fragments. An early approach that permits cloning of larger DNA fragments was based on the λ *bacteriophage*. Bacteriophages are viruses that infect and replicate in bacteria. The genome of bacteriophage λ is 48.5 kb long, approximately 15 kb of which can be deleted and replaced by foreign DNA of up to 18 kb in length. Even larger DNA fragments can be cloned into vectors based on the *P1 bacteriophage*. This phage has a larger genome and can harbor foreign DNA of up to 125 kb.

Table 7.1 Cloning vectors.

Type of vector	Insert size (kb)
Plasmid	10
Bacteriophage λ	18
Cosmid, Fosmid	40
Bacteriophage P1	125
BAC	300
PAC	300
YAC (Mega-YAC)	600 (1400)

Different types of vectors and maximal sizes of the insert in kilobases are given. BAC: bacterial artificial chromosome; PAC: P1-derived artificial chromosome; YAC: yeast artificial chromosome.

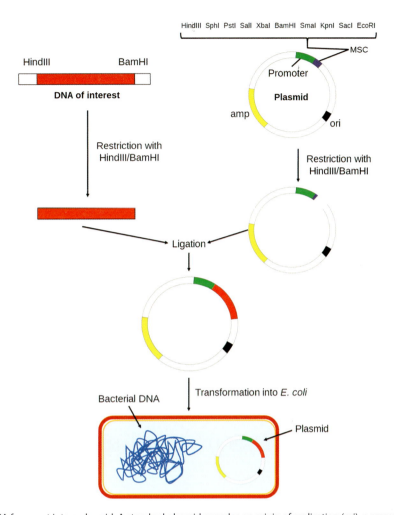

Fig. 7.1 Cloning of a DNA fragment into a plasmid. A standard plasmid encodes an origin of replication (ori), a promoter that drives the expression of the cDNA of the gene of interest, and an antibiotic resistance gene, for example, the ampicillin resistance gene. The plasmid usually harbors a multiple cloning site, which is a short region containing restriction sites of several commonly used restriction endonucleases. The DNA fragment to be cloned must contain compatible ends that can then be ligated into the opened plasmid. The recombinant plasmid containing the insert is finally transformed into bacteria (commonly *E. coli*) and amplified.

A *cosmid* is a plasmid that contains the genomic *cos* sequence of bacteriophage λ. The *cos* sequence is approximately 200 bases long and is the only region of the phage genome that is required for packaging. Cosmids can, thus, be packaged to produce phage-like particles that can infect bacterial cells. Once inside the target cell, the cosmid cannot induce the production of new phage particles like the actual phage genomes described above, but it can still replicate like a plasmid. The maximal insert size is 40 kb. A *fosmid* is a cloning vector that is very similar to a cosmid. The main difference is that a fosmid does not contain the chromosomal origin of replication (ori) of *Escherichia coli*, but the origin of replication of the F-plasmid. This plasmid, also known as the fertility plasmid, encodes information for the biosynthesis of sex pili that form a bridge between two

bacterial cells and permit transfer of genetic material between the cells. F-plasmids and the fosmids derived from them are present in *E. coli* cells in a much lower copy number than plasmids and cosmids. The limited number of fosmids in a cell reduces the potential for recombination and increases genetic stability, thus enhancing the accuracy of the genome sequencing.

Genetic engineering of artificial chromosomes permits cloning of significantly larger DNA fragments. Yeast artificial chromosomes (YACs) consist of important DNA elements of *S. cerevisiae* that are ligated into a bacterial plasmid. Therefore, they can replicate in *E. coli* and yeast. The primary components derived from *S. cerevisiae* are the autonomously replicating sequence (ARS), centromere, and telomeres (Figure 7.2). In addition, selectable marker genes are utilized to select

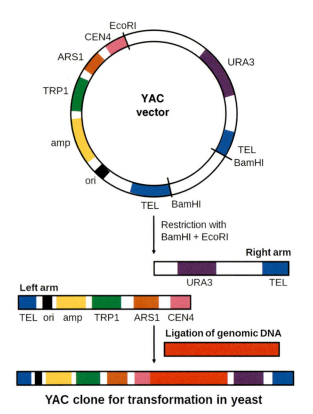

YAC clone for transformation in yeast

Fig. 7.2   Cloning into yeast artificial chromosomes. YACs consist of elements derived from *S. cerevisiae* (ARS1: autonomously replicating sequence; CEN4: centromere; TEL: telomeres; URA3 and TRP1, selection marker) and bacteria (ori: origin of replication; amp: ampicillin resistance gene). The cloning vector is initially produced in large amounts in *E. coli*. The circular plasmid is typically cut into two linear DNA molecules and the sequence of interest ligated between the two vector arms. The artificial chromosome can then be transformed into yeast cells where it is replicated and then transferred to daughter cells as an additional, 17th chromosome. The segments are not drawn to scale: The genomic sequence is much longer than the vector segments.

transformed yeast cells. The cloning procedure is described in Figure 7.2. YACs can normally carry DNA fragments of up to 300 kb, but some mega YACs can even harbor 1400 kb of foreign DNA.

Initially, the HGP used YACs for cloning of the human genome. However, YACs turned out to be unstable and up to 50% of the chimeric clones in YAC libraries were observed to consist of artificially merged segments from distinct genomic regions. Therefore, the HGP abandoned the use of YACs and switched to bacterial artificial chromosomes (BACs), which are more stable and have a lower incidence of sequencing artifacts. BACs are based on a functional F-plasmid and can be used for transforming large inserts of up to 300 kb into *E. coli*. BACs were not only employed in the HGP but have also been used in other genome projects. Finally,

P1-derived artificial chromosomes (PACs) combine properties of P1 vectors with those of BACs. PACs are transformed into bacterial cells by electroporation so that the insert size is not, like in P1 vectors, limited by the packaging capacity of the phage particle. PACs can take up foreign DNA of up to 300 kb and are comparatively stable.

At times when traditional Sanger sequencing technology was still employed in genomic projects, cloning and amplification of the DNA of interest was required. This was also true for the human genome project, as described in the next section. Some of these methods are still in use, but present-day sequencing approaches employing NGS technologies no longer require this step. As described in detail in Section 2.3, second-generation sequencing approaches include fragmentation of the genomic DNA, ligation of adapters, and amplification by PCR. Third-generation sequencing technologies are characterized by direct sequencing of the DNA without an amplification step. Omission of the cloning step significantly accelerated the process of genome sequencing.

7.1.2
Mapping and Assembly of the Genome

Even the newest sequencing technologies do not permit sequencing of large genomes like the human genome in a single read. Instead, only short segments can be analyzed; they must then be assembled into the complete sequence of the genome. Thus, comprehensive maps of the genome have traditionally been generated. Mapping methods can be divided into two categories: *genetic mapping* and *physical mapping* (Figure 7.3). A genetic map, also known as linkage map, is constructed by measuring the meiotic recombination frequencies between pairs of genetic markers. In contrast, physical maps are constructed by identifying the position of markers along the chromosome. The map of a genome eventually provides a guide for sequencing experiments by identifying the position of genes or sequence markers in the genome. In addition, maps have been used to identify the genetic basis of inherited diseases.

A genetic map contains the positions of genes and other genetic markers in a genome. Genetic mapping is based on an analysis of the transmission of heritable characteristics from parent to offspring. Genes were the first markers to be used; however, they are by no means ideal for the mapping of large eukaryotic genomes. As will be described in more detail below, protein-coding genes constitute less than 2% of the human genome. To obtain a higher marker density, other types of markers can be used. Some examples are restriction fragment length polymorphisms (RFLPs) and SNPs. A genetic

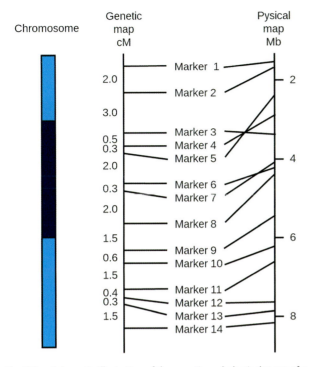

Fig. 7.3 Schematic illustration of the genetic and physical maps of a human chromosome. An ideogram of the cytogenetic band pattern is shown on the left. The genetic map is scaled in cM and compared to the physical map with distances given in mega bases (Mb). Lines join the corresponding positions of markers that are placed on both the physical and genetic maps.

map is then generated by linkage analysis of the markers. During meiosis, genetic material is recombined by crossing-over events of the two chromosomes of the diploid set. Two markers that are close together will be separated by crossovers less frequently than two markers that are more distant from one another. This means that the frequency with which the markers are unlinked by crossovers is proportional to how far apart they are on the chromosome. The distance between two markers is given in centimorgans (cM): 1 cM is defined as a recombination frequency in a single generation of 1%. A low recombination frequency (i.e., a small number of cM) indicates that two markers are close in distance. For the human genome, 1 cM equals on average ~1 000 000 bases (1 Mb). However, this can vary in different regions of the genome due to varying recombination frequencies (see below).

Genetic maps are usually insufficient for directing the sequencing phase of large genome projects, since the resolution and accuracy of genetic maps is limited. Physical maps are more precise. They describe by numbers of nucleotides the exact distance between markers. Various methods have been developed for physical mapping. For the generation of a detailed high-resolution map,

sequence tagged site (STS) mapping is the most powerful technique. A STS is a short DNA sequence of 100–500 bp that has a single occurrence in the genome and whose location is known. The STS can easily be amplified by PCR using a specific primer pair and can be detected in any genomic sequence, for example, in a YAC or BAC library.

Genetic and physical maps are not necessarily identical. As can be seen in Figure 7.3, the order of genetic markers is usually the same in both types of maps. However, their position on the chromosome may differ. One reason may be that the frequency of crossing-over events during meiosis is sometimes higher in certain regions of the chromosome than in others. Therefore, the genetic distance does not always reflect the physical distance. In some regions, the order of the markers may even be discordant, as indicated by the crossed lines.

7.1.3
Sequencing of a Large Genome

Sequencing technologies are described extensively in Section 2.3 and will not be repeated in detail here. While the chain-terminator approach developed by Frederick Sanger has been the standard technique for many years, a number of next-generation sequencing technologies have been developed in the last decade. What all these methods have in common is that they produce only short sequence reads. These reads must be assembled to give the complete genomic sequence. The most straightforward approach to this is known as *chromosome walking* (or *primer walking*, Figure 7.4a). This approach usually starts with the construction of a genomic clone library of λ or cosmid vectors. The inserted DNA from the starting clone is sequenced; its 3′ end can be used as a hybridization probe to screen the library for overlapping clones. Their inserts can then be used as probes to find the next overlapping clone and continue the walk. This technique has been used successfully to elucidate relatively small genomes. However, it cannot be applied to large genomes such as the human genome, since the human genome contains repeat sequences: A probe will hybridize to not only overlapping clones but also to nonoverlapping ones. The precision of the process can be increased by replacing the hybridization step by a PCR step, which will identify overlapping clones with primers designed to hybridize at the end of a sequenced fragment. Another alternative is the clone fingerprinting technique, which uses information such as restriction patterns of a DNA fragment to identify overlapping clones. However, none of these techniques is efficient enough to sequence large eukaryotic genomes.

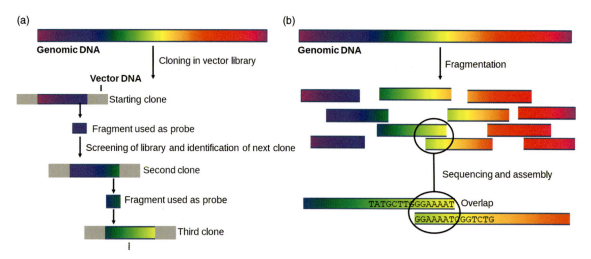

Fig. 7.4 Chromosome walking and shotgun sequencing. (a) For chromosome walking, a vector library is produced from fragmented genomic DNA. Next, the insert of the starting clone is sequenced. To identify an overlapping adjacent clone in the library, a probe is generated from the starting clone's 3′ end. After screening and successful identification of the second clone, the next section of the genome can be sequenced. This procedure continuous "walks" the complete chromosome and sequences the DNA. (b) For the shotgun sequencing approach, genomic DNA is fragmented randomly into small pieces that are sequenced. Each segment of the original DNA is covered by multiple overlapping reads, which are then assembled.

The counterpart of a well-ordered chromosome walk is shotgun sequencing (Figure 7.4b). In this approach, the DNA is randomly broken up into numerous small pieces, which are then sequenced. The library must contain multiple fragments for each segment of the original DNA. The average number of reads representing a given nucleotide is known as the *coverage*, that is, a coverage of 15 for a genome sequence means that each base in the final sequence was present, on average, in 15 reads. Finally, bioinformatics technology assembles the overlapping ends of different reads into a contiguous sequence, also denoted as *sequence contig*.

However, shotgun sequencing is prone to errors that can occur by the incorrect assembly of the overlapping sequences. Large genomes contain repeated elements known as tandem repeats. Central elements of the tandem repeats may be omitted. On the other hand, repeated sequences located at different positions of the genome may be linked together. If two genome-wide repeats are located on different chromosomes, the sequences of these two chromosomes may be mistakenly connected together. A second challenge related to the shotgun sequencing method is the complexity of the data analysis. While it is possible to assemble the fragments of a comparatively small prokaryotic genome, data analysis becomes disproportionately more complex with the substantially higher number of fragments generated from a large eukaryotic genome. The whole genome shotgun approach only became feasible for large genomes like the human genome with advancements in

computing power and bioinformatic algorithms. This will be described in the following sections.

7.2
The Human Genome

7.2.1
Sequencing of the Human Genome

In the 1980s, DNA sequencing became feasible, but it was technically challenging and was used to study only selected genes. Sequencing of the more than three billion base pairs of the whole (haploid) human genome seemed impossible. In addition, most of the genome does not contain genes and was regarded as *junk* DNA. Nevertheless, some scientists still felt that it was valuable to elucidate the sequence of the entire human genome. In 1988, the US Congress funded initiatives to explore the potential for sequencing the full human genome. In 1990, the HGP was formally initiated despite concerns that it would withdraw funding from scientifically more relevant, hypothesis-driven approaches and gather large amounts of relatively useless data. The central aim of the HGP was to sequence the entire human genome within 15 years at a cost of $3 billion, that is, at $1 per base pair. The proponents promised medical benefits from the elucidation of the human genome, including new therapies for cancer and personalized medicine. The knowledge generated and the novel technologies to be developed for the project were expected to create new jobs and fuel economic growth, justifying the large investment.

Most of the publicly funded HGP sequencing was performed in university and governmental research centers. In addition to institutions from the United States, the international consortium comprised institutes from many other countries worldwide, including the United Kingdom, France, Australia, Japan, and Germany. Sequencing efforts were not only restricted to the human genome but also included the genomes of a species of fish, worm, and plant, the fruit fly, the mouse, and various microorganisms, since the comparative genomics of model organisms was thought to be necessary to understand the human genome. In addition, the HGP aimed at educating the public and health professionals about the human genome and at raising attention about the ethical, legal, and social issues (ELSI) of the initiative, such as privacy and discrimination.

For the reasons given above, the whole genome shotgun sequencing approach did not seem feasible (lack of accuracy and need for enormous computing power) at the time the HGP was initiated. Therefore, the *clone*

contig approach (also denoted as *hierarchical shotgun* approach) was chosen for the sequencing of the human genome (Figure 7.5a). An initial goal was to produce a genetic map with a density of one marker per mega base (Mb). In 1994, an international consortium published a map that contained 5800 markers and had a density of one marker per 0.7 Mb. This map was subsequently refined by the introduction of additional markers. In parallel, a physical map was generated in 1995 that contained more than 15 000 markers with an average density of one per 199 kb. An even denser map with an additional 20 000 STSs, approaching the targeted density of one marker per 100 kb, was produced 1 year later.

As outlined above, the initial plans to use YAC libraries for the HGP had to be abandoned because of the instability of this vector system. Instead, a library of 300 000 BACs was generated. These clones were mapped onto the genome, providing the basis for the sequencing phase of the project. During this period, the insert from each BAC was completely sequenced by the shotgun approach. These efforts benefited from substantial technological developments, particularly in the automation of the sequencing process.

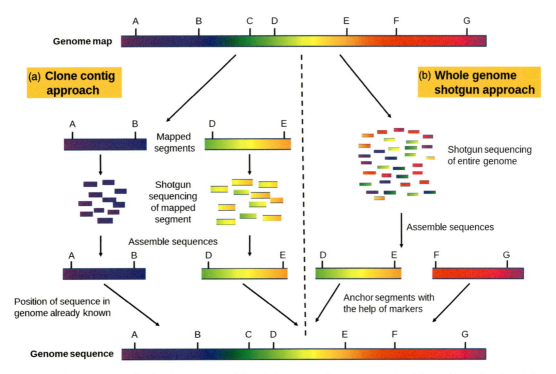

Fig. 7.5 Assembly of large genomes by the clone contig and the whole genome shotgun approaches. The positions of markers A-G were identified by mapping the genomic DNA. (a) The clone contig method starts with segments of DNA whose position between two markers is well defined. These segments are sequenced by the shotgun method and assembled. Their position between the markers is known in advance. (b) For the whole genome shotgun approach, short pieces of the entire genome are randomly sequenced. Contiguous sequences of up to several hundred kilobases can be assembled. If a segment contains a marker, it can be anchored to a specific position on the genome map. With present-day computer technology, the whole genome shotgun approach to elucidating new genomes can also be performed without mapping the genome.

However, at about the same time, the publicly funded HGP experienced unexpected competition. In 1995, Celera, a company led by the US biologist J. Craig Venter, reported the sequence of the bacterium *H. influenzae*. The striking feature of this project was that the complete genomic sequence was obtained by the shotgun approach without being informed by a genetic or physical map. Celera announced their intent to apply the whole genome shotgun sequencing technology to the human genome (Figure 7.5b). This goal, Celera believed, could be achieved with brute force, as they had approximately 300 of the most modern automated DNA sequencers and a supercomputer that was rivaled only by those of the US military. The overall cost was calculated to be $300 million, that is, one tenth of the cost of the publicly funded HGP. However, it was questionable whether the available bioinformatic resources would be able to cope with the much larger amount of data generated from the human genome compared to the bacterial genome. *De novo* assembly of the complete set of sequenced fragments would probably not have been possible, though Celera could refer to the publicly available maps and data derived from the HGP to anchor sequence contigs. This triggered an avalanche of criticism, as the company announced its intent to seek patent protection for selected genes. This, in turn, contradicted the philosophy of the government-funded HGP, which was to make all genomic information and DNA sequencing results freely available to the community without delay. Celera eventually filed preliminary patent applications on 6500 whole or partial genes, but the company's stock fell dramatically when the US authorities announced in 2000 that genome sequences could not be patented and should be made freely available.

While the clone contig method gives more accurate results, the whole genome shotgun sequencing strategy was supposed to be significantly faster. Since the HGP feared losing the sequencing race to Celera, the organizers of the public project adjusted their strategy to achieve their goals earlier. In June 2000, US President Bill Clinton, Francis Collins, the leader of the HGP, and Craig Venter, the head of Celera, jointly announced completion of the first draft of the sequence of the human genome. These studies were published in early 2001 in parallel issues in *Nature* (HGP) and *Science* (Celera), both leading scientific journals. This working draft covered approximately 90% of the genome. Most of the missing 320 million base pairs lay in constitutive heterochromatin, that is, tightly packaged regions of the chromosomes that contain few genes. In April 2003, a refined and essentially complete version of the human genome was published, 2 years earlier than planned by the HGP. However, the sequence of the human genome was still not fully analyzed and

refined, and more precise versions continue to appear. The finishing phase, when individual sequence contigs are joined by the closure of gaps, is still one of the most challenging aspects of large sequencing projects.

The initial elucidation of the human genome was exclusively carried out with the conventional chain terminator sequencing method. NGS technologies, as described in Section 2.3, have since been developed and have largely replaced conventional sequencing techniques for large genome projects.

One of the key surprising findings of the analysis of the human genome was that the number of protein-coding genes is much lower than initially thought. While calculated estimates at the outset of the HGP were approximately 100 000 genes (some researchers even estimated up to 2 000 000 genes), this number was lowered to 30 000–40 000 protein-coding genes after the initial sequencing and analysis of the human genome. Although the exact number is still unknown, there is now a broad consensus that the number of protein-coding genes is between 20 000 and 25 000. In 2012, the ENCODE project (*vide infra*) concluded that the human genome contains 20 687 protein-coding genes. This surprisingly low number of genes in humans is not far above that found in the roundworm *C. elegans*, and significantly lower than that of the weed *Arabidopsis thaliana* (Table 7.2), although humans are far more complex organisms.

Table 7.2 Genome size, protein-coding percentage of the genome, and predicted number of genes of selected organisms.

Organism	Genome length (Mb)	Protein-coding percentage (%)	Number of genes
Eubacterium *Escherichia coli*	4.6	84	4200
Fungus *Saccharomyces cerevisiae*	12.1	71	6600
Plant *Arabidopsis thaliana*	119	29	30 000
Nematode *Caenorhabditis elegans*	100	27	19 000
Insect *Drosophila melanogaster*	165	13	13 600
Mammal *Mus musculus*	2600	~2%	~20 000
Human *Homo sapiens*	3200	<2	20-25 000

In addition to protein-coding genes, the human genome harbors a large number of pseudogenes (estimates range from 11 000–20 000) that are dysfunctional relatives of protein-coding genes. They are thought to have originated by duplications of a functional gene. However, due to the accumulation of multiple mutations, they are either no longer expressed or have lost the ability to encode a protein. This can occur, for example, through mutations that led to a frameshift mutation, or to the introduction of a premature stop codon. In addition, ~18 400 genes encoding ncRNAs have been defined; these include tRNAs, rRNAs, microRNAs, and other ncRNAs.

Along with the discovery that the number of protein-coding genes is unexpectedly low came the finding that genes are more complex than initially thought. As described in more detail in Section 1.3.3, eukaryotic genes consist of protein-coding exons and intervening introns, with every human gene containing, on average, ~8 introns. The exon–intron structure of genes permits assembly of the protein-coding regions in multiple ways. This feature, known as alternative splicing, leads to the production of several hundred thousand different protein variants from the less than 25 000 genes. It has been estimated that up to 95% of all multi-exonic genes in humans are alternatively spliced. Extensive alternative splicing in mammals contributes substantially to the high complexity relative to the comparatively low number of genes. However, a dysfunctional alternative splicing process may result in disease. Single-nucleotide alterations may affect splice sites or regulatory elements. In fact, 30–60% of all disease-causing mutations have been claimed to affect splicing rather than the coding sequence.

While more than 98% of the human genome does not encode proteins, this percentage is much smaller in bacteria, whose genomes consist almost solely of protein-coding sequences. (Prokaryotes do not have introns and there is very little DNA between the genes.) In lower eukaryotes such as plants and worms, the number of protein-coding genes is on the order of that of humans, but the noncoding part of the genome is also substantially smaller (~70%, Table 7.2) than it is in humans. While for many years the noncoding DNA was regarded as "junk," the ENCODE project (Section 7.2.4) discovered that approximately three quarters of the genome is transcribed into RNAs. These RNAs may have regulatory functions even if they are not translated into protein. Approximately 80% of human DNA has been shown to fulfill this functional role. It was, therefore, suggested that the term "noncoding DNA" should be replaced by "nonprotein-coding DNA," but both terms will be used synonymously here. The deeper insights into the structure and function of genomes have thus revealed that the number of protein-coding genes is a poor indicator of the complexity of an organism. Instead, the ratio of noncoding to total genomic DNA may, in many cases, be a better reflection of complexity.

Only half of the human genome consists of those unique sequences that occur only once per haploid genome. Protein-coding exons span approximately 1.1–1.4% of the human genome; 24% of the genome consists of introns and close to 30% is composed of nonrepetitive sequences that are neither introns nor exons (Figure 7.6).

The remaining repetitive elements can be divided into four additional classes:

1) The human genome contains some 2 million inverted repeats that can fold back on themselves to form hairpin-like structures. Their function remains elusive, but they may have a regulatory role as molecular switches.

2) Segmental duplications are blocks of 1–200 kb that have been copied to one or more regions of the genome. They constitute ~5% of the human genome. The HPG found that the number of segmental duplications in the human genome was unexpectedly high. Their frequency is much higher than in lower organisms, which may indicate that novel genes specific to primates have evolved.

3) Some repetitive sequences occur in up to 10^6 copies per haploid genome and consist of segments of 100 to several thousand base pairs that are interspersed with larger blocks of unique DNA. Approximately 42% of the human genome is derived from retrotransposons, defined as mobile, transposable elements (i.e., elements that can change their position within the genome) that propagate through the intermediate synthesis of RNA. They can be grouped in three classes: (I) Long interspersed nuclear elements (LINEs) are 6–8 kb long and encode the protein that mediates their transposition. However, 99% of the LINEs have accumulated mutations that render them inactive. (II) Short interspersed nuclear elements (SINEs) are 100–400 bp long and are derived from RNA polymerase III-generated transcripts. They contain an RNA polymerase III promoter, but do not encode proteins. The most common SINEs are members of the *Alu family* (so named because they are cleaved by the restriction enzyme *Alu*I). (III) The class of long terminal repeat (LTR) retrotransposons contains LTRs flanking the *gag* and *pol* genes. They are propagated by cytoplasmic retrovirus-like particles. In addition to retrotransposons, a small fraction of the human genome comprises bacteria-like DNA transposons. Altogether, the transposable elements comprise

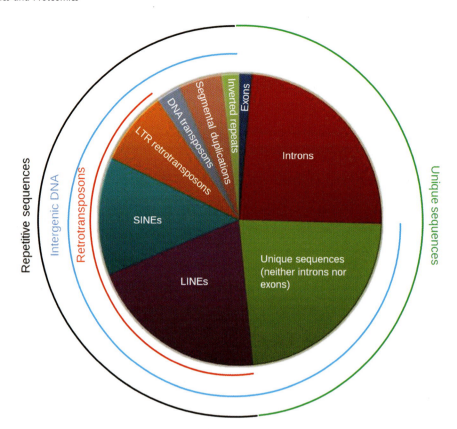

Fig. 7.6 Composition of the human genome. Approximately half of the human genome consists of unique sequences (protein-coding exons, introns, and additional unique sequences). The other half contains repetitive sequences. Around 42% of the genome consists of retrotransposons (long interspersed nuclear elements (LINEs), short interspersed nuclear elements (SINEs), and long terminal repeat (LTR) retrotransposons). However, these retrotransposons, as well as the DNA transposons, are almost entirely inactive. Additional elements in the human genome are segmental duplications and inverted repeats.

~45% of the human genome. However, they are almost entirely inactive; were this not the case, the genome would be very unstable.

4) Highly repetitive DNA consists of short sequences that are tandemly repeated, with more than 10^6 copies per haploid genome. These so-called simple sequence repeats (SSRs), also known as short tandem repeats (STRs) or satellite DNAs, cluster at telomeres and centromeres. SSRs comprise ~3% of the human genome. They are not transcribed, but may help to adjust the chromosomes during meiosis. The most frequent SSRs are repeats of CA and TA dinucleotides.

In addition to the human genome, various other mammalian genomes have been sequenced. These include the mouse, rat, dog, pig, and chimpanzee genomes. The science of comparative genomics identifies common features and differences in the genomes of different organisms. Approximately 5% of the total human genome is found in other species as well. This is higher than the protein-coding percentage of the genome, that is, an unexpectedly large noncoding percentage of the genome is evolutionarily conserved.

Chimpanzees are the closest living relatives of humans. Both species last shared a common ancestor approximately 5–7 million years ago. The difference between both genomes is only 1% (although some sources state 4%), most of which is comprised of single-nucleotide differences, although larger insertions and deletions are also present (Box 7.1).

Gene duplications account for most of the sequence differences between humans and chimpanzees. About 30% of all human genes are identical in sequence to the corresponding gene in chimpanzees; the average difference between proteins from both species is only two amino acids. Some regions of the genome, for example, the *human accelerated region 1* (HAR1), have evolved rapidly. This gene encodes a regulatory RNA that is not translated into a protein. Approximately 50 human-accelerated regions have been identified that are conserved throughout vertebrate evolution but are strikingly different in humans. Another important difference was

Box 7.1.   Of Men, Women, and Chimpanzees

It has been remarked ironically that the exchange of an X chromosome for a Y chromosome in human males (human females have two X chromosomes) results in greater genetic variation than the differences between male humans and chimpanzees. It was also proposed this may explain why men are from Mars and women . . .

found in the *FOXOP2* gene, which encodes a protein that is relevant to the development of speech. FOXOP2 differs in two amino acids between humans and chimpanzee, but is fully conserved between humans and Neanderthals (Box 7.2).

One motivation for sequencing the human genome was to help our understanding of the genetic basis of disease processes and to develop new therapeutic approaches. However, the actual medical progress realized through the elucidation of the human genome is somewhat unclear. In an interview with the German weekly news magazine *Der Spiegel* in 2010, Craig Venter stated that the medical benefits of the HGP were "close to zero." This provocative statement reflected difficulties

Box 7.2.   Ancient Genomes and the Neanderthal Genome Project

It was a long-standing dogma that it is not possible to sequence DNA from ancient species. Fragmentation of old nucleic acids into small pieces was believed to be the main hurdle to elucidating the genomes of extinct organisms. However, this belief has been revised in the last decade. As outlined in Section 2.3, NGS technologies analyze short DNA segments of at most several hundred nucleotides. Bioinformatic algorithms have been developed to assemble these fragments to produce a complete genome sequence. These techniques have also helped to assemble segments of ancient genomes. In addition, contemporary relatives of extinct species permit the elucidation of their genomes by alignment analysis.

A striking example of the successful sequence analysis of an extinct animal came in 2008, when the nuclear genome of the woolly mammoth was elucidated from a roughly 20 000-year old specimen. Comparison with the African elephant revealed a comparatively close relationship between both species, with an estimated divergence rate half of that between humans and chimpanzees. This research suggested the idea of reanimating mammoths using contemporary elephants as foster mothers. Mammoth DNA can be extracted from blood that was preserved in the permafrost. Alternatively, the DNA might be chemically synthesized based on the deciphered sequence – although the synthesis of a genome of more than 4 billion base pairs with current technologies would be an insuperable challenge (or may not even be possible). In contrast to the successful sequencing of the mammoth genome, sequencing or reanimating dinosaurs in a "Jurassic Park" manner is frankly impossible, as estimates of the half-life of nucleic acids make it highly unlikely that DNA could persist for more than ~7 million years even under optimal conditions. The dinosaurs went extinct approximately 65 million years ago.

In contrast, sequencing of the Neanderthal genome was possible, as they died out a mere 30 000 years ago. The challenge for these analyses was to prevent contamination of the Neanderthal DNA with *Homo sapiens* material from the researchers. In fact, the initially reported Neanderthal sequences were later suspected to have been contaminated with DNA from modern humans. The Neanderthal genome project, headed by the Swedish biologist Svante Pääbo at the Max Planck Institute for Evolutionary Anthropology in Leipzig, Germany, finally succeeded in obtaining the genomic sequence of Neanderthals, in addition to the sequence of a Denisovan, another Paleolithic-era member of the *Homo* species. These experiments were carried out with very small amounts of material. For example, 38 mg of bone powder from a more than 50 000-year old Altai-region Neanderthal toe phalanx were sufficient to obtain a high-quality genome sequence. Comparative data from different hominin groups permitted a recalculation of the split times of the ancestral populations. The population split time between humans on the one hand and Neanderthals and Denisovans on the other was estimated to be between 550 000 and 600 000 years ago, while Neanderthals and Denisovans split approximately 380 000 years ago (Figure 7.7).

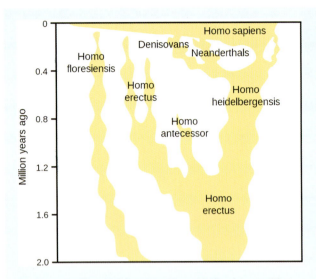

Fig. 7.7 Family tree of the genus *Homo*. Modern humans (*Homo sapiens*), Neanderthals and Denisovans, a population discovered in a cave in Siberia, are descendents from *Homo heidelbergensis*, an earlier species. Genetic analysis suggests interbreeding between the three populations. Some African populations may contain genes from another ancient group. (Adapted from Ref. [1] with kind permission from Macmillan Publishers Ltd., Copyright 2012.)

Modern humans and Neanderthals share more than 99.5% gene sequence identity. An important gene that was previously studied in early Neanderthal sequencing projects is *FOXOP2*. This gene has been implicated in the development of speech and language. It is highly conserved among mammals, but two amino acid substitutions have been fixed in humans since our split from chimpanzees. Inactivation of *FOXOP2* in humans results in deficits in speech. The *FOXOP2* variant with the two amino acid difference between chimpanzees and humans might thus have been important for the evolution of language in humans. Neanderthals carry the same *FOXOP2* allele as humans, which suggest that they may have possessed at least some language capabilities.

Another unexpected finding of the sequencing studies was the notion of gene flow between the hominin populations. Present-day humans outside Africa harbor 1.5–2.1% Neanderthal-derived DNA. Denisovans contributed approximately 5% to the genomes of the indigenous populations of Papua New Guinea and Australia. This gene flow can best be explained by interbreeding. However, some of these conclusions still remain controversial, as they are not supported by archeological evidence. Moreover, the fossil record does not conclusively place modern humans and Neanderthals in close proximity at the proposed time and place of interbreeding. Nevertheless, sequence analysis of ancient hominin DNA has contributed to our understanding of the origin and development of the modern human.

in defining just what a genomic drug is, in addition to the question of how to assess the contribution of genomic analysis to drug discovery. In fairness, however, due to the long time required for drug development (on the order of 15 years, compare Section 1.1.2), it is still far too early to draw conclusions as to whether sequencing of the human genome contributes to the development of new drugs.

Nevertheless, a survey from 2010 included in the reference list at the end of this chapter identifies 12 genomics drugs in clinical trials. These are examples that illustrate the impact of genomics data on drug discovery. For example, the monoclonal antibody (mAb) (Section 10.2.1) belimumab (trade name Benlysta) was approved by the US Food and Drug Administration (FDA) in 2011 and was the first new drug in 50 years approved to treat the autoimmune disease systemic lupus erythematosus (SLE). Development of this mAb relied on the sequencing of a cDNA library derived from neutrophil monocytes. This led to the discovery of a factor called the B-lymphocyte stimulator (BLYS), which is overexpressed in SLE and contributes to the proliferation and survival of autoimmune B cells. Binding of Benlysta to BLYS prevents its interaction with B cells and stops the autoimmune processes.

A second example of a genomic drug is the mAb romosozumab, used for the treatment of bone disorders, which entered phase III clinical trials in 2014. Genome-wide linkage analysis in families affected by osteosclerosis revealed loss-of-function mutations in the gene that encodes sclerostin. This protein, which is secreted by osteoblasts, negatively regulates bone formation; mutations in it cause bone overgrowth. The mAb inactivates sclerosin and is being developed to increase bone density for the treatment of osteoporosis.

Furthermore, knowledge of the complete human and mouse genomes has advanced the use of mouse knock-out models (Section 2.4) in drug discovery. For example, the study of phenotypes from 5000 gene knockouts led to the discovery of a new brain-specific isoform of tryptophan hydroxylase. Dysregulation of an isoform of this enzyme in the gastrointestinal tract is associated with carcinoid syndrome and irritable bowel syndrome. A mouse knockout study triggered the search for specific inhibitors of the enzyme in the gastrointestinal tract which are devoid of side effects in the nervous system. Initiated in 2007, the International Knockout Mouse Consortium (IKMC) uses genomic data to create a library of mouse embryonic stem-cell lines representing every possible gene knockout. This collection will be a valuable resource to researchers for further studies.

Also important for the development of novel therapeutics is the identification of the subset of genes that contributes to a disease phenotype and that can be modulated by small molecule drugs. The so-called *druggable genome* comprises potential targets for developing therapeutic low-molecular weight drugs. The druggability of protein families depends on the presence of protein structures that favor interactions with chemical compounds that can be used as drugs. Many proteins, however, do not have suitable binding sites and are not amenable to pharmaceutical modulation. Estimates for the number of druggable protein targets vary substantially, but are on the order of 2–3000. Many of the proteins that potentially bind a small molecule are still not suited as drug targets, as modulating their biological function will not lead to a therapeutic benefit. Thus, a potential target for a new pharmaceutical molecule must fulfill two prerequisites: It must be druggable and it must play a crucial role in a disease process. These requirements further lower the number of potential drug targets (Figure 7.8). However, many approaches in molecular medicine aim to overcome these restrictions. Biological molecules (*biologics*) substantially expand the number of potential targets, as they do not require the targeted protein to have certain specific structural features that are needed for an interaction with a low-molecular weight substance. Enzyme replacement therapies by recombinant

proteins and gene therapeutic interventions (Chapters 10 and 11, respectively) can theoretically correct the dysfunction of any protein. Likewise, anti-mRNA strategies discussed in Chapter 13 can inhibit the expression of any deleterious gene in the absence of target structural requirements.

While the direct impact of human genome knowledge on the development of new drugs is still under debate, virtually every laboratory working in biomedical research and development routinely uses genomic databases. For example, the GenBank sequence database is a freely accessible collection of all publicly available nucleotide sequences. It is maintained by the US National Center for Biotechnology Information (NCBI; www.ncbi.nlm.nih.gov/). The database contains sequences from more than 100 000 distinct organisms, including the complete sequence of the human genome. Rapid access to the sequences of every human gene has substantially accelerated life science research. It has also enabled the easy synthesis of the cDNA of every gene of interest. For most genes, clones expressing cDNAs are available in large libraries that cover the complete human genome.

In some fields, alignment searches with genomic DNA databases have become a standard procedure for the confirmation of specificity. For example, advanced algorithms used to design primers for (quantitative) PCR (Section 2.2) carry out alignment searches to avoid primer pairs that can cause amplification of targets other than the input template. RNA interference (RNAi, Section 13.3) is another field that relies directly on the knowledge of the complete sequence of the human genome. The creation of libraries targeting all human genes could only be achieved through the knowledge of their sequences. These libraries have helped to develop RNAi as a research and therapeutic strategy. In addition, genome-wide RNAi screens have also identified new targets for conventional small molecular substances for the treatment of cancer or viral infection. In addition, unspecific side effects, also known as off-target effects, are a major concern in the RNAi field. Therefore, screening of the human genome for unintended target sequences is an essential prerequisite to ensure the specificity of an RNAi approach.

A technological advance that was directly triggered by the HGP was the development of new sequencing technologies (Section 2.3). These methods would probably not exist if not for the increased interest in large-scale sequencing.

Another field that has clearly benefited from the elucidation of the human genome is the understanding of the genetic basis of inherited disorders. In the last decade, numerous disease-causing mutations have been identified. This knowledge has helped to develop genetic tests to diagnose disease or disease-relevant predispositions.

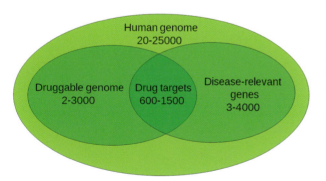

Fig. 7.8 Potential drug targets for small molecular weight drugs. Only human proteins with specific structural properties are suitable targets for small molecules. The intersection of the druggable genome and disease-relevant genes gives the number of potential drug targets for conventional pharmacological substances.

However, it is controversial as to whether genetic tests for diseases for which no therapy exists should be developed.

Since their publication, the initial sequences from the HGP and from Celera have been used as a reference genome. These sequences were obtained from multiple individuals and pools of genomes, respectively. Soon after completion of these projects, the first individual genomes were sequenced (Box 7.3). These projects, as well as a number of large, international follow-up initiatives, were begun to understand variations in the human genome and their impact on human health and disease (Table 7.3). The genetic difference between two human individuals has consistently been estimated to be on the order of 0.1%. The International HapMap Project, the

Box 7.3.   Elucidation of Individual Genomes

For the initial sequencing of the human genome, the HGP used DNA from several anonymous individuals. Large-insert genome-wide BAC and PAC libraries were created with the DNA obtained from blood or sperm of the donors. Much of the sequence of the reference genome came from an anonymous male donor from Buffalo, NY, with the code name RP11. Celera, the competing company, enrolled 21 donors. For DNA sequencing, the genomes of two males and three females were chosen, one African-American, one Asian-Chinese, one Hispanic-Mexican, and two Caucasians. Craig Venter later acknowledged that his genomic DNA was in the pool of the 21 samples, from which 5 were selected for use.

Craig Venter was also the first human whose individual genome was deciphered and published. It was analyzed by shotgun sequencing from ~32 million random DNA fragments and was sequenced by conventional Sanger dideoxy technology. Comparison of his genome with the reference database revealed approximately 3.2 million SNPs, that is, substitutions of a single nucleotide by another, and around 900 000 other types of sequence variations. These include block substitutions, short insertions and deletions (indels), inversions and larger copy number variations (CNVs). The latter comprise deletions and duplications of the genome, which range in size from about 1 kb to several mega bases. The non-SNP DNA variations account for 22% of all mutations, but involve 74% of all variant bases. The SNP number of ~3 million is in the range of what was found in subsequent sequence analyses of individual genomes. However, fewer non-SNP variations were observed. Venter's genome contained approximately 300 disease-relevant variations including predispositions for myocardial infarction, Alzheimer's disease, and alcoholism. This knowledge allowed him to take preventive measures, for example, a drug to lower his cholesterol levels. The data also revealed that the genetic variation between two individuals is much higher than previously estimated.

The first individual genome that was sequenced by an NGS technology (pyrosequencing) was that of James Watson. The approximately 3.3 million SNPs in his genome cause more than 10 000 amino acid substitutions within the coding sequence relative to the reference sequence. Large-scale gain or loss of chromosomal segments ranging from 26 000 to 1.5 million base pairs were identified. The authors of the study have claimed that the NGS technology is not only faster and less expensive than conventional sequencing approaches but it also avoids loss of genomic information by the cloning step, which is replaced by the cell-free amplification of the DNA in the newer technologies.

The elucidation of Watson's genome, which was published in 2008, also demonstrated the rapid progress made in sequencing whole human genomes. While the initial sequencing of the human genome by the HGP took 13 years and cost around $3 billion, Venter's genome was sequenced within four years and the costs were reduced to approximately $100 million. The sequencing of Watson's genome was completed within 4.5 months and cost less than $1.5 million. When Stephen R. Quake, the Stanford professor and cofounder of the company Helicos BioSciences, had his genome sequenced by true single-molecule sequencing in 2010; the price dropped to less than $50 000. The genome of the geneticist James R. Lupski was of interest, as he was affected by Charcot-Marie-Tooth disease, one of the most common inherited neurological disorders. This ailment is characterized by peripheral motor and sensory neuropathy that can lead to muscle weakness. Sequencing of his genome led to the identification of the causative genetic variations of the disorder. In the meantime, as will be described below, thousands of genomes have been sequenced. The decrease in the cost of whole genome sequencing to approximately $1000 will permit sequencing of the personal genomes of almost all (cancer) patients in wealthy health systems in the relatively near future.

Table 7.3 International initiatives studying various aspects of the human genome after completion of the HGP. SNP: single-nucleotide polymorphism.

Initiative	Start	Aim
International HapMap Project	2002	Characterize the inheritance of SNPs as haplotype blocks and identify the marker SNPs that represent all SNPs in the genome
ENCODE Project	2003	Identify all functional elements in the human genome; determine the relevance of regions in the genome that are not translated into proteins
Human Epigenome Project	2003	Identify, catalog, and interpret genome-wide DNA methylation patterns (Section 1.3.4) of all human genes in all major tissues
Personal Genome Project	2005	Sequence and publicize the complete genomes and medical records of 100 000 volunteers
Cancer Genome Atlas	2005	Catalog genetic mutations responsible for cancer, using sequencing and bioinformatics
Human Variome Project	2006	Collect and curate all data on genetic variations, their phenotypes and associated diseases; facilitate the unification of data on human genetic variation from hundreds of diverse databases worldwide and help draw useful conclusions from the large datasets
1000 Genomes Project	2008	Establish a detailed catalog of human genetic variation by sequencing the genomes of at least 1000 individuals from different ethnic groups
International Cancer Genome Consortium	2008	Large-scale study of cancer genomes in tumors from 50 different cancer types at the genomic, epigenomic, and transcriptomic level to improve prognosis and the development of new cancer therapies

1000 Genomes Project, the Personal Genome Project, and the ENCODE Project will be discussed as examples in the following sections.

7.2.2
The International HapMap Project

Any two unrelated humans share approximately 99.9% DNA sequence identity. The remaining differences may explain differing predispositions to disease. SNPs represent approximately 90% of the genetic variations in humans. A DNA fragment that contains a difference in a nucleotide between two individuals is said to occur in two *alleles*. Usually, several adjacent SNPs are inherited as a block; such a set of statistically associated SNPs is referred to as a *haplotype*. The international HapMap Project, officially started in October 2002, aimed to develop a haplotype map of the human genome to describe common patterns of genetic variation in humans. It focused on common SNPs for which each allele occurs in at least 1% of the population. The HapMap Project was based on the idea

that identification of selected marker SNPs will allow the unambiguous identification of all other polymorphic sites in this region (Figure 7.9). In other words, the presence of a particular variant at one site could predict the presence of a particular variant at another site. The number of tags necessary to represent common variants across the genome was estimated to be less than one tenth of the total number of such sites. The Hap map was believed to be helpful for investigating the genetics of common diseases.

The objective of the HapMap Project was to genotype at least one common SNP every 5 kilobases across the euchromatic portion of the genome in 270 individuals from 4 populations. At the start of the HapMap Project, the human genome was estimated to contain approximately 9–10 million common SNPs (though more recent databases contain approximately 40 million SNPs). It was expected that all SNPs can be represented by approximately 500 000 SNPs, which would facilitate whole genome association studies. In 2005, at the end of phase I of the project, a database containing more than 1 million SNPs was released. A second-generation

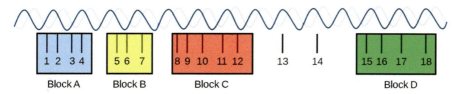

Fig. 7.9 Inheritance of haplotype blocks. The schematic representation depicts 18 SNP markers on a DNA segment. Several of these markers are linked and inherited as blocks. It is, therefore, possible to select a single-SNP marker for each of the A–D haplotype blocks. Analysis of the selected four markers is representative of the entire segment and facilitates characterization of the fragment.

haplotype map of over 3.1 million SNPs with a density of 1 SNP/kb was released in 2007. The map was further refined and released as the phase III HapMap in 2009. Despite the successful collection of common SNPs, the relevance of the findings of the HapMap Project for human health and disease, which cost more than $100 million, was considered questionable.

7.2.3
The 1000 Genomes Project and the Personal Genome Project

The 1000 Genomes Project was launched in 2008 with the aim of producing a comprehensive catalog of human genetic variations. The sequencing the genomes of more than 1000 individuals from different ethnic groups was planned. The goal was to identify variants that occur at rates in as low as 1% of the population. The variants range from SNPs over short indels to larger CNVs.

In November 2012, the 1000 Genomes Project Consortium described the genomes of 1092 individuals from 14 populations drawn from Europe, East Asia, Sub-Saharan Africa, and the Americas. The genomes were analyzed by low-coverage whole genome sequencing and deep exome sequencing (Box 7.4). A haplotype map was created that consisted of 38 million SNPs, 1.4 million indels and over 14 000 larger deletions. The results demonstrate that individuals from different populations carry

Box 7.4.   Exome Sequencing

Exome sequencing is a strategy to selectively sequence the protein-coding regions of a genome. Only a small portion of the genome is transcribed into mRNA. The primary transcripts consist of exons that are translated into protein and introns that are removed during the maturation process. There are approximately 180 000 exons in the human genome. They constitute little more than 1% (or ~30 mega bases) of the human genetic material. For many applications, it is much cheaper and more efficient to selectively sequence the exons instead of the whole genome, which consists of more than 3 billion bases. Exome sequencing is an efficient strategy that permits the sequencing of the relevant coding regions at high coverage (i.e., multiple times) to detect rare variants without the high costs of in-depth whole genome sequencing.

An important challenge of exome sequencing is to capture the exon regions and to separate them from the rest of the genome. Various methods have been developed for this purpose. The two most commonly used are described in more detail in Figure 7.10. Usually, the first step is to fragment the genomic DNA and ligate linkers for the amplification and sequencing of the fragments. Solid-phase hybridization methods make use of probes that are fixed on a solid support, usually a DNA array. The fragmented total genomic DNA is applied to the array; the sequences of interest (i.e., the exonic DNA) will hybridize to the probes. Non-hybridized fragments are washed away, and the enriched exons are subsequently eluted, amplified, and sequenced. In liquid-phase hybridization methods, the probes are not attached to a solid matrix. Instead, they carry a linker, usually biotin. Following the hybridization step, the biotinylated probes together with the captured DNA are bound to magnetic, streptavidin-coated beads and can be pulled down and washed to clear excess material. The beads are removed and the enriched exonic DNA can be amplified and sequenced by either of the NGS technologies.

Several studies have been reported in which exome sequencing has succeeded in locating the causative genes of rare inherited diseases. An example is the analysis of patients with Miller syndrome, an inherited disorder with signs and symptoms that include severe micrognathia, cleft lip and/or palate, hypoplasia or aplasia of the posterior elements of the limbs, coloboma of the eyelids, and supernumerary nipples. Genomic variations were identified by exome sequencing and filtered against known SNPs in the public databases. Two remaining variants were located in a single-candidate gene, *dihydroorotate dehydrogenase* (DHODH), which encodes a key enzyme in the pyrimidine *de novo* biosynthesis pathway. The presence of mutations in this gene was confirmed in additional families with Miller syndrome.

Several companies offer exome sequencing for clinical diagnostic purposes. However, although the exome sequencing approach is highly efficient, it has some substantial limitations. As it focuses exclusively on the protein-coding regions of the genome, it neglects the remaining >98% of the human genome that is not translated into proteins, but which also is important for cellular functioning (Section 7.2.4). Moreover, 90% of the SNPs, most of which are linked to disease, are located outside the protein-coding regions. As the cost of sequencing continues to drop rapidly, whole genome sequencing is likely to become the standard approach to investigate and understand the genetic variations found in populations.

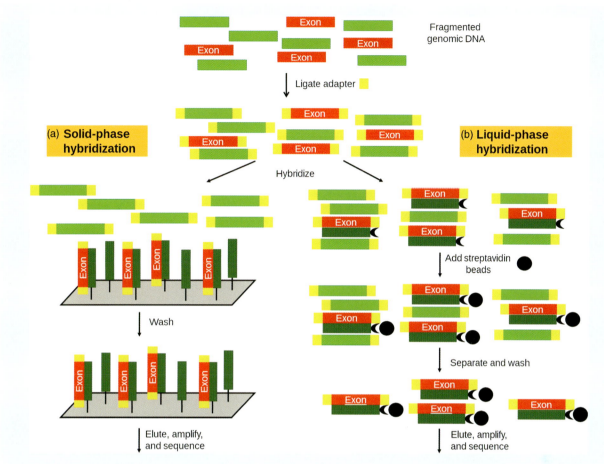

Fig. 7.10 Methods used for exome capture. (a) Solid-phase hybridization. For solid-phase hybridization methods, the targeted DNA is captured by probes fixed on an array. Unwanted material (green) is washed away and the eluted exon fragments (red) can be sequenced. (b) Liquid-phase hybridization. In liquid-phase hybridization methods, the probes are biotinylated. They bind to magnetic streptavidin beads and can be physically separated together with the hybridized exome DNA. Further details are described in the text.

different profiles of rare and common variants. Furthermore, each individual harbors hundreds of rare noncoding variants at conserved sites, such as changes that may disrupt transcription factor binding motifs. In a subsequent phase, the 1000 Genomes Project Consortium analyzed an additional 1500 genomes.

The Personal Genome Project (PGP) was launched in 2005 as a long-term study to sequence and publicize the complete genomes and medical records of 100 000 volunteers. The main focus of the PGP is to connect genetic sequences to phenotypic information. Participants in the project grant consent to making phenotypic data such as health records, personal medical histories, imaging studies, environmental exposures, nutrition, lifestyle, physical measurements, and so on, publicly available together with their genetic information. This will allow researchers to test hypotheses about the relationships between genotypes, environmental factors, and phenotypes. The aim of the PGP is to provide resources for moving toward personalized medical decisions.

In 2012, a pilot study of 10 individuals was made public. This cohort referred to as the "PGP-10" consisted of well-known researchers, including the initiator of the PGP, George M. Church. The information was interpreted by a software tool named the *Genome-Environment-Trait Evidence (GET-Evidence)* system. This algorithm automatically processes genomes and prioritizes variants for further interpretation. An unexpected finding was that presumably healthy volunteers had several gene variants that have been implicated in the literature in serious diseases. Although the late onset of a disease cannot be excluded, these analyses demonstrate the challenges connected with the use of reported variants in the clinical setting. The PGP aims to produce a system for recording and organizing evaluations per standardized evidence guidelines in order to reach consensus on the interpretation of clinically relevant variants.

An important aspect of the PGP is the participatory model of the researcher and the participant. All enrolled volunteers must give an informed "open consent" to making their data nonanonymously public. They must understand the potential risks regarding the loss of privacy not only to themselves but their relatives too. Participants may not only obtain information about their predispositions for certain untreatable diseases, but they also risk possible discrimination by insurers and employers because of their publically available data.

The field of oncology will be the first to include the analysis of a personal genome in the decision-making for an individual therapeutic strategy. Surveys have shown that the percentage of nonresponders is high for various anticancer drugs; in some cases greater than 70%. To prevent the severe side effects that can occur during cancer therapy and to avoid unnecessary treatment costs, it would be desirable to identify those patients who will benefit from a treatment while choosing an alternative strategy for those who are not likely to respond. As outlined in Section 9.3, genetic testing has already been introduced into clinical practice for some anticancer drugs. For example, the monoclonal antibody trastuzumab (trade name Herceptin) is only given after confirming that the drug target, the human epidermal growth factor receptor 2 (Her2/neu), is overexpressed in breast cancer. Inhibitors of the BRAF kinase will be suitable for the treatment of metastatic melanoma only if the target has a V600E mutation. In melanomas without the mutation, these inhibitors may actually stimulate tumor growth.

While these examples involve the analysis of only a single gene, the next step will be to expand this approach to the complete human genome. As the cost for sequencing an individual genome drops below the threshold of $1000, the analysis of a patient's genome as the basis for treatment decision becomes more realistic. The German National Center for Tumor Diseases in Heidelberg can now offer whole genome analysis to every cancer patient in the clinic. Identification of the genetic variations in the tumor is expected to assist the choice of optimal personalized therapy. Human Longevity Inc., a US-based company founded in 2014 by Craig Venter, is equipped with several of the latest-generation NGS systems. The company plans to sequence up to 40 000 human genomes per year and to rapidly scale up capacity to 100 000 genomes annually. Through a collaboration, every cancer patient of the Moores Cancer Center in San Diego, CA will have his or her own genome, as well as that of their tumor, sequenced. The company will also sequence several patients' microbiomes, that is, the genomes of all microorganisms

inhabiting a human body, to identify all the microorganisms that may be relevant to disease processes. These data will not only be used to optimize therapeutic strategies but will also be valuable for finding new therapeutic targets, pathways, and biomarkers. By the end of 2014, Illumina, a leading sequencing company, estimated that approximately 228 000 human genomes had been completely sequenced worldwide. This number is expected to double every 12 months, and by 2017, 1.6 million genomes may have been sequenced.

Though the cost for sequencing reached the magic threshold of $1000 per genome, the price is unlikely to drop much further. The bottleneck is no longer the cost, but rather the interpretation of the data due to inefficiencies in the handling of the enormous amount of data generated. Systems biology, or a more specialized branch of the discipline sometimes referred to as systems medicine, aims to develop new approaches for data integration and interpretation. Computational models will integrate genomic and epigenomic data and combine them with environmental factors, clinical records, pathology, treatment, and so on, to obtain a patient model. Thousands of datasets will be stored in a reference database. Eventually, this procedure may help to develop new therapeutics and improve individual treatment decisions (Figure 7.11).

7.2.4
Encyclopedia of DNA Elements (ENCODE)

The ENCODE project is another public research project. Launched in 2003, it aims to identify all functional elements in the human genome. As outlined above, humans encode only about 20–25 000 protein-coding genes, which account for less than 2% of the DNA in the human genome. For many years, the remaining DNA was regarded as "junk," not transcribed and with no biological function. The goal of the ENCODE project is to elucidate the role of that segment of the genome that is not translated into proteins (Figure 7.12).

In a pilot project completed in 2007, 1% of the human genome, that is, 30 million bases in 44 regions of the genome, was analyzed. The results obtained in this phase of the project revealed that much of the analyzed DNA sequence was active in some way. Approximately 75–80% of the human genome is transcribed into so-called ncRNAs. While some researchers still consider these ncRNAs as nonfunctional junk, it is now widely accepted that they may have key roles in the regulation of gene expression. Altogether, 8800 small and 9600 long ncRNAs (>200 bp long) have been identified in the complete human genome. They differ in their intracellular location; some were found in the nucleus, others

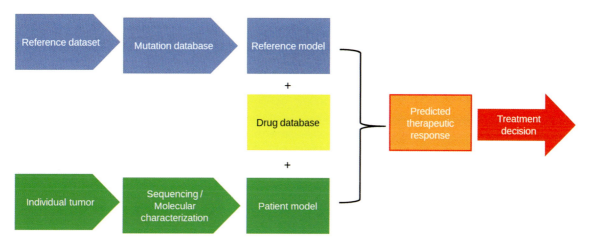

Fig. 7.11 Computer models for personalized medicine. Reference datasets and individual data are integrated with models to predict therapeutic response and aid decision-making with respect to the optimal individual treatment.

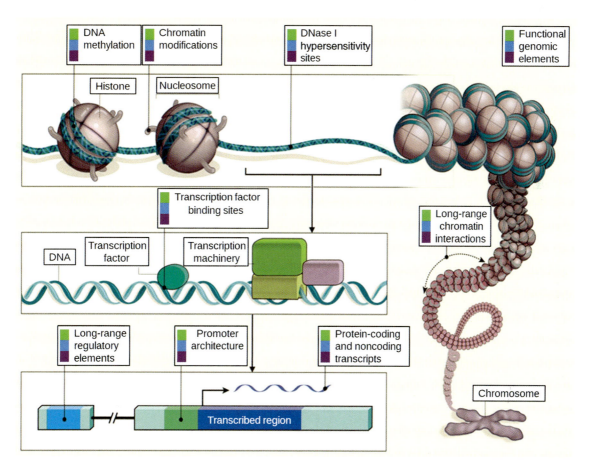

Fig. 7.12 Analysis of the functional elements in the human genome by the ENCODE project. The ENCODE project aims to study the functions of the genome beyond those contained within the DNA sequence. Epigenetic modifications (DNA methylation and the chemical modifications to histones) (Section 1.3.4) influence the rate of transcription. DNase I hypersensitive sites are accessible for the binding of transcription factors as a result of the displacement of nucleosomes. Long-range chromatin interactions, such as looping, alter the relative proximities of different chromosomal regions and may also affect transcription. In addition, the ENCODE project describes the binding of transcription factors and gene-regulatory DNA elements located either in the promoter region upstream of the point at which transcription begins, or situated in more distant regions. Another important aspect of the project is the analysis of protein-coding and noncoding RNA transcripts. (Adapted from Ref. [2] with kind permission from Macmillan Publishers Ltd., Copyright 2012.)

in the nucleolus, and some in the cytoplasm. Many ncRNAs are also transcribed in a cell-type-specific manner, reflecting their functional diversity.

After completion of the pilot phase of the project, the analysis was extended to the remainder of the genome. A larger number of cell lines and data from the 1000 genomes project were included. In addition, some model organisms such as *Drosophila melanogaster* and *C. elegans* were investigated in comparative and functional analyses (the model organism ENCODE or modENCODE project). In addition to transcription, ENCODE carried out investigations focused on the function of DNA elements in regulating the activity of genes. Maps were generated for regions of the genomic DNA that interact with transcription factors and other (regulatory) proteins. More than 40% of the genome was found to be accessible to these proteins. Examination of about 100 of the approximately 1500 transcription factors revealed that 8% of the genome falls within a transcription factor binding site. This percentage can be expected to increase further once more transcription factors have been investigated.

Approximately 90% of the SNPs in the human genome are located outside of protein-coding regions. Many of these SNPs have been linked to diseases by genome-wide association studies. Understanding the function of regions of DNA that do not encode proteins will help to elucidate the pathological relevance of these changes in the DNA sequence.

Another point of interest in the ENCODE project was the analysis of interactions between distinct DNA elements. A complex network of interactions between distant regions of a chromosome, or even different chromosomes, was found in the promoter regions of genes, emphasizing the relevance of three dimensional interactions in the genome.

When the different functional elements in the DNA are taken into consideration, 80% of the human genome is believed to have a biological function. The <2% of the DNA, which is protein-coding, and which has traditionally been the focus of biomedical research, thus represents only a small percentage of all the relevant elements. However, the interpretation of findings of the ENCODE project has not been without criticism. The ENCODE project defines functionality in such a way that anything that is transcribed must be functional. The functionality of many of the ncRNAs has been called into question, as human RNA polymerase II is known to be an unspecific enzyme and may transcribe numerous RNAs without biological meaning. Many transcribed elements do not have any known phenotypic relevance and would be considered nonfunctional by standard biological definitions.

Still, most researchers agree that the ENCODE project has revealed that the regulation of gene expression is far more complex than previously thought. The discovery of thousands of RNAs which are not translated into protein, but which are involved in the regulation of gene expression, changes the traditional gene-centric view that focused exclusively on protein-coding genes and neglected the rest of the genome. Elucidating the relevance of all the functional elements in the human genome will continue to contribute to our understanding of normal human biology and pathophysiology. It will also be of significant interest to include dynamic aspects of gene regulation in the analysis. While the results obtained so far represent a single snapshot of numerous and complex regulatory events, viewing those events over time will increase our understanding of developmental and other cellular processes.

7.3
Proteomics

The analysis of the sequence of the human genome has provided a wealth of data; however, our understanding of the function of the proteins encoded in the human genome remains insufficient. Ten years after completion of the HGP, the molecular function of roughly one third of all proteins is not known. Even for those proteins whose basic properties are known, their precise roles in physiological or disease processes are not entirely certain. Answers to these problems may be addressed via studies of *functional genomics*, which aims to understand the relationship between an organism's genome and its phenotype. In contrast to the genomic analysis of genetic information by sequencing, functional genomics also investigates dynamic aspects such as gene transcription, translation, and protein–protein interactions. Several approaches that belong to the field of functional genomics are described in other sections of the book, for example, transcriptome profiling by DNA microarrays (Section 2.1), loss-of-function studies by anti-mRNA approaches (Chapter 13), or functional studies as determined by the ENCODE project (Section 7.2.4). Here, proteomic studies, which are defined as large-scale investigations of proteins, will be described.

The term *proteome*, coined by Marc Wilkins in 1994 as a blend of *prote*in and gen*ome*, refers to the entire set of proteins and includes their posttranslational modification in any cell-type, tissue or organism. The proteome varies with time in response to environmental influences or extracellular signals. Figure 7.13 impressively illustrates the differences between the genome and the proteome. While the caterpillar and the mature swallowtail butterfly carry the identical genome, their proteome differs and determines their phenotype. Likewise, (almost) all cells of a human organism harbor the same genome. However,

Fig. 7.13 Caterpillar and imago of the swallowtail. The caterpillar and the imago carry the same genome, but differ in their proteome, which determines their phenotype. (Images of Wolfgang Schmökel and Carsten Anderssohn courtesy of the Entomological Society Orion, Berlin, Germany.)

proteins are the active molecules that differ between different cell types, for example, a liver cell and a muscle cell.

Proteomic and genomic studies are complementary approaches to understanding biological and pathological processes. Proteomics helps to overcome some of the limitations of genomics. For example, with our current state of knowledge, the presence of different protein variants arising from alternative splicing and posttranslational modification cannot be predicted by genomics, and may therefore be investigated by proteomic studies. The same is true for proteins found in extracellular fluids such as blood or urine. Quantitative proteomics also permits the determination of protein levels that do not necessarily correlate with respective mRNA levels, as determined by DNA arrays or quantitative PCR (Sections 2.1 and 2.2). This lack of correlation occurs because mRNA stability and translation efficiency varies between different transcripts, influencing the production of the encoded protein.

The Human Proteome Project (HPP) is organized by the Human Proteome Organization (HUPO), which was founded in 2001. Analogous to the HGP, it is a collaborative, international effort to study all of the proteins produced by the human genome. Individual chromosomes are covered by a working group. In addition, long-term proteomic projects study specific human tissues of clinical interest, such as the liver, brain, or plasma.

Similar to genomic research, proteomics research is based on sophisticated bioinformatic analyses and databases. A widely used research tool is the ExPASy (Expert Protein Analysis System) (www.expasy.org/) operated by the Swiss Institute of Bioinformatics (SIB). Originally, this Web site acted only as a proteomics server. However, the Web server has been extended to comprise an integrative portal that also provides resources for genomics, systems biology, structural bioinformatics, population genetics, and so on.

Proteomic analyses have widespread applications in molecular medicine. These include the identification of

potential protein targets for the development of new drugs for disease treatment. More recently, the identification of new biomarkers has become an important application of proteomic technologies. Attempts are being made to find measurable indicators of pathogenic conditions or pharmacologic responses to therapeutic interventions. Both approaches require the identification of peptides or proteins specifically expressed under certain conditions. For many years, mass spectrometry (MS) has been one of the most common technologies employed for these purposes. A well-established and widely used method is to separate proteins by two-dimensional gel electrophoresis, followed by mass spectrometric identification of differentially expressed proteins, as will be discussed in Section 7.3.1. Recently developed quantitative approaches based on isotope labeling and liquid chromatography separation of proteins prior to MS analysis will be outlined in the subsequent section.

7.3.1
Two-Dimensional Gel Electrophoresis and Mass Spectrometry

Two-dimensional (2D) gel electrophoresis is used to separate mixtures of proteins. Adding a second dimension to a conventional one-dimensional gel electrophoresis separates most proteins from each other. The resulting spots are composed of a single-protein species or a few usually very closely related species. In the commonly used version of 2D gel electrophoresis, proteins are first separated according to their isoelectric point by isoelectric focusing (IEF), followed by sodium dodecyl sulfate-polyacrylamide gel electrophoresis (SDS-PAGE). This last step separates them according to protein mass. Several thousand protein spots can be resolved by 2D gel electrophoresis simultaneously.

The standard application of 2D gel electrophoresis in molecular medicine is to identify proteins that are

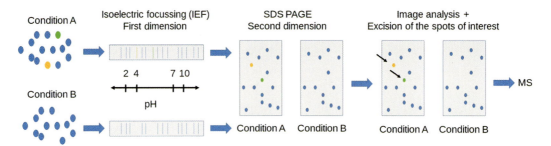

Fig. 7.14 Separation of differentially expressed proteins by two-dimensional (2D) gel electrophoresis. Protein lysates from cells (two conditions) are separated by 2D gel electrophoresis (IEF followed by SDS-PAGE). Image analysis reveals differentially expressed proteins (marked in green and orange). The spots are excised and analyzed by mass spectrometry. IEF: isoelectric focusing; SDS-PAGE: sodium dodecyl sulfate-polyacrylamide gel electrophoresis.

differentially regulated under certain conditions. For example, this can occur in cancer cells or in drug-treated cells, as compared to normal or untreated cells. A typical procedural workflow is illustrated in Figure 7.14: Protein extracts of cells are separated by IEF and SDS-PAGE. Proteins that are either up- or downregulated can be identified by comparing the 2D gels of proteins under both conditions.

However, several technical challenges to the successful application of 2D gel electrophoresis exist. Many hydrophobic proteins (e.g., membrane proteins) or very large proteins usually do not enter the gel. In addition, the wide range of protein concentrations makes it difficult to visualize all proteins. While some proteins (for example, proteins of the cytoskeleton or serum albumin) are abundant, other important proteins, such as transcription factors, may be present at very low levels and may be difficult to detect.

Another critical problem with this approach is the comparability of the two gels representing the different experimental conditions to be compared. These gels may differ in spatial resolution and spot intensities. Variations may arise from differences in protein uptake by the gels, by transfer from the first to the second dimension gel, or from local polymerization problems with the gels. These issues can be minimized by a technique known as differential in gel electrophoresis (DIGE). The basic idea of this method is to label the samples with different dyes before mixing them and running them on the same gel. For DIGE applications, cyanine dyes (Cy2, Cy3, and Cy5) are commonly used; these dyes carry reactive groups that tag them to the proteins. The samples are then allowed to comigrate in the same gel to avoid experimental variation. In the last step, the gel is illuminated with the excitation wavelengths of each of the dyes and the different fluorescent images are superimposed to identify and excise differentially expressed proteins.

Following 2D gel electrophoresis, proteins found to be differentially expressed are usually analyzed by MS, which measures the mass-to-charge (m/z) ratio for ions in the gas phase. A mass spectrometer typically consists of three main components: an ionizer, a mass analyzer, and a detector. Two approaches can be distinguished in protein mass spectrometry. In the first, referred to as a "top-down" strategy, intact proteins are ionized and introduced to the mass analyzer. In the second, "bottom-up" strategy, proteins are enzymatically digested into smaller peptides, introduced into the mass spectrometer and identified by peptide mass fingerprinting or tandem mass spectrometry (see below). Protein mass spectrometry requires soft-ionization techniques that do not destroy large macromolecules during the production of gas-phase ions. The two most common methods used for this purpose are electrospray ionization (ESI) and matrix-assisted laser desorption/ionization (MALDI) (Box 7.5).

A typical procedure for the identification of a protein in a spot identified on a 2D gel, known as *peptide mass fingerprinting*, is to digest the protein proteolytically (for example, by trypsin) and to determine the masses of the resulting peptides by MS. The protein can then be identified with the help of databases. As the complete human genome is known, it can be translated into proteins by computer programs and then digested into peptides *in silico* (trypsin is known to cleave at the carboxyl side of lysines or arginines). Experimentally obtained peptide masses can then be compared to the theoretical peptide masses of each protein encoded in the genome, thus identifying the protein.

Tandem MS (abbreviated MS/MS) has become a widely used procedure for identifying proteins by (partial) sequencing. This technology is carried out with two mass spectrometers coupled in series. In the first mass spectrometer, peptides are separated according to their m/z values. A selected peptide ion is then passed into a collision cell, where it collides with an inert gas (e.g., helium or argon). This collision causes it to fragment into smaller pieces, primarily at peptide bonds. The molecular masses of these fragments can then be

Box 7.5.   Protein Mass Spectrometry

While a variety of methods for ionization have been used in organic chemistry and material sciences for many years, the application of MS to protein science required the development of soft ionization techniques to prevent destruction of the labile biological macromolecules. The two primary methods for the ionization of whole proteins are ESI and MALDI. In the ESI technology, a solution of the protein or peptide is sprayed from a narrow capillary tube at high voltage to form a fine, highly charged aerosol from which the solvent rapidly evaporates. The gas-phase ions generated are subsequently directed into the mass spectrometer, which measures their molecular mass.

The MALDI principle is depicted in Figure 7.15a. The initial step is to cocrystallize the analyte with a UV-absorbing matrix. The energy of a UV laser beam then triggers the ablation (desorption) of the upper layer of the matrix material which, in addition to matrix molecules, contains protonated or deprotonated analyte molecules. The ionized polypeptides are accelerated in an electric field and directed into the mass analyzer.

Analysis of a protein or peptide mass is often conducted with the time of flight (TOF) principle. The typical setup of a MALDI-TOF mass spectrometer is shown in Figure 7.15b. After ionization of the particles by MALDI, the TOF analyzer accelerates ions in an electric field and measures the time they need to reach the detector. The velocity of a particle will depend on its mass; lighter ions will reach the detector first (assuming they have the same charge). The kinetic energy that particles receive in the electric field may vary slightly, which broadens the peaks of the ion signals. This distribution may be corrected by using a reflector. For ion detection, an electron multiplier is commonly used.

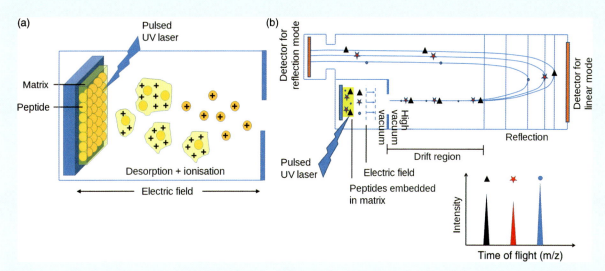

Fig. 7.15   Matrix-assisted laser desorption/ionization–time of flight (MALDI–TOF) mass spectrometry. (a) For soft ionization of macromolecules by the MALDI principle, the analyte and a matrix are cocrystallized. The matrix absorbs the energy of a UV laser beam and a layer of the material desorbs, thereby producing ionized analyte molecules that can be accelerated in an electric field. (b) Typical setup of a MALDI–TOF mass spectrometer. After ionization of the proteins or peptides by the MALDI principle followed by their acceleration in an electric field, the charged particles pass through the mass analyzer. The time they need to reach the detector is measured. A reflector is used to correct the kinetic energy distribution of the particles. An electron multiplier detects the ions.

determined by the second mass spectrometer. By comparing the masses of the various fragments generated, the identity of the amino acids removed in the collision step can be determined and the amino acid sequence of the peptide can be deduced.

A particular strength of MS-based proteomic approaches is their ability to analyze posttranslational protein modifications for example, phosphorylation, glycosylation, acetylation, or methylation. These modifications can influence the properties of proteins, but they cannot be predicted by nucleic acids-based methods such as sequencing or transcriptome analysis by DNA arrays. For example, to investigate posttranslational protein phosphorylation, phosphopeptides can be enriched by purifying the mixture over a metal resin column. The resulting peptides can then be analyzed by MS before and after treatment with phosphatase. A shift of the peptide mass by 80 Da (or multiples thereof) indicates the removal of one (or more) phosphate group by the phosphatase treatment.

7.3.2
Quantitative and Shotgun Proteomics

The techniques described so far do not permit accurate quantification of protein levels found in different proteome states. These limitations have been overcome by approaches employing stable isotope tags, which differentially label proteins from two different complex mixtures. The use of stable isotopes is suitable for quantitative proteomics; "light" and "heavy" isotopes generally exhibit identical chemical properties, but do not influence protein purification, proteolytic digestion, or MS analysis. Commonly used isotopes include ^{2}H,

^{13}C, ^{15}N, and ^{18}O. Samples are differentially labeled with light or heavy isotope tags and processed together. Comparison of the relative intensities of the heavy- and light-labeled peptides by MS allows quantification.

Metabolic labeling and chemical labeling are two strategies that have been employed (Figure 7.16). The former, also known as stable isotope-labeled amino acids in cell culture (SILAC) involves the metabolic incorporation of isotopically heavy amino acids into proteins in living cells. For this approach, two cell populations are grown in either standard media or media containing essential amino acids to which heavy atoms, such as ^{13}C or ^{15}N, have been added. After a given time, all proteins of the labeled sample will incorporate the heavy amino acid. Cells are then subjected to the desired treatment (e.g., cytokine stimulus or antisense treatment). Eventually, both, the heavy and light samples are combined into a pooled lysate. The proteins are then proteolytically digested, fractionated by liquid chromatography and

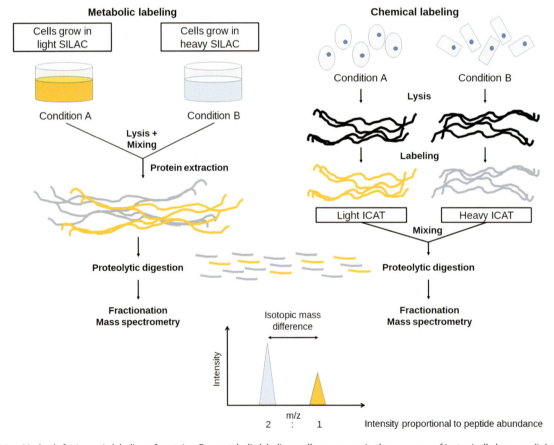

Fig. 7.16 Methods for isotopic labeling of proteins. For metabolic labeling, cells are grown in the presence of isotopically heavy or light amino acids. Both samples are then mixed and processed together. In contrast, chemical labeling is carried out after cell lysis. Both procedures result in mass spectra in which the respective peptides from both samples differ by the mass of the stable isotope labels. The intensity of the peaks in the mass spectrum is proportional to the peptide occurrence. ICAT: isotope-coded affinity tagging; SILAC: stable isotope-labeled amino acids in cell culture.

analyzed by MS. The respective peptides of both samples differ by the mass of the incorporated, labeled amino acids. The relative intensity of the peaks is proportional to peptide abundance. This permits the determination of the relative levels of proteins under both conditions.

A strategy referred to as *isotope-coded affinity tagging* (ICAT) is the most common approach for the chemical labeling of proteins. Different types of ICAT tags may be used, but they usually consist of a reactive group to covalently attach the tag to the proteins, a linker containing heavy or light isotopes, and an affinity tag used for purification of the tagged protein. As opposed to the metabolic labeling technique described above, chemical labeling is carried out post-lysis. The two samples to be compared are separately labeled with the light or heavy tag and subsequently mixed and processed as described above. In addition to the ICAT method, other labeling techniques (e.g., isobaric tags for relative and absolute quantification, iTRAQ) as well as strategies for label-free quantification have been developed.

The most recent development in the field of proteomics, referred to as *shotgun proteomics*, aims to identify proteins in complex mixtures by using a combination of liquid chromatography and MS (Figure 7.17). This approach overcomes the shortcomings of traditional proteomics as it identifies all proteins in a sample, including very small or large proteins, and membrane proteins. It also covers a broad range of proteins, from those of low abundance to highly abundant proteins as well. The challenge of resolving and detecting proteins from very complex mixtures (e.g., from whole cell lysates), has required many improvements over previous technologies. Separation of the peptides in complex mixtures is achieved by ultrahigh-performance liquid chromatography (UHPLC), which uses pumps that perform at very high pressures and columns with small bead sizes.

In addition to improvements in peptide separation by liquid chromatography techniques, high-resolution mass spectrometers that analyze peptides with very high speed are also necessary. The more recently developed Orbitrap

mass analyzer has a particularly high resolving power. It consists of an outer barrel-like electrode and a coaxial inner spindle-like electrode. Peptide masses are determined by the oscillation frequency of the peptide ions around the central spindle-shaped electrode. The mass spectrometers are usually arranged in a tandem setup, as described above, to collect three pieces of information: The first mass scan analyzes the masses of the peptides and their intensity; this can be used for quantification. The second mass spectral scan records fragment masses to determine the sequence of the peptide.

In a single experiment, shotgun proteomics can quantify thousands of proteins over concentrations of several orders of magnitude. More than 10 000 proteins have been identified in a single cell; this number thus defines the lower limit of the complexity of the human proteome. In addition, proteomics are frequently used to investigate the localization of proteins. The goal here is to produce an organellar proteome derived from subcellular structures and to quantify the proteins it contains. Quantitative proteomics helps to discriminate between true members of an organelle and low abundance background proteins that may arise from the unintended copurification of other subcellular components.

Other fields of modern proteomics include large-scale measurements of posttranslational modifications and the analysis of protein–protein interactions. For the latter, a typical approach involves affinity purification followed by MS (AP/MS). The protein-of-interest, the bait, can be pulled down by an affinity handle. The interacting proteins are then identified by MS analysis. Similar to subcellular proteomic experiments, it is of the utmost importance to identify actual interaction partners and discriminate them from background binders. The concept of quantitative AP/MS has also been adapted to investigate interactions between proteins and nucleic acids. A very recent approach has been to introduce chemical crosslinkers into living cells to fix protein–protein or protein–DNA interactions and then to analyze the covalently linked products by MS. It is anticipated that this strategy

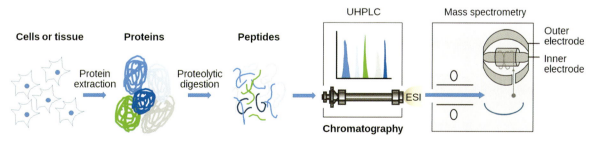

Fig. 7.17 Workflow of shotgun proteomics. In the first step, cells are extracted from tissues, fluids, or cells, followed by proteolytic digestion of the proteins into peptides. The complex peptide mixtures are separated by ultrahigh-performance liquid chromatography (UHPLC). Modern mass spectrometers use an Orbitrap mass analyzer.

will reveal a significant amount of biologically meaningful data.

Proteomic techniques are also important for the preparation and quality control of recombinant therapeutics (Chapter 10). For this application, techniques are required that identify impurities in the presence of the highly abundant main product, the recombinant protein. For example, monoclonal antibodies (mAbs) are usually produced in the Chinese hamster ovary (CHO) cell line. After harvesting and purification, the therapeutic antibody may still be contaminated with residual trace amounts of other proteins from the host cell line. Since these proteins are potential allergens, their occurrence and abundance must be monitored.

7.3.3
Structural Proteomics

Another important approach to understanding the function of a protein is to solve its molecular structure. For example, knowledge of the three-dimensional structure of enzymes has in many cases helped to elucidate the mechanism of catalysis. Structural biology may also support the development of new drugs. A prominent example is the development of the first protease inhibitor to treat HIV-1 infection, which was based on the X-ray structure of the HIV-1 protease.

The structural analysis of proteins is conventionally achieved by crystallizing a protein, followed by X-ray analysis. Atoms in a crystal cause a beam of incident X-rays to diffract in many specific directions. The diffraction pattern can be used to calculate the density of electrons within the crystal. The electron density map which is produced permits the determination of the exact structure of the protein at atomic resolution.

An alternative to X-ray crystallography is the use of multidimensional nuclear magnetic resonance (NMR) spectroscopy. NMR spectroscopy analyses the properties of atomic nuclei, which depend on their local molecular environment. These measurements provide information about the distance between two atoms and their relative movement. To facilitate these measurements, proteins are usually labeled with ^{13}C and ^{15}N isotopes as the naturally occurring isotopes are not suited for the experiments. NMR spectroscopy permits measurements in solution; the laborious crystallization step required for X-ray analysis can be omitted. In addition, there are no concerns that the protein structure in a crystal is not identical to the protein structure in solution. However, NMR spectroscopy is usually limited to proteins smaller than 35 kDa, although some larger structures have also been solved.

Solving a three-dimensional structure requires several steps. For example, X-ray crystallography includes cloning of the cDNA and expression of the gene. This is followed by purification and crystallization of the protein, collection of the X-ray diffraction patterns, and calculation of the structure. Several of these steps are well suited for automation, for example the selection of the optimal conditions of gene expression or the choice of solvents for crystallization. Structural genomics initiatives aim to develop pipelines for the high-throughput cloning and expression of genes, the purification and crystallization of proteins, and the imaging and structure determination of three-dimensional protein structures. Many of the steps of the pipeline are carried out by robots.

Several countries have founded consortia for the large-scale structural analysis of proteins. The largest of these is the Protein Structure Initiative (PSI) in the United States. Launched in 2000, it has been organized in three separate 5-year phases. As of 2014, the PSI has solved more than 6500 protein structures. The focus of the PSI is on novel structures, defined as protein structures with less than 30% sequence similarity to other proteins of known structure. According to a 2006 survey, the PSI centers have contributed more than half of all novel structures to the databases.

However, the PSI has not remained without criticism. The first two phases of the initiative cost more than $750 million. Although the costs for solving a single structure dropped from about $250 000 in 2000 to about $66 000 in 2008, critics state PSI outcomes do not justify the cost. Rather, they argue instead that scientific output would have been higher with traditional "hypothesis-driven" approaches that analyze proteins known to be of biological interest. These include large protein–nucleic acid complexes (e.g., the ribosome), macromolecular complexes (e.g., the proteasome), or membrane complexes that include important receptors for signaling cascades. In contrast, the PSI strategy can be described as "discovery based", as it primarily targets proteins from different structural classes. One of the major goals of the PSI is to have structures of representatives of as many protein families as possible. According to PSI proponents, this will permit the simulation of homologous models of closely related family members for which no physical structure exists.

References

1. Stringer, C. (2012) Evolution: what makes a modern human. *Nature*, **485**, 33–35.
2. Ecker, J.R., Bickmore, W.A., Barroso, I. *et al.* (2012) Genomics: ENCODE explained. *Nature*, **489**, 52–55.

Further Reading

Human Genome

Beyer, A., Bandyopadhyay, S., and Ideker, T. (2007) Integrating physical and genetic maps: from genomes to interaction networks. *Nat. Rev. Genet.*, **8**, 699–710.

Hall, J., Dennler, P., Haller, S. *et al.* (2010) Genomics drugs in clinical trials. *Nat. Rev. Drug Discov.*, **9**, 988.

Kelemen, O., Convertini, P., Zhang, Z. *et al.* (2013) Function of alternative splicing. *Gene*, **514**, 1–30.

Lander, E.S., Linton, L.M., Birren, B. *et al.* (2001) Initial sequencing and analysis of the human genome. *Nature*, **409**, 860–921.

Stringer, C. (2012) Evolution: what makes a modern human. *Nature*, **485**, 33–35.

Follow-Up Projects

Ecker, J.R., Bickmore, W.A., Barroso, I. *et al.* (2012) Genomics: ENCODE explained. *Nature*, **489**, 52–55.

Pennisi, E. (2012) Genomics. ENCODE project writes eulogy for junk DNA. *Science*, **337**, 1159–1161.

Regierer, B., Zazzu, V., Sudbrak, R. *et al.* (2013) Future of medicine: models in predictive diagnostics and personalized medicine. *Adv. Biochem. Eng. Biotechnol.*, **133**, 15–33.

Teer, J.K. and Mullikin, J.C. (2010) Exome sequencing: the sweet spot before whole genomes. *Hum. Mol. Genet.*, **19**, R145–R151.

Proteomics

Meissner, F. and Mann, M. (2014) Quantitative shotgun proteomics: considerations for a high-quality workflow in immunology. *Nat. Immunol.*, **15**, 112–117.

Sabido, E., Selevsek, N., and Aebersold, R. (2012) Mass spectrometry-based proteomics for systems biology. *Curr. Opin. Biotechnol.*, **23**, 591–597.

Service, R.F. (2008) Structural biology. Protein structure initiative: phase 3 or phase out. *Science*, **319**, 1610–1613.

Genetic Testing

8

Summary

- Genetic testing identifies changes in the DNA of an individual and permits the diagnosis or prediction of disorders with a genetic basis.
- A genetic test can be used to diagnose a disease (or confirm a clinical diagnosis), to predict the development of a disease, to screen newborns, to predict pharmacotherapeutic outcomes, to identify the asymptomatic carriers of mutations of recessive disorders, or to test an unborn by prenatal or preimplantation diagnosis.
- Chromosomal abnormalities can be related to the number of chromosomes (aneuploidy) or to their structure. They have conventionally been evaluated by the staining of metaphase chromosomes and microscopic analysis of the banding pattern. Fluorescence *in situ* hybridization (FISH) permits the analysis of chromosomes at much higher resolution. Array comparative genomic hybridization (aCGH) has an even higher resolution and allows genome-wide analysis of chromosomal copy number changes. This assay is based on the cohybridization of a test sample and a reference sample of human DNA and can demonstrate gain or loss of DNA at a specific genomic region.
- Many genetic disorders are caused by single-nucleotide polymorphisms (SNPs) or short insertions/deletions (indels), which can be diagnosed by molecular techniques. One of the methods most often used is the polymerase chain reaction (PCR). Many variants of this process have been developed to detect point mutations, indels, and even chromosomal aneuploidy.
- Sequencing of a sample directly identifies mutations in the genome. For monogenetic diseases, sequencing of the associated gene is common in clinical practice. Next-generation sequencing (NGS) technologies permit evaluation of complex genetic disorders. Targeted sequencing of a panel of up to several hundred genes associated with cancer, cardiac disease, or neurosensory disorders has become feasible in clinical diagnosis. More comprehensive sequencing of the whole exome or even the whole genome will still require technological advancements and – probably even more importantly – an improved ability to interpret the clinical significance of variants before it can be applied routinely.
- DNA arrays can be used to detect SNPs. They may be targeted and focus on variants of either known specific pathological relevance or genome-wide to detect/identify all variations in the genome of the tested individual. In an alternative application, DNA arrays can be employed to measure the expression levels of a set of genes that is known to be of prognostic significance. These analyses can estimate the risk of cancer recurrence and guide treatment decisions, for example, the need for a patient to receive adjuvant chemotherapy.

Contents List

Types of Genetic Tests

- Postnatal Genetic Tests
- Prenatal Genetic Tests

Chromosome Abnormalities

- Conventional Karyotyping
- Fluorescence In Situ Hybridization
- Comparative Genomic Hybridization

Molecular Diagnosis

- PCR-Based Methods
- DNA Sequencing
- DNA Microarray-Based Methods

Molecular Medicine: An Introduction, First Edition. Jens Kurreck and Cy Aaron Stein.
© 2016 Wiley-VCH Verlag GmbH & Co. KGaA. Published 2016 by Wiley-VCH Verlag GmbH & Co. KGaA.

Genetic testing, also known as DNA testing, is a type of medical test that identifies changes in the genetic material of an individual. It permits the diagnosis of vulnerabilities to inherited disorders or acquired diseases with a genetic basis. According to the GeneTest database (www.genetest.org), more than 35 000 gene tests for approximately 4000 disorders were available in 2014. Different levels of testing can be distinguished. Chromosomal genetic tests analyze whole chromosomes or the complete set of chromosomes to detect chromosomal aberrations. In contrast, molecular genetic tests study a single gene or a short DNA fragment to identify variations or mutations that lead to a genetic disorder. The tests do not necessarily have to be carried out at the genetic level. Sometimes it is easier to detect genetic disorders by the biochemical analysis of activity of an enzyme or the level of an encoded protein.

8.1
Types of Genetic Tests

Depending on the aim of the analysis, different types of medical DNA genetic tests can be distinguished (Table 8.1). The following sections will describe postnatal and prenatal diagnostic tests. Pharmacogenetic testing is extensively covered in Chapter 9. Genetic tests can also be performed to determine relatedness (e.g., paternity) or for forensic purposes.

8.1.1
Postnatal Genetic Tests

A typical application of DNA analysis is a *diagnostic test*, which is intended to diagnose or rule out a specific genetic or chromosomal condition. In many cases, the genetic test is used to confirm a clinical diagnosis based on physiological parameters and symptoms. The test can be carried out at any time during a person's lifetime and can guide the management of the disease. An example is the diagnosis of hemochromatosis, a hereditary disease characterized by excessive intestinal absorption of dietary iron, resulting in a pathological increase in total body iron stores. The accumulated excess iron can disrupt the normal organ function of, for example, the liver, adrenal glands, heart, skin, gonads, joints, and pancreas. The disorder is inherited in an autosomal recessive pattern. Transferrin saturation is commonly used as screening for hemochromatosis. Formerly, the only way to confirm an initial diagnosis was a liver biopsy. A genetic test of the *HFE* (*H*igh for *Fe*) gene encoding the human hemochromatosis protein can be performed as a noninvasive alternative. The most common mutation found among those of Northern European ancestry is the substitution of tyrosine by cysteine at position 282 of the protein. The disorder can be treated by phlebotomy, iron-chelating compounds, or by diet modification.

In contrast to diagnostic tests, a *predictive test* is carried out in an asymptomatic individual to detect the risk for the subsequent development of a genetic disorder.

Table 8.1 Classification of different medical genetic tests.

Category of the test	Aim of the test	Examples
Diagnostic testing	Diagnosis of a disease or confirmation of a clinical diagnosis	*HFE* gene for hemochromatosis
Predictive testing	Test of an asymptomatic individual for a genetic disorder that may appear later in life	(A) Presymptomatic: *huntingtin* gene for Huntington's disease. (B) Predictive: *BRCA1* gene for breast cancer. (C) Predispositional: *APOE* gene encoding the apolipoprotein E; the E4 isoform has been implicated in atherosclerosis, Alzheimer's disease, and impaired cognitive function
Newborn screening	Routine test just after birth for genetic disorders that can be treated early in life	Phenylketonuria, congenital hypothyroidism
Pharmacogenetic testing (Chapter 9)	Testing of the effect of the genotype on an individual's drug response	Cytochrome P450 isoforms for drug metabolism
Carrier testing	Test used to identify heterozygous carriers of a recessive disorder to determine the disease risk for offspring	*CFTR* gene for cystic fibrosis and other recessive disorders
Prenatal diagnosis	Testing of a fetus for genetic disorders	Trisomy 21, monogenetic disorders
Preimplantation genetic diagnosis	Genetic profiling of embryos after *in vitro* fertilization prior to implantation	Duchenne muscular dystrophy and other monogenetic (recessive or dominant) disorders

CFTR: cystic fibrosis transmembrane conductance regulator.

The onset of disease symptoms appear after birth, most often later in life. These types of tests can be helpful to individuals with family members with a genetic disorder who are asymptomatic at the time of testing. Information about a person's risk of developing a disorder can help with lifestyle and medical care decision-making.

Different classes of predictive tests can be distinguished based on the levels of penetrance of the disorder, that is, the probability that it will eventually develop: *Presymptomatic tests* diagnose a disorder that is (almost) certain to develop in an individual's lifetime. An example is the test for Huntington's disease (HD, Section 3.1.1). This neurodegenerative disorder usually develops between the ages of 35 and 45. Young asymptomatic offspring of affected individuals may choose to have a presymptomatic test. A high number of repeats of the cytosine–adenine–guanine trinucleotide sequence in the *huntingtin* gene are indicative of a 100% probability of developing HD at later age. This type of testing is psychologically and ethically problematic, particularly for diseases (such as HD) for which no cure exists (Section 15.2.1).

The second class of predictive test refers to evaluating an asymptomatic individual to predict the risk of developing a genetic disorder where a mutation in a specific gene confers increased risk, although not everyone with the mutation develops the disorder. An example is the *breast cancer 1, early-onset* gene (*BRCA1*). *BRCA1* is a tumor suppressor gene; the encoded protein is part of a complex that repairs double-stranded breaks in DNA. Certain variants in the *BRCA1* gene lead to an increased risk for breast and ovarian cancers as part of the hereditary breast–ovarian cancer syndrome. Mutations in the *BRCA1* and the related *BRCA2* gene account for approximately 7% of all breast cancers and 11–15% of all ovarian cancer cases. Hundreds of mutations in these genes have been identified that are associated with an increased risk of cancer. Depending on the type of mutation, women with mutated *BRCA* genes have a 36–85% risk of developing breast cancer and a 25–55% risk of developing ovarian cancer. The company Myriad Genetics offers a test for detecting mutations in *BRCA1* and *BRCA2* based on PCR amplification followed by (conventional) sequencing of the amplified genes (see below). The high cost of the test (several thousand dollars) and company's patent-protected monopoly were heavily criticized and were the subject of legal controversy. In June 2013, the US Supreme Court ruled that isolated genes are not patentable, invalidating the *BRCA* gene patents held by Myriad Genetics. Regardless of cost, many women with a family history of breast cancer decide to have this test performed. If variations in the *BRCA* genes are diagnosed that are associated with increased breast cancer risk, the affected women may choose to increase mammography frequency to detect the early development of a tumor. Many mutation carriers who have inherited *BRCA* variants that confer a very high risk of developing cancer choose to have a prophylactic mastectomy (Box 8.1).

A third class of predictive test is *predispositional* or *susceptibility testing*. This refers to tests of asymptomatic individuals that predict the risk of developing a disease with a complex genetic (and environmental) basis. Therefore, the predicted risk is usually low. An example is apolipoprotein E (ApoE), a protein found in chylomicrons, and intermediate-density lipoproteins (IDLs), particles that transport triglycerides and cholesterol in the

Box 8.1. Angelina Jolie and *BRCA1* Mutations

In May 2013, the popular actress Angelina Jolie announced that she had undergone a preventive double mastectomy earlier that year. Her mother had breast cancer and died from ovarian cancer at the age of 56; her maternal grandmother had ovarian cancer and died aged 45; and a maternal aunt died at age 61 of breast cancer. This family history prompted a genetic evaluation that revealed mutations in the *BRCA1* gene associated with an 87% risk of developing breast cancer and a 50% risk of developing ovarian cancer. This diagnosis led her to undergo preventive double mastectomy, a procedure that lowered her risk of developing breast cancer to under 5%. Two years later, in March 2015, she also underwent preventive oophorectomy due to the increased risk of developing ovarian cancer.

Angelina Jolie's announcement of her mastectomy drew extensive public attention. Her decision was widely discussed in the media. While some experts welcomed the publicity that raised awareness of the genetic testing options available to those at risk, others feared widespread overestimation of BRCA mutations and a misunderstanding of the risk by those who tested positive. The public discussion of the issue resulted in increased worldwide interest in genetic testing, a phenomenon some referred to as the "The Angelina Effect".

bloodstream. *ApoE* is polymorphic, the three major isoforms being *ApoE2, ApoE3,* and *ApoE4.* The E4 isoform is found in approximately 14% of the population and has been implicated in various diseases, such as atherosclerosis, Alzheimer's disease (AD), and impaired cognitive function. Homozygous Caucasian and Japanese carriers of the E4 allele have between 10 and 30 times the risk of developing AD by the age of 75 as compared to individuals not carrying any E4 allele. Asthma and diabetes mellitus are further examples of complex diseases that include a genetic component, as described in Section 3.2.

A widespread routine application of genetic testing is in *newborn screening*, used just after birth to identify genetic disorders that can efficiently be managed when treatment is initiated early in life. In many countries, infants are routinely tested for phenylketonuria (Section 3.1.2) and congenital hypothyroidism, a disorder of the thyroid gland. Here, the test is not carried out by direct DNA sequence analysis, but rather by measuring enzyme activities or metabolite levels that suggest the presence of a genetic variation. Newborn screening is the most common application of (indirect) genetic testing, as it can help to prevent the development of severe symptoms later in life.

8.1.2
Prenatal Genetic Tests

Genetic testing also plays an important role in family planning and prenatal diagnosis. *Carrier testing* is offered to people who are potential heterozygous carriers of a mutation of a recessive disorder. It is used both when individuals have a family history of a genetic disorder and for people in ethnic groups with a known increased risk of having a specific genetic condition. If both partners are tested, the couple's risk of having a child with the respective genetic condition can be determined. The test is useful for all recessive disorders, several examples of which are described in Section 3.1.2. For a typical autosomal recessive disorder, cystic fibrosis for example, the risk of two heterozygous carriers of a mutated allele having an offspring that is homozygous for the mutated allele and who thus will develop the disease phenotype is 25%. This information may influence family planning or the decision to carry out prenatal or preimplantation diagnosis. Tay–Sachs disease is a special case of a genetic disorder with a particularly high incidence in Ashkenazi Jews. Some orthodox Jewish organizations offer genetic testing and mate selection evaluation to decrease the incidence of the disease in the population.

Prenatal diagnosis evaluates genetic disorders in a fetus or embryo before it is born. This procedure is well known for detecting chromosome abnormalities,

Trisomy 21 (Down syndrome) for example, but can also be used for testing hundreds of other monogenetic disorders. Diagnostic prenatal procedures can be classified as being invasive or noninvasive. Typical noninvasive techniques include ultrasonography and maternal serum screens. The most common application of an ultrasound scan is in the measurement of nuchal thickness, used to diagnose chromosomal conditions such as Down syndrome in a fetus. The test can be combined with maternal screening for certain biomarkers. The so-called triple test measures serum levels of alpha fetoprotein (AFP), estriol, and β-human chorionic gonadotropin (β-hCG). Sometimes inhibit A is added to the panel in the "quad test." Together, these biomarkers have a sensitivity of 70–80% and a false-positive rate of ~5% for detecting Down syndrome.

Until recently, more reliable genetic prenatal testing required invasive sampling (Figure 8.1). Chorionic villus sampling (CVS) can be performed between 10 and 12 weeks gestation to obtain placental tissue for genetic analysis. Amniocentesis is usually done at a later stage when a woman is between 14 and 20 weeks pregnant. The procedure collects a small amount of amniotic fluid, which contains fetal cells whose DNA can be examined for genetic abnormalities. Both invasive procedures are associated with a risk of miscarriage that is slightly higher for CVS (~1% versus ~0.5% for amniocentesis). At a later stage of gestation (from week 18), percutaneous umbilical cord blood sampling (PUBS), also called cordocentesis, examines blood from the fetal umbilical cord to detect fetal abnormalities or infection.

The great disadvantage of all invasive, prenatal testing techniques is the risk of fetal loss. However, in the late 1990s, small amounts of cell-free fetal DNA (cff-DNA) were discovered in maternal plasma. This laid the basis for noninvasive prenatal testing (NIPT), sometimes also called noninvasive prenatal diagnosis (NIPD). The cff-DNA most likely originates from apoptotic trophoblastic cells from placental tissue. Initially, the low concentrations of cff-DNA, only about 2–10% of the total DNA in the plasma, did not permit reliable diagnoses. This obstacle was resolved with the technical development of next-generation sequencing (Section 2.3). Massive parallel sequencing of DNA fragments of a sample of 10 ml maternal blood now permits the reliable diagnosis of chromosomal aberrations. Bioinformatic analysis then separates fetal and maternal DNA and assigns the cff-DNA to the respective chromosomes. An increase in the prevalence of reads from one chromosome indicates that it is present in three copies instead of two. Commercial tests exist for the detection of Trisomy 21, Trisomy 18, and Trisomy 13 as well as Trisomy of the sex chromosomes. These tests are highly sensitive and have a 0.2%

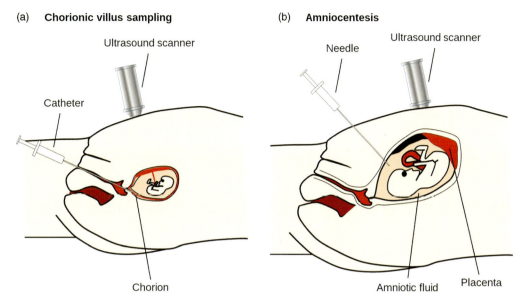

(a) Chorionic villus sampling

Ultrasound scanner

Catheter

Chorion

(b) Amniocentesis

Needle

Ultrasound scanner

Amniotic fluid   Placenta

Fig. 8.1   Sampling of cells for prenatal diagnosis. (a) Chorionic villus sampling is carried out by inserting a catheter through the cervix to obtain a sample of the chorionic villi from the placenta. The chorionic villi contains fetal cells. Sampling can be accomplished by inserting a needle through the abdominal wall into the placenta. (b) In amniocentesis, the doctor inserts a needle through the abdominal wall into the amniotic fluid and a sample of the fluid containing fetal cells is withdrawn for analysis. Both procedures are guided by ultrasonography.

false positive rate. Diagnosis of single-gene disorders, such as Huntington's disease and myotonic dystrophy, are being validated for translation into clinical practice. Whole exome sequencing (WES) and whole genome sequencing (WGS) (Section 7.2.3, including Box 7.4) is still impractical for routine prenatal diagnosis due to cost and turn-around time. Even with the expected decrease in cost and handling time, major problems remain with the annotation and interpretation of variant information, which limits the amount of useful information for prenatal diagnosis. In addition, important ethical issues may emerge if variants that are beyond the scope of conventional medical diagnosis are identified.

During the 1990s, *preimplantation genetic testing* became increasingly popular for the genetic profiling of an embryo after *in vitro* fertilization (IVF) and prior to implantation. Evaluation can be performed as preimplantation genetic screening (PGS) or preimplantation genetic diagnosis (PGD). Genetic testing can be done at three stages in the development of the embryo: the polar bodies can be biopsied, one or two cells can be taken from the cleavage-stage embryo, or several cells can be taken from the trophectoderm of the blastocyst.

For PGS, the embryo is obtained from parents with presumed normal karyotypes and without suspected severe genetic disorders. However, chromosomal aneuploidy, that is, an abnormal number of chromosomes (see below), is a major cause of failure of the implantation procedure. Approximately half of the embryos obtained by IVF had chromosomal abnormalities that decreased the success rate. On day 3 after fertilization, the chromosomes of the embryo can be analyzed by fluorescence *in situ* hybridization or chromosomal microarrays (see below) to select embryos without chromosomal aberrations. However, the benefit of PGS, that is, an increased success rate, has not yet been convincingly demonstrated. The performance of PGS on cleavage-stage embryos has been seriously questioned, as the analyzed blastomere may not be representative of the entire embryo, a phenomenon known as mosaicism.

PGD is available for a large number of monogenetic diseases and can be used to investigate a specific genetic disorder if there is a proven family history. A typical example is that in which both parents are carriers of the mutated allele of a recessive genetic disorder (e.g., cystic fibrosis or β-thalassemia). In this case, the test can be used to select embryos that are not homozygous carriers of the mutated allele. The embryonic DNA is analyzed by PCR-based methods or sequencing. Depending on the condition, FISH and chromosomal microarrays may also be used. PGD may not only be employed to evaluate recessive genetic disorders but can also be performed for dominant or X-linked disorders. While PGD was initially developed for severe, early-onset genetic disorders, the American Society for Reproductive Medicine more recently considered PGD ethically justifiable for serious adult-onset conditions for which no adequate treatment exists, for example, Huntington's disease or cancer susceptibility genes (*BRCA1* and *BRCA2*). Additional applications of PGD are still under debate. One example is the selection of an embryo with a human leukocyte antigen (HLA) that matches a sick sibling, making the child available for stem cell donation. This type of selection of a "savior sibling" is permitted by law in several countries. The use of PGD for prenatal sex discrimination is even more controversial. While this application can be useful to test for sex-linked monogenetic disorders, such as X-linked diseases, ethical uncertainties have arisen if used for nonmedical reasons (Section 15.2.3).

8.2
Chromosome Abnormalities

The complete set of chromosomes in a cell or an individual is known as the *karyotype*. The term also refers to the number and appearance of chromosomes in the nucleus. The normal human karyotype contains 23 pairs of chromosomes: 22 pairs of autosomal chromosomes and 1 pair of sex chromosomes. In females, the normal karyotype contains two X chromosomes; males have one X and one Y chromosome. *Chromosome abnormalities* or *aberrations* can be anomalies in the number or the structure of chromosomes.

An abnormal number of chromosomes is called *aneuploidy*. It occurs when an individual is either missing a chromosome from a matched pair (monosomy) or has more than two chromosomes of a matched pair (trisomy). Since different species have different numbers of chromosomes, the term aneuploidy refers to the number of chromosomes that is typical for that species: Humans have 46 chromosomes in a diploid complement, chicken have 80 chromosomes, and the Indian muntjac, a deer, only 6, although its genome consists of almost as many nucleotides as the human genome. Aneuploidy originates during meiotic cell division when the chromosomes do not separate properly between the two daughter cells. The failure to separate, also known as chromosomal nondisjunction, occurs more frequently with the increasing age of the mother, which translates to a higher risk of trisomies in the fetal set of chromosomes. In most cases, trisomies result in miscarriage or failure to implant in the uterus. The most common autosomal trisomies that survive to birth in humans are Trisomy 21 (Down syndrome), Trisomy 18 (Edwards syndrome), and Trisomy 13 (Patau syndrome). These conditions are associated with intellectual disability and shortened life expectancy. Trisomy of the sex chromosomes include the XXX (Triple X syndrome), XXY (Klinefelter syndrome), and XYY genotypes. These types of trisomies normally have less severe consequences and many individuals have few or no symptoms, are fertile, and have a normal life expectancy. The only full monosomy seen in humans is the Turner syndrome in women, who have a single X chromosome instead of the usual two sex chromosomes. The affected individuals have characteristic physical abnormalities (e.g., short stature, webbed necks), are sterile, and suffer from various other health problems. While all other full monosomies are lethal during embryonic development, some fetuses with partial monosomies caused by a deletion of only a part of a chromosome survive to birth.

Structural abnormalities of chromosomes can take several forms (Figure 8.2). These mutations can affect a single chromosome or two chromosomes. The major types of single chromosome mutations are deletions, duplications, and inversions. While the total amount of DNA remains unaltered by inversions, it changes after duplications or deletions. Examples of short deletions are the Prader–Willi syndrome (PWS) and the Angelman syndrome (AS) (Chapter 3, Box 3.1) that are (usually) caused by a deletion of several genes located on the long arm of chromosome 15. Duplication of the gene encoding the peripheral myelin protein 22 on chromosome 17 is the cause of the Charcot–Marie–Tooth disease type 1A.

Mutations involving two chromosomes include insertions and translocations. In the former, a portion of one chromosome has been removed from its normal place and inserted into another chromosome. In reciprocal translocations, segments from two different chromosomes have been exchanged. Translocations may either be balanced, that is, an even amount of genetic material is exchanged, or unbalanced when the exchange of chromosomal material is unequal and genes are extra or missing. Translocations can result in gene fusions, when two otherwise separated genes are joined. The expression level of a translocated gene may also be altered when it finds itself in a different genetic context. In many cases, individuals with a balanced translocation are generally healthy, but infertile, as fetuses conceived by such people are not viable. A prominent example of an acquired translocation that leads to the development of cancer is the Philadelphia chromosome, a reciprocal translocation between chromosomes 9 and 22 (Section 4.7.2). This translocation is found in 95% of all patients with chronic myelogenous leukemia (CML) and 25–30% of the patients with acute lymphoblastic leukemia (ALL). The translocation results in the creation of an oncogenic fusion gene by juxtaposing the *Abl1* (*Abelson murine leukemia viral oncogene homolog 1*) gene to a sequence

Single chromosome mutations

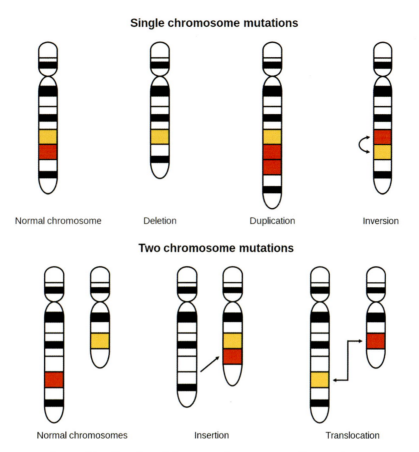

Normal chromosome   Deletion   Duplication   Inversion

Two chromosome mutations

Normal chromosomes   Insertion   Translocation

Fig. 8.2 Structural chromosome abnormalities. Alterations of chromosomal structure may affect one chromosome or two chromosomes. While the total amount of DNA changes through deletions or duplications, it remains unaltered in inversions, insertions, and (balanced) translocations.

of the *BCR* (*breakpoint cluster region*) gene. Another special case of translocation is the Robertsonian translocation, in which an entire chromosome attaches itself to another chromosome at the centromere.

Various tests are routinely employed to detect chromosome abnormalities. Typical applications are in prenatal diagnosis following amniocentesis or CVS as described above, or in tumor diagnostics. Some of the most commonly used methods will be described below. While we will focus on more recently developed methods such as fluorescence *in situ* hybridization and comparative genomic hybridization (CGH), well-established methods like Southern blotting in combination with restriction fragment length polymorphism (RFLP), also continue to be used in clinical diagnosis.

8.2.1
Conventional Karyotyping

The study of the karyotype is a standard analysis and has been used for more than 50 years to detect large chromosomal abnormalities such as translocations,

deletions, or duplications. The most common application is in prenatal diagnosis, where it is used to detect aneuploidy. In a typical procedure, a sample of cells is fixed and stained (e.g., by Giesma stain) to create a typical light and dark banding pattern. The chromosomes are then analyzed under a microscope. Each chromosome has a characteristic banding pattern that helps to uniquely identify it. The standard format, in which the chromosomes are depicted in pairs ordered by size and centromere position for chromosomes of the same size, is known as a *karyogram* or *ideogram*. An example of a karyogram of a female with Trisomy 21 (Down syndrome) is shown in Figure 8.3a. A shortcoming of this method is that it can only be used to detect large chromosomal abnormalities, since a band of a Giesma-stained chromosome represents approximately 10 Mb.

8.2.2
Fluorescence In Situ Hybridization

Fluorescence *in situ* hybridization (FISH) is a cytogenetic technique that was developed in the early 1980s and is

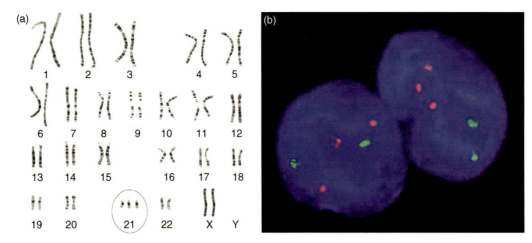

Fig. 8.3 Trisomy 21 diagnostic methods. (a) Karyogram of a Trisomy 21 female. The G-banded karyogram obtained by Giesma staining demonstrated the presence of three copies of chromosome 21. (b) Fluorescence *in situ* hybridization (FISH) of a Trisomy 21 fetus. FISH was carried out with interphase nuclei. In each cell there are two green spots originating from a probe for chromosome 13 and three red spots from a probe for chromosome 21. (Adapted from Ref. [1] with kind permission from Macmillan Publishers Ltd., Copyright 2004.)

used to detect and localize the presence or absence of specific DNA sequences on chromosomes. This technique has higher resolution than the banding method described above and can detect deletions and duplications at a resolution of approximately 2 Mb for metaphase FISH and with improved resolution for interphase FISH. FISH uses probes that can be fluorescently labeled. These probes bind specifically to regions of the chromosome with which they have a high degree of sequence complementarity. Binding of the probe to its target DNA can be visualized by fluorescence microscopy. In addition to its use in chromosomal analysis for genetic counseling, FISH also has research applications. These include, for example, the detection of specific RNA targets in cells or tissue samples.

There are different ways to design a FISH probe. Short probes can be used to hybridize and detect or locate a specific target. An example is the clinical diagnosis of heterozygous female carriers of the mutation that causes Duchenne muscular dystrophy (DMD) (Section 3.1.3). This recessive X-linked neuromuscular disorder can be caused by a deletion in the *dystrophin* gene. FISH experiments employing a probe for the *dystrophin* gene can be used to identify asymptomatic female carriers of the deletion whose male offspring are at a 50% risk of being affected through inheritance of the mutant maternal allele (Figure 8.4).

Long probes or mixtures of probes that hybridize along an entire chromosome can be used to count the numbers of specific chromosomes, to visualize translocations, or

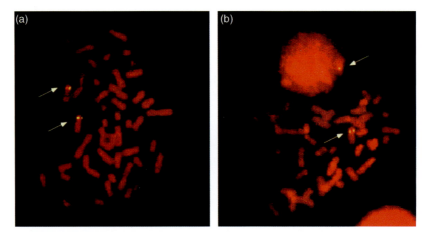

Fig. 8.4 FISH analysis for the *dystrophin* gene. (a) In a normal control, the dystrophin probe binds to and generates a signal from both X chromosomes in metaphase. (b) For a heterozygous carrier of the deletion, only one signal is detected, either in interphase or in metaphase (the two dots in metaphase originate from the same chromosome). (Adapted from Ref. [2] with kind permission from Elsevier.)

to identify extrachromosomal fragments. This approach is often called "whole chromosome painting" and can, for example, be used to diagnose Trisomy 21. Figure 8.3b shows the presence of three copies of chromosome 21 (red), while the reference chromosome 13 (green) occurs only twice.

FISH can also be used in cancer diagnosis. For example, the Philadelphia chromosome, which is frequently found in leukemic cells in CML and ALL, can be detected in this manner (see above and Section 4.7.2). This abnormality is a reciprocal translocation between chromosome 9 and chromosome 22. FISH with differently colored probes for the *BCR* and *ABL1* DNA, respectively, reveals the fusion gene (Figure 4.18 in Section 4.7.2).

The development of multiple fluorescent dyes has broadened the spectrum of FISH applications. A mixture of differently colored probes permits the simultaneous visualization of all human chromosomes in different colors, which considerably facilitates karyotype analysis. Since there are more chromosomes than there are easily distinguishable fluorescent dye colors, probe mixtures can be used to create *secondary* colors. Simultaneous staining of all human chromosomes is also known as spectral karyotyping (SKY) or multiplex fluorescence *in situ* hybridization (M-FISH). Both techniques use a combinatorial labeling scheme with spectrally distinguishable fluorochromes, but employ different detection methods. While image acquisition in SKY is based on the measurement of the entire emission spectrum with a single exposure, M-FISH captures separate images for each of the fluorochromes and combines them with special software. In both techniques, unique pseudo-colors are assigned to the chromosomes based on their specific fluorochrome signatures. SKY and M-FISH have many applications for the detection of chromosomal abnormalities in clinical diagnosis and biomedical research. These abnormalities include aneuploidy, subtle translocations, and complex rearrangements. Figure 8.5a shows the spectral karyogram of a normal female. Chromosomes can easily be identified by their respective color. In cancer cells, many numerical and structural abnormalities are frequently observed (Figure 8.5b).

8.2.3
Comparative Genomic Hybridization

The resolution of a cytogenetic analysis is further improved by comparative genomic hybridization (CGH). This technique is used for analyzing copy number variations (CNVs), that is, genomic variations that result in copy number changes in specific chromosomal regions. This method was originally developed for the evaluation of chromosomal changes in tumors. CGH compares two genomic DNA samples, most commonly a test and a reference sample. In conventional (or chromosomal) CGH, both samples are independently labeled with different fluorophores (e.g., Cy-5 (red) and Cy-3 (green)), denatured, and hybridized to a normal metaphase spread of chromosomes. The labeled DNA samples will hybridize to their locus of origin. Signals are detected by fluorescence microscopy and differentially colored spots indicate extrachromosomal DNA or deletions in the test DNA. A neutral color (yellow when the fluorophore labels are red and green) indicates no difference between the two samples in that location. CGH can detect unbalanced chromosomal abnormalities; however, the resolution of CGH is still limited and the method is technically challenging.

A more advanced variation of this method that has a simpler procedure and improved resolution is known as array comparative genomic hybridization (Figure 8.6). Array comparative genomic hybridization (aCGH) can detect structural chromosome variations of only a few hundred base pairs and can be used for a genome-wide, high-resolution analysis of chromosomal copy number changes. The basic principle for aCGH is the same as for conventional CGH; in both techniques, DNA from a reference and a test sample are differentially labeled and cohybridized. However, while the samples are hybridized to a reference metaphase spread in conventional CGH, they are hybridized to target pieces of human genomic DNA, usually oligonucleotides or cloned segments of DNA (e.g., bacterial artificial chromosomes (BACs), plasmids, or cDNA) that in aCGH have been spotted onto a glass slide.

Depending on the design, aCGH may cover the entire genome. It can also be targeted to a specific chromosome or chromosomal segment to evaluate specific DNA dosage abnormalities in individuals with suspected microdeletions. While the genome-wide design has advantages for research applications, it may be difficult to interpret for routine diagnostic applications. An example of a successful research application of aCGH is the identification of the genetic cause of the CHARGE syndrome (the term CHARGE is derived from "*C*oloboma of the eye, *H*eart defects, *A*tresia of the nasal choanae, *R*etardation of growth and/or development, *G*enital and/or urinary abnormalities, and *E*ar abnormalities and deafness"). A genome-wide array was used to identify a potential chromosomal region that is causative for the CHARGE syndrome. The approach identified nine candidate genes and eventually led to the identification of mutations in the *CHD7* gene encoding the chromodomain-helicase-DNA-binding protein 7, the genetic cause of the disease.

aCGH has many applications as a clinical diagnostic tool in the detection of aneuploidy, microdeletions and microduplication syndromes, and unbalanced chromosomal

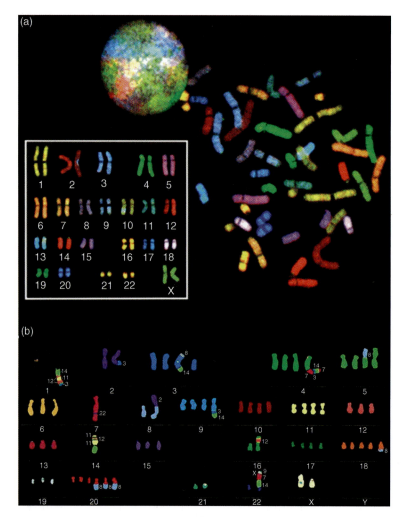

Fig. 8.5 Fluorescence karyotyping of normal and cancer cells. (a) Spectral karyotype (SKY) of a female. (b) Multiplex fluorescence *in situ* hybridization (M-FISH) of a non-small cell lung cancer cell line. (Part (a) courtesy of the National Human Genome Research Institute. Part (b) adapted from Ref. [3] with kind permission from Macmillan Publishers Ltd., Copyright 2000.)

rearrangements. The advantages of aCGH are high resolution, simplicity, and short turnaround time, since there is no need to culture cells as in classical cytogenetic methods. Another area of clinical use of aCGH is in the diagnosis of genetic disorders. As described above and in Box 3.1 of Section 3.1, the Prader–Willi syndrome and the Angelman syndrome are, in the majority of cases, caused by 3–5 Mb deletions in the long arm of chromosome 15. Due to its high resolution, these aberrations can be detected with aCGH, although they are missed using conventional cytogenetics or CGH. The clinical application of aCGH also led to the discovery of (small) chromosomal imbalances in individuals with mental retardation, autism, and other congenital anomalies. Other methods were unable to detect any etiology for these problems.

In addition, aCGH can be used to detect genetic alterations and rearrangements that occur frequently in cancer. These data can assist in cancer classification and prognostification. For example, a study demonstrated the sensitive and specific detection of a 13q14 deletion by aCGH in chronic lymphocytic leukemia (CLL). This deletion correlates with a favorable prognosis in the absence of other chromosomal abnormalities. aCGH can also be used in preimplantation genetic screening or prenatal genetic diagnosis to detect chromosomal abnormalities.

A main limitation of aCGH is that it can only detect gains and losses of genetic material relative to the normal level. It cannot detect chromosomal aberrations without copy number changes, such as balanced translocations or inversions. Nevertheless, aCGH has become

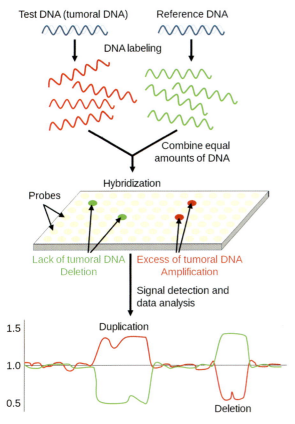

Test DNA (tumoral DNA)   Reference DNA

DNA labeling

Combine equal
amounts of DNA

Hybridization

Probes

Lack of tumoral DNA   Excess of tumoral DNA
Deletion   Amplification

Signal detection and
data analysis

Duplication

Deletion

Fig. 8.6 Array comparative genomic hybridization (aCGH). Whole genomic DNA from a test sample (e.g., a tumor specimen) and a reference are differentially labeled with two different dyes. Repetitive DNA sequences are commonly blocked with Cot-1 DNA (placental DNA enriched with repetitive sequences) to improve the signal. Subsequently, the labeled samples are cohybridized onto the microarray. The fluorescence ratio is determined for each spot. The ratio between reference and test DNA can be plotted to visualize variations.

a routine clinical diagnostic tool, one that has gradually been replacing older cytogenetic methods.

8.3
Molecular Diagnosis

While the methods described so far are used to analyze large chromosomal alterations, many genetic disorders are caused by SNPs or short insertions/deletions (indels). The detection of these types of mutations requires molecular techniques rather than the cytogenetic and chromosome-wide approaches discussed above. Various methods have been used to scan a segment of DNA for small changes by detecting alterations in the mobility or chemical reactivity of the target DNA compared to normal DNA. These approaches include denaturing high-performance liquid

chromatography (dHPLC), analysis of single-stranded conformation polymorphisms (SSCP), denaturing gradient gel electrophoresis (DGGE), and the chemical cleavage of mismatch (CCM). All these techniques reveal the presence of a change in the DNA, but cannot detect the precise type of mutation or its pathological relevance. More advanced methods such as PCR, DNA sequencing, and DNA microarray-based methods have become cheaper in recent years. They are also more informative, as they reveal the precise change in the DNA sequence. As a result, these techniques have largely replaced the former ones and will be discussed in more detail in the following sections.

8.3.1
PCR-Based Methods

PCR is a cyclic process that can be used to amplify DNA concentrations by several orders of magnitude. The basic principle of the technique and its quantitative variant (quantitative or qPCR) are extensively described in Section 2.2. Since it is a robust and comparatively cheap and simple method (although prone to contamination), it is widely used in clinical molecular diagnostics. A large number of variants of the general principle of PCR have been developed that can be used for specific applications in genetic testing.

Two straightforward, PCR-based approaches can be illustrated for the diagnosis of factor V Leiden. This inherited disorder of blood clotting is associated with an increased risk of thrombosis during, for example, surgery or pregnancy. The predominant genetic cause of the disorder is a G to A substitution in position 1691 of the *coagulation factor V* gene. This mutation can be detected with the so-called amplification refractory mutation system (ARMS), which uses an allele-specific PCR primer (Figure 8.7a). One of the two primers is designed to overlap with the SNP location. The 3'-terminal base of the primer is complementary to a normal template DNA so that a PCR product may be generated. In contrast, the primer cannot bind to the mutated DNA and no PCR product will be generated.

An alternative approach combines PCR amplification with the analysis of a RFLP (Figure 8.7b). In this procedure, the site that potentially harbors the mutation is first amplified by PCR with primers that bind at a distance. The normal sequence is cleaved by a restriction endonuclease, for example, MnlI. A mutation (G → A in the case of factor V Leiden) removes the target site of the restriction endonuclease so that the fragment is not cleaved by treatment with the MnlI enzyme. The difference can easily be visualized by gel electrophoresis.

PCR can also be used to analyze the length of a specific segment (Figure 8.7c). An example is the diagnosis

PCR-based genetic testing

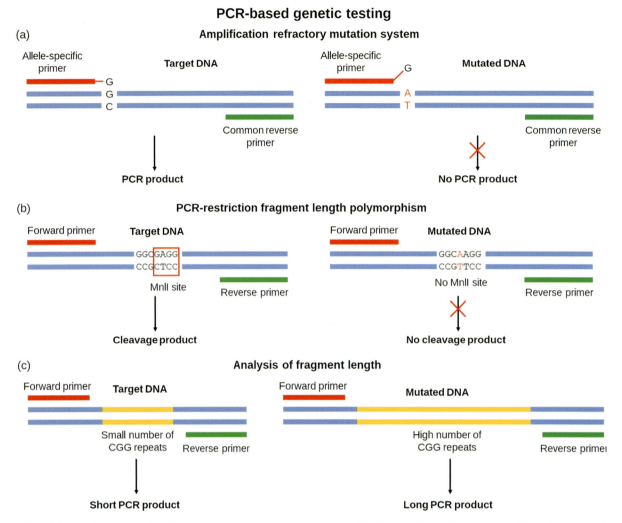

Fig. 8.7 PCR-based genetic testing. (a) Amplification refractory mutation system (ARMS). An allele-specific primer amplifies only the normal DNA sequence in a PCR reaction. In the presence of a mutation, the primer cannot bind and the segment is not amplified by PCR. (b) PCR-restriction fragment length polymorphism (PCR-RFLP). After amplification of the DNA segment that harbors the potential mutation, the fragment is cleaved by a specific restriction endonuclease (in the illustrated case MnlI). The mutation removes the target sequence of the enzyme so that the fragment is not cleaved by enzymatic treatment. (c) Analysis of fragment length. Disorders that are caused by insertions or deletions can be diagnosed by analyzing the length of the PCR-amplified fragment. The fragile X syndrome, for example, is caused by an extension of CGG trinucleotide repeats. The length of a PCR fragment obtained with primers flanking the repeats can be analyzed by simple gel electrophoresis, usually complemented by a methylation-sensitive Southern blot.

of the fragile X syndrome, a neurodegenerative disease that is characterized by autism, intellectual disabilities, and additional physical and behavioral characteristics. The disorder occurs as a result of a mutation in the *fragile X mental retardation 1* (FMR1) gene on the X chromosome. Typically, the number of CGG trinucleotide repeats in the 5′ untranslated region (UTR) of affected individuals is extended. The *FMR1* gene normally contains 6–54 CGG repeats. Extension to 55–200 repeats is considered a premutation that is not associated with phenotypic consequences. However, the premutated allele is unstable and can be further extended

during meiosis to the fully mutated variant with over 200 repeats, which virtually ensures that the offspring will develop the disease. Extension of the CGG repeats causes hypermethylation of the promoter, leading to inhibition of *FMR1* gene expression.

To diagnose fragile X syndrome, the 5′ UTR of the *FMR1* gene is amplified by PCR with primers flanking the repeat. The number of CGG repeats can be deduced from the length of the PCR product. This analysis is accompanied by a Southern blot, a procedure that combines the separation of DNA fragments by electrophoresis followed by transfer to a membrane and then

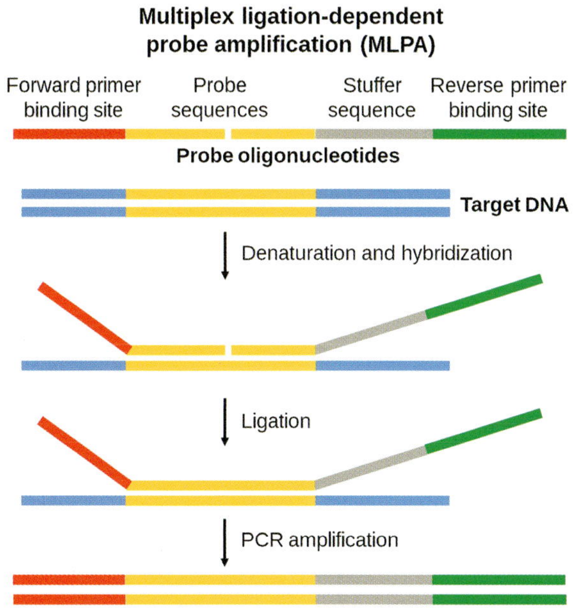

Multiplex ligation-dependent probe amplification (MLPA)

Fig. 8.8 Multiplex ligation-dependent probe amplification (MLPA). MLPA is carried out with two probe oligonucleotides that bind to adjacent positions of the target DNA. Each of the oligonucleotides consists of a binding site for one of the PCR primers as indicated in red and green, respectively. One of the probes also harbors a stuffer sequence (gray). If the target DNA is present in a sample, the probes are ligated and amplified by a subsequent PCR reaction. The stuffer sequences may be varied in length so that multiple targets can be amplified and separated in a single experiment.

detection by hybridization with a complementary probe. The procedure can both analyze the length of the fragment (particularly important as the fully mutated, hyperextended variant is often resistant to PCR amplification) and determine the methylation status of the gene. For this analysis, the DNA is treated with a restriction endonuclease that is methylation sensitive. For example, the nuclease *HpaII* requires that a cytosine within its target site is methylated. The hyperextended and methylated variant of the *FMR1* gene, but not the unmethylated normal fragment, will therefore be cleaved by the restriction enzyme.

Another widely used variant of PCR technology is multiplex ligation-dependent probe amplification (MLPA). This approach represents a further refinement of conventional multiplex PCR and uses multiple primer sets within a single PCR mixture to produce amplicons of multiple DNA sequences, thereby analyzing various regions of the genome in a single reaction. However, conventional multiplex PCR requires laborious optimization of the annealing temperatures and amplicon sizes for each of the primer sets. These limitations are overcome by MLPA analysis, as only a single set of PCR primers is required for the amplification reaction. The

salient point of this method is that the amplification of a probe is made dependent on a preceding ligation step (Figure 8.8): Two adjacent probes are designed that contain binding sites for the forward and reverse primers. One of the two probes contains a "stuffer" sequence whose length can be varied. The probes are hybridized against the target DNA and are ligated only if the target DNA is present in the sample. Subsequent PCR amplification occurs only if ligation has taken place. The forward primer is fluorescently labeled so that the amplicons can be visualized by capillary gel electrophoresis. The amount of PCR product is proportional to the amount of target DNA present in the sample. Thus, the technique is suitable for quantitative measurements, for example, in the diagnosis of aneuploidy (e.g., Trisomy 21). MLPA is also used to detect genomic deletions and insertions in cancer diagnosis, the *BRCA1* and *BRCA2* breast cancer genes being examples. By varying the length of the stuffer sequences, up to 50 targets can be amplified in a single experiment.

Another PCR-based test used for the detection of aneuploidy is known as quantitative fluorescence PCR (QF-PCR). This assay analyzes the levels of several chromosomes in a single PCR experiment. In the most common assay format, aneuploidy of medically relevant chromosomes 13, 18, and 21 and the sex chromosomes can be analyzed (see above). Multiple short tandem repeats (STRs) for each of the chromosomes are amplified by PCR. The amplicons are detected by the incorporation of fluorochromes. In contrast to quantitative PCR (qPCR) as described in Section 2.2, the PCR products are not detected during the cyclic amplification procedure, but rather by electrophoretic analysis following PCR amplification. Aneuploidy can be identified by determining the ratios of the levels of the different markers. For example, if the amount of the amplified STRs of chromosome 21 occurs in a 3:2 ratio compared to the other markers, this chromosome will be present in three copies instead of the normal two copies (Trisomy 21).

8.3.2
DNA Sequencing

DNA sequencing is another cornerstone of genetic testing. PCR amplification of genetic material followed by sequencing of the region that potentially harbors the disease-relevant mutation or PCR amplification and sequencing of all the exons of the respective gene is already a widely used approach. The disease-targeted sequencing of a panel of multiple genes associated with a complex genetic disorder has also become a valuable strategy in genetic testing. Genome-wide or at least

Sequencing of PCR product

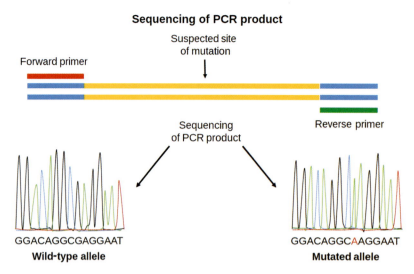

Fig. 8.9   Sequencing of PCR fragments. Either the segment that carries a potentially disease-causing mutation or all the exons of the respective gene can be PCR amplified and sequenced to directly identify the presence of a mutated allele.

exome-wide sequencing of an individual is likely to gain increasing relevance in the near future.

A common diagnostic procedure for monogenetic disorders includes PCR amplification of the fragment harboring the suspected mutation, followed by sequence analysis (Figure 8.9). In the example of the diagnosis of factor V Leiden discussed above, a segment can be amplified that includes position 1691 of the gene. Sequencing then reveals whether the tested individual carries a wild-type G in this position or a disease-causing A.

However, in most cases, a genetic disorder can be caused by multiple sequence errors: Testing for a single mutation is not adequate. Sequencing of the complete gene (or at least all the exons of the gene) identifies all the mutations present in an individual's gene, although this procedure is significantly more laborious and expensive. For example, cystic fibrosis is an autosomal recessive disorder that is caused by mutations in the *cystic fibrosis transmembrane conductance regulator* (*CFTR*) gene. These mutations result in the abnormal transport of chloride ions across the epithelium, leading to organ, particularly lung, damage (Section 3.1.2). The coding sequence is very long (more than 4000 bp); DNA tests often focus on the most common mutations, which are present in about 88% of the affected individuals. The observed rate will be even lower if the test, which is optimized for Caucasians, is applied to individuals with different ethnic background who may have other frequently occurring mutations. More elaborate sequencing of the complete coding sequence (perhaps including intron borders and the promoter region) will give a more complete picture and help to uncover all possible mutations independent of the genetic background of the

patient. PCR amplification, either of a suspected region or of all exons followed by sequence analysis, is a widely established approach in clinical practice and is available for virtually every inherited disease for which a causative mutation is known.

For complex genetic disorders such as cardiac disease or cancer, the causative genetic variation will frequently not be found by single gene testing. However, with the advent of NGS, it has become possible to sequence multiple genes or even the whole genome of a patient. The technological background of NGS is covered extensively in Section 2.3. Three NGS-based approaches can be distinguished: disease-targeted sequencing of a panel of genes that are associated with a disease phenotype; sequencing of the whole exome, that is, all protein-coding parts of the genome; and sequencing of the whole genome.

Currently, only disease-targeted sequencing has a firm place in clinical diagnostics. Commercial and academically affiliated clinical laboratories offer several dozen disease-targeted tests, each of which typically includes ten to several hundred genes. Examples include tests for cancer, cardiac disease, immune, neurological, neuromuscular, metabolic, and sensory disorders. In some cases, retinitis pigmentosa for example, mutations in several genes produce nearly identical symptoms. In other cases, mutations in different genes lead to different phenotypes, for example, dilated, hypertrophic, and arrhythmogenic cardiomyopathies, which require different treatments.

In broad panel tests, genes can be included that make only a rare contribution to a phenotype. An example is the inclusion of the *alpha-galactosidase* gene in tests for the evaluation of hypertrophic cardiomyopathy (HCM), the most common cause of sudden cardiac death in individuals under the age of 25. While the gene was previously not routinely screened, its inclusion in panel tests revealed that 2% of patients with assumed HCM carry a pathogenic variant of the *alpha-galactosidase* gene, which is associated with Fabry's disease. This diagnosis has permitted a significant change in clinical management, as patients with an impaired *alpha-galactosidase* gene can receive enzyme replacement therapy (Section 10.2.7).

Cancer is another field in which disease-targeted sequencing has begun to be employed to direct treatment decisions. For example, the Center for Genomics and Transcriptomics (CeGaT GmbH, Tubingen, Germany) offers a germline tumor panel for the diagnosis of hereditary cancer syndromes. This panel includes the sequencing of 97 genes that have been reported to be associated with a significantly increased risk of developing a malignancy. Depending on family history, subpanels can be analyzed that include genes relevant for specific tumor types. In addition, CeGaT offers a somatic tumor panel comprising more than 550 genes with known mutations that can have an impact on tumor development. The analysis is usually carried out with tumor and normal tissue samples to detect somatic mutations. Knowledge of the causative mutations in a tumor may alter treatment decisions.

With the declining cost of sequencing, more comprehensive analyses are becoming possible. As described in Box 7.4 in Section 7.2.3, whole exome sequencing (WES) focuses on the analysis of all exons of the human genome. WES analyzes all protein-coding regions, but these regions comprise only approximately 1% of the human genome, thus facilitating the experimental procedure. WES will undoubtedly reveal more mutations than the disease-targeted sequencing of a defined set of genes, and has already become commercially available from several laboratories. Medical geneticists make use of WES, particularly for unexplained genetic disorders. However, the clinical benefits are still limited as the interpretation of the data remains difficult. In addition, as discussed above, most SNPs that are linked to disease are located outside the protein-coding region, and will therefore be overlooked by WES. This requires the analysis to be extended to the complete genome. Fortunately, due to the latest technological developments, whole genome sequencing (WGS) has become economically feasible and can be expected to supersede WES in the coming years.

Sporadic case studies demonstrating the successful application of WGS have already been reported. An example is the WGS of two 14-year old fraternal twins diagnosed with dopamine responsive dystonia (DRD) at the age of 5. Despite treatment with L-DOPA, their condition deteriorated. WGS revealed two mutations in the *SPR* gene encoding sepiapterin reductase, which is required for the synthesis of the neurotransmitters dopamine and serotonin. Based on this finding, L-DOPA therapy was supplemented with 5-hydroxtryptophan, a serotonin precursor, to bypass the *SPR* gene defect. This adjusted treatment resulted in clinical improvement in both twins. Some academic laboratories, such as the German National Center for Tumor Diseases in Heidelberg and commercial enterprises (e.g., Human Longevity Inc., San Diego, CA), offer WGS for cancer patients, as outlined in more detail in Section 7.2.3.

However, before WGS becomes a routine clinical procedure, several challenges must be met. In addition to economic factors, technological improvements must be made. An important point is coverage. Although called "whole" exome or "whole" genome sequencing, these approaches have typical coverages of 90–95% (in exome sequencing). This means that important regions of the gene may be missed with these approaches. For example, a survey of publicly available data revealed that four genes known to be causative for nonsyndromic hearing loss have an especially low coverage in exome sequencing. This may be due to various technical reasons, including probe preparation, the presence of repetitive sequences, and poorly performing probes. In contrast, disease-targeted tests as described above can have a higher (often complete) coverage, as they focus on the relevant genes and may fill in missing NGS content with supplemental Sanger sequencing or other sequencing technologies.

While it is almost certain that the technological problems that impact the accuracy of sequencing information will be solved, the interpretation of the data obtained by WES or WGS is likely to remain a challenge for many years. As each individual carries up to several million variants in his or her genome compared to the reference database, data filtering must prioritize variants and make assumptions about their potential role in disease. While some variants are directly associated with a specific phenotype, each person's genome carries thousands of variants of unknown significance (VUSs). Genetic filters usually filter out variants that are frequently found in a control population. However, this approach may miss interactions between genes (or their encoded proteins). Modifier alleles, that is, variants that do not cause a pronounced phenotype by themselves, but may be of relevance in combination with a specific genetic constellation, are often excluded even though in combination they may have profound phenotypic effects. For the interpretation of WES and WGS data, it is often advantageous to sequence the patient and both parents (and additional relatives). This approach permits the analysis of patterns of inheritance in multiple family members and the identification of genetic markers. If the parents are unaffected, *de novo* mutations in the affected individual can be distinguished from inherited variations. Although the analysis of biological relatives may be helpful in some cases, significant bioinformatic progress will be required before the results of WES or WGS can be converted into meaningful clinical information that supports treatment decisions.

Finally, even if WGS reveals the causes of the development of a specific tumor, in many cases no drugs will currently be available to target this mutated protein. These drugs may not be developed until the exact function of the protein is deciphered. Only in those rare cases where the driver mutation of a patient's tumor can be identified by sequencing the tumor genome, and by coincidence there exists a targeting drug that is active in usually another tumor type, is personalized treatment possible. Unfortunately, such coincidences are extremely rare. In most tumor entities, there is such a great variety of mutations and so much heterogeneity that WGS does not currently provide new therapeutic options. It is difficult to predict when and for which cancer types WGS will meaningfully expand our repertory of personalized treatment options.

8.3.3
DNA Microarray-Based Methods

As outlined in Section 2.1, DNA microarrays consist of up to several hundred thousand DNA segments immobilized on a solid support. A DNA or RNA sample is hybridized to the array either to determine a specific genotype or to measure the expression levels of multiple genes. While the use of aCGH has been described in Section 8.2.3, the employment of arrays for the detection of disease-relevant SNPs and for the prognostic analysis of gene expression signatures will be outlined here.

For SNP arrays, the sample DNA is hybridized to the array platform and the presence or absence of specific DNA sequence variants is evaluated. The SNP analysis uses a single labeled test DNA that is hybridized to the array to search for known sequence variations spotted on the array. This is in contrast to array CGH discussed above, in which two DNA samples are labeled with fluorescent dyes of different colors and are cohybridized to a single array to obtain a direct comparison between the samples.

The SNP analysis can detect homozygosity or heterozygosity with respect to a specific allele. SNP microarrays may either be targeted or genome-wide. In the former case, they carry a panel of SNP probes that have been associated with a specific condition. Such phenotype-specific panels have been produced for many

disorders. Although this targeted approach can offer high-quality molecular diagnoses, it obviously lacks information about new causal alleles and genes, which are continuously being discovered.

This problem is less pronounced in genome-wide SNP genotyping. Latest generation genome-wide human SNP arrays comprise up to one million SNPs. A single test can, therefore, assess the risk of multiple common genetic disorders. However, as described for the genome-wide sequencing approach, genome-wide SNP analysis is also likely to identify differences that have an uncertain or no clinical consequence. SNP tests can be employed as diagnostic tools for a variety of conditions, including certain cancers and ophthalmologic, cardiac, renal, and neurological disorders. SNP arrays can also be combined with aCGH to detect point mutations as well as larger chromosomal rearrangements and deletions. Array-based assays are also widely used in direct-to-consumer genetic testing, that is, the marketing of such tests by commercial companies directly to consumers without involving a health care professional (Box 15.1 in Section 15.2.1).

As an alternative microarray-based approach to SNP testing, an analysis of gene expression levels can be used to assess the risk of a cancer patient experiencing tumor recurrence and/or the development of metastases after surgical removal of the tumor (Figure 8.10). In these approaches, panels of genes are tested that classify patients into low or high risk groups. The outcomes of these analyses can guide treatment decisions. For example, they may be used to improve prognostic criteria for decision-making regarding adjuvant chemotherapy treatment for breast cancer patients. Based on traditional prognostic factors such as histopathology or clinical tumor characteristics, approximately 85% of patients do not benefit from adjuvant cytotoxic chemotherapy with respect to 10-year breast cancer-specific survival. Improving prognostic abilities will spare many patients potentially toxic and ineffective therapies.

The first widely distributed gene expression assay (approved by the FDA in 2007) is the MammaPrint assay marketed by Agendia Inc., Irvine, CA. This array is based on the breast cancer study described in Section 2.1, in which patterns of gene expression levels were used to distinguish patients with a high probability of metastasis-free survival at 5 years after diagnosis from patients at high risk of developing distant metastases. The study identified 70 prognostic marker genes, which were included in the DNA array. Several validation studies confirmed the prognostic value of the array analysis. While this demonstrated the *prognostic* value of the assay, that is, the ability to predict temporally distant recurrence, it was also shown to be of *predictive* significance, that is, it can assess the potential benefit of

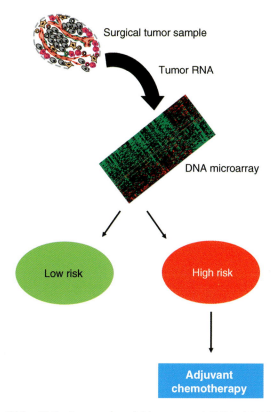

Surgical tumor sample

Tumor RNA

DNA microarray

Low risk

High risk

Adjuvant chemotherapy

Fig. 8.10 DNA microarray-based risk assessment. RNA isolated from a surgical tumor sample is hybridized to a DNA array that analyzes the levels of multiple genes that have been associated with tumor recurrence and/or the development of metastases. According to the assay outcome, the patient is classified as having a low or high risk for recurrence. In the latter case, adjuvant chemotherapy can be recommended as a treatment option.

adjuvant chemotherapy in addition to hormonal therapy. Patients who were classified by the MammaPrint assay as belonging to the high risk group had improved distant recurrence-free survival after treatment with chemotherapy, while no benefit was seen with chemotherapy in the low risk group. A similar gene expression test is currently available to assess recurrence rates and guide treatment decisions for colon cancer.

As previously discussed in Chapter 2, array-based approaches and quantitative PCR are two complementary methods that can both be employed to measure the levels of gene expression. The Oncotype DX® test from Genomic Health Inc., Redwood City, CA is an alternative prognostic test for breast cancer patients that measures gene expression by qPCR. This array analyzes a panel of 21 genes within a tumor to determine a recurrence score that corresponds to the likelihood of breast cancer recurrence within 10 years of the initial diagnosis. The score is not only of prognostic significance but also has predictive value as to whether or not a patient will benefit from adjuvant chemotherapy. In fact, meta-

analyses have shown that treatment recommendations were changed in approximately one third of the patients after determination of the Oncotype DX risk score. Some patients given an initial recommendation for hormone therapy alone instead received chemotherapy, while adjuvant chemotherapy was spared for a high percentage of patients. However, no impact on survival has been shown in these decision-impact studies.

Traditionally, tumor grading, an important prognostic indicator, has been based on histopathology. However, histopathologic grading often suffers from observer variability and may not be optimal for the guidance of treatment decisions. Another array-based test, the Genomic Grade Index (GGI), was therefore designed to make breast cancer tumor grading more objective. GGI is based on a 97-gene microarray signature, which mainly captures the expression of proliferation genes. The assay has been recommended as a complement to histologic grading and divides patients into low and high risk groups, instead of the traditional 1, 2, and 3 histologic grades. This array has been shown to be of value for the prediction of recurrence in endocrine-treated patients and for predicting the benefit of adjuvant chemotherapy in early-stage breast cancer.

A major limitation of any DNA array-based approach is that it can only detect known sequences and those mutations that are printed on a chip. Novel mutations that are specific for a given individual (or a family), sometimes referred to as *private mutations*, cannot be detected by this method. It has been speculated that NGS approaches are likely to replace array methods in the future. However, it is not yet clear when this transition will happen, or if sequencing-based approaches will completely replace, or only complement, array-based methods. Similarly, the analysis of gene expression signatures will probably shift from array-based approaches to qPCR assays.

References

1. Antonarakis, S.E., Lyle, R., Dermitzakis, E.T. *et al.* (2004) Chromosome 21 and down syndrome: from genomics to pathophysiology. *Nat. Rev. Genet.*, **5**, 725–738.
2. Rosenberg, C., Navajas, L., Vagenas, D.F. *et al.* (1998) Clinical diagnosis of heterozygous dystrophin gene deletions by fluorescence *in situ* hybridization. *Neuromuscul. Disord.*, **8**, 447–452.
3. Speicher, M.R., Petersen, S., Uhrig, S. *et al.* (2000) Analysis of chromosomal alterations in non-small cell lung cancer by multiplex-FISH, comparative genomic hybridization, and multicolor bar coding. *Lab Invest.*, **80**, 1031–1041.

Further Reading

Babkina, N. and Graham, J.M., Jr. (2014) New genetic testing in prenatal diagnosis. *Semin. Fetal. Neonatal. Med.*, **19**, 214–219.

Go, A.T., van Vugt, J.M., and Oudejans, C.B. (2011) Non-invasive aneuploidy detection using free fetal DNA and RNA in maternal plasma: recent progress and future possibilities. *Hum. Reprod. Update*, **17**, 372–382.

Katsanis, S.H. and Katsanis, N. (2013) Molecular genetic testing and the future of clinical genomics. *Nat. Rev. Genet.*, **14**, 415–426.

Mahdieh, N. and Rabbani, B. (2013) An overview of mutation detection methods in genetic disorders. *Iran J. Pediatr*, **23**, 375–388.

Chromosomal Abnormalities

Shinawi, M. and Cheung, S.W. (2008) The array CGH and its clinical applications. *Drug Discov. Today*, **13**, 760–770.

Fluorescence *In Situ* Hybridization

Tsuchiya, K.D. (2011) Fluorescence *in situ* hybridization. *Clin. Lab Med.*, **31**, 525–542.

Sequencing

Rehm, H.L. (2013) Disease-targeted sequencing: a cornerstone in the clinic. *Nat. Rev. Genet.*, **14**, 295–300.

Array Analysis

Schaaf, C.P., Wiszniewska, J., and Beaudet, A.L. (2011) Copy number and SNP arrays in clinical diagnostics. *Annu. Rev. Genomics Hum. Genet.*, **12**, 25–51.

Sinn, P., Aulmann, S., Wirtz, R. *et al.* (2013) Multigene assays for classification, prognosis, and prediction in breast cancer: a critical review on the background and clinical utility. *Geburtshilfe Frauenheilkd*, **73**, 932–940.

Pharmacogenetics/ Pharmacogenomics

9

Summary

- The terms pharmacogenetics and pharmacogenomics can be used synonymously for a discipline that elucidates the effects of genotype on an individual's drug response. Genetic variations influence the pharmacokinetics (uptake and metabolism) of a drug as well as its pharmacodynamics (effects of a drug on the target).
- Inherited differences in the uptake and distribution of a compound influence its bioavailability. P-glycoprotein (PGP), encoded by the *multidrug resistance gene* (*MDR*), is an important transporter that pumps numerous substances (e.g., xenobiotics) out of cells.
- Cytochrome P450 (CYP) proteins are the major class of drug-metabolizing enzymes (DMEs) and affect most marketed drugs. Different CYP genotypes may result in poor, intermediate, extensive, or ultrarapid metabolization of a drug or chemical.
- Mutations of the target and changes in its expression level may substantially alter the effect of any drug and must, therefore, be analyzed prior to treatment with certain drugs.
- Pharmacogenomic testing is an important constituent of personalized medicine and has already made its way into clinical practice. The choice of a suitable drug and the adjustment of its dosage will help to ensure treatment efficacy and prevent adverse drug responses. In addition, pharmacogenomics will help to improve the process of drug development by selecting patient subpopulations with a suitable genotype for clinical trials.

Contents List

A drug that has proven effective in many individuals in large populations often fails in others. In other cases, the medication may demonstrate activity, but causes side effects. These variations have inspired efforts to tailor individual optimal pharmacotherapy. Many factors influence the effects of a substance on an organism, including patient-specific parameters such as age, gender, general physiology, interactions with other drugs, alcohol intake, or cigarette smoking. An important contribution to an individual's response to pharmacotherapy is defined by his or her genetic makeup. Although attempts to monitor drug therapy have been made since the 1950s, it was not until the molecular era that specific genes could be identified that influence the effects of a drug on an individual. In recent years, increasing knowledge has been gained about the genetic variations that influence the uptake, transport, and metabolism of drugs and their effects on molecular targets.

Pharmacogenetics is a discipline that elucidates the effects of the genotype on an individual's drug response. This approach usually includes a genotype–phenotype connection. Much clinically significant and meaningful data on the influence of genetic variations on the outcome of a therapeutic intervention have been obtained in the few last decades. More recently, the focus has shifted to genome-wide integrative analyses that permit a global view of the relevance of the genotype on the reactions to a given drug. This change of paradigm is sometimes referred to by the term *pharmacogenomics*. However, the relevance of these large data sets for

Molecular Medicine: An Introduction, First Edition. Jens Kurreck and Cy Aaron Stein.
© 2016 Wiley-VCH Verlag GmbH & Co. KGaA. Published 2016 by Wiley-VCH Verlag GmbH & Co. KGaA.

pharmacological intervention still remains to be clarified. In addition, the significance of newly identified variations derived from genomic studies needs to be understood. Since there is no consensus on the precise definition and differences between the terms pharmacogenetics and pharmacogenomics, both will be used interchangeably when referring to the effects of the genotype on the response to drugs. They are uniformly abbreviated as PGx in this chapter.

As of 2014, the FDA listed more than 150 drugs with pharmacogenomic labeling in their package inserts, a selection of which is summarized in Table 9.1. The complete FDA list refers to more than 50 genetic biomarkers that are relevant to the effects of these drugs. Most of the genetic variations influence individual variability with respect to uptake, metabolism, transport, distribution, and elimination of the drug and in its interactions with its target. The majority of the drugs for which PGx consideration has been recommended are indicated for the treatment of cancer, cardiovascular, infectious, and psychiatric diseases. Inclusion of pharmacogenetic data in oncology is particularly important and at the same time challenging. Chemotherapeutics can only be used in a very narrow dose range and often display severe and sometimes even life-threatening side effects. In addition, PGx in anticancer treatment involves the analysis of different genomic systems, the germline genome, and the tumor genome, which usually differs from the inherited genome. The tumor genome itself may further be divided into the genome of the primary tumor and that of the metastases. Modern tumor therapy using molecular targeted agents has achieved significant progress in cancer treatment. However, as will be outlined in more detail below, it is essential to separate responders from nonresponders, which can be accomplished by pharmacogenetic testing.

The large number of drugs for which genetic testing prior to therapeutic intervention has been recommended demonstrates that PGx has penetrated clinical practice. It is an important aspect of the goal of "individualized" or "personalized medicine," that is, the attempt to tailor medical treatment to the individual characteristics of each patient. Genomic information (as well as other personalized data) can be used to improve targeted pharmacotherapy. However, PGx not only influences individual treatment but can also facilitate the discovery and clinical testing of new products.

The aim of pharmacotherapy is to achieve the maximal therapeutic effect of a drug in the absence of toxicity. The range of dosages of a drug that treat disease effectively without causing adverse effects is defined as the *therapeutic window* of the drug (Figure 9.1a). The standard dosage of a drug is chosen to fulfill these requirements. However, individual variability may lower the overall efficacy of a drug. For example, almost 10% of the population lacks the enzyme that converts the prodrug codeine, which is used to treat mild to moderate pain and for other indications, into the active form, morphine. These patients are considered nonresponders and must be treated with alternative drugs. In other instances, the maximal efficacy of a drug can be reached, but only at a higher dose than the standard dosage (Figure 9.1b). Individual factors may also alter the toxicity dose curve (Figure 9.1c) causing patients to experience treatment side effects, called adverse drug reactions (ADRs). Millions of people experience ADRs every year

Table 9.1 Selected drugs for which pharmacogenetic tests are indicated.

Drugs	Therapeutic area	PGx biomarker	Variation type	FDA recommendation
Abacavir	Infectious diseases	HLA-B allele	SNP	Mandatory
Cetuximab and panitumumab	Oncology	EGFR	Overexpression	Recommended
		KRAS	Mutation	Recommended
Codeine	Anesthesiology	CYP2D6	SNP	Recommended
Imatinib, nilotinib, and dasatinib	Oncology	PhChr	Translocation	Recommended
Mercaptopurine	Oncology	TPMT	SNP	Mandatory
Trastuzumab	Oncology	Her2/neu	Overexpression	Recommended
Vemurafenib and dabrafenib	Oncology	BRAF	Mutation	Recommended
Warfarin	Cardiology or hematology	CYP2C9	SNP	Proposed
		VKORC1	SNP	

The list summarizes examples from different therapeutic areas, which are discussed in this chapter. CYP: cytochrome P450; EGFR: epidermal growth factor receptor; Her2: human epidermal growth factor receptor 2; HLA: human leukocyte antigen; PhChr: Philadelphia chromosome; SNP: single-nucleotide polymorphism; TPMT: thiopurine-*S*-methyltransferase; VKORC1: vitamin K epoxide reductase complex subunit 1.
(Based on Ref. [1] and http://www.fda.gov/drugs/scienceresearch/researchareas/pharmacogenetics/ucm083378.htm).

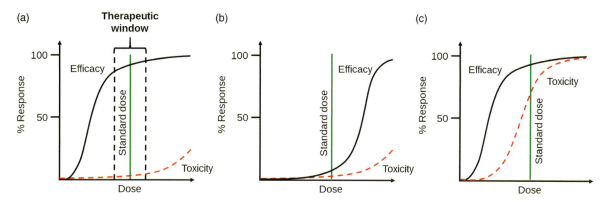

Fig. 9.1 Dose curves of drug response. (a) All drugs have dose-dependent therapeutic efficacy and toxicity. The therapeutic window is defined as the dose range in which the drug is effective without causing toxicity. Individual factors can influence the efficacy (b) or the toxicity curve (c). If maximal efficacy is reached at a higher dose of the drug, treatment will be ineffective at the standard dose and a higher dose is required. However, it is also possible that optimal efficacy will not be achieved even at high dose (not shown). A shift in the toxicity curve will result in adverse drug reactions (ADRs) at the standard dose.

and although the estimated numbers of fatal ADRs vary greatly, the side effects of pharmacotherapies are the cause of death in, at a minimum, ten thousand cases. The term ADR refers only to injury caused by the drug when given at a standard dose; the numbers of adverse drug effects due to overdose are much higher.

Efficacy of drug treatment and possible adverse side effects are largely influenced by the *pharmacokinetic* and *pharmacodynamic* properties of the substance. Pharmacokinetics describes the fate of an administered substance in a living organism. This includes drug uptake and tissue distribution, drug metabolism and degradation, and its subsequent elimination from the body. These features will be described in Sections 9.1 and 9.2. In contrast, pharmacodynamics is the study of the effects of the drug on the human body, particularly on the drug target, as described in Section 9.3. An important aspect of pharmacodynamics is the mechanism of interaction between the drug and its target (e.g., a cellular receptor) and the relationship between drug concentration and effect. All of these aspects may be greatly influenced by the individual genotype.

Genetic variations are usually inherited as germline variants, but may also be acquired as somatic mutations. The latter case is particularly relevant in cancer therapy as tumor genomes frequently are enriched in genetic alterations. The analysis of somatic mutations is often used to optimize the choice and efficacy of a chemotherapeutic agent. The most important genetic variations relevant to PGx are single-nucleotide polymorphisms (SNPs), but may also be insertions or deletions of nucleotides or gene duplications. PGx also analyzes changes in the expression level of a drug target. While these variations have been extensively investigated in

pharmacogenetic studies for many years, the regulation of gene expression by microRNAs (Section 13.4) and epigenetic modifications have only more recently been recognized as important factors in pharmacotherapy.

9.1
Uptake and Transport of Drugs

Drugs are usually not distributed in tissue simply by passive diffusion, but rather are actively transported across biological membranes by transport proteins. Uptake and efflux transporters are distinguished by the direction of drug transport.

Uptake transporters are further categorized as *Organic Anion Transporting Polypeptides* (OATP), *Organic Cation Transport* (OCT) proteins, and *PEPtide Transporters* (PEPTs). OATPs are expressed in many tissues of the human body. In the liver, they extract foreign chemical substances including drugs from the blood. For example, the transport protein OATP-C is exclusively found in the basolateral membrane of hepatocytes and is involved in the uptake of pravastatin, a member of the drug class of statins, which are used for lowering cholesterol. They do this by inhibiting the enzyme hydroxymethylglutaryl-CoA (HMG-CoA) reductase. Several SNPs in the OATP-C gene have been discovered that influence the plasma concentration of pravastatin after ingestion. In some individuals, tyrosine in position 521 of the protein is substituted by cysteine. This mutation is abbreviated by the one-letter codes of the amino acids; the position of the substitution is thus T521C. This variation, also denoted as the haplotype OATP-C*5, results in delayed hepatocellular uptake of pravastatin. On the other hand,

A388G (OATP-C*1b) seems to accelerate uptake of the drug. These findings may help to adjust the dosage of the drug according to genotype.

The best-characterized efflux transporter is the P-glycoprotein (PGP). This protein belongs to the family of ATP-binding cassette transporters (ABC transporters) and is encoded by the multidrug resistance gene 1 (MDR1), which was originally identified as the factor that confers tumor cell resistance to chemotherapeutic agents. PGP pumps many amphiphilic substances out of the cell, including numerous drugs, such as chemotherapeutic agents (e.g., vincristine and topotecan), cardiac drugs (e.g., digoxin, see below), HIV-protease inhibitors (e.g., saquinavir and ritonavir), and immunosuppressant drugs (e.g., cyclosporine A). ABC transporters function as dimeric proteins (Figure 9.2). Each monomer consists of a bundle of six transmembrane helices and a cytoplasmic domain harboring the ATP-binding cassette. ABC transporters utilize the energy of ATP hydrolysis to change conformation, which facilitates the transporting of substances across membranes. Drugs that inhibit substrate binding or the ATP hydrolytic activity of PGP may block the efflux transporter and can maintain the sensitivity of tumor cells to chemotherapeutic agents.

Several SNPs in the MDR1 gene that influence the activity of PGP have been identified. A well-studied example is the substitution of cytosine in position 3435 for thymine. Although this substitution is silent, that is, it does not change the amino acid sequence of the encoded protein, it modulates gene expression levels. While the expression of PGP in enterocytes (intestinal absorptive cells) is low in individuals that are homozygous for the T-allele, it is significantly enhanced in the genotype with a homozygous C-allele. The cardiac glycoside digoxin, which is used in the treatment of various heart conditions, is an example of a drug that is transported by PGP. Reduced PGP levels in individuals with the homozygous T-genotype lead to reduced efflux of digoxin from enterocytes into intestinal lumen. As a

consequence, intestinal digoxin uptake is increased and plasma levels of the drug are elevated.

9.2
Drug Metabolism

Biochemical modification of a drug substance and/or xenobiotic metabolism is another important process that influences the efficacy and toxicity of a pharmacological intervention. Because many therapeutics are given as inactive prodrugs, their efficient conversion into the active form is important for treatment success. DMEs are also involved in inactivating the active compound to prepare it for elimination from the cell and excretion. These reactions must be well adjusted to maintain the desired level of the drug. Inefficient activation of a prodrug or too rapid clearance of the active compound may reduce efficacy. In contrast, incomplete detoxification may result in an ADR due to increased drug levels.

Drug metabolism is widely divided into phase I and phase II reactions. Phase I reactions are usually minor modifications of the substance, for example, oxidation, reduction, or hydrolysis. Conversion of a C−H bond to a C−OH group is a typical phase I oxidation that may convert a prodrug into a pharmacologically active compound. Phase II reactions are synthetic reactions that frequently involve the transfer of moieties such as methyl-, acetyl-, or sulfhydryl groups or the conjugation of the tripeptide glutathione (GSH). While phase I reactions often produce active metabolites, phase II reactions usually inactivate substances.

9.2.1
Cytochrome P450 Enzymes

The major class of phase I DMEs is the cytochrome P450 superfamily (abbreviated by the acronym CYP). In fact, most approved drugs are metabolized by one or more members of this class of enzymes. CYPs either catabolize drugs to inactive metabolites, bioactivate prodrugs into their active form, or produce a reactive metabolite for further modification. The name cytochrome P450 arises from their characteristic 450 nm peak in the visible absorption spectrum.

CYP enzymes most commonly catalyze the oxidation of organic substances. The general feature of all cytochromes is that they contain a heme group that catalyzes redox reactions. The human genome encodes 57 CYPs that are grouped within families denoted by an Arabic numeral. Most of the DMEs are found within families 1, 2, and 3. CYPs are further specified by a letter that indicates the subfamily, and an another number that denotes

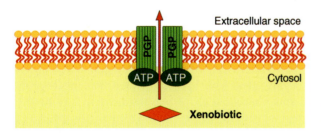

Fig. 9.2 P-glycoprotein. PGP homodimer pumps xenobiotics, including many drugs, out of the cell. Each monomer consists of six membrane-spanning helices and an intracellular ATP-binding domain. Hydrolysis of ATP results in a conformational change, which allows it to pump the substance into the extracellular space.

a specific member within the subfamily, for example, CYP2D6.

CYPs are monooxygenases that are embedded primarily in the membrane of the endoplasmic reticulum. They most commonly catalyze the insertion of one atom of oxygen into the aliphatic position of an organic substrate (RH):

$$RH + O_2 + NADPH + H^+ \rightarrow ROH + H_2O + NADP^+$$

While some CYP enzymes are specific for one or a few substrates, others metabolize multiple substrates. Altogether, CYPs are involved in metabolizing approximately 75% of all marketed drugs. Of the 57 human CYPs, 5 enzymes (CYP1A2, CYP2C9, CYP2C19, CYP2D6, and CYP3A4) are involved in 95% of these reactions. CYPs are not only the major DMEs, but are also involved in metabolizing toxins and may play a role in cancer development.

CYPs exist in numerous genetic variants with significant consequences for drug metabolism. DME polymorphism may result in the bimodal distribution of individuals into fast and slow metabolizers or even into more complex multimodal responders. An example of the latter is the inactivation of debrisoquine, an antihypertensive drug to treat high blood pressure, through 2-hydroxylation by CYP2D6. The population can be grouped into poor metabolizers (PMs), who have inactive CYP2D6, intermediate metabolizers (IMs) with low residual enzyme activity, extensive metabolizers (EMs) with a normal rate of metabolism and ultrarapid metabolizers (UMs), who have multiple copies of CYP2D6 with very high enzyme activity (Figure 9.3). The delay in clearance in PMs results in very high drug levels that may have adverse effects. Accordingly, PMs should receive a lower dose of the drug. In UMs, in contrast, a standard dose results in low drug exposure and lack of efficacy. Here, the dosage should be increased to achieve the desired plasma drug concentration. Other adjustments are required for IMs and EMs.

The frequency of the different phenotypes varies strongly depending on ethnic background. While 7% of the Caucasian population are PMs with respect to CYP2D6, only 1% of Asian people are PMs. Approximately 50 and

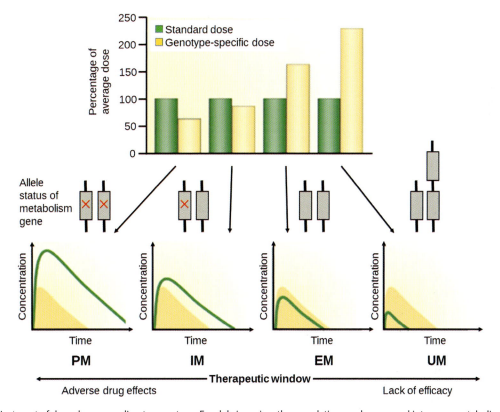

Fig. 9.3 Adjustment of drug dose according to genotype. For debrisoquine, the population can be grouped into poor metabolizers (PMs), intermediate metabolizers (IMs), extensive metabolizers (EMs), and ultrarapid metabolizers (UMs). Metabolic activity is determined by the genotype (allele status), as indicated. Depending on the clearance activity of the metabolizing enzyme, the concentration of the drug may be too high, causing adverse drug effects, or too low, diminishing drug efficacy (green lines). Accordingly, the standard dose should be adjusted to a genotype-specific dose to obtain the optimal concentration, indicated in orange. (Adapted from Ref [2] with kind permission from Macmillan Publishers Ltd., Copyright 2005.)

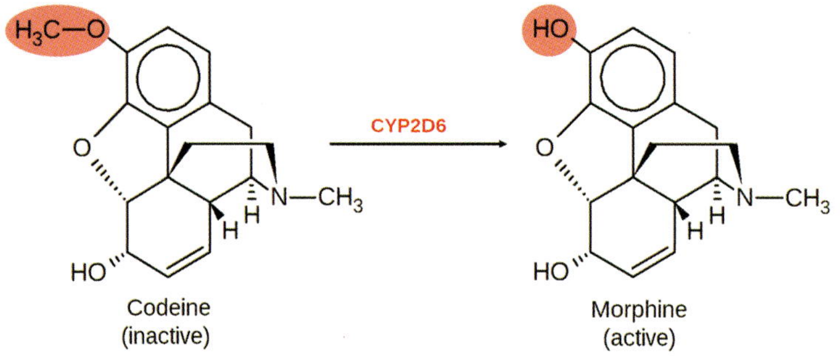

Fig. 9.4 Conversion of codeine to morphine. CYP2D6 catalyzes the demethylation reaction of the inactive prodrug codeine to give the active compound, morphine.

40% of Caucasians are IMs and EMs, respectively, and another 1–2% of this population are UMs. In contrast, 30% of Ethiopians have significantly increased enzyme activity. Ethnic differences pose additional challenges to the process of drug development. Regulators in some countries may request the empirical testing of new drugs in different populations. Although there is a clear scientific basis for these requirements in many cases, the argument may also be abused.

CYP2D6 is also involved in the activation of prodrugs. As mentioned in the introduction to this chapter, codeine is an inactive substance that is metabolized to produce the pharmacologically active morphine. The demethylation reaction is catalyzed by CYP2D6 (Figure 9.4). As expected, this reaction does not occur in individuals with CYP2D6 deficiency, in whom codeine lacks efficacy. In contrast, UMs may experience ADRs due to their above-average rate of morphine production.

Drugs may be substrates of several CYP proteins. Tamoxifen is a widely used anticancer drug in the treatment of breast cancer, which often requires the hormone estrogen to bind to and activate the estrogen receptor. While tamoxifen itself is an inactive prodrug, its metabolites 4-hydroxytamoxifen and endoxifen bind to the estrogen receptor with high affinity, but do not activate it. Thus, they act as competitive antagonists and eventually inhibit cell proliferation. Tamoxifen is predominantly metabolized by two CYP enzymes: CYP2D6 and CYP3A4 (Figure 9.5). Polymorphisms in CYP2D6 influence response to tamoxifen: PMs with CYP2D6 deficiency experience significantly reduced treatment efficacy, while UMs with multiple copies of CYP2D6 functional alleles have high plasma concentrations of endoxifen, which may result in ADRs. Although evaluation of CYP activity is not always carried out in clinical practice due to the low number of tamoxifen ADRs, this example demonstrates that the evaluation of

a patient's genotype for CYP expression has the potential to optimize pharmacological treatment (Box 9.1).

In addition to genotype, environmental factors may influence the activity of CYP3A4 by modulating its expression. Many drugs are inducers or inhibitors of CYP3A4 expression and may cause interference after simultaneous administration. Studies on the interactions of newly identified pharmacological substances with the CYP3A4 promoter are frequently included in the process of drug development. Of interest is a widely recognized study that described grapefruit juice as a potent inhibitor of CYP3A4; this interaction can affect the metabolism of a number of drugs. This was the first reported food–drug interaction and has been known to be fatal in a small number of cases.

While the pharmacogenetic implications of CYP2D6 and CYP2C19 polymorphisms are well understood in many cases, the situation is less clear for CYP3A4, the most abundant CYP in the human liver. This CYP is involved in the metabolism of more than 50% of all approved drugs. More than 20 variants of CYP3A4 have been identified, many of which alter enzyme activity. However, the clinical relevance of these polymorphisms is minimal and unclear in many cases. One reason for these uncertainties may be that many CYP3A4 substrates are also metabolized by CYP3A5. Although in most cases metabolism by CYP3A5 is significantly slower than by CYP3A4, the dual pathways obscure the significance of the CYP3A4 polymorphisms.

9.2.2
Other Drug Metabolizing Enzymes

In addition to the CYP enzymes, many other DMEs play a critical role in the conversion of drugs. One of the earliest studied phase II DMEs is arylamine *N*-acetyltransferase 2.

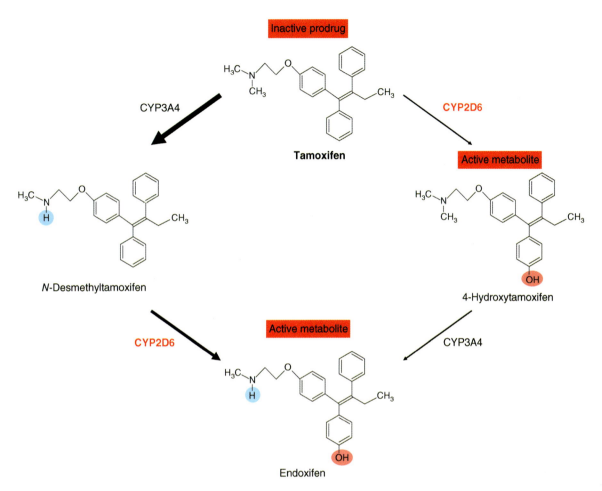

Fig. 9.5   Metabolism of tamoxifen. Tamoxifen is an inactive prodrug. Two CYP proteins, CYP3A4 and CYP2D6, metabolize tamoxifen into the intermediate *N*-desmethyltamoxifen and the two active compounds 4-hydroxytamoxifen and endoxifen. CYP2D6 is the rate-limiting enzyme in the activation of tamoxifen and its polymorphisms have a significant impact on its anticancer efficacy.

Variations in this gene divide people into "slow acetylators" and "fast acetylators." As a consequence, half-lives and plasma concentrations of important drugs such as the antiarrhythmic agent procainamide vary significantly. This may have dramatic consequences as ADRs may occur in the group of slow acetylators, a group to which approximately half of the Caucasian population belongs. In contrast, less than 20% of the East Asian population are slow acetylators.

Thiopurine-*S*-methyltransferase (TPMT) is another example of a phase II DME. This enzyme catalyzes the methylation of thiopurine compounds such as azathioprine, 6-mercaptopurine, and 6-thioguanine, which are commonly used as chemotherapeutic agents and immunosuppressive drugs. TPMT transfers a methyl group to its substrates, inactivating them (Figure 9.7). Due to the high rate of polymorphisms in the TPMT gene, enzyme activity varies greatly in the population.

More than 20 genetic variants of TPMT have been identified, most of which result in reduced enzyme activity. While enzyme activity is high in approximately 90% of Caucasians, 10% inherit intermediate activity and 0.3% have low or no enzyme activity. Individuals lacking TPMT need less than 10% of the standard dose of the drug. If treated with the full dose, the functional deficiency in TPMT leads to an increased level of cytotoxic thiopurine nucleotides that can result in severe (and sometime fatal) hematological toxicity. Therefore, the FDA recommends genotyping of the three major SNPs in TPMT prior to the use of 6-mercaptopurine; these variations account for 95% of individuals with reduced TPMT activity. If the patient is carrier of one of the SNPs, the dose of the drug must be substantially reduced.

Another common phase II reaction is the conjugation of the reduced tripeptide glutathione (GSH) to a

Box 9.1. Analysis of the Cytochrome P450 Genotype

The AmpliChip™ CYP450 genotyping test manufactured by Roche was the first array-based pharmacogenetic test. It was approved by the FDA in 2004 (Figure 9.6). The device analyzes genotypes for CYP2D6 and CYP2C19, which are involved in the metabolism of approximately one fourth of all prescribed drugs, including representatives of many important classes such as β-blockers, monoamine oxidase inhibitors, analgesics, antidepressants, antiarrhythmic agents, antidiabetic agents, and antiestrogens. The DNA array is produced by the photolithography method described in Section 2.1. It tests for three CYP2C19 alleles, classifying individuals into EMs and PMs, in addition to evaluating 27 CYP2D6 alleles for classification into the UM, EM, IM, and PM phenotypes. The analysis requires a sample from the patient (e.g., blood) to amplify the gene by PCR. The PCR product is then fragmented and labeled prior to hybridization on the microarray. The results are obtained after the scanning of the chip and bioinformatic analysis. This test has been extensively employed in psychiatry where, for example, it may help to adjust the dose of the antipsychotic drug risperidone according to the genotype.

Fig. 9.6 AmpliChip CYP450 array by Roche. The DNA chip evaluates genotypes of CYP2D6 and CYP2C19. It is produced by photolithography. (Image courtesy of Roche Diagnostics, Mannheim, Germany.)

pharmacological agent by glutathione-*S*-transferases (GSTs). A clinical example of such a conjugation can be found in the metabolism of paracetamol, in Europe still a widely used analgesic and antipyretic whose main mechanism of action is the inhibition of cyclooxygenase-2 (COX-2). Paracetamol is metabolized primarily in the liver by several metabolic pathways. One of the pathways involves the generation of the toxic intermediate *N*-acetyl-*p*-benzoquinone imine (NAPQI) by three CYP enzymes (Figure 9.8). Although CYP2D6 metabolizes only a comparatively small fraction of the total amount of paracetamol, it contributes significantly to paracetamol toxicity due to its pronounced polymorphism. Thus, the rate of NAPQI formation depends on the CYP2D6 genotype. At standard doses, NAPQI is detoxified by the phase II reaction involving the conjugation of GSH by GST. However, following an overdose and also in EMs and UMs who produce elevated levels of NAPQI by CYP2D6, the detoxification pathway may become saturated. NAPQI will then accumulate, causing liver and renal toxicity. In addition, depletion of GSH itself may be hepatotoxic, since the antioxidant function of the tripeptide prevents

damage to cellular components by reactive oxygen species.

9.3
Drug Targeting

Polymorphisms in genes encoding drug targets may directly influence the response to a pharmacological treatment. Figure 9.9 illustrates how changes in the drug

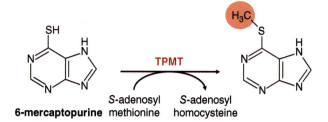

Fig. 9.7 Inactivation of 6-mercaptopurine. The cytostatic drug 6-mercaptopurine is inactivated by the transfer of a methyl group. This reaction is catalyzed by thiopurine-*S*-methyltransferase (TPMT), which uses *S*-adenosyl methionine as methyl group donor.

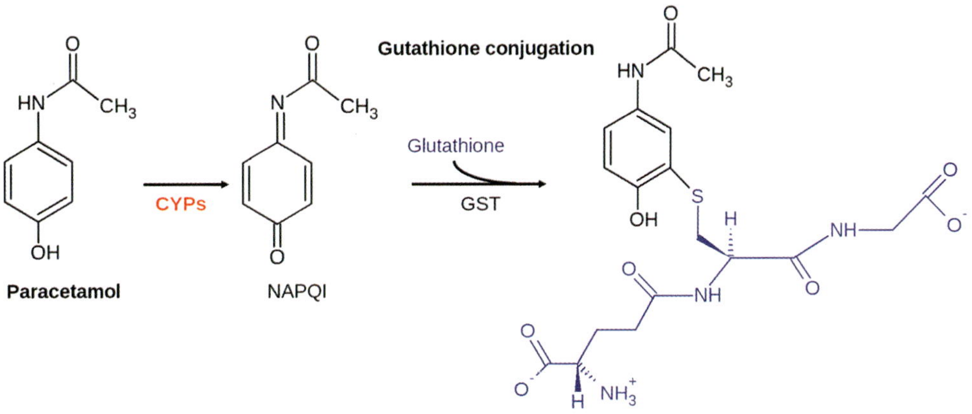

Gutathione conjugation

Glutathione

GST

Paracetamol

NAPQI

Fig. 9.8 Metabolization of paracetamol. The three hepatic CYP proteins CYP2E1, CYP1A2, and CYP2D6 metabolize paracetamol into *N*-acetyl-*p*-benzo-quinone imine (NAPQI). This intermediate is detoxified by conjugation of reduced glutathione by the enzyme glutathione-*S*-transferase (GST). Saturation of this detoxification pathway causes liver and renal toxicity.

target can modulate the efficiency of a drug: Mutations in a target gene may alter protein structure to prevent drug binding to its target, hence inactivating it. In addition, the response to treatment may be significantly reduced (or completely abolished) if the expression of the drug target is low. As will be described below, some anticancer drugs are recommended only if their target is expressed at high levels in the tumor.

An example of the loss of efficacy due to changes in the drug target sequence is a mutation in an oncogene that prevents binding of the anticancer drug and consequently leads to drug resistance. As described in Section 4.7.1, translocation of the Abelson kinase (ABL) to produce the breakpoint cluster region (BCR)-ABL fusion protein, also known as the Philadelphia chromosome, is commonly associated with chronic myeloid leukemia (CML). The tyrosine kinase inhibitors (TKIs) imatinib, nilotinib, and dasatinib, which block the activity of the BCR/ABL fusion protein, are used for the treatment of Philadelphia chromosome-positive CML patients.

However, the T315I mutation in the protein leads to a loss of binding of the TKI and renders the cancer cells resistant to treatment. Patients with this genotype will, therefore, not respond to treatment with TKIs.

The importance of the level of expression of the drug target for the success of a therapeutic intervention is nowhere more clearly seen than for the monoclonal antibody (mAb) trastuzumab (trade name Herceptin), which has been used successfully for the treatment of breast cancer (Section 10.2.1). This antibody binds to the human epidermal growth factor receptor 2 (Her2/neu) and inactivates its target, which regulates cell growth and division. Her2 is overexpressed in approximately 30% of breast cancers and results in accelerated cell proliferation (Section 4.1). Since trastuzumab will have no activity and may cause an ADR if the cancer cells do not overexpress Her2, the level of its expression must be determined prior to treatment. Several methods have been developed to test for Her2 expression in patient biopsies: For immunohistochemical (IHC) analysis, the tissue is fixed and treated with an antibody against the target protein (Figure 9.10). Direct detection by one-step staining involves the use of a labeled antibody, while the indirect two-step method makes use of an unlabeled primary antibody that binds to the target, followed by tagging with a labeled secondary antibody. The latter method is more sensitive due to amplification of the signal. As an alternative, fluorescence *in situ* hybridization (FISH) analyzes amplification of *Her2* at the genetic level. The assay is carried out with a DNA probe complementary to the target gene that is labeled with a fluorescent reporter for visualization under fluorescence microscopy, as described in more detail in Section 8.2.2. The current standard often involves a combination of IHC and FISH to help determine a

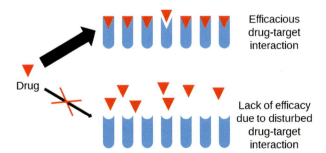

Efficacious drug-target interaction

Drug

Lack of efficacy due to disturbed drug-target interaction

Fig. 9.9 Variations in the drug target. The drug (red triangle) binds to its target (blue protein) and induces a response. Mutations in the target protein that prevent binding of the drug will modulate drug response.

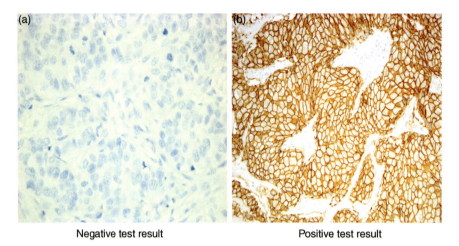

(a) Negative test result   (b) Positive test result

Fig. 9.10   Analysis of Her2 expression levels. Only those breast cancers overexpressing Her2 respond to treatment with trastuzumab; levels of expression must be analyzed prior to the treatment. A negative (a) and a positive (b) result is shown for an immunohistochemical (IHC) assay. (Image courtesy of Roche Austria, Vienna, Austria.)

treatment, but additional methods have already been approved or are in development.

Similar to the example above, the endothelial growth factor receptor (EGFR; Her1) is overexpressed in many colon (Section 4.4) and head and neck cancers. Cetuximab and panitumumab, mAbs targeted against the EGFR protein, are used to inhibit the growth and survival of tumor cells overexpressing *EGFR*. However, these mAbs were ineffective in some patients even though their tumor cells overexpressed the target. Mutations in the *KRAS* gene were identified that confer treatment resistance. KRAS is a G-protein that activates factors in the signal cascade downstream of EGFR; an example is BRAF (Figure 9.11). Two mutations in *KRAS* have been identified that lead to aberrant activation of *KRAS*; here inactivation of EGFR will be irrelevant. Therefore, the FDA recommends pharmacogenetic testing and use of the anti-EGFR antibodies only for patients with *EGFR* overexpression *and* mutant negative (i.e., wild-type) *KRAS*.

The signal cascade depicted in Figure 9.11 can also be blocked further downstream: Vemurafenib (trade name Zekboraf) and dabrafenib (trade name Tafinlar) are inhibitors of BRAF that are FDA approved for the treatment of metastatic melanoma. However, these drugs can only be used if BRAF contains a V600E mutation, which is present in 60% of melanomas. Treatment with the inhibitors is inefficient in melanoma cells without this mutation and may even promote tumor growth by stimulating wild-type BRAF.

In some cases, polymorphic DMEs combined with genetic variants of the drug target may determine response to pharmacotherapy. An example is warfarin,

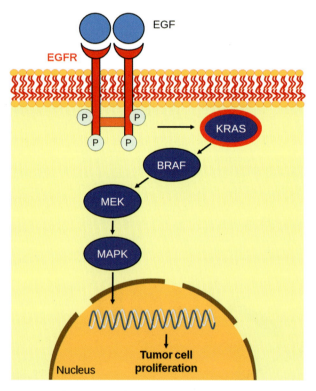

Fig. 9.11   The EGFR signaling cascade. Binding of the epidermal growth factor (EGF) to its receptor (EGFR) leads to activation of a signal cascade. Overexpression of EGFR results in uncontrolled tumor cell proliferation. Inhibition of EGFR by anticancer drugs can fail to block the signaling pathway if mutated *KRAS* activates factors downstream of the pathway independent of EGFR. MEK: mitogen-activated protein kinase kinase; MAPK: mitogen-activated protein kinase.

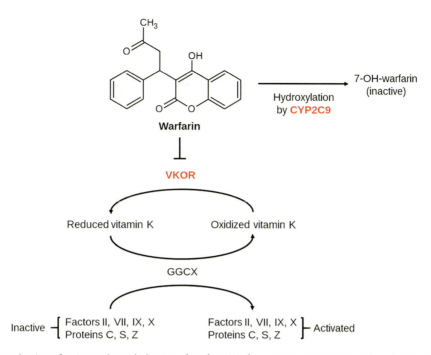

Fig. 9.12 Simplified mechanism of action and metabolization of warfarin. Warfarin targets vitamin K epoxide reductase (VKOR) to inhibit vitamin K reduction. Depletion of reduced vitamin K results in decreased activation of coagulation factors II, VII, IX, and X and proteins C, S, and Z by γ-glutamyl carboxylase (GGCX). Loss of function of these factors leads to slowing of blood coagulation. CYP2C9 inactivates warfarin by hydroxylating one carbon atom.

the most commonly used oral anticoagulant worldwide. This drug inhibits vitamin K epoxide reductase (VKOR), an enzyme that reduces vitamin K (Figure 9.12). The reduced form of vitamin K is required by γ-glutamyl carboxylase (GGCX) to activate coagulation factors II, VII, IX, and X and the regulatory proteins C, S, and Z, thus inducing blood coagulation. By targeting the VKOR complex subunit 1 (VKORC1), warfarin prevents the formation of reduced vitamin K, which eventually leads to slowing of blood coagulation.

The therapeutic window for warfarin is narrow and interindividual response to the drug is highly variable. The dose required for efficient prevention of thrombosis and embolism may vary 20–30-fold from patient to patient. An incorrect dosage can lead either to severe bleeding or failure to prevent clotting. Polymorphisms in two genes account for approximately 40% of the dose variation between patients. Genotyping of patients treated with warfarin has revealed several common polymorphisms in VKORC1, the drug target. The two main variants are a low-dose haplotype group (A) and a high-dose haplotype group (B). These polymorphisms also explain why African-Americans are, on average, relatively resistant to warfarin, since the proportion of group B haplotypes is higher. In contrast, Asian-Americans are generally more sensitive due to a higher proportion of group A haplotypes. A second, somewhat smaller, impact on drug response is due to polymorphisms in CYP2C9. This enzyme catalyzes the 7-hydroxylation of warfarin, inactivating the drug. Two variants with single amino acid substitutions, CYP2C9*2 and CYP2C9*3, lead to reduced warfarin clearance. Individuals with these genotypes have a two- to threefold increased risk of experiencing adverse effects and thus require reduced daily doses of the drug. Evaluation of the CYP2C9 and VKORX1 polymorphisms may help to lower the risk of bleeding complications from warfarin treatment. Indeed, initial studies have indicated lower hospitalization rates in patients treated with individually adjusted doses.

However, warfarin pharmacogenetics is complicated. Warfarin is a racemic mixture of two active enantiomers: the *S*-enantiomer is a significantly more potent anticoagulant than the *R*-enantiomer. Each is cleared by a different pathway: The hydroxylation reaction by CYP2C9 described above is specific for the *S*-form, while the less active *R*-form is inactivated by three polymorphic DMEs: CYP1A2, CYP1A1, and CYP3A4. In addition, proteins involved in the uptake and transport of warfarin (e.g., PGP) and polymorphisms in other proteins such as GGCX, which catalyzes the γ-carboxylation of the vitamin K-dependent clotting factors, also play a role in the variability of treatment response. The relevance of these factors for bleeding complications after warfarin treatment has not yet been fully elucidated. Moreover, the current approach that tests for two genotypes (CYP2C9 and VKORX1) will provide only limited information about an individual's response to the drug.

9.4
Drug Toxicity and Hypersensitivity

The adverse effects of a pharmacotherapeutic agent may originate from the interaction of the drug and its target or from a protein different from the therapeutic target. An example for the former, also known as *on-target* toxicity, is the risk of excessive bleeding by treatment with warfarin due to drug overdosing, as described previously. *Off-target* toxicity has, for example, been observed for lumiracoxib (trade name Prexige), which was developed for the symptomatic treatment of osteoarthritis and acute pain by inhibiting cyclooxygenase-2. However, due to severe hepatotoxicity, lumiracoxib has never been approved by the US FDA and was withdrawn from the market in other countries. Detailed analyses revealed that SNPs in MHC class II regions were associated with the observed ADRs, for example, hepatotoxicity.

Adverse effects of drugs that resemble allergic reactions are denoted as *drug hypersensitivity reactions*. These may require prolonged hospitalization and changes in drug treatment, and may in some cases be life-threatening. Although the mechanisms underlying drug hypersensitivity reactions are not fully understood, genetic polymorphisms may predispose individuals to drug allergy. An example is abacavir (trade name Ziagen), a nucleoside analog reverse-transcriptase inhibitor (NRTI) used to treat HIV infection (Section 5.4.1). The drug is well tolerated by more than 90% of patients, but the remaining patients develop an "abacavir hypersensitivity syndrome" (AHS). This ADR is strongly associated with a SNP in the human leukocyte antigen (HLA). Individuals with the common histocompatibility allele *HLA∗5701* are at risk of developing AHS. Abacavir binds with high specificity to the mutant protein, changing the shape of the antigen-binding cleft. This change modulates immunological tolerance and induces activation of abacavir-specific T cells and a systemic reaction that can be severe and in some cases fatal. Genetic testing for the *HLA∗5701* polymorphism is mandatory before starting treatment with abacavir. Here, pharmacogenetic testing enables a definitive decision as to whether or not antiviral treatment should be initiated with this drug. In contrast, the analysis of how genetic variability influences pharmacokinetics, as described in the previous sections, often does not result in a specific

recommendation. This is because the guidelines for dose adjustments that can compensate for gene-based differences are in many cases not well defined.

9.5
Drug Development and Individual Pharmacotherapy

PGx underlies the genetic component of personalized medicine. The knowledge of causal relations between genetic variations and drug response can, hopefully, better aid physicians in prescribing the appropriate medication for the correct disease target. In addition, patients can be optimally dosed to achieve maximal therapeutic benefit with minimal ADRs. In contrast, physicians currently usually prescribe a standard dose of a drug and adjust it by observing its efficacy and adverse events during treatment. To date, only a limited set of drugs requires genotyping before treatment is initiated. Otherwise, pharmacogenetic testing is performed only when patients are refractory to therapy or where severe adverse events supervene. In the future, PGx may play an increasingly important predictive role. Genotyping may routinely guide drug dosing. This paradigm shift is further illustrated by warfarin therapy example. Due to a narrow therapeutic window that is bounded by the risk of failing to control blood clotting with a too low a dose, and extensive bleeding with too high a dose, patients must be closely monitored by conventional methods during treatment. Currently, physicians determine the time required for blood to clot after patients have been treated with the drug. They then adjust the dose to balance between the desired therapeutic effect and the prevention of severe side effects. An individualized medicine approach would incorporate pharmacogenetic testing for relevant polymorphisms, particularly VKORC1 and CYP2C9, before determination of the optimal dose prior to treatment. For example, it is known that individuals homozygous for the 1173T allele of VKORC1 require 50% of the warfarin dose compared with patients carrying the wild-type allele. However, these two genotypes account for only approximately one fourth of the dose variations, while other factors (genetic determinants as well as nongenetic factors) that have an impact on the response to warfarin treatment are not well understood.

Since many proteins, particularly CYPs (but transporters as well), are involved in the metabolism of a large number of pharmacologically relevant drug substances, a pharmacogenetic certificate has been proposed based on the collection of individual pharmacogenetic information. For example, knowledge of an individual's genotype for the major CYPs (e.g., CYP1A2, CYP2C9, CYP2C19,

CYP2D6, and CYP3A4) would facilitate the physician's choice of a suitable drug and dosage. However, exhaustive pharmacogenetic testing may raise important privacy issues. Per definition, pharmacogenetic markers are relevant only to an individual's response to drug treatment and not to disease-related predispositions. However, several variants studied by PGx are also associated with common diseases. An example is the apolipoprotein E4 allele, a marker for reduced responsiveness to statin treatment for the lowering of cholesterol levels. In addition to its pharmacogenetic relevance, this allele has also been associated with an increased risk of developing Alzheimer's disease. The privacy of pharmacogenetic data must be ensured because the information it provides is as personal as any other genetic information.

Despite the predictive power of PGx in some cases, many challenges remain for optimizing pharmacotherapy. As exemplified by warfarin, multiple factors may be involved in the uptake, transport, metabolism, and action of a compound. A complex network of proteins influences the activity of every drug. This requires the analysis of multiple polymorphisms and a full understanding of their interactions. Modern sequencing technologies (Section 2.3) will soon permit physicians to obtain a comprehensive picture of a patient's genome; several companies already offer the sequencing of complete tumor genomes. However, collecting the appropriate data and the complex bioinformatic analysis required to generate therapeutic strategies make this approach complicated and expensive.

In addition, the outcome of a pharmacologic intervention is not only influenced by genetic background but also strongly depends on other individual factors such as individual physiology, age, environment, and nutritional status. Environmental chemicals, such as coadministered drugs, cigarette smoke, and alcohol are known to induce or inhibit CYPs and other DMEs, altering drug efficacy and toxicity. Personalized medicine cannot be restricted to the analysis of inherited markers only, but must also take into consideration the many factors that influence an individual's response to a drug.

PGx not only aims to optimize individual pharmacotherapy but also helps to improve the process of drug development. Less than 10% of the pharmaceutical candidates that enter clinical trials are approved for marketing by the FDA. The majority of the candidate drugs fail because of unacceptable side effects or insufficient efficacy. PGx may help to stratify those patients in the larger population that are suitable candidates for a prospective drug. Knowledge of the genetic variants that confer resistance to treatment or are associated with an increased risk of experiencing ADRs can help to exclude

carriers of that genotype. This stratification strategy can be expected to produce clear data on therapeutic outcomes and unwanted side effects in treated patients.

The inclusion of PGx data in the process of drug testing will initially add cost as suitable biomarkers must be identified and validated. However, this extra expense may be compensated for by improved trial design. Through genotype stratification, a more homogenous population will allow inclusion of fewer patients in a study. A reduction in the number and severity of side effects will also accelerate the approval process by the regulatory authorities. Studies may also be designed to test different doses in PM, IM, EM, and UM patients. However, though this strategy has already proven successful in oncology, success may be more difficult to achieve in other fields.

In addition to the development of new agents, PGx may give a second chance to candidates that have failed in clinical trials due to toxicity or limited efficacy. In addition, based on pharmacogenetic stratification, approved drugs that were withdrawn from the market (usually because of severe ADRs) may be re-investigated for suitable subpopulations. Where treatment failure can be associated with a specific genotype, approval may come with a label requiring mandatory pharmacogenetic testing prior to treatment.

References

1. Weng, L., Zhang, L., Peng, Y. *et al.* (2013) Pharmacogenetics and pharmacogenomics: a bridge to individualized cancer therapy. *Pharmacogenomics*, **14**, 315–324.
2. Kirchheiner, J., Fuhr, U., and Brockmoller, J. (2005) Pharmacogenetics-based therapeutic recommendations: ready for clinical practice? *Nat. Rev. Drug Discov.*, **4**, 639–647.

Further Reading

Cascorbi, I., Bruhn, O., and Werk, A.N. (2013) Challenges in pharmacogenetics. *Eur. J. Clin. Pharmacol.*, **69** (Suppl. 1), 17–23.
Kirchheiner, J., Fuhr, U., and Brockmoller, J. (2005) Pharmacogenetics-based therapeutic recommendations: ready for clinical practice? *Nat. Rev. Drug Discov.*, **4**, 639–647.
Ma, Q. and Lu, A.Y. (2011) Pharmacogenetics, pharmacogenomics, and individualized medicine. *Pharmacol. Rev.*, **63**, 437–459.
Weng, L., Zhang, L., Peng, Y. *et al.* (2013) Pharmacogenetics and pharmacogenomics: a bridge to individualized cancer therapy. *Pharmacogenomics*, **14**, 315–324.

Cytochrome P450

Guengerich, F.P. (2008) Cytochrome p450 and chemical toxicology. *Chem. Res. Toxicol.*, **21**, 70–83.

Recombinant Protein Drugs   10

Summary

- The development of recombinant DNA technologies has provided a path for the production of protein drugs in expression organisms. This procedure improves the purity and safety of proteins that were previously isolated from animal or human sources. Furthermore, the pharmacological properties of these proteins can be altered by the introduction of mutations. Human proteins that have previously not been available can easily be expressed based on knowledge obtained from human genome sequencing. Novel proteins can also be designed and synthesized.
- Large-scale protein production can be easily achieved in *Escherichia coli*. However, proteins that require posttranslational modifications, for example, the formation of disulfide bridges or glycosylation, must be produced in eukaryotic cells such as yeast, insect, or mammalian cells or in transgenic organisms.
- In the three decades following approval of the first recombinant protein in 1982, approximately 250 protein-based drugs have been approved for the treatment of diseases. These ailments include cancer, cardiovascular disease, inflammatory disease, and infectious and genetic diseases.
- Monoclonal antibodies (mAbs) are a particularly successful class of recombinant protein drugs and are employed for the treatment of a large number of diseases. Insulin is probably the best-known hormone to be synthesized recombinantly and is produced mainly in bacteria. Growth factors such as erythropoietin can be used to treat anemia. An important representative of the cytokine class is interferon-β, used for the treatment of multiple sclerosis (MS). While blood coagulation factors improve clotting times in patients with hemophilia, thrombolytic proteins dissolve blood clots after

Contents List

Production of Recombinant Proteins

- Bacteria
- Yeast and Other Fungi
- Insect Cells
- Mammalian Cells
- Transgenic Animals and Plants

Classes of Recombinant Drugs

- Monoclonal Antibodies
- Hormones
- Growth Factors
- Fusion Proteins
- Cytokines
- Blood Coagulation Factors: Anticoagulants and Thrombolytics
- Therapeutic Enzymes
- Recombinant Vaccines

---

myocardial infarction or stroke. A growing number of vaccines are now produced by recombinant expression. Additional classes of recombinant protein drugs are fusion proteins and therapeutic enzymes, which are used to treat inherited genetic disorders.

The development of new drugs based on the recombinant expression of proteins has become an advanced area of molecular medicine. Peptides and proteins, (e.g., insulin) obtained from animal sources had been used for therapeutic purposes for decades, but the advent of gene technology has provided new processes for the generation of protein-based drugs. Recombinant drugs can be reproducibly obtained from bacteria or eukaryotes in large quantities in high purity. Currently, the vast

Molecular Medicine: An Introduction, First Edition. Jens Kurreck and Cy Aaron Stein.
© 2016 Wiley-VCH Verlag GmbH & Co. KGaA. Published 2016 by Wiley-VCH Verlag GmbH & Co. KGaA.

majority of insulin used worldwide is now obtained via expression in *E. coli* or yeast. These new technologies provide improved sources for the production of older therapeutic agents. In addition, a large number of human proteins that were previously not available can now be obtained by recombinant expression. Gene technology also allows the modification and pharmacological optimization of naturally occurring proteins. Monoclonal antibodies are the dominant commercial representatives of this class of biopharmaceuticals.

The development of molecular cloning technologies, which allows the amplification of a defined DNA fragment, is the basis for the recombinant production of proteins as drugs. In this process, genes can be transferred from one species, for example, humans, into another species, for example, *E. coli* bacteria. This procedure became possible only after the discovery of restriction enzymes in the early 1970s. These enzymes, which are also known as restriction endonucleases, recognize specific sequence motifs in DNA and cleave them by endonuclease activity. This enables the novel combination of genetic material and subsequent heterologous expression of a gene in a chosen host organism (usually bacteria, yeast, mammalian cells, animals, or plants). The process is initiated by the cloning of the cDNA of a gene-of-interest into a plasmid, which, in addition to the cDNA, contains an origin of replication, a promoter, and an antibiotic resistance gene (Figure 10.1). The plasmid will then be transferred into the host and the desired protein expressed.

Recombinant production has improved the safety of protein drugs, which had been extracted from other sources. For example, insulin had formerly been isolated from cattle and pigs. Since animal insulin differs in its amino acid composition from human insulin, and cannot be isolated in pure form from animal sources, it

occasionally caused immune responses and allergic reactions. These problems were overcome by the use of human insulin obtained by recombinant expression in *E. coli* or yeast cells. Another example is human growth hormone (HGH), which in the past was isolated from the pituitary glands of human cadavers. This procedure, however, is associated with the risk of transmission of Creutzfeldt–Jacob prion disease. The replacement of HGH harvested from human cadavers by the recombinant protein led to a substantial improvement in product safety. A third example is coagulation factor VIII, which is used for the treatment of patients with hemophilia. Factor VIII was originally concentrated from donated blood plasma. However, the transfer of plasma products led to the transmission of the hepatitis B, hepatitis C, and human immunodeficiency viruses. The recombinant production of the drug now prevents disease transmission during replacement therapy.

In addition, many protein drugs were not available before the advent of recombinant DNA technology. Widely used hormones and enzymes, such as erythropoietin (EPO), granulocyte colony-stimulating factor (G-CSF), tissue-type plasminogen activator (tPA), or interferon-β (INF-β) became available only after the cloning and recombinant expression of the human genes. The production of monoclonal antibodies, a highly successful class of biopharmaceuticals, depends on the heterologous expression of a protein that does not exist naturally in any organism.

The initial aims of the generation of recombinant drugs were to produce proteins with a sequence identical to that of the respective human protein. The methods of gene technology, however, also permit the directed modification of protein-encoding DNA sequences. This strategy of protein engineering can be used to improve the pharmacokinetic and/or pharmacodynamic properties of proteins, for example, their stability, solubility, or specificity. Modifications can be restricted to a single amino acid (point mutation), but deletions, insertions of larger fragments, or the fusion of protein sequences can also be produced. Mutated proteins are sometimes referred to as *muteins*.

The properties of recombinant proteins have also been improved by posttranslational chemical modification. One of the most widely used approaches is the covalent attachment of polyethylene glycol (PEG) chains to proteins or peptides (Figure 10.2a). This polymer is neither toxic nor immunogenic and has been used in a number of FDA approved drugs. PEG conjugation improves the properties of the recombinant protein in several ways (Figure 10.2b): PEGylation increases the size of the drug and prolongs its circulation time by reducing renal clearance. In addition, it masks the protein from the host's immune system and reduces or eliminates its immunogenicity. PEGylation also

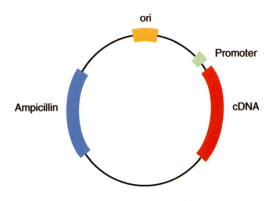

Fig. 10.1 Plasmid for gene expression in *E. coli*. At a minimum, standard plasmids encode an origin of replication (ori), a promoter that drives the expression of the cDNA of the gene-of-interest, and an antibiotics resistance gene, for example, the ampicillin resistance gene.

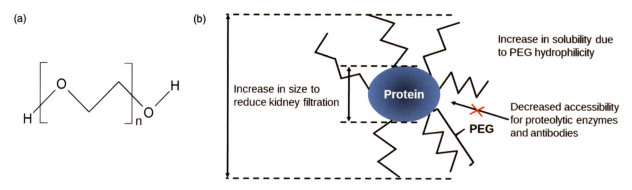

Fig. 10.2 PEGylation of proteins and peptides. (a) Chemical structure of polyethylene glycol. (b) Advantages of PEGylation: In this example, several PEG chains are linked to a protein. PEGylation shields the protein from degrading enzymes and antibodies, increases drug size to reduce renal clearance, and may increase the solubility of the protein. (Reprinted from Ref. [1] with kind permission from Elsevier.)

protects the protein from degradation by metabolic enzymes and can enhance the water solubility of proteins that are hydrophobic. The first PEGylated protein, approved by the FDA in 1990, was PEG-adenosine deaminase (trade name Adagen). This drug is used to treat patients with severe combined immunodeficiency (SCID) syndrome. Since then, a number of PEGylated protein and peptide pharmaceuticals have also been FDA approved (Table 10.1).

The PEG moiety is conjugated to a previously synthesized protein or peptide by chemical methods. The first step of the PEGylation process is the activation of the PEG polymer at one or both of its termini. Activated PEG will then form a covalent linkage with certain reactive amino acids, including lysine, cysteine, histidine, arginine, aspartic acid, glutamic acid, serine, threonine, and tyrosine. Furthermore, either the N-terminal amino group or the C-terminal carboxylic acid group can be employed in the conjugation step. The PEGylated product is then purified by chromatographic methods. While

the initial PEG polymers were linear chains, the latest generation of PEGylation agents is branched. An alternative strategy is to fuse the therapeutic protein with conformationally disordered polypeptide sequences composed of the amino acids proline, alanine, and serine. This is called "PASylation." Like PEGylation, PASylation significantly increases the size of the protein and prevents renal clearance. In contrast to PEGylation, PASylation does not require a posttranslational chemical coupling step.

In recent years, biosimilars (also known as follow-on biologics in the United States) have gained importance as recombinant drugs. The term "biosimilar" refers to a protein-based follow-on product that is obtained by recombinant expression in living organisms following patent expiry of the innovator product. Since biologics generally exhibit higher molecular complexity than small-molecule drugs, they may be very sensitive to changes in the manufacturing process. In most cases, the follow-on manufacturers do not have access to the originator's molecular clone or fermentation and

Table 10.1 Selection of PEGylated proteins and peptides in clinical practice.

Conjugate	Trade name	Indication	Approval year
PEG-adenosine Deaminase	Adagen	Severe combined immunodeficiency (SCID)	1990
PEG-asparaginase	Oncaspar	Leukemia	1994
PEG-interferon-α2b	PegIntron	Hepatitis C infection	2000
PEG-interferon-α2a	Pegasys	Hepatitis C infection	2001
PEG-human growth hormone mutein antagonist	Somavert	Acromegaly	2002
PEG-G-CSF	Neulasta	Neutropenia	2002
PEG-erythropoietin	Mircera	Anemia associated with chronic kidney disease	2007
PEG-anti-TNF-Fab'	Cimzia	Rheumatoid arthritis and Crohn's disease	2008
PEG-uricase	Pegloticase; Krystexxa	Chronic gout	2010
PEG-erythropoietin-analog	Omontys	Anemia associated with chronic kidney disease	2012

Based on Ref. [2] with kind permission from Elsevier.

purification processes. As a consequence, biosimilars may have different properties and side effects than the original product. Biologics produced in independently generated cell lines may differ in glycosylation patterns, impurities, and breakdown products. These differences may result in significant clinical consequences. Because copies of biologics might perform differently than the original version, specific approval procedures differing from those for small-molecule generic drugs have been instituted. For biosimilars, the comparability of the similar product to the existing approved product must be demonstrated, as opposed to merely showing chemical identity as for small-molecule drugs. This process includes the evaluation of clinical data on the efficacy and safety of the biosimilar drug. The complex regulations for the approval of biosimilars have tempered the initial hype in this field. Estimates indicate that it takes 7–8 years to bring a biosimilar to the market, at a cost of $100–250 million. As a consequence, as of 2015, only 11 different biosimilar products have been approved in Europe and only one in the United States.

Another concept receiving much recent attention is the development of biobetters. A biobetter is a recombinant protein drug that is in the same class and has improved properties when compared to an existing biopharmaceutical, but is not identical to it. The biobetter has the same target as the original drug, but superior properties such as extended activity, higher efficiency, lower immunogenicity, or an improved safety profile.

These improvements can be obtained by protein engineering and/or glycoengineering.

10.1
Production of Recombinant Proteins

Recombinant proteins may be produced in many hosts (Table 10.2). Since the early 1980s, biopharmaceuticals have been obtained from *E. coli*. However, they may also be synthesized in single-cell eukaryotes, such as yeast. A growing number of proteins are produced in cultured cells from multicellular eukaryotes such as insects or mammals. The latest developments involve the production of biopharmaceuticals in transgenic organisms, particularly plants and animals that have been genetically modified to produce the desired protein.

The choice of the production organism depends on the cost, the purification strategy, and the desired properties of the product (Figure 10.3). Bacterial expression systems are the ones most widely used, since they permit relatively easy production of proteins at a comparatively low cost. However, many human proteins require posttranslational modifications; thus, eukaryotic cells must be used for heterologous expression. For example, glycosylation is an important posttranslational modification required for the functioning of many proteins, especially for membrane and secretory proteins. Since bacteria do not glycosylate proteins, they must be

Table 10.2 Example of host organisms chosen for the recombinant expression of protein drugs.

Protein	Indication	Production organism
Insulin	Diabetes mellitus	*E. coli*, *S. cerevisiae*
Interferon-α	Kidney cancer	*E. coli*
Interferon-β-1b	Multiple sclerosis	*E. coli*
Granulocyte colony-stimulating factor (G-CSF)	Neutropenia	*E. coli*, CHO
Tissue-type plasminogen activator (Reteplase)	Myocardial infarction, stroke	*E. coli*
Hepatitis B surface antigen	Hepatitis B vaccine	*S. cerevisiae*
Human papilloma virus L1 protein	Human papilloma virus vaccine	High Five insect cells, *S cerevisiae*
Erythropoietin (EPO)	Anemia	CHO, BHK
Interferon-β-1a	Multiple sclerosis	CHO
Tissue-type plasminogen activator (Alteplase, Tenecteplase)	Myocardial infarction, stroke	CHO
Glucocerebrosidase	Gaucher disease	CHO, carrot cells
Monoclonal antibodies	Various, including cancer and inflammatory diseases	CHO, BHK
Blood clotting factors	Hemophilia	CHO, BHK
Antithrombin-α	Antithrombin deficiency	Transgenic goat

Heterologous genes are commonly expressed in bacteria such as *E. coli*, yeast (usually *S. cerevisiae*), or mammalian cells such as Chinese hamster ovary (CHO) or baby hamster kidney (BHK) cells. More recently, transgenic plants (or plant cells) and animals have been created for biopharming. Protein drugs can be obtained from different sources, a fact that can influence their properties.

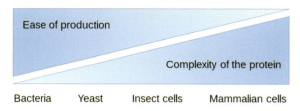

Fig. 10.3 Advantages and disadvantages of host organisms for the recombinant expression of proteins. While proteins can easily be produced in bacteria at a comparatively low cost, more complex proteins (e.g., glycosylated proteins) must be produced in more demanding eukaryotic cell expression systems.

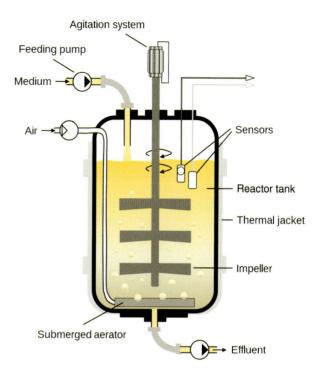

Fig. 10.4 Bioreactor. Microorganisms are usually cultivated in a sterile, closed system. Important parameters such as pH, oxygen concentration, and nutrition content are measured by sensors and required substances can be added continuously. The solution is stirred by an impeller.

expressed in eukaryotic cells, for example, yeast. However, the protein glycosylation pattern produced in yeast cells still differs significantly from that produced in humans: This may have a negative impact on the function or pharmacokinetic properties of the protein. Therefore, production of the protein in higher eukaryotic (mammalian) cells may be required. However, this is a much more demanding process than the expression of a recombinant protein in bacteria or yeast. Production organisms that have been widely used, well characterized, and for which no safety issues have been observed are categorized as "generally recognized as safe" (GRAS) species. Their large-scale use to produce recombinant drugs has been approved by the regulatory authorities in a straightforward manner.

10.1.1
Bacteria

In 1977, the mammalian peptide hormone somatostatin was successfully expressed in *E. coli*. This milestone experiment achieved the *in vitro* expression of a foreign gene in bacterial cells. Since then, *E. coli* has been widely employed for the recombinant expression of proteins. Its advantages as an expression system include the rapid growth rate of the bacterium, the capacity for continuous fermentation (the bacteria can be cultured in large bioreactors that maintain optimal conditions (Figure 10.4)), and the relatively low cost compared to other production organisms. Yields in the gram per liter range can routinely be obtained for secreted proteins.

The human gene is usually encoded on an expression vector with expression controlled by an inducible promoter. The most widely used systems are based on the T7-polymerase promoter, requiring an *E. coli* strain to express the T7 RNA polymerase originating from the T7 bacteriophage. The expression of the recombinant protein is often repressed during the exponential growth phase of the bacteria. When the suspension culture has reached a certain density, protein expression is induced while bacterial culture remains in a growth-arrested phase.

Three expression strategies can be distinguished:

1) Fusion proteins can be produced by fusing the gene-of-interest to a sequence encoding a tag. Examples of widely used tags are the glutathione-*S*-transferase (GST) tag, the maltose-binding protein (MBP), the FLAG tag (an octapeptide), and a poly-histidine tag. The tags enhance the solubility of the protein and allow its purification by affinity chromatography. Histidine tags, for example, have a high affinity for nickel ions, which can be immobilized with a chelator on a stationary phase. Proteins carrying the histidine tag will be retained on the column, while other proteins will be washed away. For product elution, imidazole can be used as a competitor to bind to the metal ion and release the tagged protein. The inclusion of sequences for site-specific proteases, such as the Tobacco Etch Virus protease, allows removal of the affinity tag from the recombinant protein.

2) The expression of secreted proteins can in itself simplify the purification procedure. For example, the protein of interest may contain a signal sequence that directs the protein-synthesizing machinery to secrete the protein into the periplasmic space. The

(a)   **(b)**

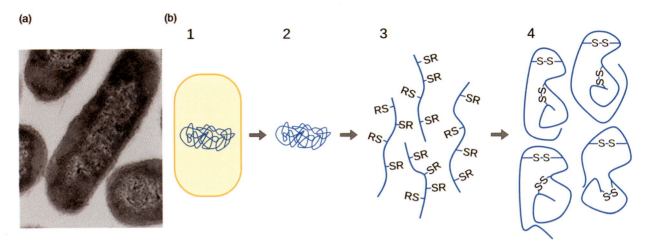

Fig. 10.5 Inclusion bodies. (a) Electron micrograph of inclusion bodies in *E. coli* cells. (b) Renaturing of recombinant proteins in inclusion bodies. (1) The recombinantly expressed proteins form an inclusion body in *E. coli*. (2) After cell lysis, the inclusion body is denatured by chaotropic reagents. (3) The solubility of the protein can be improved by reversible modification (R) of free SH groups. (4) Finally, the protein is refolded to its native conformation under controlled conditions. (a) Reprinted from Ref. [3] with kind permission from Elsevier.)

signal sequence is then removed by a bacterial protease.

3) High-level expression of proteins in *E. coli* often leads to misfolding and the formation of *inclusion bodies* (Figure 10.5). These aggregates facilitate extraction of the protein by centrifugation. However, the isolated, denatured proteins are in many cases nonfunctional and must be solubilized and refolded. This is accomplished by dissolving the protein in a solution of, for example, a chaotropic agent such as urea or guanidinium. The denaturant is then removed by dialysis or ultrafiltration and the protein is slowly allowed to renature.

A major disadvantage of *E. coli* as an expression system is its inability to introduce posttranslational modifications such as glycosylation and the inability to form disulfide bridges or to acetylate fatty acids. These modifications may influence proper folding of the protein and may thus have an effect on many of its properties, including bioactivity, solubility, stability, and protease resistance. In many cases, the formation of disulfide bonds and glycosylation are essential for the correct functioning of the protein, precluding its recombinant expression in a prokaryotic system.

Additional problems may arise because some proteins expressed in *E. coli* are also misfolded due to the absence of appropriate chaperones in the producer cell. This problem can be addressed by renaturing procedures as described above, or by the coexpression of these chaperones in the bacterial cell. Furthermore, codon usage differs between human and bacterial cells. Low copy numbers of certain bacterial tRNAs may, therefore, become limiting

for the expression of a human recombinant protein. Known solutions for this problem include the adjustment of the codons employed in the expression vector to the codons preferred by the bacterial cell, or to artificially supply tRNAs for codons that are rarely used in *E. coli*. A further problem is that the endotoxins of Gram-negative bacteria, which may remain associated with the purified protein, can produce the severe immune responses in humans that are often associated with septic shock.

In addition to *E. coli*, alternative bacterial expression systems are being developed. The soil bacterium *Bacillus subtilis* has several advantages. Being a Gram-positive bacterium, it does not produce lipopolysaccharides, which are a common contaminant of the proteins purified from *E. coli*. This bacterium can also secrete proteins directly into the medium, facilitating the purification process. Another expression system, the Gram-negative bacterium *Ralstonia eutropha*, can be used to overcome the problem of the formation of insoluble inclusion bodies at high levels of expression. However, additional knowledge about the cultivation of these microorganisms is required before they can be routinely employed for the large-scale production of recombinant proteins.

10.1.2
Yeast and Other Fungi

Yeast is a unicellular eukaryotic organism that has been employed for millennia to produce bread and ferment grains and grapes. In modern life sciences, the baker's yeast *Saccharomyces cerevisiae* has been employed for the expression of heterologous genes since the early 1980s and is well characterized and generally regarded as safe. In addition to *S. cerevisiae*, *Pichia pastoris*, and *Schizosaccharomyces pombe* are commonly used for the production of recombinant proteins. Additional yeast strains are being developed for heterologous expression. Yeast is an attractive expression organism since it combines two desirable features: As a eukaryote, yeast possesses the machinery for the secretion of proteins and can produce correctly folded proteins that have undergone posttranslational modification (particularly *N*- and *O*-glycosylation and the formation of disulfide bridges). At the same time, yeast as a microorganism is suitable for large-scale fermentation and can rapidly grow in low-cost media. Yeast cells are also easier to culture and manipulate than mammalian cells.

Foreign genes are usually transformed into yeast cells by shuttle vectors, which contain all the features necessary for replication in two different host species (*E. coli* and yeast). The gene of interest can thus be inserted into the vector *in vitro*, and the ligation product is then amplified in *E. coli*. Only after confirmation of the correctness of the construct will it be transformed into yeast. Several protein drugs, some of which are listed in Table 10.2, have been produced in yeast.

One limitation of the production of protein drugs in *S. cerevisiae* is its tendency to hyperglycosylate the proteins it synthesizes. This strain is thus not suited for the production of proteins that require a glycosylation pattern that is identical to the one in mammalian cells for their biological activity (e.g., erythropoietin). While the problem of hyperglycosylation is less pronounced in *P. pastoris*, its pattern of glycosylation differs sufficiently enough from the human pattern that some proteins need to be produced in mammalian cells, both to attain full functionality and to prevent immunologic reactions.

In addition to yeast, filamentous fungi have been exploited for the recombinant production of proteins. However, these systems are less well characterized and many challenges must be met before they can be employed for the commercial production of proteins.

10.1.3
Insect Cells

Insect cells are relatively easy to handle and manipulate and can be cultured at comparatively low cost. Proteins produced from these cells are correctly folded and many posttranslational modifications are similar to those produced by mammalian cells. Cell lines commonly used for the expression of recombinant proteins are derived from the *Spodoptera frugiperda* armyworm (e.g., Sf9 and Sf21 cells), and from the cabbage looper *Trichoplusia ni* (High Five cells). These cells can be genetically modified to express the gene of interest by transfection with a vector and subsequent selection of a stably transfected clone. However, a more commonly used procedure employs the infection of insect cell lines with recombinant baculovirus, which contains the foreign gene in its genome. Since baculoviruses are infectious only for insects and are not harmful to humans, they belong to the GRAS category. The insect cell/baculovirus system is often well suited for the recombinant production of glycoproteins; the synthesized protein is conformationally similar to the native conformation and has a proper posttranslational modification pattern. Furthermore, there is no risk of contamination by endotoxins. However, subtle differences in glycosylation patterns exist between proteins synthesized in insect and human cells, particularly in the length of the oligosaccharides and in the mannose content. These deviations from the natural product may have a negative impact on protein bioactivity and may also produce immunogenicity.

10.1.4
Mammalian Cells

Mammalian cells are employed for the expression of human proteins that require correct processing, folding, and posttranslational modification for their bioactivity. These cells usually secrete glycosylated proteins, which facilitates their harvesting and purification. Human tissue-type plasminogen activator (tPA) was the first recombinant protein produced in mammalian cells and was approved by the FDA in 1986. Since then, numerous additional protein drugs, such as erythropoietin (EPO), clotting factor VIII, follicle-stimulating hormone, and numerous monoclonal antibodies have followed. In fact, more than half of the biopharmaceuticals approved between 2006 and 2010 are produced in mammalian cells.

For the large-scale expression of proteins, stably transfected cell lines are usually employed that can be grown in suspension culture. The most commonly used host for the commercial production of therapeutic proteins are recombinant Chinese hamster ovary (CHO) cells. Although additional cell lines such as baby hamster kidney (BHK) cells and the HEK293 human embryonic kidney cell line have been developed for the recombinant expression of proteins, CHO cells account for the production of nearly 70% of all recombinant therapeutic proteins currently generated in mammalian cell lines. In particular, monoclonal antibodies are commonly produced in CHO cells. CHO cells have been used for the production of therapeutic protein drugs for more than two decades and have been demonstrated to be safe hosts. Improvements in the cultivation of CHO cells in the last two decades have helped to overcome the problem of low productivity, a major disadvantage of using mammalian cells for protein production. Employing CHO cells for gene amplification, optimizing feeding strategies, and the development of serum-free media have improved yields more than 100-fold. Furthermore, suspension cultures of CHO cells can be grown in stirred tank bioreactors with capacities of over 10 000 l. Although the glycosylation patterns of proteins produced in CHO cells differs slightly from the natural human product, these minor deviations usually do not adversely impact the bioactivity of the protein. Nevertheless, these differences are still sufficient to distinguish recombinant EPO from the naturally produced protein in doping tests, as described in Section 10.2.3.

Compared to bacterial cultures, mammalian cell cultures are more complicated and more expensive to maintain. In addition, they are susceptible to contamination, for example, with viruses. In 2009, the biotechnology company Genzyme was forced to temporarily shut down a plant after a bioreactor contamination with Vesivirus 2117. Although the virus is not known to cause human disease, it can inhibit the growth of CHO cells. During the decontamination period, the drugs Cerezyme and Fabrazyme could not be produced, which resulted in a supply shortage of these enzyme replacement therapies for rare diseases (Gaucher disease and Fabry disease, respectively). Despite these difficulties, mammalian cells are the most suitable cell types for the recombinant expression of numerous proteins.

10.1.5
Transgenic Animals and Plants

A comparatively novel approach is the use of transgenic animals and plants for the production of pharmaceutically relevant proteins, a procedure also referred to as biopharming. The production of proteins in milk is currently the most mature biopharming process, since proteins can be produced at high concentrations and purification is relatively easy. Furthermore, heterologous proteins in milk are unlikely to negatively impact the producing animal's health. Blood, egg white, and urine are also possible sources as well, but have been used to a lesser extent. Recombinant proteins can be produced from many transgenic animal species, including pigs, cows, sheep, and goats. Hamsters and rabbits have also been used because of their faster breeding, but they produce limited amounts of milk. However, glycosylation in rabbits produces carbohydrate units that resemble those incorporated into human proteins, while oligosaccharides produced by goats, cows, and sheep contain carbohydrate monomers that are virtually absent in native human proteins.

A widely used method for the generation of transgenic farm animals is the microinjection of genetic material into the male pronucleus of a fertilized egg (Figure 10.6). Since the nuclei from the oocyte and sperm must still be separate for efficient transfer of the transgene, this procedure is also known as pronuclear injection. The genetic material is typically linear DNA encoding the gene of interest. For the production of the transgene in milk, the expression of the foreign cDNA is controlled by a mammary-specific promoter, for example, the promoter of the beta-lactoglobulin (BLG) gene. Problems associated with this technology are the random integration of the injected DNA into the genome and the low efficiency of the gene transfer, which varies among species. To overcome these hurdles, methods have been developed for the transfer of the gene of interest into cultured cells and the subsequent cloning of transgenic animals by somatic cell nuclear transfer (SCNT) (Box 10.1).

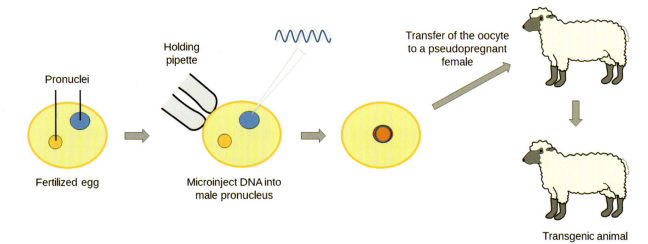

Fig. 10.6 Pronuclear microinjection. The genetic material is directly injected into the male pronucleus of a fertilized egg. Subsequently, the oocyte is implanted into a pseudopregnant female to create the transgenic animal.

Box 10.1. Dolly, Polly, and Molly

Dolly is probably the world's most famous sheep, as she was the first mammal to be cloned from an adult cell by the method of somatic cell nuclear transfer (SCNT). This procedure involves the transfer of a cell nucleus from an adult cell into a denucleated egg (Figure 10.7). The fusion cell is then stimulated to divide by an electric shock and begins to develop into an embryo. At the blastocyst stage, it is implanted into the uterus of a foster mother, where it can develop normally into a lamb.

Dolly, named after country music singer Dolly Parton, was cloned by Ian Wilmut, Keith Campbell, and colleagues at the Roslin Institute and PPL Therapeutics, a Scottish biotechnology company. The sheep was born in 1996 and later gave birth to six lambs by natural breeding. Dolly had to be euthanized because she developed a progressive lung disease and severe arthritis at the age of 6, half of the average life expectancy of the species. Whether or not her early death was connected with being a clone has been heavily debated. The chromosomes in the nucleus of the donor cell, which was used for the transfer into the denucleated egg, were 6 years old and had shortened telomeres, a typical sign of aging.

The original attempt to clone animals by SCNT was motivated by the desire to overcome problems associated with the microinjection procedure. The transgenic sheep Molly and Polly are clones that express clotting factor IX in their mammary glands. Their creation began with the introduction of the human factor IX cDNA linked to the mammary-specific BLG promoter into cultured primary fibroblasts. In addition, an antibiotic resistance gene was transferred to select cells that took up the transgene. After confirmation of successful genetic modification, embryos were generated by nuclear transfer as described above, and Polly and Molly were born. After induced lactation, both sheep began to produce factor IX in their milk.

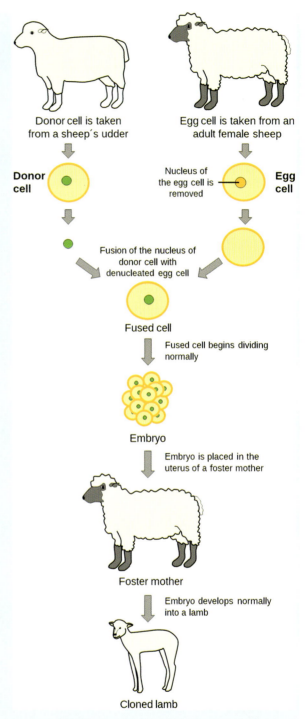

Fig. 10.7 Cloning by somatic cell nuclear transfer (SCNT). The nucleus of a somatic cell, for example, from the udder, is transferred to an egg cell, whose nucleus was removed. After stimulation with an electric shock, the fusion cell begins to divide and an embryo evolves, which is placed in the uterus of a foster mother. The embryo will develop into a lamb. It should be noted that in practice, the whole donor cell instead of the isolated nucleus is usually transferred into the denucleated egg.

In 2009, the FDA approved the first transgenic animal drug: ATryn is recombinant human anti-thrombin-α produced in the milk of transgenic goats. This anti-coagulant is indicated for individuals with hereditary antithrombin deficiency, who are at high risk of developing blood clots during surgery and childbirth. Originally, ATryn-producing goats were created by microinjection, but subsequently animals were bred by nuclear transfer to ensure that all animals were transgenics. In 2014, the FDA approved conestat alfa (trade name Ruconest) as the second protein obtained from transgenic animals. Ruconest is a recombinant version of the human C1 esterase inhibitor protein and is used for the treatment of patients with hereditary angioedema. The protein is produced in the milk of transgenic rabbits.

Plants have been a source for drugs for millennia; more than 25% of the medicines used nowadays contain ingredients derived from them. A more recent development is the production of protein therapeutics from transgenic plants. Various plant species have been used for this purpose, including tobacco, *Arabidopsis thaliana*, maize, spinach, potato, and lettuce. While the generation of transgenic plants is laborious, the large-scale production of proteins from an engineered plant is safe, simple, and economic. Plants can be cultivated en mass, which presents an opportunity to supply low-cost drugs and vaccines for developing countries. However, soil and weather conditions are variable and precise growth conditions are difficult to control in the open field. Furthermore, crops expressing pharmacologically active proteins may contaminate the food supply or cross-pollinate with conventional crops. Genetically modified plants can be grown in greenhouses to avoid these problems. Hydroponics may also be used to improve control of growth conditions.

However, these measures limit the scale of production; plant cell cultures have been developed as an alternative. Examples are carrot cell cultures and cultures of the moss *Physcomitrella patens*, which can easily be genetically modified by homologous recombination. These cultures are grown in disposable bioreactors, which permits easy upscaling. A great advantage of plant systems is the low risk of toxic contamination or of human pathogens becoming associated with the final product. However, there are differences between plant and animal glycosylation patterns, although it is possible to obtain a typical animal glycosylation pattern with genetically engineered plants. The inactivation of endogenous glycosyltransferases and the expression of heterologous glycosyltransferases can even be used to improve the bio-properties of a recombinant protein by modifying its glycosylation pattern, a process called glycoengineering.

Several classes of plant-derived pharmaceutical proteins have reached the clinical evaluation stage, especially in the production of antigens for vaccination and enzymes. Plant-based systems have also been used to produce antibodies; these are sometimes referred to as plantibodies. An example is CaroRX; this antibody binds *Streptococcus mutans*, the bacteria that cause tooth decay, and prevents its adherence to teeth. In 2012, the first plant-made biologic was approved by the FDA. Protalix BioTherapeutics, an Israeli company, produces a recombinant form of human glucocerebrosidase in carrot cell bioreactors (i.e., not in whole plants). This protein has a glycan and amino acid structure similar to its natural counterpart produced in human cells. The drug, known as taliglucerase alfa (trade name Elelyso), is used to treat patients with Gaucher disease, a rare genetic disorder in which affected individuals do not produce glucocerebrosidase, and as a consequence, accumulate lipids that damage multiple organs. Elelyso is an alternative to Genzyme's imiglucerase (trade name Cerezyme), whose production was disrupted after contamination of the producing CHO culture as described above.

The advantages and disadvantages of the different expression systems for recombinant protein drugs are summarized in Table 10.3. A steady increase in the prominence of mammalian-based over nonmamalian-based expression systems used for the production of approved biopharmaceuticals has been a long-term trend. While 25 years ago two thirds of these products were produced in nonmamalian systems, now 60% of all protein drugs are generated in mammalian expression systems. Quantitatively, however, microbial production still predominates. In terms of protein mass, approximately two thirds of all recombinant proteins are produced in microbial systems, with insulin constituting the bulk of the products.

10.2
Classes of Recombinant Drugs

Insulin was the first recombinant protein drug and was FDA approved in 1982. In the 30 ensuing years, more than 200 drugs produced by heterologous expression have been approved for the treatment of a large number of diseases. By the end of 2014, 246 biopharmaceuticals had been approved in the United States and European Union. However, only 166 of these 246 products have distinct active ingredients. In addition, 34 products had to be withdrawn after approval. The market for biopharmaceuticals reached $140 billion in 2013, accounting for approximately one fifth of the overall pharmaceutical market.

The major categories of biological drugs are monoclonal antibodies, hormones, growth factors, fusion proteins, cytokines, blood coagulation factors and thrombolytics, therapeutic enzymes, and recombinant vaccines (Table 10.4).

Table 10.3 Comparison of different prokaryotic and eukaryotic host organisms for the recombinant expression of proteins.

Host organism	Advantages	Disadvantages
Bacteria (*E. coli*)	• Well characterized • Great choice of expression vectors • High yields • Low cost and easy handling	No posttranslational modifications Formation of inclusion bodies May be immunogenic May contain bacterial impurities, for example, endotoxins derived from Gram-negative bacteria
Yeast	• No endotoxins detectable • Fermentation is well established and comparably cheap • Allows glycosylation and the formation of disulfide bonds	Glycosylation is not identical to the pattern observed in mammalian cells Protein expression is more difficult to control than in *E. coli*
Insect cells	• Posttranslational modifications (e.g., glycosylation) similar to that in mammalian cells • Biologically safe	More demanding culture conditions Mechanism of glycosylation not fully understood Function and immunogenicity may be problematic
Mammalian cells	• Highest level of posttranslational protein processing, particularly glycosylation patterns and formation of disulfide bonds	Demanding and expensive cell culture Lower yields than in microorganisms
Transgenic plants and animals	• Posttranslational modifications • Large scale production	Very complex and time-consuming generation of transgenic lines

While some of these agents are indicated for the treatment of a few hundred to a thousand patients with rare diseases, others are used for common diseases such as diabetes. Due to the large number of existing protein drugs, it will not be possible to describe them all in detail. Rather, representative examples of each category will illustrate the features that recombinant protein drugs have in common.

10.2.1
Monoclonal Antibodies

Monoclonal antibodies are the dominant class of biologic therapeutics. Between 2010 and 2014, they represented approximately one fourth of all approvals in the field. The total mAb sales of $63 billion accounted for more than 40% of the total market value of biopharmaceuticals in 2013. mAbs bind to their targets with high affinity and specificity. In contrast to polyclonal antibodies, which are made from several different immune cells, mAbs are homogeneous because they are produced by identical immune cells that are clones of a unique parent cell. The first mAb was approved for therapeutic purposes in humans by the FDA in 1986. In the years since, approximately 50 therapeutic mAbs have been marketed (for a selection, see Table 10.5). Hundreds of mAb therapeutics are currently undergoing clinical trials and many may be approved in the near future.

As outlined in greater detail in Section 1.5, antibodies are glycoproteins belonging to the immunoglobulin superfamily. They are grouped into five classes. Therapeutic mAbs typically belong to the IgG isotype (Figure 10.8b). They consist of two heavy and two light chains that are connected by disulfide bridges. In addition, glycosylation of the heavy chains is essential for

Table 10.4 Classes of recombinant protein drugs and examples.

Class of recombinant drug	Examples
Monoclonal antibodies	Anti-Her2, anti-TNF, anti-VEGF
Hormones	Insulin Glucagon-like peptide-1
Growth factors	Erythropoietin Granulocyte colony-stimulating factor
Fusion proteins	TNF-receptor-Fc
Cytokines	Interferon-α, interferon-β, and interferon-γ Interleukin-2
Blood coagulation factors, anticoagulants, and thrombolytics	Factor VIII, VIIa, IX Tissue-type plasminogen activator
Therapeutic enzymes	Glucocerebrosidase
Vaccines	Human papilloma virus L1 protein Hepatitis B surface antigen

Her2: human epidermal growth factor receptor 2; TNF: tumor necrosis factor; Fc: constant, crystallizable segment of immunoglobulin G; VEGF: vascular endothelial growth factor.

Table 10.5 Selected examples of marketed therapeutic mAbs.

Scientific name	Trade name	Origin and isotype	Target	Year of first approval	Indication	Special remarks
Muromonab-CD3	Orthoclone OKT3	Murine IgG2a	CD3	1986	Transplant rejection	First mAb ever approved
Ibritumomab tiuxetan	Zevalin	Murine IgG1	CD20	2002	Non-Hodgkin's lymphoma	Conjugated with radioactive isotope
Infliximab	Remicade	Chimeric IgG1	TNFα	1998	Autoimmune diseases (RA, Crohn's disease, etc.)	
Catumaxomab	Removab	Chimeric IgG2a/b	CD3/ EpCAM	2009	Malignant ascites	First approved bispecific mAB
Brentuximab vedotin	Adcetris	Chimeric IgG1	CD30	2011	ALCL, Hodgkin lymphoma	Antibody drug conjugate (ADC)
Trastuzumab	Herceptin	Humanized IgG1	Her2	1998	Her2 positive breast cancer	
Trastuzumab emtansine	Kadcyla	Humanized IgG1	Her2	2013	Her2 positive breast cancer	ADC of trastuzumab with a cytotoxin
Palivizumab	Synagis	Humanized IgG1	RSV F protein	1998	Prevention of RSV infection in neonates	
Ranibizumab	Lucentis	Humanized igG1	VEGF	2006	Macular degeneration	Fab fragment of Avastin
Bevacizumab	Avastin	Humanized IgG1	VEGF	2004	Colorectal, lung, breast cancers	
Certolizumab pegol	Cimzia	Humanized IgG1	TNFα	2008	RA, Crohn's disease	PEGylated Fab fragment produced in *E. coli*
Adalimumab	Humira	Human IgG1	TNFα	2002	RA, Crohn's disease, plaque psoriasis	First approved human mAb; obtained by phage display
Panitumumab	Vectibix	Human IgG2	EGFR	2007	Colorectal cancer	
Obinutuzumab	Gazyva	Humanized IgG1	CD20	2013	Chronic lymphocytic leukemia	First approved glycoengineered antibody

ADC: antibody drug conjugate; ALCL: anaplastic large-cell lymphoma; EGFR: epidermal growth factor receptor; EpCAM: epithelial cell adhesion molecule; Her2: human epidermal growth factor receptor 2; RA: rheumatoid arthritis; TNF: tumor necrosis factor; VEGF: vascular endothelial growth factor.

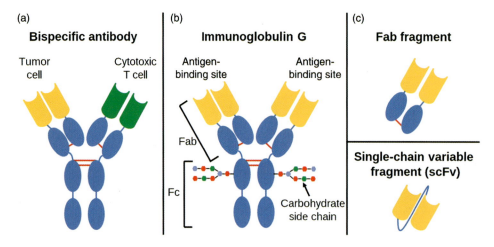

Fig. 10.8 Types of mAb therapeutics. Therapeutic mAbs are typically immunoglobulin G molecules (b) that consist of an antigen-binding fragment (Fab) and a crystallizable fragment (Fc) formed by two heavy and two light chains connected by disulfide bridges (red lines). Antibodies require carbohydrate side chains for their functionality. Bispecific antibodies (a) are variants that bind to two different antigens, for example, a tumor cell and a cytotoxic T cell. Fab fragments are connected by disulfide bridges, while single-chain variable fragment (scFv) antibodies consist of the two variable domains of the heavy and light chains, which are tethered by a short peptide linker (c).

their functionality; thus, antibodies must be produced in mammalian cells, usually CHO or BHK cells. IgGs can be separated into an antigen-binding fragment of variable sequence (Fab) and a crystallizable fragment of constant sequence (Fc). While the variable region determines the specificity of a mAb, the Fc region is important for mediating effector function, particularly complement-dependent cytotoxicity (CDC) and antibody-dependent cellular cytotoxicity (ADCC). The latter is triggered by binding of the Fc region of the antibody to Fc receptors expressed on immune effector cells, eventually resulting in the killing of a target cell.

Antibodies targeted against a given molecule can be obtained by experimental immunization of an animal. The injected antigen induces the generation of a large set of antigen-specific antibodies. However, these antibodies are polyclonal and heterogenous. Since lymphocytes do not grow continuously in culture, it is unfortunately not possible to isolate and clone a single lymphocyte from an injected animal to produce a homogeneous immuno-globulin molecule. A major breakthrough, recognized with the Nobel Prize in Physiology or Medicine in 1984, came when Georges Köhler and César Milstein developed a method to immortalize lymphocytes. Their key idea was to fuse antibody-producing B-cells to immortalized mye-loma cells that had lost their ability to secrete antibodies (Figure 10.9). This process is facilitated by the addition of PEG to promote the fusion of adjacent cell membranes. The cells were then grown on a selective medium con-taining hypoxanthine, aminopterin, and thymidine, called HAT. The aminopterin blocks the *de novo* synthesis of purines. Since unfused myeloma cells have a deficiency in the salvage pathway of nucleic acids, they cannot grow on this selection medium. Lymphocytes isolated from the mouse spleen have a limited life span; thus, only fused hybridoma cells can grow indefinitely in the medium. The mixture of cells is then separated and single-cell clones are grown in individual wells of microtiter plates and screened for the presence of antibodies to the original antigen. The selected clone can then be used for the large-scale mAb production by cultivation in bioreactors.

Initially, high hopes were placed in antibody technology to provide new therapeutic options. In 1986, the first mAb was approved by the FDA as a drug for patients: Muromonab-CD3 (trade name Orthoclone OKT3) is a murine IgG2a antibody targeting the CD3 receptor on the surface of T cells. It is indicated as an immuno-suppressant to reduce acute rejection in recipients of organ transplants. However, a major problem with the early mAbs that were developed in the 1980s is that they were produced in mice. Their murine origin had signifi-cant drawbacks, particularly the induction of antimouse antibodies, since the mAbs were recognized as an

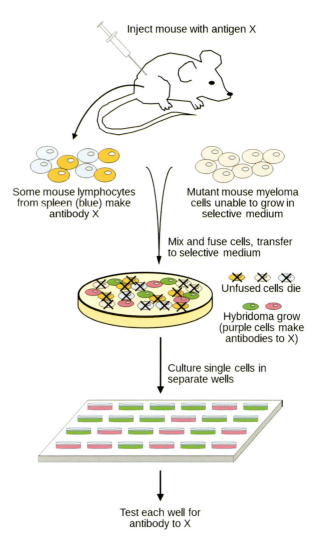

Fig. 10.9 Generation of monoclonal antibodies. The injection of antigen X into a mouse induces the generation of antibodies. Antibody-producing mouse lymphocytes are isolated from the spleen and fused to myeloma cells. Only hybridomas will grow in culture on selective media. The culture is separated into single-cell clones in separate wells and the clone that produces an antibody against the antigen is selected. (Adapted from Ref. [4] with kind permission from John Wiley & Sons, Inc.)

extrinsic antigen by the human immune system. Murine mAbs also caused allergic reactions and their circulation half-life in man was relatively short. They were also poor inducers of effector functions such as antibody-dependent cellular cytotoxicity and complement-dependent cyto-toxicity. To overcome these problems, mAbs were devel-oped by recombinant technologies to more closely resemble human antibodies (Figure 10.10).

The first step in the humanization of mABs was the grafting of the entire antigen-specific variable domain of a mouse mAb into the constant domains of a human antibody. The resulting molecules are approximately

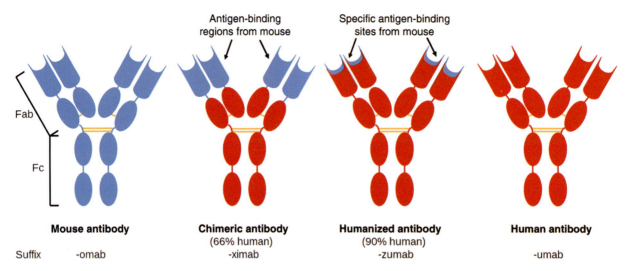

Fig. 10.10   Design of improved mAbs. The first mAbs were of murine origin and induced an immune reaction in patients. Regions of the molecule were therefore replaced by human sequences. Chimeric mAbs are composed of approximately two thirds human sequence, while in humanized mAbs only the hypervariable regions remain of murine origin. Finally, it became possible to obtain human mAbs, either by phage display or from transgenic mice expressing human antibodies. The suffix in the antibody name indicates its origin.

two-thirds human and exhibit an extended half-life in man as well as reduced, but by no means absent immunogenicity. To prevent this, variants were developed that are approximately 90% human in character. Only the hypervariable region of these so-called humanized mAbs remained of murine origin. The ultimate solution was the creation of human mAbs that can be obtained either *in vitro* by phage display technology or from transgenic

mouse strains expressing human antibodies. For the latter approach, the murine immunoglobulin genes are inactivated by a knockout strategy and human immunoglobulin genes are introduced. As a consequence, these transgenic mice generate fully human antibodies.

Phage display technology is a laboratory method for the identification, from large libraries, of proteins that interact with a given ligand (Figure 10.11). This technology is

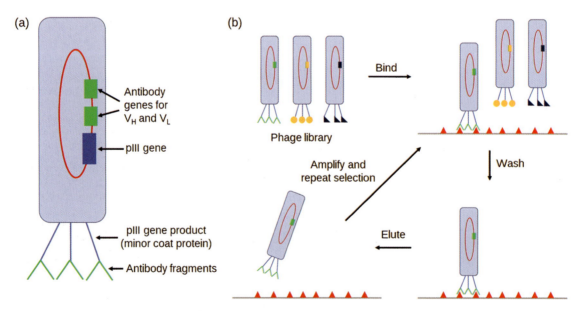

Fig. 10.11   Basic principles of phage display. (a) Antibody genes for the variable fragments are fused to a phage gene of, for example, pIII, a minor coat protein. The fusion construct is expressed and presented on the tip of the phage. (b) For selection, a library is generated that expresses the fusion proteins of the polypeptides and the phage coat protein. To identify high-affinity binders, the library is incubated with the immobilized target. Nonbinding phages are removed by washing, while bound phages are subsequently eluted and amplified in *E. coli*. After several rounds of the *in vitro* selection procedure, phages that express a polypeptide with high affinity to a given target ligand are enriched.

based on the connection of phage phenotype and geno-type. A fusion protein consisting of a polypeptide and a bacteriophage coat protein is displayed on a phage parti-cle that contains the respective encoding genes. In the first step, a library consisting of $>10^{11}$ different phages that express these fusion proteins is generated. These libraries can be amplified by passage through a bacterial host and proteins binding the target ligand can be identi-fied by *in vitro* selection procedures. For the selection of mAbs by phage display, human variable gene segments isolated from B cells of a donor are introduced into the phage genome. A widely used approach is to fuse the antibody genes to the phage gene encoding pIII, a minor coat protein. Several copies of the fusion protein are dis-played on the tip of the phage. During *in vitro* selection, antibody fragments with high affinity to the target are enriched. After several rounds of selection, antibody frag-ments binding to the target can be identified by sequenc-ing the phage genome. The selected variable region can be joined to the Fc fragment of a human antibody, or it can be used as an Fab or as a single-chain fragment varia-ble (scFv) antibody (see below). The antibody adalimumab (trade name Humira), targeted against tumor necrosis fac-tor α (TNFα), was the first approved human mAb that was obtained by phage display.

The different generations of antibodies can be distin-guished on the basis of a systematic nomenclature. Gen-erally, therapeutic mAbs are designated with the suffix "-mab." Additional letters in the suffix provide informa-tion about the similarity of the mAb to human antibod-ies (Figure 10.10).

Therapeutic mAbs can be employed for numerous indications, such as cancer and infectious and inflamma-tory diseases. Selected examples of each class will be subsequently discussed.

The humanized mAb palivizumab (trade name Syna-gis) is used for the prevention of respiratory syncytial virus (RSV) infection. This is a serious infection in pre-mature newborns and infants with other medical prob-lems such as congenital heart disease. Palivizumab binds to the F protein of RSV and inhibits its entry into the cell, thus preventing infection. Treatment with the mAb was shown to reduce the risk of hospitalization due to RSV infection. Treatment with palivizumab is recom-mended for premature newborns and infants that are otherwise at risk during the RSV season.

A very successful example of a mAb for the treatment of solid tumors is trastuzumab (trade name Herceptin), which is used to treat breast cancers (Section 4.1). This antibody is a humanized mAbs and binds to human epi-dermal growth factor receptor 2, also known as Neu (and abbreviated as Her2/neu). The receptor mediates signal-ing across the cell membrane and stimulates proliferation.

Her2/neu is overexpressed in approximately 30% of breast cancers and results in uncontrolled proliferation. Its overexpression is associated with very aggressive forms of breast cancer and with a poor prognosis. Trastuzumab binds to the receptor and leads to cell cycle arrest in the G1 phase, reducing proliferation. Since this treatment is beneficial only in Her2/neu overexpressing cancers, trastuzumab is administered to patients after analysis of tumor samples for Her2/neu amplification (Figure 9.10 in Section 9.3). This proce-dure also avoids adverse effects and ineffective, expen-sive treatment. Testing for eligibility for trastuzumab treatment is an example of personalized medicine. In appropriate cases, the mAb improves survival and reduces the risk of cancer recurrence after surgery. However, treatment with trastuzumab can often lead to the development of resistance; this problem pro-pelled the creation of a new variant of the antibody coupled to a cytotoxic molecule (see below).

A common mechanism of action of mAbs for the treatment of autoimmune and other inflammatory dis-eases is the binding to a soluble messenger molecule, for example, a cytokine, to prevent its interaction with its cellular receptor. The chimeric mAb infliximab (trade name Remicade) is directed against TNFα, a cytokine that is an important immune modulator. Inactivation of TNFα has been proven to be important in downregulat-ing inflammatory reactions associated with autoimmune diseases, including rheumatoid arthritis, psoriasis, Crohn's disease, and ulcerative colitis. Remicade was the first mAb for which biosimilar products were approved in the European Union (in 2013). However, the launch of the follow-on products Remsima and Inflectra was delayed until 2015, when patent protection of the origi-nal product expired.

Due to the problems with antibodies that contain regions of murine origin, the development of a fully human mAb against TNFα was a major advance. Adali-mumab (trade name Humira for "human monoclonal antibody in rheumatoid arthritis") was the first approved fully human mAb. It was derived from a phage display and is the best-selling biopharmaceutical worldwide, with sales of $11 billion in 2013. Humira is marketed in preloaded syringes and pen devices that are used for subcutaneous home injection by patients. Several other mAbs, as well as the fusion protein etanercept (Section 10.2.4), also bind to TNFα and are used for similar indications. A possible adverse effect of all TNFα inhibitors is the weakening of the immune response to infection, since the cytokine has important immunosti-mulatory functions.

Papain and other proteases cleave the immunoglobulin molecule into the Fc and two Fab fragments (Figure 10.8c), which maintain antigen-binding capacity. A promising variation of this approach has been the development of the scFv format. These single-chain antibodies have full antigen-binding activity, but consist of only the variable regions of the heavy (V_H) and light chain (V_L). The two domains are tethered by a flexible peptide linker, usually comprising stretches of glycine and serine residues. The advantages of scFv antibodies include the possibility of discovery by page display and production in bacterial cells. Due to their small size, both Fab fragments and scFv antibodies penetrate tumors better than the intact immunoglobulin molecule. They are also more rapidly cleared from the blood and have a shorter retention time in nontargeted tissue and reduced immunogenicity. These "downsized" antibodies can also be designed as bispecific molecules, and can

deliver a cargo such as a cytotoxin to a specific cell type (see below). An additional reduction in size is achieved with so-called *nanobodies*; these molecules have been isolated from camels and related animals, such as dromedaries, llamas, and alpacas (collectively known as camelids) (Box 10.2).

An example of an approved Fab fragment in clinical use is ranibizumab (trade name Lucentis), which is derived from the parent antibody bevacizumab (trade name Avastin); both are directed against VEGF-A. The affinity of the Fab fragment has been increased by *in vitro* evolution to achieve tighter binding to its target, VEGF; the molecule thus has a pronounced antiangiogenic effect. Bevacizumab is licensed to treat various metastatic cancers, while ranibizumab has been approved for the treatment of the wet form of age-related macular degeneration (AMD), a major

Box 10.2. Nanobodies

In addition to conventional heterotetrameric antibodies composed of two light and two heavy chains, the sera of camelids contain antibodies that are devoid of the light chains (Figure 10.12). Furthermore, instead of three constant domains and a dedicated variable domain, the heavy chains of these camelid antibodies are composed of only two constant domains and a variable domain referred to as a V_{HH} domain (variable domain of the H chain of heavy-chain antibodies). The V_{HH} domain alone possesses antigen-binding capacity, and can easily be identified by cloning the repertoire of B cells circulating in blood after a camelid is immunized. Due to its small size (in the nanometer range) it was named a nanobody. The camelid V_{HH} domains share a high degree of identity with human V_H domains and do not induce a marked immune response in humans. Several amino acids that differ between the human V_H and camelid V_{HH} domains can be replaced to further humanize the nanobodies.

Compared to conventional mAbs, nanobodies have several advantages: For example, they are encoded by single genes and can efficiently be produced in bacteria and yeast. Furthermore, nanobodies are very stable and do not lose their antigen-binding capacity at room temperature over long periods of time. Another important advantage is their small size; they may reach hidden epitopes of pathogens and can penetrate tumor tissue, which is inaccessible to larger antibodies. Several nanobodies have been evaluated in clinical studies. An example is the nanobody caplacizumab that binds the von Willebrand factor and has antithrombotic activity. This antibody was developed

to treat patients with a rare disease named thrombotic thrombocytopenic purpura (TTP), which is characterized by the accumulation of ultra-large multimers of the von Willebrand factor in the plasma that may result in the formation of life-threatening blood clots. Nanobodies against IL-6R and TNFα have also been used to treat inflammatory diseases.

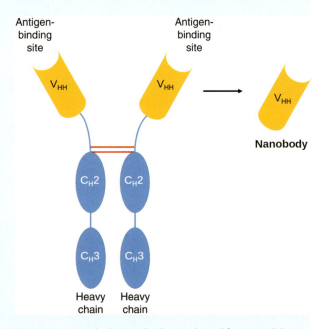

Fig. 10.12 Nanobody. Nanobodies are derived from camelid antibodies that consist of heavy chains only. The V_{HH} domain called a nanobody is the smallest functional antigen-binding fragment.

reason for vision loss in older adults. AMD is caused by abnormal blood vessel growth and can be prevented by antiangiogenic treatment. However, the cost of the treatment has been the subject of a long-lasting debate. Lucentis (the Fab fragment) is expensive, costing approximately 50 times more than Avastin for the treatment of AMD, even though the off-label use of Avastin seems to be equally effective as Lucentis. However, it is uncertain if the division of Avastin into smaller doses for the treatment of wet AMD is associated with an increased risk of infection.

In 2008, the first Fab fragment produced in *E. coli* was approved for the treatment of Crohn's disease and rheumatoid arthritis: Certolizumab pegol (trade name Cimzia) is a TNFα-binding mAb Fab fragment that is also PEGylated to prevent rapid clearance of this comparatively small molecule and to extend its circulation time.

Another type of mAb with potential in cancer therapy is the bispecific antibodie class. These are composed of fragments of two monoclonal antibodies and bind to two different antigens (Figure 10.8a). Bispecific mAbs bring cytotoxic effector cells into proximity with those tumor cells that present a specific antigen on their surface. The first-generation bispecific mAbs were composed of two light and two heavy chains like a typical IgG molecule. However, the two Fab fragments bind to two different antigens, while the Fc region forms a third binding site for cells with an Fc receptor. Bispecific mAbs are thus trifunctional. In 2009, the first bispecific mAb was approved. Catumaxomab (trade name Removab) is indicated for the treatment of malignant ascites in patients with human epithelial cell adhesion molecule (EpCAM)-positive carcinomas. The antibody simultaneously binds EpCAM on tumor cells and CD3 on cytotoxic T cells, bringing them in close contact to facilitate killing of the tumor cells. Furthermore, the Fc region also induces ADCC. In addition to the first-generation bispecific mAbs, other smaller molecules are being developed; these include chemically linked bispecifics that consist of only two Fab fragments. An example is Amgen's immunotherapy drug blinatumomab (trade name Blincyto) that was approved by the FDA in late 2014 for the treatment of B-cell acute lymphoblastic leukemia (B-cell ALL). Blinatumomab is a single-chain bispecific antibody that consists of two variable fragments that bind to CD3 on T cells and CD 19 on the targeted B cells. The drug works by linking the two cell types and activating the T cell to exert its cytotoxic activity on the malignant B cells.

Further applications of mAbs include their use for the delivery of a cargo; this, for example, includes the combination of an mAb linked to a potent cytotoxic agent that can bind to a tumor-specific antigen. These linked molecules are called antibody–drug conjugates (ADC). The chemical moiety is attached to the mAb using a linker that maintains stability in the circulation and releases the cytotoxin when the mAb is bound to or internalized by the cancer cell. The linker used for the conjugation may either be cleavable or noncleavable. To avoid toxic side effects, the cytotoxic substance must remain bound to the mAb until internalization into the cell. In 2011, the ADC brentuximab vedotin (trade name Adcetris) was approved by the FDA for the treatment of Hodgkin lymphoma and anaplastic large-cell lymphoma. The antibody targets the cell surface protein CD30 and is linked to three to five molecules of the antimitotic agent monomethyl auristatin E. In clinical trials, brentuximab vedotin produced high rates of tumor shrinkage and, in some cases, complete remissions. Another example of an ADC is trastuzumab emtansine (trade name Kadcyla), in which the mAb trastuzumab is covalently linked to the cytotoxic agent maytansine. This drug is used to treat patients with breast cancer who have become resistant to trastuzumab. The mAb mediates specific delivery of the cytotoxin to Her2/neu overexpressing cells. The maytansine then enters the cells and destroys them by binding to tubulin. Treatment with the cytotoxic agent alone is limited by toxicity, but the safety profile, as well as progression-free and overall survival of the treated patients is improved versus maytansine alone.

In addition to cytotoxic agents, mAbs can also be linked to radioactive isotopes. An example is ibritumomab tiuxetan (trade name Zevalin), a murine mAb used for the treatment of B cell non-Hodgkin's lymphoma. The mAb is congugated with a chelator that binds a radioactive isotope (either yttrium-90 or indium-111). The conjugated mAb targets the CD20 antigen found on the surface of normal and malignant B cells (but not B cell precursors). The bound radioisotope then irradiates and kills the cancer cell. Comparative studies have shown that treatment with ibritumomab tiuxetan was superior to treatment with a comparable mAb targeted against CD20 without the linked radioisotope. A more recent application of mAbs involving the cell-type specific delivery of small interfering RNAs is outlined in Section 13.3.3.

10.2.2
Hormones

The peptide hormone insulin was the first recombinant drug approved for therapeutic purposes in humans. Insulin is a peptide hormone that is central to regulating an organism's carbohydrate and fat metabolism. It is produced by β-cells of the pancreas and causes the cells in the liver, skeletal muscles, and fat tissue to take up glucose from the blood. Endogenous insulin is released into the circulation in response to elevated blood glucose levels after food intake (prandial insulin). In addition, basal insulin is continuously released at low levels. The body's failure to produce sufficient amounts of insulin or an inadequate response of the target cells to insulin results in elevated levels of glucose in the blood, known as *diabetes mellitus* (Section 3.2). This disease affects hundreds of millions of people worldwide. Diabetes mellitus increases the risk of long-term damage of the blood vessels, which can lead to cardiovascular diseases such as ischemic heart disease, stroke, and peripheral vascular disease. Furthermore, damage to the capillaries can cause disease of the eye, kidney, and nervous system. Diabetes mellitus is the third leading cause of death in the United States.

There are two major types of diabetes mellitus, insulin-dependent and non-insulin-dependent. The insulin-dependent form is also known as juvenile-onset or type 1 diabetes mellitus (T1DM), since it often strikes in childhood. It is caused by the inability of pancreatic β cells to produce sufficient insulin. This disease is usually treated by insulin injection. In contrast, non-insulin-dependent type 2 diabetes mellitus (T2DM) usually develops gradually at an older age. It may be caused by a deficiency of insulin receptors or of insulin receptor signal transduction in addition to decreased insulin production.

Mature insulin consists of two polypeptide chains, the A- and B-chains, which are linked together by disulfide bonds (Figure 10.13). Initially, a single polypeptide called preproinsulin is synthesized and translocated into the endoplasmic reticulum. The signal peptide is cleaved, forming proinsulin. After its transfer to the Golgi apparatus, proinsulin undergoes maturation into active insulin through the activity of cellular peptidases, which release the central fragment, known as the C-peptide.

For decades, insulin obtained from cow and pig pancreases was used for the treatment of individuals having diabetes mellitus. Since insulin from these sources differs from human insulin in only a few amino acids (one amino acid in porcine and three amino acids in bovine insulin), it is effective in humans. However, these differences, as well as the impurities in the product, may produce allergic reactions. Recombinant DNA technology permits expression of "human" insulin in bacteria. The pharmaceutical company Eli Lilly first marketed insulin produced in *E. coli* in the early 1980s.

While the initial aim of recombinant technology was to produce insulin with the identical amino acid sequence as human insulin, a number of insulin analogs with improved properties have also been developed. Recombinantly produced human insulin tends to form hexamers that must dissociate into dimers and monomers to be biologically active. The time needed for the dissociation delays the onset of its activity. Several muteins have been developed to achieve a more rapid dissociation and onset of action.

In 1996, the first insulin analog, insulin lispro, became commercially available. In this mutein, the lysine and proline residues B29 and B28, found on the C-terminus of the B-chain, are reversed. Since B28 proline is critical for the formation of hexamers, this alteration results in a more rapidly dissociation and onset of action. Other analogs belonging to the class of rapid-acting insulin variants are insulin aspart and insulin glulisine. In insulin aspart, the B28 proline is replaced by aspartic acid, while insulin glulisine contains two mutations, lysine

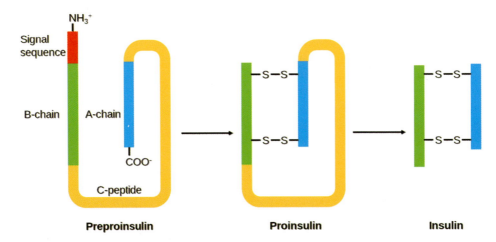

Fig. 10.13 Biosynthesis of insulin. The first step in insulin biosynthesis is the translation of a single polypeptide (preproinsulin) that undergoes several maturation steps. The active form of insulin consists of two chains, which are connected by disulfide bridges.

instead of asparagine at B3 and glutamate instead of lysine at B29. These mutations also prevent the formation of hexamers. These rapid-acting insulin variants provide prandial coverage with an onset of action of 15 min or less. They are active for up to 3 h, compared to regular human insulin, which has a 30 min onset of action and remains active for about 8 h.

Intermediate-acting insulin NPH (or neutral protamine Hagedorn) was developed to limit the number of injections required to achieve glycemic control. It has an onset of action of 3–4 h and a duration of activity of about 15 h. In NPH, the small, positively charged protein protamine is added to regular human insulin to stabilize the hexameric structure, resulting in a longer duration of action. Additional insulin analogs with even longer activity attempt to mimic basal insulin release. Insulin glargine differs from human insulin by three amino acids, producing a more stable hexamer structure. Once injected, insulin glargine forms a microcrystalline precipitate resulting in a depot form of the drug not readily absorbed into the circulation. Insulin detemir is synthesized in yeast (*S. cerevisiae*). Its prolonged duration of action is achieved by removal of threonine from the B30 position and the binding of myristic acid, a fatty acid, to B29 lysine. Long-lasting insulin preparations have a duration of action of up to 24 h. New, ultra-long acting analogs are being developed that have a duration of action of up to nearly 2 days, reducing the subcutaneous injection frequency to three times a week. Different insulin variants can be combined so that the injection mixture can have both a rapid onset and a long duration of activity.

A major problem with current diabetes treatment is insulin injection discomfort. In 2006, the first inhaled insulin product, Exubera, was approved, but it was taken off the market within a year due to problems with the inhalation device and concerns over the potential for compromised lung function and an increased risk of lung cancer. In 2014, a new inhaled insulin product, Afrezza, was approved that uses a much improved inhaler device. Development of alternative delivery routes is a general challenge in the field, as parenteral administration remains the mainstay of most of the approved drugs of this class. Several companies have initiated programs to develop oral insulin (and other therapeutic proteins that can be administered orally).

Another category of biopharmaceutical for the treatment of diabetes mellitus is glucagon-like peptide-1 (GLP-1) and its analogs. These peptides have multiple modes of action. Their known physiological functions include increasing insulin secretion from the pancreas in combination with decreasing glucagon secretion and increasing insulin sensitivity of its target cells.

10.2.3
Growth Factors

Approximately a dozen growth factors produced by recombinant expression are available in the United States, most of them colony-stimulating factors (CSF) that fall into the category of erythropoiesis-stimulating agents (ESAs) or granulocyte colony-stimulating factors (G-CSFs). CSFs are secreted glycoproteins that bind to receptors on the surface of hematopoietic stem cells and induce their proliferation and differentiation into a specific type of blood cell.

ESAs stimulate red blood cell production, a process known as erythropoiesis. They are used for the treatment of anemia resulting from chronic kidney disease or cancer treatment. Erythropoietin (EPO) is the most widely used ESA. In the human body, EPO is primarily produced in the kidney under hypoxic conditions. An important feature of the hormone is its extensive glycosylation. Alterations of the glycosylation pattern by recombinant technologies permit an increase in EPO's plasma stability, reducing the frequency of injection. EPO binds to the erythropoietin receptor on the surface of red cell progenitor cells and induces their maturation.

While insulin was obtained from natural sources prior to its biotechnological production, EPO could only be produced after cloning and expression of the EPO gene by the methods of genetic engineering. All available recombinant EPO products differ from the endogenous human protein in their glycosylation pattern. The different EPO variants are grouped according to their composition and denoted as "Epoetin" followed by a Greek letter. Since glycosylation is essential for the function of recombinant EPO, it is produced in CHO, BHK, or other mammalian cells.

The first EPO product, Epogen, was marketed by Amgen in 1989 (also marketed by other companies under the trade name Procrit within and Eprex and other names outside the United States). This drug belongs to the Epoetin-α class. Treatment of cancer and dialysis patients with recombinant EPO was initially demonstrated to produce fewer side effects than standard blood transfusion therapy. Soon after, an Epoetin-β variant was developed, which has a higher molecular mass and a different glycosylation (sialylation) pattern compared to the α-form. Epoetin-β was reported to have a higher bioactivity than Epoetin-α, at least in animal models.

Due to the success of EPO, additional variants with improved pharmacological properties have been developed by protein engineering. A major aim was to increase the molecular size in order to extend serum half-life and duration of activity. Darbepoetin-α is a genetically modified erythropoietin that contains additional oligosaccharide

Box 10.3.   Erythropoiesis-Stimulating Agents in the Doping of Athletes

Stimulation of erythropoiesis is one of the most widely used methods of doping and has been used by endurance athletes. Drugs that were developed to treat patients suffering from anemia are often misused to enhance physical performance. EPO was among the first protein drugs obtained by recombinant expression. It was added to the list of forbidden substances of the World Anti-Doping Agency (WADA) in 1990, but it took another 10 years until adequate methods for the detection of EPO were available. Doping with erythropoiesis-stimulating agents in sports is not only considered unethical but it may also have severe consequences for athletes. Higher levels of red blood cells reduce blood flow due to increased viscosity, which may lead to thrombosis and stroke.

To control doping, detection of EPO in the urine must differentiate between the endogenous hormone and the recombinant protein used for the doping. While both forms of EPO are identical in their amino acid sequence, they differ in their posttranslational modifications. In particular, the glycosylation pattern of endogenous human EPO generated in the kidney differs from that of the recombinant protein produced in cultured CHO cells originating from hamsters. Highly sensitive methods have been developed that detect the low levels of EPO in urine. The most widely used protocol is composed of a three-step detection procedure (Figure 10.14). First, the urine is concentrated by ultrafiltration, followed by isoelectric focusing (IEF), a method that separates proteins according to their isoelectric points in a polyacrylamide gel that contains a pH gradient. Due to differences in their glycosylation patterns, natural and recombinant EPO migrate differently in these gels. In the third step, EPO is detected by immunoblotting with a specific mAb. To overcome problems with the nonspecific binding of the secondary antibody used for the immunodetection of EPO, a secondary blotting step under acidic conditions is required in which only the mAbs, but no other proteins are transferred. Finally, the anti-EPO mAbs are detected with a secondary antibody. This procedure results in different observed patterns for natural and recombinant EPO. As new EPO variants are continuously being developed and in some cases even used for doping prior to approval for medical purposes by the regulatory authorities, improved methods for the detection of erythropoiesis-stimulating agents are also regularly required for antidoping control.

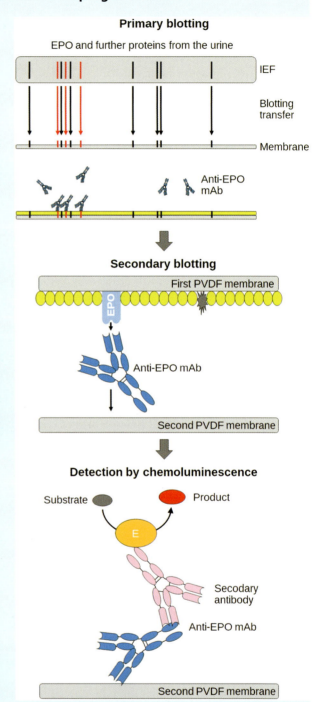

Fig. 10.14   Immunodetection of EPO. Concentrated proteins from the urine are separated by isoelectric focusing (IEF) and transferred to a membrane by blotting. A specific monoclonal antibody (mAb) binds to EPO. In a secondary blotting step, only the mAb is transferred to a second membrane and detected with a secondary antibody.

chains as a consequence of the substitution of five amino acids. The first approved chemically modified ESA, Mircera, is a long-acting agent whose size was increased by the addition of polyethylene glycol (PEG). Further EPO variants are currently being developed. Some are fusion proteins (see Section 10.2.4), while others are EPOs produced in cultured human cells rather than in animal cells. Since EPO was one of the first recombinant drugs to be marketed, key intellectual property rights have expired. Epoetin alfa Hexal was the first biosimilar on the European market.

For many years, EPO was the most commercially important recombinant drug. However, sales have fallen dramatically due to a decline in prescriptions, new financial arrangements with dialysis centers, the development of competing drugs, and new evidence of severe side effects. Studies have shown that for unclear reasons, EPO increases mortality in cancer patients. Furthermore, EPO is not only an important drug to treat anemia but also has a history of misuse as blood doping agent in endurance sports such as cycling, long-distance running, and cross-country skiing (Box 10.3).

The second major class of recombinant growth factors, G-CSF, stimulates the proliferation and maturation of precursor cells into granulocytes. Human G-CSF is a glycoprotein consisting of 174 amino acids. Many cancer patients suffer from chemotherapy-induced neutropenia, characterized by a depressed number of circulating neutrophils, the cells that serve as a primary defense against infection. Recombinant G-CSF is effective in chemotherapy patients because it increases the number of circulating neutrophils. It is also used to mobilize hematopoietic stem cells from the bone marrow into the peripheral blood for use in hematopoietic stem cell transplantation before collection by leukapheresis (Section 12.6.1).

G-CSF (trade name Filgrastim) was initially produced in *E. coli*. Therefore, it lacks the oligosaccharide chains present in the endogenous human protein. A variant of this protein is PEGylated (trade name Pegfilgrastim) and has a longer half-life, eliminating the need for daily injection. The newer product Lenograstim is produced in CHO cells to ensure glycosylation of threonine 133 as in natural human G-CSF. A number of biosimilar CSFs have also been launched in recent years.

10.2.4
Fusion Proteins

Fusion proteins or chimeric protein drugs are constructed from two different parental proteins, ideally to combine the best properties of both components. In a broader sense, chimeric antibodies are fusion proteins,

since the antigen-binding domains derived from mice are fused to a human antibody. Specifically, the term fusion protein refers to chimeric proteins composed of two completely unrelated proteins. Commonly, a soluble extracellular receptor domain is fused to the Fc segment of immunoglobulin G1 (IgG1). The receptor domain provides specificity for the drug target, while the Fc segment adds to the stability and ability to deliver the active domain. Many biologically active proteins have short serum half-lives due to rapid renal clearance, a problem that can be solved by fusing them to the Fc domain. Furthermore, Fc-fusion proteins frequently permit easier protein expression and enable protein A-affinity purification. Since the Fc domain contains disulfide bridges as well as posttranslational glycosylation, this type of fusion protein is commonly produced in CHO cells.

The most widely used chimeric protein drug is etanercept (brand name Enbrel) (Figure 10.15), a fusion protein composed of the extracellular domain of the human p75 TNFα receptor and the Fc segment of human IgG1. The cytokine TNFα is a major regulator of inflammation. Similar to the anti TNFα mAbs described in Section 10.2.1, etanercept acts as a decoy that binds TNFα,

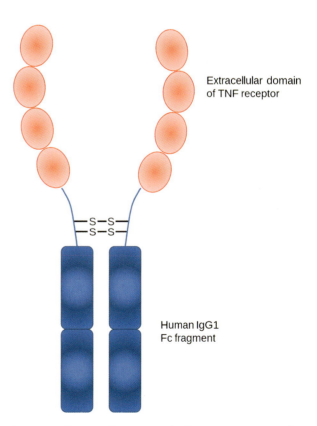

Extracellular domain of TNF receptor

Human IgG1 Fc fragment

Fig. 10.15 Structure of etanercept. The fusion protein consists of the extracellular domain of the human TNFα receptor and the Fc portion of human IgG1. It acts as a decoy to bind TNFα.

preventing it from binding to its cellular receptor. Inhibition of TNFα activity by etanercept reduces the inflammatory response and is useful for the treatment of autoimmune diseases. The drug is approved for rheumatoid arthritis, psoriasis, and other inflammatory diseases.

Aflibercept (brand name Elyea) is a fusion protein consisting of the extracellular domains of the VEGF receptors 1 and 2 and the Fc portion of IgG1. It is approved for the treatment of neovascular AMD and is also being evaluated in late-stage trials for diabetic macular edema. The drug, which was approved in 2011, has become a competitor for other VEGF binders such as Lucentis, the antigen-binding fragment of anti-VEGF antibody (see above).

Another example of an approved fusion protein is belatacept (brand name Nulojix), which is composed of the extracellular domain of the cytotoxic T-lymphocyte antigen 4 fused to an Fc segment. This chimeric protein binds to CD molecules on the cell surface, thus preventing T cell stimulation. Nulojix is indicated for the prophylaxis of organ rejection in adult kidney transplant recipients. As of 2012, approximately a dozen fusion proteins have been introduced to the market.

10.2.5
Cytokines

Cytokines are a large family of proteins that are secreted by cells and act as intercellular signaling molecules. They regulate the growth and differentiation of cells and often have an immunomodulatory function. Four varieties of cytokine are currently available as recombinant drugs: interferon (IFN)-α, IFN-β, IFN-γ and the interleukins. IFNs are a major class of cytokine and were named after their ability to "interfere" with viral replication. In addition, they can activate immune cells such as natural killer cells and macrophages. IFNs thus not only have antiviral activity but they also help the body to recognize and destroy tumor cells.

Activating the immune system and fighting viral infections has made IFNs attractive candidates as antiviral agents. A combination of IFN-α and a low-molecular-weight antiviral drug, usually ribavirin, has long been the standard treatment for hepatitis B and hepatitis C infection (Sections 5.4.4 and 5.4.2, respectively). While conventional IFN-α must be administered three times per week, a single weekly injection of PEGylated IFN-α is therapeutically sufficient due to its extended duration of action. However, only ~60% of the patients respond to treatment with combination therapy, particularly those infected with HCV genotype 1. IFN-α and IFN-γ both are also used for the treatment of various forms of cancer.

The first recombinant IFN-β product, IFN-β 1b (trade names Betaseron in North America and Betaferon elsewhere), was approved by the FDA in 1993 for the treatment of relapsing remitting multiple sclerosis. MS is an inflammatory disease in which the myelin sheath around the axons of the brain and spinal cord are damaged. Since myelin insulates axons, the resulting demyelination and scarring affects the ability of the nerve cells to conduct action potentials. Furthermore, the blood–brain barrier (BBB) is frequently impaired in patients with MS. The damage is caused by the body's own immune system, which attacks and destroys the myelin sheath. The causes of the disease are still uncertain, but seem to involve both genetic predisposition and viral infection. MS is associated with a variety of neurological symptoms, can lead to physical and cognitive disability, and may reduce life expectancy. Multiple variants of the disease are known, with new symptoms occurring either in discrete attacks (relapsing disease) or as disease that progresses slowly over time (progressive disease).

Treatment of MS patients with IFN-β reduces relapse rates by about one third and the appearance of new MRI lesions by about two thirds. Furthermore, IFN-β was shown to reduce disability progression and the rate of brain atrophy. The mechanism of action of IFN-β in the treatment of MS is not fully understood. IFN-β binds to a heterodimeric cell surface receptor, which sends the signal through distinct JAK/STAT phosphorylation cascades resulting in the transcriptional regulation of more than 1000 genes. It is believed that IFN-β slows MS progression by its anti-inflammatory properties. It probably also improves the integrity of the blood–brain barrier. However, as with other protein treatments, IFN-β may induce the formation of antibodies in some patients. This immune response will reduce the efficacy of further treatment.

Natural human IFN-β is a small protein that is glycosylated at a single site. Interestingly, this carbohydrate chain is important for the stability of the protein and its solubility and biodistribution, but it is not required for receptor binding. It was thus possible to use *E. coli* to express IFN-β-1b that is functional despite its lack of the carbohydrate moiety. The IFN-β-1b variant also contains an amino acid substitution that minimizes erroneous disulfide bridge formation. More recently approved IFN-β-1a variants are produced in CHO cells. They contain the wild-type human amino acid sequence and a carbohydrate moiety very similar to that found in the natural product. The IFN-β-1a drugs on the market (brand names Rebif, Avonex) differ mainly in formulation. IFN-β-1a and IFN-β-1b have very similar drug profiles and similar efficacy at the approved doses. In some countries, IFN-β biosimilar products are available. Furthermore, a PEGylated IFN-β is being developed that has a longer half-life, reducing the dosing frequency.

Interleukins are a subclass of cytokine. The term interleukin is derived from their major function, which is to mediate communication between (lat. *inter*) white blood cells (*leuko*cytes). Recombinant interleukins can be used as therapeutic agents to activate the immune system against cancer cells (or viral infection). They are usually given in combination with additional anticancer drugs, for example, chemotherapeutic agents. A recombinant variant of interleukin-2, aldesleukin, is used to treat kidney cancer and melanoma. This drug contains mutations at several positions of the amino acid sequence and is not glycosylated since it is produced in *E. coli*. However, the use of aldesleukin has decreased more recently due to toxic side effects and the development of more active agents.

10.2.6
Blood Coagulation Factors: Anticoagulants and Thrombolytics

The process by which blood forms clots is known as coagulation or thrombogenesis and is essential for stopping blood loss from damaged vessels. However, the coagulation system must always be in balance. Enhanced blood clotting can lead to thrombus formation, possibly resulting in myocardial infarction or stroke. Several genetic disorders that can increase either the risk of bleeding (hemorrhage) or clotting (thrombosis) are known. Protein drugs are available to treat affected individuals.

Damage of blood vessels induces the coagulation cascade, which leads to the formation of an insoluble fibrin network that stops further blood loss. Classically, the coagulation cascade is divided into three pathways: the intrinsic and the extrinsic pathway, both merging into the final, common, pathway (Figure 10.16). Most of the coagulation factors are proteases, which act by cleaving and activating downstream factors. The components of the coagulation cascade are assigned Roman numerals; the suffix "a" denotes an active factor.

The extrinsic pathway is initiated when a membrane protein called tissue factor is exposed to the bloodstream by tissue damage and forms a complex with circulating factor VII or VIIa. This complex converts factor

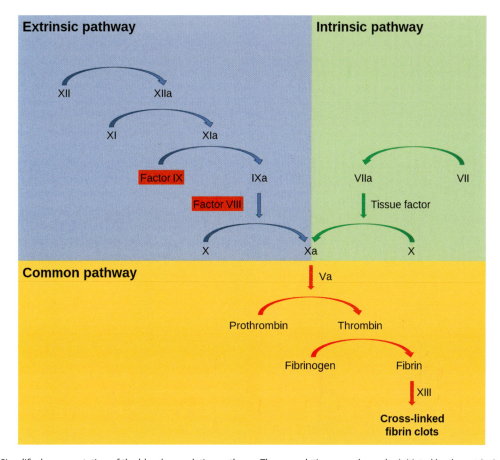

Fig. 10.16 Simplified representation of the blood coagulation pathway. The coagulation cascade can be initiated by the extrinsic or intrinsic pathway, which come together in a common pathway. The two major forms of coagulopathy result from congenital defects in factor VIII (hemophilia A) or factor IX (hemophilia B).

X to the active enzyme factor Xa, which converts pro-thrombin to thrombin. The activated thrombin cleaves fibrinogen to form fibrin, which is finally cross-linked by factor XIII. The intrinsic pathway is initiated by the activation of factor XII upon contact with anomalous surface structures as a result of endothelial damage. This initiates a cascade also known as the contact activation pathway; it leads to the cleavage of prothrombin in the common pathway as described above. More recent research has revealed that the interactions between the coagulation factors are more complex than the simple schematic representation suggests.

Hemophilia is a group of hereditary genetic disorders that impair the body's ability to control blood clotting. The most common forms are hemophilia A and B, which are caused by a deficiency of factors VIII and IX, respectively. Since both forms of hemophilia are recessive X-linked genetic disorders, the disease occurs predominantly in males. The affected individuals suffer from abnormal bleeding. Prior to the development of modern treatments, many hemophilics died before they reached adulthood. While the life expectancy was increased to 50–60 years by the 1980s, today hemophilics receiving adequate treatment have a near normal quality of life, although their average lifespan is still about 10 years less than that of unaffected males.

Management of hemophilia involves regular infusions of the deficient coagulation factor, that is, factor VIII for hemophilia A or factor IX for hemophilia B. Two categories of standard care, prophylactic and on-demand, can be distinguished. Prophylactic care involves the infusion of the deficient coagulation factor on a regular schedule to keep levels sufficiently high enough to prevent spontaneous bleeding. Additional on-demand treatment is given when needed, for example, directly after trauma or prior to surgery.

For many years, hemophilics received pooled plasma or isolated factors from human blood serum. Although this treatment improved disease management, the use of blood-derived products was associated with a high risk of contracting blood-borne viruses, especially HIV or HCV. Almost 10 000 hemophilics in the United States (almost half of the total number) were estimated to have become infected with HIV after receiving tainted clotting products, and an even higher proportion was infected with HCV. When it became possible to produce recombinant coagulation factors by heterologous expression, safety dramatically improved.

Producing the deficient clotting factor by heterologous expression is the most straightforward way to obtain the drug. Recombinant human factor VIII for the treatment of hemophilia A became available in the early 1990s. Since factor VIII is a large protein with multiple glycosylations and other posttranslational modifications required for full activity, it can only be produced in mammalian cell culture systems (e.g., CHO or BHK cells). Interestingly, a mutein of factor VIII (moroctocog alfa, brand name ReFacto) is more stable due to the deletion of one of the protein domains. In the late 1990s, recombinant factor IX became available. Like factor VIII, the major factor IX product (nonacog alpha, brand name BeneFIX) is also produced in CHO cells. In 2014, the first recombinant factor XIII was approved in the United States for patients with congenital factor XIIIA-subunit deficiency, a rare autosomal recessive clotting disorder.

Though treatment of hemophilics with coagulation factors has been considered safe and successful, some patients develop inhibitory antibodies to the transfused clotting factor, reducing treatment efficacy. One way to resolve this problem is to bypass the factor VIII/IX complex of the coagulation cascade by treating the patient with activated factor VIIa. This recombinant enzyme (eptacog alpha, brand name Novoseven, and its biosimilar follow-on product AryoSeven) induces coagulation by direct proteolytic activation of factor X.

The new generation of blood coagulation factors has a longer half-life that reduces the injection frequency for patients. Prolonged activity is achieved by either fusing the factor to an Fc fragment (see Section 10.2.4), coupling it to albumin, or by PEGylation.

In contrast to coagulation factors, anticoagulant and thrombolytic drugs are used to prevent clotting and to dissolve blood clots, respectively. Several of these blood thinners have been used for decades; these include warfarin, a vitamin K antagonist, and heparin, a highly sulfated glycosaminoglycan. A more recently developed drug is the antithrombin protein itself (brand name ATryn), which is produced in the milk of genetically modified goats (Section 10.1.5). The recombinant protein has been approved for the thromboprophylaxis of individuals with hereditary antithrombin deficiency before, during, or after surgery or childbirth.

Thrombolytic drugs dissolve blood clots and are used for the treatment of myocardial infarction, stroke, thrombosis, and pulmonary embolism. The aim of treatment is to clear blocked arteries and prevent permanent tissue damage. Optimal treatment time is at the earliest possible point after the onset of ischemic symptoms. The most prominent representative of this class of drugs is tissue-type plasminogen activator (tPA), a serine protease that catalyzes the conversion of plasminogen to plasmin (Figure 10.17a). After this activation step, plasmin lyses clots by degrading fibrin.

Alteplase was the first recombinant tPA on the market and is identical to native human tPA. The drug is

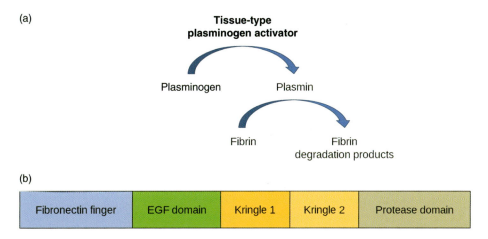

(a)

**Tissue-type
plasminogen activator**

Plasminogen   Plasmin

Fibrin   Fibrin
degradation products

(b)

| Fibronectin finger | EGF domain | Kringle 1 | Kringle 2 | Protease domain |

Fig. 10.17   Tissue-type plasminogen activator (tPA). (a) Simplified illustration of clot lysis (fibrinolysis) by tPA. tPA converts plasminogen to plasmin, which then lyses blood clots by degrading fibrin. (b) Domain structure of tPA. The domains are not drawn to scale. The two Kringle domains are named after the Scandinavian pastry that these structures resemble. EGF: epidermal growth factor.

produced in CHO cells and contains several carbohydrate side chains. tPA is composed of five classical domains (Figure 10.17b): The fibronectin finger mediates binding to fibrin. Elimination of tPA by hepatocytes occurs via the epidermal growth factor domain (EGF domain), while the Kringle 1 domain is responsible for elimination by hepatic endothelial cells. Fibrin-stimulation of the protease, the catalytic entity that cleaves plasminogen to plasmin, is mediated by the Kringle 2 domain.

Two variants of tPA alteplase are reteplase and tenecteplase. Reteplase is much shorter in length than native tPA and consists only of the Kringle 2 and the protease domains. This is one of the rare times when expression as an unglycosylated protein in *E. coli* has therapeutic advantages. The lack of carbohydrate side chains and the deletion of the domains that are required for efficient hepatic clearance (EGF domain and Kringle 1) increase the plasma half-life to 14–18 min

compared to 3–4 min for alteplase. This enables the drug to be injected as a bolus rather than being given by continuous infusion. Due to the absence of the fibronectin finger, the binding of reteplase to fibrin is reduced compared to alteplase. Tenecteplase is a mutated version of alteplase that is expressed in CHO cells. It contains the same domains as the native tPA, but harbors several point mutations. One of the point mutations eliminates the glycosylation site that facilitates hepatic clearance, while a second mutation introduces a new glycosylation site for the purpose of enlarging the molecule. This further reduces its elimination and prolongs its plasma half-life. Additional mutations reduce the inhibitory effects of plasminogen activator inhibitor. Tenecteplase has a relatively long half-life of 17 min and since no domains are missing, its fibrin selectivity is relatively high. Desmoteplase is yet another tPA in clinical development (Box 10.4).

Box 10.4.   Desmoteplase

The source of a new tPA in clinical development was the saliva of the common vampire bat *Desmodus rotundus* (Figure 10.18). Bats constitute almost a fourth of all mammalian species. Only three of the 1100 species of bats have adapted to blood feeding; one of these is *D. rotundus*. Blood-feeding animals have evolved the ability to prevent clotting; the saliva of the vampire bat was shown to harbor a fibrinolytic agent as long ago as the 1930s. Cloning of the four

fibrinolytic proteases from the saliva of *D. rotundus* in the 1990s revealed similarities to human tPA.

One of the four salivary plasminogen activators of *D. rotundus*, now known as desmoteplase, was developed as a thrombolytic agent. The basic structure of desmoteplase is similar to that of human tPA, except that it lacks the Kringle 2 domain. Furthermore, compared to human alteplase, desmoteplase has a 180-fold higher selectivity for fibrin, fewer toxic side effects, and a half-life of several hours rather than a few minutes. The clinical advantage expected from the development of desmoteplase is the extension of the treatment time window. Currently available tPAs must be given within 4.5 h of the onset of symptoms, since the risk of brain hemorrhage outweighs any benefits at later time points. Desmoteplase was expected to extend this time window to 9 h post-stroke onset, which would allow a much larger percentage of stroke patients to receive thrombolytic treatment. However, a large phase III trial failed to demonstrate treatment efficacy. Additional studies are designed to investigate therapeutic benefit in selected cohorts of patients with severe acute ischemic stroke.

Fig. 10.18 Image of the common vampire bat *D. rotundus*. (Reprinted with permission from the Natural History Museum Vienna.)

10.2.7
Therapeutic Enzymes

Recombinantly generated therapeutic enzymes are mostly employed for the treatment of (ultra-)rare genetic disorders. These so-called orphan diseases sometimes affect less than 1000 individuals worldwide. Given the small number of patients that need the product, it is financially risky for pharmaceutical companies to conduct the expensive development and clinical evaluation of a new drug. Special regulations for the approval of orphan medicinal products have been created to facilitate drug development for these groups of patient. Due to the small patient numbers, treatment of a single patient suffering from an orphan disease can cost several hundred thousand dollars per year.

Among the most successful treatments available for orphan diseases is enzyme replacement therapy for Gaucher disease. The affected individuals have an inborn error of glycosphingolipid metabolism due to the deficiency of the lysosomal enzyme acid β-glucocerebrosidase. The disorder is characterized by an enlarged spleen and liver. Symptoms may also include the painful progressive deformity of the skeleton and impaired organ function. Patients may be more susceptible to infection and may have a shortened life expectancy. Since the early 1990s, the disease has been treated with glucocerebrosidase purified from human placenta. However, thousands of placentas were required for the production of enough enzyme to treat a single patient. This was not only laborious and expensive but also left the patients at considerable risk for pathogen transmission. Substantial progress was made when human glucocerebrosidase was successfully expressed in genetically engineered CHO cells. The

enzyme product, named imiglucerase (brand name Cerezyme), must be further modified by exoglycosidases to expose glycan residues, which are required for selective uptake of the protein in macrophage-rich tissues.

Two biosimilar protein products for the treatment of Gaucher disease have subsequently been introduced into the market. Velaglucerase-alpha (brand name VPRIV) is generated by activation of the endogenous human *glucocerebrosidase* gene in an immortalized human fibrosarcoma cell line. The second follow-on agent, which was previously described in Section 10.1.5, is taliglucerase-alpha (brand name Elyseo). This drug is produced in carrot cells, thus avoiding the risk of contamination of susceptible mammalian cell cultures.

Another example of a recombinantly expressed enzyme for the treatment of a genetic disorder is α-galactosidase, a glycoside hydrolase that cleaves the terminal α-galactosyl moiety from glycolipids and glycoproteins. Mutations in the α-galactosidase gene cause Fabry's disease, a metabolic disorder that produces a wide variety of symptoms and shortens life expectancy. Two recombinant forms of human α-galactosidase (brand names Replagal and Fabrazyme), differing in the structure of their oligosaccharide side chains, are available as treatment. Additional therapeutic enzymes are currently being developed or have already been approved (Figure 10.19).

10.2.8
Recombinant Vaccines

The concept of producing antigens for vaccination by recombinant expression of viral surface or capsid proteins is compelling. In contrast to the use of attenuated

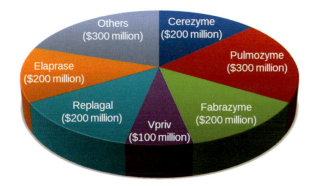

Fig. 10.19 Therapeutic enzymes. The slices of this pie chart indicate the US sales of each protein drug. Cerezyme and Vpriv are recombinant versions of glucocerebrosidase used to treat Gaucher disease, Pulmozyme is a recombinant deoxyribonuclease used for patients with cystic fibrosis, Fabrazyme and Replagal are recombinant α-galactosidase proteins used for the treatment of Fabry's disease and Elaprase is used for the enzyme replacement therapy of Hunter's disease [5].

or inactivated microorganisms, the entire intact pathogen is not needed, either for the production of the vaccine or for the vaccination itself. This strategy, also known as subunit vaccination, is thus safer and not associated with the risk of vaccination-induced infection. A disadvantage of the approach is that the subunit vaccines are sometimes less immunogenic than the intact virion. Immunogenicity may be enhanced by the addition of an adjuvant.

As described in more detail in Section 5.2.4, a recombinant vaccine to prevent infection with the human papilloma virus (HPV) entered the market in 2006. Several types of HPV (e.g., HPV 16 and HPV 18) have been implicated in cervical cancer, the third most common cancer by incidence among women worldwide. Additional HPV types, particularly HPV 6 and HPV 11, cause genital warts. Recombinant HPV vaccines contain the major capsid protein of HPV, L1, which can spontaneously self-assemble into virus-like particles (VLPs) that resemble HPV virions. The VLPs trigger an antibody response and confer protection against HPV but do not induce cancer since they lack the viral DNA. Vaccination is recommended for young females prior to sexual activity, since it can prevent nearly 100% of the precancerous cervical cell changes caused by the various types of HPV included in the vaccine.

The two major HPV vaccines are named Cervarix and Gardasil. Cervarix is a bivalent vaccine consisting of the L1 protein of HPV types 16 and 18, which account for 70% of newly diagnosed cervical cancer cases. The proteins are produced by a Baculovirus expression system in High Five insect cells. The vaccine also contains aluminum hydroxide as an adjuvant to boost the immune

response. The alternative product, Gardasil, is produced in yeast cells (*S. cerevisiae*) and is a quadrivalent vaccine containing VLPs from the L1 proteins of HPV types 6, 11, 16, and 18. Thus, it not only protects against cervical cancer but also against genital warts.

The second recombinant vaccine widely used since the mid-1980s was developed for the prevention of hepatitis B virus infection (brand names Recombivax HB, Engerix-B etc.). It replaced vaccines previously produced by extracting the viral surface protein from the blood of chronically infected patients. The subunit vaccine contains the hepatitis B surface antigen (HBsAg), one of the viral envelope proteins. The vaccine is produced in yeast; individuals are vaccinated by a course of two to three intramuscular injections. Approximately 10–15% of adults fail to respond or respond poorly to vaccination. Here too, an adjuvant can be employed to enhance the immunogenicity of the viral antigen. Additional subunit vaccines have been approved or are in development for the prevention of influenza and HIV infection.

References

1. Veronese, F.M. and Pasut, G. (2005) PEGylation, successful approach to drug delivery. *Drug Discov. Today*, **10**, 1451–1458.
2. Pasut, G. and Veronese, F.M. (2012) State of the art in PEGylation: the great versatility achieved after forty years of research. *J. Control. Release*, **161**, 461–472.
3. Chen, Y., Song, J., Sui, S.F. *et al.* (2003) DnaK and DnaJ facilitated the folding process and reduced inclusion body formation of magnesium transporter CorA overexpressed in *Escherichia coli. Protein Expr. Purif.*, **32**, 221–231.
4. Voet, D., Voet, J.G., and Pratt, C.W. (2013) *Principles of Biochemistry – International Student Version*, 4th edn, John Wiley & Sons, Inc., Hoboken.
5. Aggarwal, R.S. (2014) What's fueling the biotech engine-2012 to 2013. *Nat. Biotechnol.*, **32**, 32–39.

Further Reading

Recombinant Drugs

Aggarwal, R.S. (2014) What's fueling the biotech engine-2012 to 2013. *Nat. Biotechnol.*, **32**, 32–39.
Beck, A. (2011) Biosimilar, biobetter and next generation therapeutic antibodies. *MAbs*, **3**, 107–110.
Morrison, C. (2015) Fresh from the biotech pipeline-2014. *Nat. Biotechnol.*, **33**, 125–128.
Pasut, G. and Veronese, F.M. (2012) State of the art in PEGylation: the great versatility achieved after forty years of research. *J. Control. Release*, **161**, 461–472.
Veronese, F.M. and Pasut, G. (2005) PEGylation, successful approach to drug delivery. *Drug Discov. Today*, **10**, 1451–1458.
Walsh, G. (2014) Biopharmaceutical benchmarks 2014. *Nat. Biotechnol.*, **32**, 992–1000.

Expression Systems for Foreign Genes

De Muynck, B., Navarre, C., and Boutry, M. (2010) Production of antibodies in plants: status after twenty years. *Plant Biotechnol. J.*, **8**, 529–563.

Houdebine, L.M. (2009) Production of pharmaceutical proteins by transgenic animals. *Comp. Immunol. Microbiol. Infect. Dis.*, **32**, 107–121.

Kim, J.Y., Kim, Y.G., and Lee, G.M. (2012) CHO cells in biotechnology for production of recombinant proteins: current state and further potential. *Appl. Microbiol. Biotechnol.*, **93**, 917–930.

Ma, J.K., Chikwamba, R., Sparrow, P. *et al.* (2005) Plant-derived pharmaceuticals: the road forward. *Trends Plant Sci.*, **10**, 580–585.

Yin, J., Li, G., Ren, X. *et al.* (2007) Select what you need: a comparative evaluation of the advantages and limitations of frequently used expression systems for foreign genes. *J. Biotechnol.*, **127**, 335–347.

Monoclonal Antibodies

Ahmad, Z.A., Yeap, S.K., Ali, A.M. *et al.* (2012) scFv antibody: principles and clinical application. *Clin. Dev. Immunol.*, **2012**, 980250.

Buss, N.A., Henderson, S.J., McFarlane, M. *et al.* (2012) Monoclonal antibody therapeutics: history and future. *Curr. Opin. Pharmacol.*, **12**, 615–622.

Muyldermans, S. (2013) Nanobodies: natural single-domain antibodies. *Annu. Rev. Biochem.*, **82**, 775–797.

Growth Factors

Leuenberger, N., Reichel, C., and Lasne, F. (2012) Detection of erythropoiesis-stimulating agents in human anti-doping control: past, present and future. *Bioanalysis*, **4**, 1565–1575.

Fusion Proteins

Beck, A. and Reichert, J.M. (2011) Therapeutic Fc-fusion proteins and peptides as successful alternatives to antibodies. *MAbs*, **3**, 415–416.

Cytokines

Key, N.S. and Negrier, C. (2007) Coagulation factor concentrates: past, present, and future. *Lancet*, **370**, 439–448.

Medcalf, R.L. (2012) Desmoteplase: discovery, insights and opportunities for ischaemic stroke. *Br. J. Pharmacol.*, **165**, 75–89.

Nordt, T.K. and Bode, C. (2003) Thrombolysis: newer thrombolytic agents and their role in clinical medicine. *Heart*, **89**, 1358–1362.

Rudick, R.A. and Goelz, S.E. (2011) Beta-interferon for multiple sclerosis. *Exp. Cell Res.*, **317**, 1301–1311.

Therapeutic Enzymes

Cox, T.M. (2010) Gaucher disease: clinical profile and therapeutic developments. *Biologics*, **4**, 299–313.

Recombinant Vaccines

Markowitz, L.E., Tsu, V., Deeks, S.L. *et al.* (2012) Human papillomavirus vaccine introduction: the first five years. *Vaccine*, (30 Suppl. 5), F139–F148.

Gene Therapy

11

Summary

- Gene therapy is a molecular approach involving the transference of genetic material into the cells of a patient with the intent to treat a disease.
- The most straightforward gene therapy strategy is to replace a deficient gene of a monogenetic disease by an intact version of that gene. In addition, gene therapy can be used to treat cancer and cardiovascular and infectious diseases in addition to many other disorders.
- The choice of an appropriate vector for gene transfer is of major importance. Widely used vehicles are retroviral vectors, adenoviral vectors, and vectors based on adeno-associated viruses (AAVs). The vectors differ in their properties (e.g., duration of action, tissue distribution, and efficiency) and in their side effects (insertional mutagenesis, immunogenicity, etc.).
- While viral vectors employed in gene therapy are usually replication deficient, that is, they can only transduce the patient's cells once but cannot disseminate, oncolytic viruses specifically replicate in cancer cells and can destroy a tumor.
- Although more than 2000 clinical trials have been conducted since the early 1990s, the number of approved gene therapies is currently very limited due to serious problems with efficacy and safety. However, advances in vector technology are likely to overcome these obstacles in the near future.

Contents List

The term gene therapy refers to the transfer of genetic material into the cells of an individual's body with the intention of treating a disease. Classically, the affected patient carries a defective gene, the cause of the disease. The methods of molecular biology can be used to transfer the functional form of the gene to hopefully correct the defect (Figure 11.1). Examples of gene therapy include correction of metabolic functions, the replacement of missing blood coagulation factors, or the transfer of tumor suppressor genes for the treatment of cancer. In addition, gene therapy can be used to create cells with novel properties. For example, the body's own defense mechanisms against tumor cells can be triggered by the use of genes encoding cytokines. Genes for enzymes can be transferred, which convert the inactive precursor of a drug (prodrug) into the active form. DNA vaccination involves transferring viral genes, which trigger an immune response against the virus. Indeed, any transfer of nucleic acids designed to manipulate genes or to regulate gene expression may also be considered gene therapy. Examples also include antisense and RNA interference therapies, which will be addressed in Chapter 13.

Molecular Medicine: An Introduction, First Edition. Jens Kurreck and Cy Aaron Stein.
© 2016 Wiley-VCH Verlag GmbH & Co. KGaA. Published 2016 by Wiley-VCH Verlag GmbH & Co. KGaA.

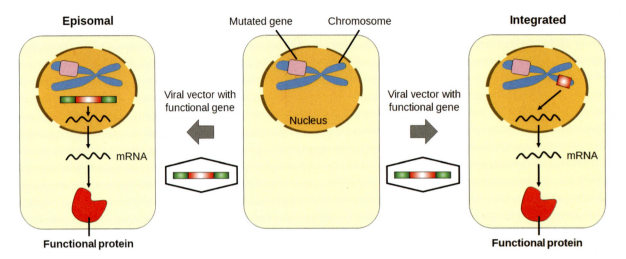

Fig. 11.1   Principles of gene therapy. Due to a defective gene, a protein may not be produced in the cells of the affected patient. With the aid of vectors, the intact version of the gene can be introduced into these cells. Whether or not the therapeutic gene is integrated into the genome of the host or remains episomal is determined by the choice of vector. The functional gene can then be expressed so that the active protein is synthesized and the defect corrected. The defective gene usually remains in the genome with no consequences.

Gene therapy expands our treatment paradigms to encompass the idea of molecular medicine, which involves treatment of disease on the molecular–cellular level. The discovery of restriction enzymes in the early 1970s allowed scientists to combine segments of DNA in novel ways and transfer genes between organisms. The idea of replacing a defective gene with a functional one became experimentally feasible once this technology was available. The hope of using gene therapy to treat severe hitherto untreatable diseases were correspondingly high. The first evaluations of gene therapy began in 1990, but it soon became apparent that there were substantial hurdles to overcome. In addition to the limited efficiency of gene transfer, significant safety concerns were raised. Multiple deaths as a result of gene therapy were serious setbacks for the field. Nevertheless, the resulting efforts to improve the properties of the vectors for gene transfer have brought continuous progress. Between 1990 and 2014, approximately 2150 clinical studies involving gene therapy have been performed. Roughly, two thirds were done in the United States, 10% in Great Britain, and 4% in Germany. The first gene therapy product was approved for therapeutic use in 2003 in China, but it was not approved in the United States or in Europe. In 2012, Glybera™ became the first gene therapeutic product to receive approval in Europe. It is used for the treatment of lipoprotein lipase deficiency, a metabolic disease, and costs the staggering sum of approximately $1.6 million per treatment, which will make it the most expensive medicine in the world. Several further gene therapeutics are being considered for approval in the near future.

11.1
Types of Gene Therapy

There are two fundamental types of gene therapy: Germ cell therapy and somatic cell therapy. The intention of germ cell therapy is to genetically modify ova or sperm cells. As a result, after fertilization, every cell in the body, including future germ cells, will include the inserted gene. Therefore, the patient's children could also benefit from the intact form of the gene. Via this approach, genetic diseases could, theoretically, be permanently corrected. However, this could also mean that any side effects of the therapy could also manifest themselves in the children of treated patients. As a result, germ cell therapy has not been pursued for ethical reasons, although opinions differ as to whether this might be a temporary situation to be revisited when all the technical problems have been resolved. On the other hand, germ cell therapy may be fundamentally nonviable due to its associated irreversible risks.

Somatic cell therapy, in contrast, transfers the therapeutic gene into somatic cells. Here, any genetic change is restricted to the patient and is not passed to his or her children. There is broad consensus that this type of gene therapy should be further developed.

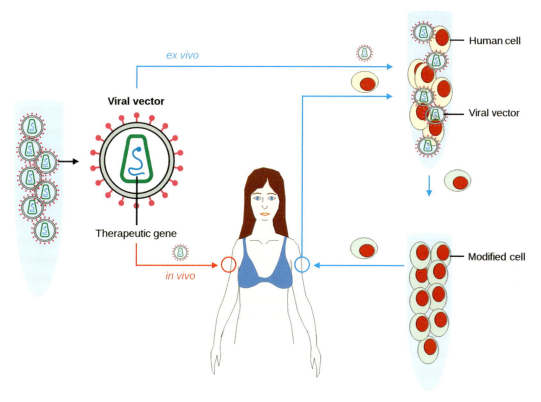

Fig. 11.2 *Ex vivo* and *in vivo* gene therapy. In *ex vivo* gene therapy, cells are removed from the patient and treated with the vector transferring the therapeutic gene. The genetically modified cells are transferred back into the patient. In contrast, *in vivo* gene therapy requires directly treating the patient with the vector.

Somatic cell gene therapy is further divided into *ex vivo* and *in vivo* gene therapy (Figure 11.2). In *ex vivo* gene therapy, cells are removed from the patient and are then genetically modified with the vector in tissue culture before being reimplanted. In contrast, *in vivo* gene therapy involves treating the patient directly with the vector. Depending on the technique used to transfer the nucleic acid, the therapeutic gene can either be stably integrated into the host's genome or remain episomal (Figure 11.1). Therapeutic benefit can be of limited duration or it can be permanent.

11.2
Methods of Gene Transfer

The efficient transfer of the therapeutic gene into the target cells of the patient is an important challenge for successful gene therapy. This is referred to as "delivery." Two major methods, one biological and the other chemical/physical, are used to transfer genes. The latter method includes the transferring of DNA into cells by means of, for example, liposomes. Physical means, such as the bioballistic use of a "gene gun," have also been employed. Another physical method of gene transfer involves opening pores in cell membranes for a short time by means of a weak electric shock. This allows the plasmids to enter cells and is referred to as electroporation. Chemical/physical strategies have been employed in approximately one fourth of gene therapy studies, while the remainder employ biological methods (Figure 11.3).

Biological approaches take advantage of the fact that viruses have been evolutionarily optimized to efficiently infect cells. In most cases, however, viruses are harmful to their hosts and their pathogenic properties must be eliminated before their use as therapeutic agents. This is accomplished by removing the essential viral genes from the genome and replacing them with a therapeutic gene. It is especially important that the ability of the virus to replicate is destroyed. This creates viral vectors that are referred to as "replication deficient." These vectors can enter the target cell once and express their genetic material, but are not capable of reproducing. The property of replication deficiency is implied in the term

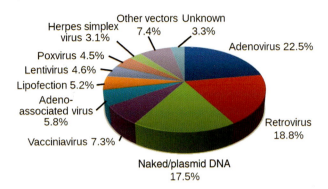

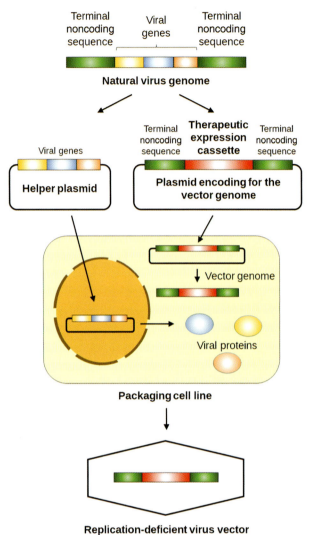

Fig. 11.3 Gene therapy vectors used in clinical studies. The numbers were derived from an international database of gene therapy studies, which is updated regularly (data as of March 2015; http://www.wiley .com/legacy/wileychi/genmed/clinical/).

"viral vector," as opposed to an intact and usually harmful "virus."

To produce viral vectors, helper DNA is required in what is known as a packaging cell line. This cell line contains the essential genes and – as a second component – the backbone of the vector. The vector backbone consists of the terminal repeat elements of the viral genome and the expression cassette of the therapeutic transgene (Figure 11.4). For example, capsid proteins, which are necessary for the assembly of virus particles, are expressed by the helper DNA. The vector backbone and the therapeutic gene are then packaged into the viral particles. The viral proteins of the helper DNA are usually spread across a number of plasmids to minimize the risk that an intact virus can be created by recombination.

The first vectors used in gene therapeutic studies were retroviruses. Along with adenoviruses, retroviruses comprise almost half of the vectors employed in clinical studies (Figure 11.3). As a result of the variety of problems that have been observed for both of these types of viral vectors (see below), in addition to complex production problems, they have lost much of their previous appeal. Although adeno-associated viruses are used in a small proportion of current clinical studies, they play an important role in medical and preclinical research. They will subsequently be explained in more detail, along with retroviral and adenoviral vectors. In addition, numerous other vectors (e.g., vaccinia virus, pox virus, and herpes simplex virus (HSV)) have been the basis for the development of gene transfer vectors.

11.2.1
Retroviral Vectors

Retroviruses contain a single-stranded RNA genome, which is transcribed in infected cells into double-

Fig. 11.4 Composition and production of viral vectors. For gene transfer, the native central gene-coding segment of the viral genome is removed and replaced with the therapeutic gene. The vector genome is then packaged by use of a packaging cell line, which contains the necessary viral genes spread over several plasmids for safety reasons. The resulting viral vector contains only the expression cassette for the transgene, while the essential virus genes are missing. Further viral replication is impossible. (Adapted from Ref. [1] with kind permission from Wiley-Blackwell.)

stranded DNA and subsequently integrated into the host's genome. Retroviruses were discovered as a result of their tumorigenic properties in animals. They were thus designated oncornaviruses, fusing the oncogenic and the RNA genome aspects of the virus into a single name. The family of the Retroviridae is divided into two subfamilies, the Orthoretrovirinae and the Spumavirinae. The Orthoretrovirinae subfamily, in turn, contains six genera of retroviruses (α-, β-, γ-, δ-, and ε-retroviruses and lentiviruses). Further information about the biology

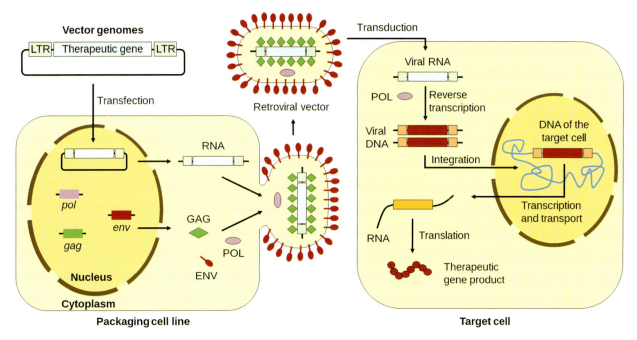

Fig. 11.5   Use of retroviral vectors for gene therapy. The retroviral genes are replaced by the therapeutic transgene. The vector particles are assembled in a packaging cell line, which contributes the viral *pol*, *gag*, and *env* proteins. The vector particles can be used to transduce the target cells. Once inside these cells, it is reverse transcribed, integrated into the host genome, and expressed.

of the human immunodeficiency virus (HIV) as the most prominent representative of the retrovirus family is given in Section 5.1.1. Of special significance for gene therapy are the murine leukemia viruses (MLV, γ-retroviruses), which were the source of the first vector employed in clinical studies. Lentiviral vectors are also becoming increasingly important.

The retroviral vectors contain all the *long-terminal repeat* (LTR) region sequences necessary for integration in addition to the Ψ element, which is necessary for packaging. The regions normally coding for the viral proteins are replaced by the therapeutic gene provided it does not exceed 7 kb in size. To package the gene, the viral polymerase (*pol*), the group specific antigen (*gag*), and the envelope protein (*env*) genes, usually from separate plasmids, are expressed in the packaging cell line. The RNA genome along with the LTRs and the foreign gene are then packaged into virus particles and exit the cell (Figure 11.5).

The vector particles produced are collected, purified, and used to transduce the patient's cells. The transgenic RNA is reverse transcribed into double-stranded DNA in the cell cytoplasm and integrated into the host genome, leading to the expression of the therapeutic gene. Since the genetic information for the viral genes is missing, no new virus particles can be created. The stable integration of the proviral DNA into the host chromosome

guarantees the long-term expression of the transgene and its transfer to daughter cells when the cell divides.

However, the use of retroviruses is associated with a major disadvantage: The integration of the proviral DNA into the host chromosome cannot be directed. As a result, essential genes can either be destroyed or cellular genes can be overexpressed due to the proximity of the powerful viral promoter. This danger was demonstrated when several children whose immune deficiency was initially successfully treated later developed leukemia because of these vectors. Over several clinical trials, a total of five patients developed T-cell leukemia, four of whom were successfully treated with chemotherapy. A more detailed analysis showed that the vectors integrated near the proto-oncogene *LMO-2*, where the strong viral promoter led to its overexpression. However, long-term studies have suggested that not only insertional mutagenesis but also the therapeutic gene itself (the interleukin-2 receptor γ-chain) may have contributed to the genesis of the T-cell lymphomas.

A major advantage to employing retroviral vectors is their high transduction efficiency (often up to 100%). Unfortunately, the use of these vectors is limited to the transduction of dividing cells, since they can only integrate into the host genome while the nuclear membrane is dissolved during cell division. As a result, they are often used for the *ex vivo* gene therapy of proliferating

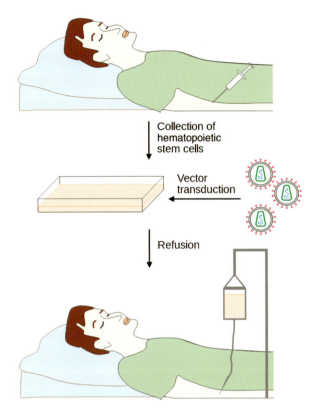

Fig. 11.6 *Ex vivo* gene therapy with retroviral vectors. Hematopoietic stem cells are collected from the patient. They are then treated *ex vivo* with the retroviral vector. The cells, which now express the correct version of the originally defective gene, are reinfused back into the patient.

cells. An example is their use for the treatment of severe combined immunodeficiency (SCID) (Section 11.4.1). Here, hematopoietic stem cells are removed from the patient and are treated with retroviral vectors in tissue culture. The modified cells are then reinfused back into the patient (Figure 11.6).

In addition to oncoretroviral vectors, lentiviruses are another genus of the retrovirus family currently being employed in gene therapy. In comparison to vectors that are derived from mouse leukemia viruses, lentiviruses (e.g., HIV) have the advantage of efficiently transducing mitotically quiescent, that is, nondividing, cells, since they assemble a preintegration complex consisting of the viral genome and host proteins. This complex is actively imported into the cell nucleus. As a result, lentiviral vectors can be used for diseases that require the therapeutic treatment of nondividing cells. In addition, the safety of newer retro/lentiviral vectors has been improved in several ways, including the use of self-inactivating LTRs and the use of cellular rather than viral promoters.

11.2.2
Adenoviral Vectors

Adenoviruses contain a linear, double-stranded DNA genome and primarily cause disease of the respiratory tract. In addition, they infect the conjunctiva of the eye and the gastrointestinal tract. Adenoviruses are complex viruses and contain groups of genes that are expressed in a specific temporal order. These genes are characterized as early genes (E1–E4) or late genes (L) depending on whether they are expressed before or after replication of the viral genome has begun. For gene therapy, the preferred vectors are derived from the human serotype Ad5. An important difference from retroviruses is that the adenoviral genome is not integrated into the host genome, but remains episomal. As a result, transgene expression is temporally limited and the genetic modification is not passed on to daughter cells. Also, the risk of insertional mutagenesis described previously is absent. Adenoviral vectors also efficiently transduce resting (nondividing) cells.

However, a major problem with the use of adenoviral vectors is that they trigger a strong immune response. As a result, vector-transduced cells are rapidly removed by the immune system, which shortens the duration of the gene therapy. Unfortunately, the immune response can also be fatal. In 1999, a young man died as a result of an attempt to employ gene therapy to correct his ornithine transcarbamylase deficiency. Treatment with a high dose of the adenovirus vector triggered a massive immune response, which led to multiple organ failure. This was a serious setback for the field of gene therapy, and subsequent investigation revealed that the approved treatment protocol was violated and that previously observed side effects had not been reported to the appropriate regulatory bodies.

The problems with the use of adenoviral vectors have led to intensive efforts to make this type of vector safer and more efficient. The first-generation adenoviral vectors lacked only the E1 and E3 genes, which eliminated the ability of the virus to replicate. The second generation also removed E2 and E4 (Figure 11.7). Finally, "gutless" vectors were developed, which no longer contain any viral genes. Instead, they consist merely of the inverted terminal repeats (ITRs) and the packaging signal, Ψ. Very large foreign genes of up to 36 kb in length can be inserted between these viral genome elements. To produce gutless vectors, the viral proteins must be provided by a helper adenovirus. Gutless vectors trigger a much weaker immune response compared to adenoviral vectors of the first and second generations. This increases patient safety and extends the duration of therapy.

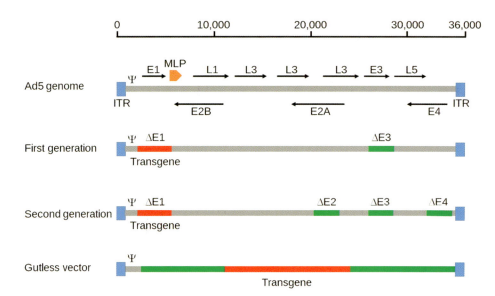

Fig. 11.7 Development of adenoviral vectors. In first generation adenoviral vectors, only the E1 and E3 genes were deleted; in the second vector generation, the early genes E2 and E4 were removed. The newest generation of adenoviral vectors is referred to as "gutless" vectors because they no longer code for any viral genes; instead, they consist only of the ITRs, the packaging signal Ψ, and the transgene. (Adapted from Ref. [2] with kind permission from Macmillan Publishers Ltd., Copyright 2005.)

The first gene therapeutic method that was approved by the regulatory authorities was based on an adenoviral vector. The gene therapeutic agent known as Gendicine™ transfers the tumor suppressor gene p53 into transduced cells and was approved for the clinical treatment of head and neck squamous cell carcinomas (HNSCC) at the end of 2003 by the Chinese State Food and Drug Administration. However, it was not approved in Europe or the United States because subsequent evaluation did not meet the required standards.

A special case of gene therapy involves use of oncolytic viruses, which, in contrast to viral vectors, are replication competent in tumor cells and may destroy their host cells (Box 11.1). Adenoviruses belong to the most-frequently used classes of virus for this purpose.

Box 11.1. Oncolytic Viruses

Oncolytic viruses represent a special instance of the treatment of cancer by viruses. In contrast to viral vectors for gene therapeutic use, oncolytic viruses are replication-competent. Using this treatment modality, known as virotherapy, viruses are employed that preferentially replicate in cancer cells or are tumor specific as a result of genetic modification. In this way, normal cells are spared while the viruses replicate in the tumor cells and destroy them (Figure 11.8).

Adenoviruses are among the most advanced oncolytic viruses. In 2005, the *H101* adenovirus became the first approved oncolytic virus in China for the treatment of tumors. Adenoviruses can only replicate in the absence of the tumor suppressor p53. Normally, the protein product of the adenovirus *E1B-55K* gene destroys the host's p53 protein. The *E1B-55K* gene has been removed from the oncolytic virus so that it cannot replicate in normal cells. In many tumor cells, p53 is defective; *H101* can replicate and destroy these p53-deficient cells. Alternatively, the replication of oncolytic viruses can be restricted to tumor cells by the use of tumor-specific promoters. Here, the early genes, for example, can be placed under the control of a tumor-specific promoter so that the virus cannot enter into its replication cycle in healthy cells.

In newer generations of oncolytic viruses, tumor specificity is increased by microRNAs (miRNAs) (Section 13.4). For example, a binding site for a liver-specific miRNA was added to the essential early adenovirus gene E1A so that it cannot be

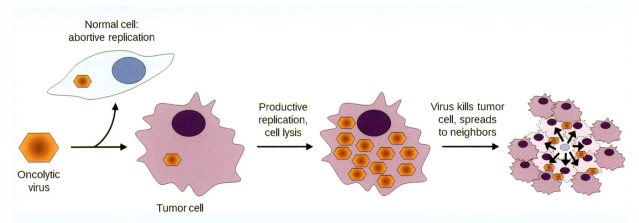

Fig. 11.8 Oncolytic viruses. The viruses can infect normal and tumor cells, but can replicate only in tumor cells. These cells are destroyed by cell lysis and the viruses can then infect and kill neighboring cells. (Adapted from Ref. [3] with kind permission from Macmillan Publishers Ltd., Copyright 2001.)

expressed in this organ. This reduces the liver toxicity of the oncolytic virus without reducing its antitumor effects. The efficiency of oncolytic viruses is further increased when they are equipped with additional elements such as cytokine, cytotoxic, or proapoptotic genes, which kill the cancer cell. This procedure is referred to as "arming." An important challenge for the use of oncolytic viruses is to ensure that they do not recombine on replication and that no viruses with novel properties emerge.

11.2.3
Adeno-Associated Virus Vectors

An additional type of vector, currently clinically evaluated, is based on the adeno-associated virus (AAV). AAVs are small, nonenveloped viruses with a single-stranded DNA genome. To replicate, they require helper viruses (e.g., adenoviruses) that provide the genes necessary for replication. In the absence of the helper virus, AAVs integrate into a specific locus on human chromosome 19. AAVs are attractive as gene transporters since no disease has been associated with infection by wild-type virus.

The viral genes are usually removed from the AAV vectors so that only the therapeutic gene is found between the two ITRs. As a result, the vectors are not only replication-deficient but they have also lost the ability to integrate into the host chromosome. Thus, there is virtually no danger of insertional mutagenesis. Since AAVs are single-stranded, after the cells are infected the opposite strand must be synthesized. As a result, the expression of the transgene contained in AAV vectors is delayed. The development of self-complementary AAV (scAAV) vectors, which contain a double-stranded genome, represents a significant leap forward since it enables the strong and rapid expression of a foreign gene (Figure 11.9). Unfortunately, the already limited 4.7 kb packaging capacity of the AAV vectors is diminished by 50% in scAAV vectors. This limits its utility when a large foreign gene must be transferred.

For gene therapeutic purposes, efficient transduction of the target cells is challenging. To reduce side effects,

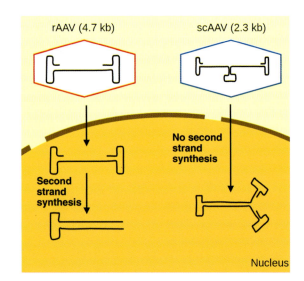

Fig. 11.9 Self-complementary adeno-associated virus (AAV) vectors. Single-stranded recombinant AAV vectors (rAAV) must synthesize the opposite strand, while self-complementary AAVs (scAAV) fold into a double-stranded genome and bypass second-strand synthesis. This rapidly leads to the strong expression of the foreign gene.

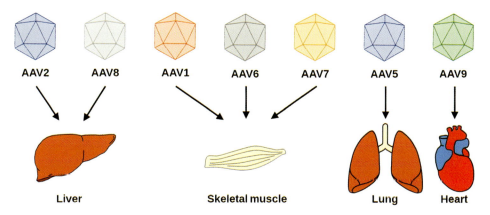

Fig. 11.10 Tissue tropism of AAV vectors derived from distinct capsids. AAV vectors with capsid proteins derived from serotypes 2 and 8 preferentially target the liver, while AAV vectors based on serotypes 1, 6, and 7 are specific for skeletal muscle. AAV-5 and AAV-9 vectors efficiently transduce the lung and heart, respectively. (Adapted from Ref. [4] with kind permission from John Wiley & Sons, Inc.)

it is also advantageous not to transduce the foreign gene into cells where it should not be expressed. More than 100 different serotypes of AAV with differing tissue tropisms have been isolated from human and nonhuman primates. Initially, the AAV-serotype 2 was the preferred vector, but serotypes 1–10 are also widely used. In addition, artificial capsid structures that expand the spectrum even further have been created by artificial selection. Often, so-called pseudotyped AAV vectors are employed. This involves combining the ITRs of the standard serotype 2 AAV with capsid proteins of other serotypes to direct tissue tropism. Pseudotyped AAV vectors permit specific tissue targeting for gene transfer (Figure 11.10).

In many cases, it is necessary to administer the vectors multiple times. However, an immune reaction is often triggered that is directed against the vector, which, after repeated applications, can be inactivated by neutralizing antibodies. This problem can also be addressed by pseudotyping as the serotype is changed after each injection. Thus, the new capsid protein is not recognized by the immune system. However, all of the serotypes employed must transduce the target cells.

The most important characteristics of the different vector types (retro-/lentiviral vectors, adenoviral vectors, and AAV vectors) are summarized in Table 11.1. The production of viral vectors for clinical use is very complex; this is the reason that gene therapeutic treatments are very expensive. The vectors must be free of contamination (for example, from the producing cell line) and no replication competent viruses must have been created by recombination events. Upscaling the production of viral vectors for broad use represents a major challenge at the present time.

Several additional requirements have been identified for the successful application of gene therapy. These include (i) an increase in the efficiency of vector uptake, transport, and uncoating; (ii) the prolonged persistence of episomal vector genomes; (iii) the maintenance of sustained transgene expression, which in some cases is extinguished by epigenetic modifications of the vector genome; and (iv) the prevention of a host immune response either against the transgene product or the vector particles. It also seems likely that further improvements in vector properties will significantly enhance gene therapeutic efficiency.

Table 11.1 Summary of the main properties of viral vectors.

	Retrovirus vector	*Adenovirus vector*	*Adeno-associated virus vector*
Transduction of proliferating cells	Yes	Yes	Yes
Transduction of quiescent cells	No – oncoretrovirus Yes – lentivirus	Yes	Yes
Genomic integration	Yes	No	No (or as a rare event)
Replication	No	No – adenovirus vector Yes – oncolytic adenovirus	No
Potential risks	Insertional mutagenesis	Immune reaction, cytotoxicity	Cytotoxicity

11.2.4
Nonviral Gene Transfer

Due to the difficulties with viral vectors (limited capacity for expression of the transgene, complex production requirements and safety concerns), other strategies for gene transfer not involving viral vectors are being pursued. Since negatively charged nucleic acids must transit the hydrophobic cell membrane, positively charged cationic lipids are often used to introduce RNA and DNA into cells. In combination with cationic lipids, nucleic acids either form liposomes, which carry the nucleic acid in their aqueous interior, or form complexes called lipoplexes. The concentration of the cationic lipid must be kept low to reduce toxicity. Additional progress can be expected based on the use of nanoparticles, which, to a degree, can specifically transfer nucleic acids into selected cell types (e.g., tumor cells). Cell specificity can be achieved by coupling ligands that bind to receptors expressed by the target cells to the surface of the nanoparticles. For example, nanoparticles carrying transferrin on their surface are taken up by receptor-mediated endocytosis into those tumor cells which express the transferrin receptor at high levels on their surface.

The simplest and most frequently used strategy to transfer genes without employing viruses is the use of naked plasmid DNA. This approach has been selected for almost 20% of all gene therapy studies (Figure 11.3). The use of a naked plasmid is not likely to trigger serious adverse effects; however, the efficiency of gene transfer is limited. The two standard methods of DNA transfer involve either injection via syringe or the use of a *gene gun*. This device transfers DNA attached to microparticles into cells at high velocity without severely damaging the tissue (Figure 11.11).

The low efficiency of gene gun gene transfer inhibits its broad use, since successful treatment of tumors or viral infection essentially requires that all affected cells are reached. On the other hand, the application of naked DNA is suitable for immunization, as the plasmid codes for an antigen that only needs to reach a limited cell population. The viral gene is then expressed in these cells and the synthesized protein exits the cell. The immune system recognizes it as foreign and induces the production of antibodies and the activation of killer cells. Thus, the immunized individual develops protection from subsequent infection by the virus. An important advantage of DNA vaccination is its great flexibility: When a new variant of the virus emerges, the plasmid can be altered, a process that is much simpler and faster than the production of a new vaccine. Although, in the narrow sense of the word, a vaccination is not a therapy

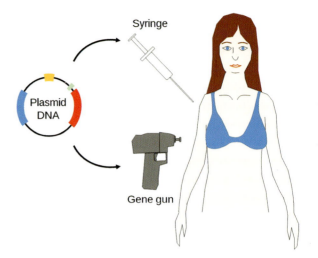

Fig. 11.11   Application of plasmid DNA. The DNA is either injected via syringe or "shot" into the skin at high velocity with a gene gun.

but a form of prophylaxis, DNA immunization is usually still viewed as a form of gene therapy since it involves the transfer of nucleic acids into a host.

An additional development of the plasmid approach is the use of so-called minicircle DNA vectors that are devoid of dispensable bacterial genes. The minicircles are produced in genetically modified *Escherichia coli* bacterial strains (Figure 11.12). In comparison to a standard plasmid, minicircle DNAs produce more persistent gene expression at very high levels (generally 100–1000 times greater than a standard plasmid). The reason for the increase in duration and level of expression is not fully understood, but minicircles can be expected to play an increasingly important role in gene therapeutic applications.

The main disadvantage of employing plasmids in gene therapy is the rapid degradation of extrachromosomal DNA in human cells. This problem has in part been circumvented by an innovative approach of nonviral gene transfer involving the use of transposons, which integrate the expression cassette into the host's genome. As a naturally occurring component of human cells, these mobile genetic elements can change their position in the genome. Characteristic elements of transposons are inverted repeat sequences on their ends. The enzyme transposase, which cleaves the inverted repeats out of the genome and inserts them elsewhere, is also included. For example, the *Sleeping Beauty* transposon system has been developed for gene therapeutic use. It consists of the transposon combined with the expression cassette for the therapeutic gene, in addition to the transposase (Figure 11.13). These elements can be found on a single plasmid, or in some cases on two plasmids. In comparison to viral vectors,

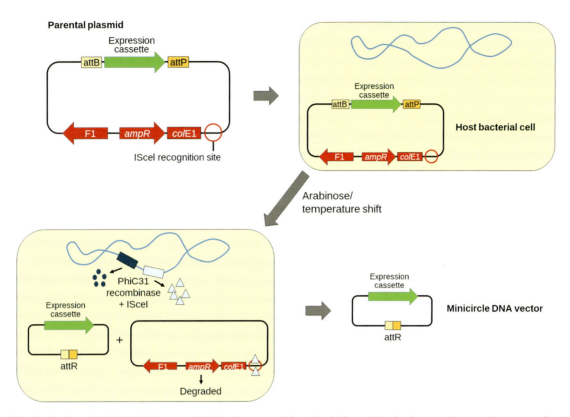

Fig. 11.12 Production of minicircle DNA vectors. Initially, the parental plasmid, which contains the therapeutic expression cassette flanked by attB and attP sequences and multiple IScil recognition sites of the bacterial plasmid backbone, is amplified in the same manner as a standard plasmid. However, the host bacterial cell contains ~20 modifications that carry out inducible recombination events. After the amplification step, the recombination program is induced and results in the formation of a minicircle and an undesired bacterial plasmid. The latter is cleaved and degraded by the IScel restriction enzyme, while the minicircle is purified and used for the gene therapeutic application. The minicircle vector no longer contains any components of the bacterial plasmid and has a low CpG content.

plasmids can be produced simply and inexpensively. In addition, they are stable in long-term storage and can be used safely.

In early studies, the low activity of the transposase was a problem. However, by a combination of molecular *in vitro* evolution and selection, a transposase variant (SB100X) was discovered which has greatly increased efficiency compared to the natural enzyme. However, the Sleeping Beauty system integrates into a random spot in the genome. Thus, as with the integration of viral vectors, the danger of insertional mutagenesis exists. To prevent this, spacers are employed to avoid activating nearby genes. A further safety feature is the limited duration of the activity of the transposase, which minimizes the risk that cellular transposons will be mobilized.

As with the use of plasmid DNA, cell delivery remains a significant challenge for the use of transposons. In some cases, viral vectors have been used to transfer the transposons. However, the advantages of nonviral

delivery (ease of use, safety) are lost. For *ex vivo* gene transfer into immune cells, transposon plasmids can be transferred by lipofection or electroporation. In the first clinical use of the Sleeping Beauty transposon system, the chimeric antigen receptor (CAR) was transferred *ex vivo* into T cells from patients with B-lymphoid malignancies. The goal of this genetic modification was to modulate the specificity of the T cells, stimulating them to destroy the cancer cells.

11.3
Tissue Specificity of Gene Transfer and Gene Expression

The expression of foreign genes in cells can produce toxicity, requiring restriction of transgene expression to the target cells. As previously described, the choice of suitable capsid proteins can direct the tropism of the

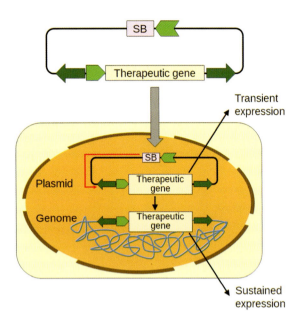

Fig. 11.13 Sleeping Beauty transposon system. The expression cassette of the therapeutic gene is placed between two inverted repeats. The transposase (SB) cleaves out this section and integrates it into the host genome. While plasmid DNA leads only to transient gene expression, integration guarantees the long-term expression of the therapeutic gene. (Adapted from Ref. [5] with kind permission from Oxford University Press.)

vector. Transductional targeting, as this strategy is sometimes called, may influence organ distribution but it normally does not provide sufficient cell specificity. For example, AAV vectors of serotype 9, used to transfer genes into heart cells, also transduce skeletal muscle and liver cells with high efficiency. As a result, additional strategies have been developed to increase transgene specificity. One approach involves expressing the transgene under the control of an organ-specific promoter (transcriptional targeting). For example, the promoter for α_1-antitrypsin is specific for the liver, while the promoter for myosin light chain is specific for the heart.

In addition to directing organ specificity through the appropriate choice of promoter elements, a gene can be expressed under special circumstances only. An example is the gene therapy vector *Repoxygen*™, which had been in development to treat anemia until the program was terminated by Oxford Biomedica. This vector transfers the human *erythropoietin (EPO)* gene, whose gene product stimulates the production of erythrocytes. The *EPO* gene is under the control of a hypoxia responsive element (HRE) and it is expressed only under hypoxic conditions (low oxygen concentrations). Interestingly, the first suspected abuse of gene therapeutic methods for doping involved the use of Repoxygen to enhance athletic endurance (Box 11.2).

Box 11.2.   Gene Doping by Athletes

No Tour de France and no Olympiad take place without news that athletes have employed prohibited means to enhance their competitiveness, a process often referred to as doping. Methods to improve performance are becoming increasingly sophisticated and the procedures to detect them more laborious. It is not surprising that genetic manipulation is used not only for therapeutic purposes but also to enhance physical fitness. This problem has been recognized by the World Antidoping Agency (WADA) and "the nontherapeutic use of cells, genes, genetic elements, or of the modulation of gene expression, having the capacity to enhance athletic performance, is prohibited."

The viral vectors that are used to transfer a gene for therapeutic purposes can also be used to transfer a gene to increase performance. For example, in animal experiments, it was possible to increase the muscle mass of an animal almost fourfold by overexpressing *follistatin* and knocking out *myostatin*, a gene that limits the growth of muscles (Figure 11.14). While

such discoveries could be used to develop new therapies for muscle diseases, genetic manipulation can simultaneously be abused by body builders and other athletes.

Abuse of the *EPO* gene is also a problem. Recombinant manufactured EPO is used on a large scale to improve endurance. In bicycle racing, and also in other sports, the abuse of EPO has been widespread. The laboratory test for EPO doping takes advantage of the fact that the pattern of glycosylation of EPO in humans differs from that of EPO produced recombinantly, usually in hamster cells (Box 10.3 in Section 10.2.3). Gene doping, on the other hand, generates human EPO in the cells of the doper, which cannot be differentiated from the body's own EPO. In 2006, the case of the German trainer Thomas Springstein emerged. He expressed interest in Repoxygen and was later convicted of the (conventional) doping of a young runner. However, it has not been proven that gene therapy has actually ever been successfully used in doping.

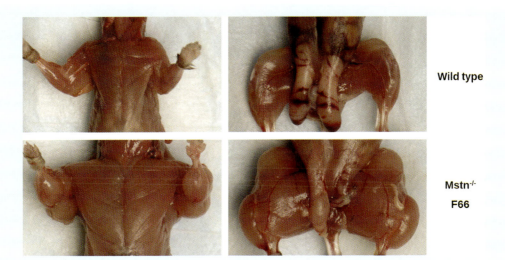

Fig. 11.14   Muscle growth in mice. Shown is a comparison of wild-type mice with mice that have had the gene for myostatin inactivated (Mstn$^{-/-}$) and a gene for follistatin (F66) added [6].

Conventional doping has been associated with serious health risks. The risks of gene doping are significantly higher since the technology is not mature. In the 1980s, the US physician Bob Goldman discovered that more than half of the top athletes surveyed would take a medicine that guaranteed a gold medal, even if they knew it would kill them within 5 years. The acceptance of such risk by athletes pointedly demonstrates the temptation to employ immature gene therapy approaches to increase performance.

The necessity of restricting gene expression to target cells is particularly critical in the case of the expression of cytotoxic genes designed to destroy tumor cells, as healthy tissue must be spared. Therefore, the transgene can be expressed under the control of a promoter that is specific for tumor cells. For example, telomerase is only active in tumor (and stem) cells, and its promoter appears to be suitable for the tumor-specific expression of genes. Oncolytic virus-mediated gene expression (Box 11.1) can also be limited to tumor cells by restricting the expression of genes essential for viral replication to these cells only.

The use of specific promoters has several limitations. Many of these promoters are at times large enough to exceed viral packaging capacity, particularly of AAV vectors. Shortened promoter variants are often not strong enough to produce sufficient amounts of the therapeutic gene. In addition, the so-called "tissue-specific" promoters are often not completely specific; they are often active in their target tissue, but are weakly expressed in other organs as well. These difficulties are probably the reason why only a few different promoters have been employed in gene therapeutic studies. In most cases, these are the ubiquitously active viral promoters such as the cytomegalovirus immediate early promoter.

The discovery of microRNAs (miRNAs) (Section 13.4) has led to the development of the novel concept of posttranscriptional targeting. This approach takes advantage of the fact that many miRNAs are expressed in a cell-specific manner. If a target sequence for an miRNA is introduced into the 3′UTR downstream of the cDNA encoding the therapeutic gene, it blocks transgene expression in those cells that express the corresponding miRNA. For example, AAV vectors of serotype 9 are suitable for gene transfer into cardiomyocytes. As previously described, they also transduce the liver, which in turn specifically expresses miRNA-122 at a high level. The target sequences of this miRNA can prevent the expression of the transgene in the liver without influencing its expression in cardiomyocytes.

The miRNA-dependent regulation of transgene expression can also help to solve the problem of the immune reaction against therapeutic genes. Attempts to treat hemophilia B with gene therapy using coagulation factor IX have been underway for many years. However, an immune reaction against factor IX is triggered, rendering the therapeutic benefit of short duration. In order to avoid this immune response, the target site of an miRNA, which is specifically expressed in hematopoietic cells, was added to the transgene's cDNA in the viral

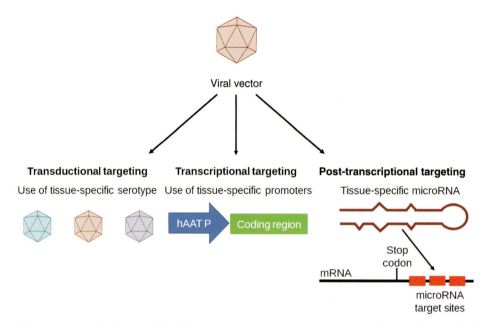

Fig. 11.15 Specificity of gene transfer. The specificity of the expression of a therapeutic gene can be increased by transductional targeting, transcriptional targeting, and post-transcriptional targeting. For transductional targeting, serotypes of the viral vector are used, which are specific for the target organ. Since the vector often targets more than one organ, specificity can be increased by the use of a tissue-specific promoter. An example would be a liver-specific promoter, for example, the promoter of human alpha 1-antitrypsin (hAAT). Finally, expression of the transgene in specific organs can be avoided by adding target sites for miRNAs that are expressed only in that organ.

vector. This prevented direct stimulation of the immune cells, although the expression of factor IX in the liver was not affected. A sustained therapeutic effect became possible in animal studies.

Post-transcriptional silencing, however, can only inhibit expression of transgenes in those cells in which it is not desired. No efficient miRNA-based system has yet been developed to induce the expression of a gene in a defined cell type.

Taken together, there are three different strategies available to increase the specificity of gene transfer and corresponding protein expression (Figure 11.15): For transductional targeting, suitable serotypes of the virus can be chosen that efficiently transduce the target cells. Pseudotyping can be used for other vector types, in addition to AAV. For example, clinical studies often employ lentiviral vectors that are pseudotyped with the G-glycoprotein of the vesicular stomatitis virus (VSV-G). This is intended to expand tissue tropism since these lentiviruses are able to infect almost any cell type. For other uses, lentiviral vectors have been created with surface proteins that are cell type specific. Nevertheless, transductional targeting is in many cases not sufficiently specific for a given tissue. As a result, a promoter can be chosen that is primarily active in the target cell (transcriptional targeting). Since the promoters are often not sufficiently specific and can lead to gene expression in

nontarget cells (though frequently at low levels), post-transcriptional targeting with miRNA target sequences can be employed to silence gene expression in these cells. These three strategies can be combined *ad libitum*. They have not yet been employed in clinical studies yet, but are undergoing continuous development.

11.4
Applications of Gene Therapy

In principle, gene therapy can be used for any disease in which the negative consequences of a gene defect can be overcome by the introduction of the intact form of the gene. In addition, via these induced genetic changes, cells can assume those novel properties that are desirable for disease treatment. Gene therapy is, therefore, complementary to silencing strategies (Chapter 13), where the expression of a harmful gene can be diminished.

Gene therapy was originally clinically evaluated for the treatment of monogenic diseases. This involves replacing the defective gene with the functional gene, though "replacement" usually means that the defective gene remains in the genome and the correct gene is either integrated into another position in the chromosome or remains episomal. As monogenic diseases are

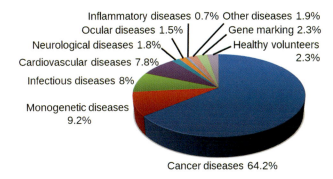

Fig. 11.16 Clinical targets of gene therapy. Approximately two thirds of the gene therapy studies are for cancer. Monogenic, cardiovascular, and infectious diseases each approach 10% of the total. (Data as of March 2015; http://www.wiley.com/legacy/wileychi/genmed/clinical/).

comparatively rare, common diseases, such as cancer, cardiovascular diseases, and infectious diseases have gained in importance for gene therapy studies. Currently, two thirds of clinical gene therapy studies are for cancer, followed by monogenetic diseases, cardiovascular diseases, and infectious diseases, which each comprise almost 10% of the total (Figure 11.16).

11.4.1
Gene Therapy of Monogenic Diseases

Due to the uncertainties inherent in the use of any new form of treatment, the first gene therapy studies chose a disease that is particularly severe and for which no suitable therapy exists. SCID is produced by an inherited lack of appropriate immune system responsiveness. Typically, this disease involves T-lymphocytes that are either defective or too few in number. In addition, B-lymphocytes or natural killer cells can also be affected. While SCID is very rare among the general population, affecting perhaps 1 birth in 50 000–100 000, it is much more common in certain ethnic groups, such as the Navajo and Apache people, where it can affect as many as 1 birth in 2000.

Two forms of SCID have been studied for gene therapeutic applications. These include either defects in the *adenosine deaminase* gene (ADA-SCID) or in the gene encoding the γ-chain of the interleukin-2 (IL-2) receptor (X-SCID), located on the X chromosome. Left untreated, SCID leads to an early death, since the carriers lack important protection from infection. SCID is also known as the "bubble boy disease," since several so-affected children spent their lives in a sterile plastic tent. SCID can be treated with a bone marrow transplant if a suitable donor is available. Alternatively, patients can be treated with enzyme replacement therapy, which involves injecting the

recombinant ADA enzyme. However, lifelong treatment is required and it is not always successful.

In 1990, the first children with ADA-SCID were treated with gene therapy. The transfer of the intact *ADA* gene took place *ex vivo* with the aid of a retroviral vector. A temporary positive effect was observed in a single patient. Since she simultaneously received enzyme replacement therapy, the utility of the gene therapy could not be proven. Several years later, a study that treated X-SCID by the retroviral transfer of the *IL-2 receptor γ-chain* gene initially appeared to be successful, since the immunodeficient phenotype could be ameliorated in many patients. It was thus a major, tragic setback when five of the children later developed leukemia and one died as a result. The integration of the retroviral vector had activated a proto-oncogene that triggered unchecked proliferation of T cells (Section 11.2.1). Nevertheless, even taking into account these severe adverse events, gene therapy produced superior clinical results versus the alternative treatment, bone marrow transplantation. In addition, another study performed in parallel involving gene therapy of ADA-SCID patients resulted in restoration of a permanently functional immune system in 8 of 10 patients without negative side effects. This success was featured in a 2009 *New England Journal of Medicine* article entitled "Gene therapy fulfilling its promise." A follow-up more than 10 years later demonstrated that 18 of 20 treated X-SCID patients and all 27 treated ADA-SCID patients were still alive. Furthermore, the immune deficiency was corrected in 17 of the X-SCID and 19 of the ADA-SCID patients. Additional progress using newer vectors with added safety features has been achieved, through, for example, the use of self-inactivating LTRs and cellular promoters and the use of nonmurine lentiviral vectors.

In addition to retroviral vectors, AAV-based vectors have also shown promising results in the clinical setting, as demonstrated by the development of gene therapy for the treatment of lipoprotein lipase (LPL) deficiency. In this rare genetic disease, the affected patients are missing a key enzyme in their lipid metabolism pathways. The high blood triglyceride levels can lead to severe, life-threatening pancreatitis. With the aid of an AAV vector, the intact LPL gene was transferred into muscle cells. Biologically active LPL protein was increased and the frequency of pancreatitis was reduced. This AAV-based treatment of LPL was the first gene therapy approved in Europe.

Leber's congenital amaurosis (LCA) is a rare inherited eye disease that leads loss of vision. LCA type 2 is caused by mutations in the *RPE65* gene. In late 2014, treatment of the disorder with an AAV2 vector delivering the intact *RPE65* gene via subretinal delivery led to a significant

Table 11.2 Selection of monogenic diseases that have been treated in gene therapy studies.

Disease
Cystic fibrosis
Severe combined immunodeficiency (SCID)
Leber's congenital amaurosis
Duchenne muscular dystrophy
Becker muscular dystrophy
Alpha-1-antitrypsin deficiency
Chronic granulomatous disease
Ornithine transcarbamylase deficiency
Lipoprotein lipase deficiency
Hemophilia A and B

improvement in the eyesight of a majority of the patients. A breakthrough therapy designation was received from the FDA, which will expedite further the development of this gene therapeutic approach.

An additional genetic disease for which gene therapeutic treatment has been attempted is cystic fibrosis (CF). This autosomal recessive inherited disease is characterized by a malfunction of chloride channels, which leads to hyperviscosity of the secretions of the bronchi and other organs (Section 3.1.2). Patients suffer from repeated lung infections and severe lung inflammation, which shorten life expectancy. At the molecular level, the cause of CF is the mutation in the *cystic fibrosis transmembrane conductance regulator* (*CFTR*) gene, which encodes the chloride channel. In various clinical studies, transferring the intact *CFTR* gene was attempted. Since the disease primarily affects the bronchi, the vector chosen was an adenovirus, which naturally infects the bronchi. Unfortunately, the efficiency of the gene transfer was insufficient because the

viral vector could not penetrate the viscous bronchial mucus. As a result, other vectors were also evaluated.

All told, more than 20 monogenic diseases in almost 200 clinical trials involving gene therapy have been targeted. A selection is summarized in Table 11.2.

11.4.2
Gene Therapy of Cancer

In addition to the use of oncolytic viruses (Box 11.1), three different strategies for the gene therapeutic treatment of cancer are being pursued: The replacement of defective tumor suppressor genes, the activation of prodrugs, and gene-based immunotherapy. Different tumor types have been treated in clinical studies, including lung, gynecologic, skin, urologic, neurologic, and gastrointestinal tumors, as well as hematologic malignancies and pediatric tumors.

Tumor suppressors are responsible for controlling the cell cycle, preventing the unchecked proliferation typical of cancer cells (Box 4.1 in Chapter 4). The best-known tumor suppressor is p53, which, due to its central role in the control of the cell cycle, is also referred to as "the guardian of the genome." In normal cells, p53 is inactivated by its negative regulator mdm2. DNA damage or cellular stress leads to the activation of p53 (Figure 11.17). This triggers cell cycle arrest and initiates DNA repair processes. If the damage is sufficiently great, programmed cell death (apotosis) occurs. In approximately 50% of human tumors, p53 is defective due to mutations. Therefore, there have been numerous gene therapy attempts to reintroduce p53 into tumor cells to stop their uncontrolled proliferation. As already described, an adenoviral vector that introduced p53 into tumor cells was the basis of the first approved gene therapy in China.

In *suicide gene therapy*, genes that convert an inactive pharmacological substance (the prodrug) into a functional

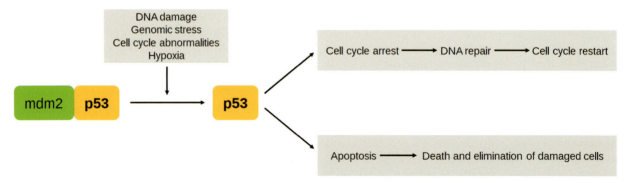

Fig. 11.17 Function of the tumor suppressor p53. Under normal conditions, p53 is blocked by the negative regulator mdm2. It is only activated as a result of DNA damage or cellular stress. p53 either halts cell cycle progression to enable the repair of DNA damage, or triggers cellular apoptosis.

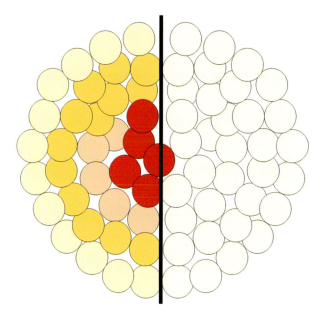

Fig. 11.18 Bystander effect. If there were no intercellular connections, the prodrug would be activated only in those cells that were transduced by the viral vector (right). Therapeutic efficacy would be low, since not every cell can be reached. However, tumor cells exchange small molecules via gap junctions, thus, the activated Ganciclovir can also reach neighboring cells (left). The color code signifies that a gradient exists between the transduced cells and more remote cells.

chemotherapeutic agent are introduced into tumor cells. Thymidine kinase from the HSV is often used. This enzyme activates the prodrug Ganciclovir™, a guanosine analog that cellular kinases subsequently convert into the triphosphate (Section 5.4). Ganciclovir triphosphate is then incorporated into newly synthesized DNA by DNA polymerase. Since further nucleotides cannot be attached to the analog, premature termination of DNA strand synthesis is followed by cell death.

Gene therapeutic vectors can never reach all of the tumor cells. Ganciclovir can be activated only in the transduced cells and the remaining unaffected cells can re-grow the entire tumor. However, this problem is ameliorated by the so-called bystander effect (Figure 11.18): The cells reached by the vector can pass along the activated toxic drug via gap junctions into neighboring cells. Thus, tumor cells that have not directly received the suicide gene are also affected by the toxin it produces.

The aim of gene-based immunotherapy, the third form of gene therapy against cancer, is to strengthen the usually weak humoral and cellular immune response against antigens on the surface of tumor cells. Different strategies have been entertained. Tumor cells can be provided with immunostimulatory factors, which induce a stronger immune response to all the cells that carry the

corresponding tumor antigens on their surface. Alternatively, a vaccine can be administered by viral vectors that lead to the expression of tumor antigens. This directs an immune response against the tumor cells. Vaccination with dendritic cells, which express the tumor antigen after manipulation *ex vivo*, has also proven to be promising (Box 5.2 in Section 5.2).

11.4.3
Other Diseases

Most of the gene therapeutic studies for the treatment of cardiovascular disease aim to induce angiogenesis to improve blood flow in ischemic regions. This includes treatment of both myocardial ischemia caused by coronary artery disease and lower extremity ischemia caused by peripheral artery disease. Genes stimulating angiogenesis can be introduced: These include fibroblast growth factor (FGF), vascular endothelial growth factor (VEGF), or hypoxia inducible factor (HIF). An important cause of heart failure is the disruption of calcium homeostasis, which should in principle be correctable by gene therapy. In a clinical study examining the treatment of patients with heart failure, AAV1 vectors encoding the sarcoplasmic reticulum Ca^{2+}ATPase pump (SERCA2a) were used for gene transfer, which led to an improvement in or stabilization of the disease. In addition, the risk of major cardiovascular events was significantly reduced. Positive results persisted after 24 months.

For the treatment of infectious diseases, genes are transferred which block viral replication. These include ribozymes or short hairpin RNAs (Chapter 13). These approaches have been evaluated for HIV infection, as well as for infection with the hepatitis C or respiratory syncytial viruses. In additional clinical gene therapeutic studies, a coreceptor of HIV was inactivated by means of directed genome engineering (Box 11.3). Gene therapeutic methods for the treatment of neurodegenerative diseases are also in development. Parkinson's disease is characterized by the death of the dopamine-producing cells of the *substantia nigra*. This results in reduced levels of the neurotransmitter dopamine, which leads to motor defects as well as to psychological and neurological problems. AAV vectors encoding neurturin were also able to halt the degeneration of dopaminergic neurons. Although neurologic function improved, the effects were not impressive since too few cells were transduced by the vector. In an alternative approach, the three genes involved in dopamine synthesis (*tyrosine hydroxylase, aromatic amino acid decarboxylase*, and *GTP cyclohydrolase*) were transferred. In other clinical studies, the potential for gene therapy to treat inflammatory diseases is also being investigated.

Box 11.3.   Genome Engineering

In recent years, several tools have been developed for the sequence-specific manipulation of genetic material. These tools are based on cellular systems that can be adapted for directed genome engineering. Most techniques are based on zinc finger nucleases (ZFNs), transcription activator-like effector nucleases (TALENs), and a newer method guided by clustered regulatory interspaced short palindromic repeats (CRISPR) and their associated *cas* genes.

ZFNs are artificial restriction enzymes that consist of a DNA-binding zinc finger domain and a DNA-cleaving restriction domain. They are designed to cleave only a specific DNA sequence in the genome. This permits the specific inactivation of a gene or the incorporation of foreign DNA at a specific locus. ZFNs are active as heterodimers, each consisting of a different DNA-binding domain in addition to the DNA-cleaving FokI-domain. They are joined by a linker.

The DNA-binding domain usually consists of three to six zinc finger motifs, which each recognize a DNA sequence of approximately three base pairs. The binding of the zinc finger to the DNA is largely responsible for determining the activity and specificity of the ZFN. If the DNA-binding domain does not have sufficient affinity for the target sequence, the ZFN will either fail to find its target or it will bind and cleave other sequences in the genome. This can produce toxicity. For the assembly of DNA-binding domains, either zinc fingers with known binding sequences can be combined or new binding domains can be generated using *in vitro* selection strategies. The DNA-binding domain is joined to the FokI restriction enzyme, which cleaves both DNA strands when it dimerizes with a second monomer. The modular nature of a ZFN theoretically allows it to target any gene.

A double-strand break can be used for two approaches in gene therapy. It can either be used to functionally knockout a harmful gene or to stimulate the targeted correction of a gene defect. In the first case, the repair of a double-strand break employs *nonhomologous end joining* (NHEJ), which can produce insertions or deletions at the break point. This can alter the reading frame or destroy the coding region and inactivate the gene. By combining two ZFN pairs, entire segments of genomic DNA can be deleted. For the accurate repair of a gene, donor DNA with homologous sequences to the target gene must be used; these are incorporated via homologous recombination, permitting the correction of a mutated gene. ZFNs can also be used for the targeted insertion of a therapeutic expression cassette into the so-called safe haven for the insertion of transgenes in the human genome.

An important hurdle for the practical use of artificial nucleases is the efficient transfer of the ZFN-coding cDNA into the target cells. Since the strong, long-term expression of the ZFN can be cytotoxic due to non-specific nuclease activity, integrating vector systems such as retroviral vectors are unsuitable for ZFN gene transfer. To reduce toxicity, short-term episomal systems such as plasmid DNA or AAV vectors are more suitable.

The potential of ZFNs as a novel tool to deactivate genes has been investigated in clinical trials. In two studies involving the treatment of HIV patients, the ZFN was directed against the *CCR5* receptor gene on T cells. HIV employs CCR5 as a coreceptor to invade cells. As a result of a mutation, individuals who lack CCR5 are protected from infection by HIV without any known health consequences (Section 5.4.1). Based on this discovery, ZFNs have been developed that delete a section of the *CCR5* gene, thus inactivating it. In the first clinical trial of this therapy, the ZFN coding cDNA was transferred by an adenoviral vector into T cells

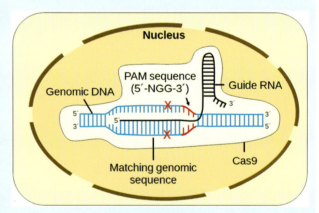

Fig. 11.19   Targeted genome editing with RNA-guided Cas9. Cas9 is a DNA endonuclease found in many bacteria. The enzyme has two active sites that each cleave one strand of double-stranded DNA. The enzyme is guided to the target DNA by a guide RNA that contains a sequence that matches the sequence to be cleaved, which is demarcated by the so-called PAM sequence. The RNA-guided Cas9 activity creates site-specific double-stranded DNA breaks that are then repaired by nonhomologous end joining, thereby introducing random mutations. Alternatively, the break can be repaired by homologous recombination, which permits a donor DNA to insert a new specific DNA sequence at the break site.

isolated from patients. The modified T cells, which no longer express functional CCR5, were reinfused and were shown to be protected against infection with HIV by the loss of the coreceptor. The aim of another clinical study was to use a ZNF to inactivate the glucocorticoid receptor to strengthen the T cell-based immunologic response against cancer.

In addition to ZFNs, other artificial restriction enzymes have been created that produce targeted cuts in the genome. A promising strategy consists of fusing the DNA-binding domain of a transcription activator-like effector (TALE) with a DNA-cleaving domain. The resulting specific restriction enzymes, which work like ZFNs, are known as transcription activator-like effector nucleases (TALENs). They are considered particularly efficient and specific.

A rapidly emerging technique for genome engineering that may soon replace these methods is based on the CRISPR/Cas9 system (Figure 11.19). Its natural function is to serve as an adaptive immune mechanism to protect bacteria from infection by bacteriophages. CRISPRs are DNA loci containing short repetitive sequences interspaced by sequences derived from previous exposures to a virus. The Cas9 protein is an endonuclease that naturally cleaves pathogenic DNA guided by the RNA transcribed from the CRISPRs. As the Cas9 protein can be retargeted by redesigning the RNA, it can be used to cleave virtually any DNA sequence. Accordingly, the CRISPR/Cas system has successfully been used to introduce directed mutations into the genomes of human cells. This approach can also be used to inactivate a harmful gene.

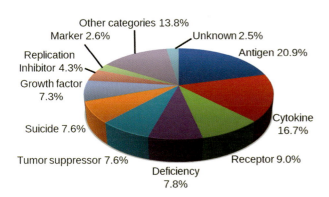

Fig. 11.20 Gene types transferred in gene therapy clinical trials. (Data as of March 2015; http://www.wiley.com/legacy/wileychi/genmed/clinical/).

11.5
Future Prospects

After the initial euphoria surrounding the promise of gene therapy for the treatment of severe disease, disappointment set in as it was realized that gene transfer was often not efficient and that it could be associated with toxicity. After the year 2000 and a period of consolidation, the first notable successes were achieved. In these cases, too, there were problems with vector-induced side effects. The risk, however, needed to be weighed against that of alternate methods of treatment. In particular, the *ex vivo* gene therapy of hematopoietic stem cells for the treatment of severe immune defects has progressed dramatically. In parallel, significant improvements were achieved for the direct *in vivo* applications of viral vectors and nonviral methods of delivery.

An important challenge for gene therapy is the choice of the most suitable viral vector. Intensive exploration of numerous viral vectors has clearly shown that a standard vector that is optimal for all purposes will most likely never be found. For the *ex vivo* transduction of hematopoietic cells, modern retro- or lentiviral vectors with added safety features have proven to be adequate. For lung diseases such as cystic fibrosis, adenoviruses were employed because they naturally infect the respiratory tract. Finally, for the transfer of genes into the heart, pseudotyped AAV vectors are highly efficient. Vectors that stably deliver their genetic material into the host chromosome are preferentially used to transduce

Since the first clinical trials in the early 1990s, gene therapeutic strategies for the treatment of a large variety of diseases have been developed. Many different genes have been transferred. In Figure 11.20, these classes of transferred genes are summarized: Antigens for the stimulation of an immune response, cytokine genes, tumor suppressor genes, and suicide genes together comprise approximately 50% of the total and have been used mainly for the treatment of cancer. The "deficiency" genes represent the replacement of defective genes in monogenic diseases, and growth factors have been used almost exclusively for the treatment of cardiovascular diseases.

dividing cells, since vectors that remain episomal are lost from rapidly proliferating cells. These vectors are, therefore, better suited to cell types that either do not divide or divide only very slowly. In addition to the efficiency of gene transfer, issues surrounding safety and complexity of production are important criteria for the choice of a vector. The death of patients in a clinical trial using a vector that was based on the murine leukemia virus produced a shift toward employing lentiviruses and AAV vectors. Additional advantages of these vectors lie in the fact that they are easier to produce and use than the vector types previously employed.

However, due to the risk of severe side effects, in the foreseeable future, gene therapy will only be considered for use against very serious diseases and will remain restricted to diseases with no alternative forms of treatment. For example, several clinical studies for the gene therapeutic treatment of hemophilia have been performed. However, these approaches must compete with simpler and established treatment with recombinantly produced coagulation factors.

The replacement of a defective gene to treat a monogenic disease is a more straightforward application of gene therapy. The number of patients suffering from these severe inherited diseases is small, so for commercial reasons multifactorial diseases with larger numbers of affected patients have been and are currently being intensively investigated. Taken together, the developments of the last few years, in particular improvements in vector technology, will presage increasing success for the clinical use of gene therapy in the near future. The technology will also be aided by combining it with stem cell therapy, as described in Section 12.6.3.

References

1. Kurreck, J. (2009) RNA interference: from basic research to therapeutic applications. *Angew Chem., Int. Ed.*, **48**, 1378–1398.
2. Alba, R., Bosch, A., and Chillon, M. (2005) Gutless adenovirus: last-generation adenovirus for gene therapy. *Gene Ther*, (12 Suppl 1.), S18–S27.
3. McCormick, F. (2001) Cancer gene therapy: fringe or cutting edge? *Nat. Rev. Cancer*, **1**, 130–141.
4. Arruda, V.R. and Xiao, W. (2007) It's all about the clothing: capsid domination in the adeno-associated viral vector world. *J. Thromb. Haemost.*, **5**, 12–15.
5. Aronovich, E.L., McIvor, R.S., and Hackett, P.B. (2011) The Sleeping Beauty transposon system: a non-viral vector for gene therapy. *Hum. Mol. Genet.*, **20**, R14–R20.
6. Lee, S.J. (2007) Quadrupling muscle mass in mice by targeting TGF-beta signaling pathways. *PLoS One*, **2**, e789.

Further Reading

Gene Therapy

Ginn, S.L., Alexander, I.E., Edelstein, M.L. *et al.* (2013) Gene therapy clinical trials worldwide to 2012: an update. *J. Gene Med.*, **15**, 65–77.
Kay, M.A. (2011) State-of-the-art gene-based therapies: the road ahead. *Nat. Rev. Genet.*, **12**, 316–328.
Lewis, R. (2014) Gene therapy's second act. *Sci. Am.*, **310**, 52–57.
Sheridan, C. (2011) Gene therapy finds its niche. *Nat. Biotechnol.*, **29**, 121–128.
Tani, J., Faustine, and Sufian, J.T. (2011) Updates on current advances in gene therapy. *West Indian Med. J.*, **60**, 188–194.

Viral Vectors

Alba, R., Bosch, A., and Chillon, M. (2005) Gutless adenovirus: last-generation adenovirus for gene therapy. *Gene Ther.*, (12 Suppl. 1), S18–S27.
Fechner, H. and Kurreck, J. (2011) Viral vectors for RNA interference applications in cancer research and therapy, in *Drug Delivery in Oncology* (eds F. Kratz, P. Senter, and H Steinhagen), Wiley-VCH-Verlag GmbH, Weinheim, pp. 1415–1442.
Russell, S.J., Peng, K.W., and Bell, J.C. (2012) Oncolytic virotherapy. *Nat. Biotechnol.*, **30**, 658–670.
Thomas, C.E., Ehrhardt, A., and Kay, M.A. (2003) Progress and problems with the use of viral vectors for gene therapy. *Nat. Rev. Genet.*, **4**, 346–358.

Transposons

Aronovich, E.L., McIvor, R.S., and Hackett, P.B. (2011) The Sleeping Beauty transposon system: a non-viral vector for gene therapy. *Hum. Mol. Genet.*, **20**, R14–R20.
Hackett, P.B., Largaespada, D.A., and Cooper, L.J. (2010) A transposon and transposase system for human application. *Mol. Ther.*, **18**, 674–683.

Genome Engineering

Gaj, T., Gersbach, C.A., and Barbas, C.F., 3rd (2013) ZFN, TALEN, and CRISPR/Cas-based methods for genome engineering. *Trends Biotechnol.*, **31**, 397–405.
Handel, E.M. and Cathomen, T. (2011) Zinc-finger nuclease based genome surgery: it's all about specificity. *Curr. Gene Ther.*, **11**, 28–37.
Hsu, P.D., Lander, E.S., and Zhang, F. (2014) Development and applications of CRISPR-Cas9 for genome engineering. *Cell*, **157**, 1262–1278.
Wilkinson, R. and Wiedenheft, B. (2014) A CRISPR method for genome engineering. *F1000Prime Rep.*, **6**, 3.

Stem Cells

12

Summary

- Stem cells are defined by the property of self-renewal, that is, unlimited growth, and the ability to differentiate into more specialized cell types. Much hope has been placed on the development of novel stem cell therapeutic approaches for the treatment of heretofore untreatable diseases.
- Embryonic stem cells (ESCs) are obtained from the inner cell mass of a blastocyst. These cells are pluripotent, that is, they can develop into cells from the three germ layers and also into germline cells. Patient-specific ESCs can be obtained by therapeutic cloning. This procedure involves the transfer of the nucleus of a donor cell into a denucleated egg, cultivation of the fused cell to the blastocyst stage, and the generation of ESCs from the inner cell mass.
- Adult stem cells exist in special niches throughout the postnatal organism and are responsible for tissue renewal and for repair following damage. They are multipotent and can differentiate into many of the cells of the tissue from which they originate. Their plasticity in differentiating into those cells required for cell therapy is under intensive investigation.
- Various methods have been developed to reprogram differentiated somatic cells into so-called induced pluripotent stem cells (iPSCs). These pluripotent cells can be generated without destroying an embryo and can be obtained directly from a patient. Thus, autologous transplantation can be performed using these cells without the risk of inducing an immune reaction.
- Another recent development in this field is the direct reprogramming of differentiated cells into a desired cell type while bypassing the pluripotent state. Cellular switching can be achieved by treatment of these cells with transcription factors, microRNAs, or small molecules. This approach may provide a fast and straightforward alternative for the generation of patient-specific cells with low tumorigenic potential.
- Protocols have been established to differentiate pluripotent stem cells (ESCs or iPSCs) into virtually any and every cell type of medical interest. Most protocols for directed differentiation require the sequential addition of inducing factors that would normally direct cell differentiation during embryonic development.
- Hematopoietic stem cell transplantation (HSCT) for the treatment of blood cancer is the only form of stem cell therapy that is widely practiced. However, other novel applications of stem cells for treatment purposes can be expected in the near future. Pluripotent cells, either ESCs or iPSCs, can be used for two purposes: the design of disease

Contents List

Molecular Medicine: An Introduction, First Edition. Jens Kurreck and Cy Aaron Stein.
© 2016 Wiley-VCH Verlag GmbH & Co. KGaA. Published 2016 by Wiley-VCH Verlag GmbH & Co. KGaA.

models or direct cell therapy. Several patient-specific iPSCs have been produced and used to establish cellular models to study mechanisms of disease and to screen for novel therapeutic compounds. Clinical trials have been initiated that employ stem cell-derived cells to treat diseases such as spinal cord injury or vision loss. Heart disease, neurodegenerative diseases, diabetes, and muscular disorders are additional areas of advanced stem cell research.

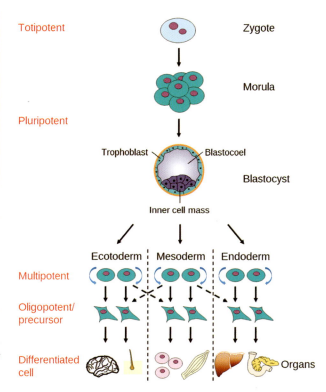

The development of new therapeutic options based on stem cells has been among the most intensively discussed topics in molecular medicine in recent years. Stem cells are undifferentiated cells that can differentiate into specialized cells; they can also divide to produce more stem cells. Intensive research efforts are underway to explore the potential of stem cells to treat or prevent disease. However, bone marrow transplantation is the only form of stem cell therapy routinely used in practice. All other envisaged applications of stem cells, for example, in the treatment of neurodegenerative diseases, diabetes, and heart disease, are still in their infancy and will require time before becoming clinically routine.

The most common types of stem cells are either embryonic stem cells, which are usually isolated from the inner cell mass of blastocysts, or adult stem cells, which are found in several tissues. More recently, specialized adult cells have been reprogrammed to assume a stem cell-like state; this new type of stem cell is called the induced pluripotent stem cell. Despite substantial differences, all types of stem cells are characterized by two properties: (i) They can self-renew, that is, they can divide indefinitely while remaining in an undifferentiated state. This property permits the *in vitro* culturing of stem cells indefinitely. (ii) Stem cells can differentiate to form specialized cell types. Stem cells are, therefore, distinctly different from immortalized cell lines, such as HeLa, HEK 293, or CHO cells, and from primary cells. The latter cannot divide indefinitely and, thus, cannot be cultured *in vitro* for an extended period. Immortalized cell lines usually do not differentiate into other, more specialized cell types and their ability to divide indefinitely has been acquired at the price of mutations, which makes them unstable and somewhat artificial.

The potential to differentiate into more specialized cell types, known as potency (Figure 12.1), varies between different types of stem cells. Cells with the greatest differentiation potential are known as *totipotent*. This designation applies to cells that possess the ability to divide and

Fig. 12.1 Cell potency. Fertilization of an egg by a sperm cell generates the zygote, a totipotent cell that can differentiate into all cells of the embryo and construct a complete, viable organism. Cells produced by the first few divisions of the zygote are also totipotent. Cells of the inner cell mass of the blastocyst are pluripotent, that is, they can differentiate into the three germ layers: the endoderm, mesoderm, and ectoderm. Multipotent stem cells can differentiate into a number of closely related cells. Under certain conditions, differentiation of multipotent stem cells into cells of other lineages has been observed (dashed lines). Oligopotent stem cells can differentiate into only a few cell types.

produce all of the cells in an organism as well as embryonic components that are required for development but do not become tissues of the adult body (i.e., the placenta). Totipotent stem cells are produced when a sperm cell fertilizes an egg, resulting in a zygote. In the first hours after fertilization, this zygote divides into identical totipotent cells. The 16-cell stage is named the *morula*. At this stage, the cells lose their totipotency by differentiating into cells that will eventually become part of either the blastocyst's inner cell mass or the outer trophoblasts. Four to five days postfertilization, the blastocyst consists of approximately 150 cells, which have not yet implanted into the uterus, and is composed of three primary components: the inner cell mass, which subsequently forms the embryo and will become the adult organism, the trophoblast, which becomes the placenta, and a fluid-filled space named the blastocoel.

The cells of the inner cell mass of a blastocyst are *pluripotent*. Pluripotency, in its original sense, describes a cell that can differentiate into each of the approximately 220 different cell types found in the adult. These cell types originate from three germ layers: (i) The *endoderm* is the inner layer of embryonic cells that will develop into the lungs, digestive organs, liver, pancreas, and other organs. (ii) The *mesoderm* is the middle layer of embryonic cells that will develop into muscle, bone, blood, kidneys, connective tissue, and related structures. (iii) The *ectoderm* is the outer layer of embryonic cells that will develop into skin, the nervous system, sensory organs, tooth enamel, the lens of the eye, and other structures. These layers are determined by the physical position of the cells in the gastrula, the stage of embryonic development following the blastocyst stage. Pluripotent cells, as has been discovered in the last decade, can also differentiate into germ lines and can give rise to an entire organism.

Multipotent cells form the next layer in the stem cell hierarchy. They can differentiate into multiple mature cells, usually a related group of differentiated cell types. For example, hematopoietic stem cells (HSCs) (Figure 1.26 in Section 1.5) are multipotent cells. They can differentiate into myeloid or lymphoid progenitor cells; however, a multipotent hematopoietic stem cell cannot differentiate into brain cells, bone cells, or other nonhematopoietic cell types. Multipotent cells have also been found in adipose tissue, the heart, and the bone marrow. Mesenchymal stromal cells (MSCs) are also multipotent and can differentiate into osteoblasts, chondrocytes, and adipocytes.

Myeloid and lymphoid progenitor cells are examples of *oligopotent* cells and have the ability to differentiate into a small number of cell types. For example, a lymphoid progenitor cell can differentiate into natural killer cells, T cells, B cells, or dendritic cells, but it cannot give rise to red blood cells (erythrocytes).

Research in recent years has revealed that these potency levels represent a continuum. Pluripotent cells have been found to possess properties that were originally attributed to totipotent cells. Thus, it has been proposed that the distinction between totipotency and pluripotency should be abandoned. Furthermore, fully differentiated cells are nowadays routinely dedifferentiated into pluripotent cells and different types of differentiated cells can be converted into one another. Taken together, these findings blur the strict hierarchy of stem cell potency that was originally proposed.

12.1
Embryonic Stem Cells

12.1.1
Generation and Properties of Embryonic Stem Cells

Embryonic stem cells are pluripotent stem cells that usually arise from the inner cell mass of a blastocyst. They can grow without limit in culture or can be induced to differentiate into all the cell types of the three germ layers. The first pluripotent ESCs from mouse blastocysts were isolated in 1981 by Martin Evans and Matthew Kaufman at the University of Cambridge, UK, and by Gail Martin at the University of California, San Francisco. The most common procedure used to obtain ESCs involves isolation of the inner cell mass (ICM) of a blastocyst (Figure 12.2). The cells are transferred to a culture dish containing an appropriate medium. The surface of the culture dish is typically coated with a layer of *feeder cells* that have been

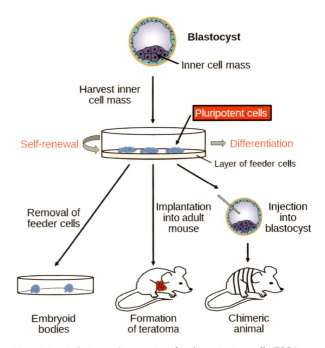

Fig. 12.2 Isolation and properties of embryonic stem cells (ESCs). ESCs are typically isolated from the inner cell mass of a blastocyst. They are transferred to a culture dish and grown on a layer of feeder cells. These stem cells can be grown without limit (self-renewal) or they can be differentiated into a specialized cell type. Upon removal of the feeder cells, ESCs will form three-dimensional embryoid bodies. When implanted into adult immunodeficient mice, they will form teratoma-type tumors. ESCs can also be implanted into a mouse blastocyst where they will give rise to a chimeric animal.

inactivated using X-irradiation or drug treatment to prevent them from dividing. The feeder cells are necessary to supply the ESCs with nutrients, growth factors, and extracellular materials. They produce leukemia inhibitory factor (LIF), an interleukin-6 family cytokine which is essential to maintain ESCs in an undifferentiated, pluripotent state.

Like any type of stem cell, ESCs can be grown without limit in cell culture or can alternatively be differentiated into a specialized cell type. Due to their pluripotency, they possess several additional properties (Figure 12.2): When mouse ESCs are plated without feeder cells on nonadhesive dishes, they form three-dimensional aggregates known as *embryoid bodies*. Over 2–3 weeks, embryoid bodies generate the different cell types found in a normal embryo (however, they do not form a complete embryo). ESCs can also be implanted into an adult immunosuppressed animal, where they will form a special type of tumor called a *teratoma*. These tumors contain a wide mixture of tissue types, often consisting of cells from all three germ layers. Another feature of ESCs is their ability to integrate into the inner cell mass when injected into another mouse blastocyst. There, they contribute to the development of all fetal tissues; the organism is thus a chimera consisting of a mixture of genetically distinct cells. Implanted ESCs can even contribute to the germ cells of these chimeric embryos. It is thus possible to breed offspring from the chimeric animal that carry the genetic background of the ESCs that were injected into the blastocyst. The ability to form embryoid bodies, giving rise to teratomas and contributing to the developing organism when injected into a blastocyst, have become important criteria in demonstrating the pluripotency of a cell line.

ESCs have been intensively characterized. Their pluripotent character is maintained by a network of transcription factors. Critical transcription factors for the maintenance of pluripotency include Oct4, Sox2, and Nanog. These factors activate and repress genes to prevent the cell from entering the early steps of differentiation, thus keeping it in a stable pluripotent state.

The first human embryonic stem cell line (hESC) was established in 1998 by James Thomson and coworkers at the University of Wisconsin, Madison. These experiments have generated controversy, since they require the destruction of a human embryo. Typically, hESCs are isolated from surplus preimplantation embryos generated during an *in vitro* fertilization (IVF) procedure. The ethical implications of ESC research will be discussed in Section 15.2.2. To avoid these ethical issues, researchers developed a method to generate hESCs without embryo destruction. Single blastomeres were removed from the embryos with a technique similar to

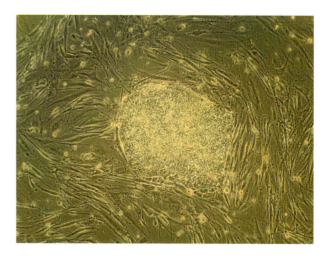

Fig. 12.3 Human embryonic stem cell (hESC). The image shows a colony of hESC, from the H9 cell line. The colony contains a few hundred small cells and is surrounded by a feeder layer of large elongated fibroblasts.

that used in preimplantation genetic diagnosis. The remaining embryos were capable of growing to the blastocyst stage. To improve the efficiency of the generation of hESC lines, the isolated blastomeres were cultured in a specific environment that resembled their niche in the inner cell mass of the blastocyst.

Figure 12.3 shows a colony of hESCs. Human and mouse ESCs are very similar at a basic level, despite some differences in gene expression, behavior, and appearance. Both types of ESCs possess the defining properties of stem cells, that is, self-renewal and the potential to differentiate into specialized cell types. There is general consensus that it is highly unethical to inject hESCs into a human blastocyst in an attempt to develop a chimeric (human) organism. However, hESCs have been demonstrated to form embryoid bodies *in vitro* and to generate teratomas when engrafted into immunodeficient mice that do not reject them. These types of evaluations are commonly performed to confirm the pluripotency of newly established hESC lines.

Most of the hESC lines that have been established to date cannot be used for therapeutic purposes in humans since they were exposed to animal cells. For many years, all hESC lines were established and maintained with mouse embryonic fibroblasts as feeder layers and/or with medium conditioned with murine components to promote their propagation in the undifferentiated state. However, exposure of the hESC lines to animal components risks contamination with pathogens (e.g., retroviruses) that can be transmitted to a human patient (and potentially to a wider human population). Therefore, protocols were developed to establish and maintain hESCs in the absence of all animal-derived products. These approaches include the use of human feeder systems and feeder-free systems. This is achieved by the use of xeno-free media (i.e., media free of animal components) and special matrices to replace the feeder cells. Although it is possible to cultivate hESCs using these procedures, their efficacy is not yet adequate. Further improvements in the cultivation of hESCs under completely feeder layer- and serum-free conditions can be expected to solve the problems associated with the use of hESCs in the treatment of human patients.

12.1.2
Therapeutic Cloning

hESCs have another important disadvantage that limits their utility in regenerative medicine: Since the hESC lines are not patient-specific, transplants generated from them are subject to immune rejection by the recipient. *Therapeutic cloning* is an approach aimed at solving this problem. Cloning describes the creation of genetically identical copies of an organism. While *reproductive cloning* aims at creating an entire organism that is genetically identical to an existing donor (Box 12.1), therapeutic cloning is intended to produce cells for medical use, for example, transplants for regenerative medicine. The first step of the cloning procedure is to obtain donor cells from the patient, for example, from a biopsy (Figure 12.4). A nucleus from a donor cell is then fused to a denucleated egg. This procedure is identical to the somatic cell nuclear transfer (SCNT) procedure described in animals in Box 10.1 of Section 10.1.5. The

fused cell is cultivated to give rise to a blastocyst from which hESCs can be obtained as described above. After differentiation of the pluripotent hESCs into the desired type of cell (Section 12.5), they are transplanted into the patient. As this procedure is an autologous transplantation (i.e., transplantation of a graft to the same person from whom the cells (or their genetic material) had been taken), an immune reaction will not be induced.

Therapeutic cloning is ethically controversial, since it involves the destruction of an embryo, as does any experiment that generates an hESC line. In addition, it requires the development of technologies that may be misused for reproductive cloning, which is generally considered unethical. Another problem with therapeutic cloning is the need for human oocytes. Since only a small minority of the reconstituted eggs develops sufficiently to establish an hESC line following SCNT, a large number of human oocytes will be required if this method becomes a standard medical procedure. These cells can be obtained from female volunteers who undergo hormonal stimulation and laparoscopic harvesting of their oocytes. The egg donors are usually paid, but the procedure involves some health risk. However, the shortage of oocytes may yet be overcome, as it has been shown that oocytes can be derived from ESCs.

In 2004 and 2005, the Korean stem cell researcher Hwang Woo-suk reported successfully generating 11 different patient-specific hESC lines from a cloned human blastocyst using the SCNT technology. However, 1 year later he was forced to retract all his publications as much of his stem cell research had been faked.

Box 12.1. Unverified Claims of Reproductive Cloning of Humans

There is broad consensus that reproductive cloning of humans is unethical. However, some individuals have challenged this view. Severino Antinori is an Italian expert in IVF. He extended this technique to women who have gone through menopause. From the 1990s on, he has assisted women over 60 years old to become pregnant. He later argued that reproductive cloning of humans represents an important opportunity for infertile couples, who cannot employ other methods. In 2002, he announced that he had successfully used cloning to induce pregnancy in three women. However, he refused to provide the identities of the clones and his claims could never be confirmed.

Another group that supports reproductive cloning of humans is the Raëlian movement, a religion that

believes in UFOs and that life on earth was scientifically created by a species of extraterrestrial. The company Clonaid, led by the Raëlian Bishop Brigitte Boisselier, claimed in 2002 that an American woman underwent a cloning procedure that led to the birth of her daughter Eve. Raëlians consider this the first step not only to reproduce but also to improve human beings, for example, by the acceleration of the rate of growth into adulthood. In addition, as per the Raëlian vision, reproductive cloning offers the opportunity to transfer the mind and personality of the donor into the clone to achieve eternal life. The Raëlian claim that a human baby was conceived through cloning technology could never be independently verified and is not considered credible.

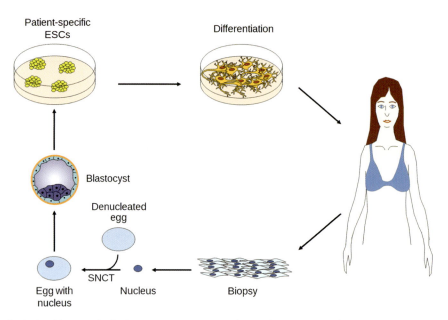

Fig. 12.4 Therapeutic cloning. Therapeutic cloning aims to generate autologous (or patient-specific) embryonic stem cells (ESCs). The procedure starts with the isolation of nuclei from a patient biopsy. The nuclei are transferred to denucleated eggs, a process known as somatic cell nuclear transfer (SCNT). The fused cell is cultivated to the blastocyst stage to isolate ESCs from the inner cell mass. Eventually, the ESCs are differentiated into the desired cell type and transplanted into the patient.

The generation of hESCs after SCNT has proven more difficult than expected, even though various animals including mice, goat, sheep, pigs, and monkeys had successfully been cloned by SNCT. Numerous attempts to generate hESCs from human cells failed due to early embryonic arrest. In 2011, scientists at the New York Stem Cell Foundation finally succeeded in generating hESC lines. Using standard SCNT procedures, they observed that the cells stopped dividing at the 6–10-cell stage. The researchers then placed the diploid donor nucleus into an oocyte that retained its haploid nucleus. These cells developed into blastocysts from which hESCs could be derived and that had the characteristic property of developing into the cells of all three germ layers. However, all cells contained three genome copies, one from the haploid oocyte and two from the diploid donor cell. This procedure was thus unsuitable for medical applications.

In 2013, the group of Shoukhrat Mitalipov from the Oregon Health & Science University was the first to obtain diploid hESCs using SCNT. They used somewhat different culture conditions compared to the standard procedure, which enabled the cells to develop into blastocysts from which they could derive hESCs. However, in these experiments, fetal or infant somatic donor cells were used. One year later, a research group reported the successful creation of hESCs by SCNT from adult donor cells using dermal fibroblasts from 35- and 74-year-old males.

Although the generation of patient-specific hESCs by SCNT is promising, it remains to be seen whether it will be of much importance in stem cell research. Ethical issues, the shortage of oocytes, and a technically challenging procedure remain hurdles for routine use. The progress made in the generation of patient-specific pluripotent cells by the reprogramming of somatic cells (Section 12.3) and the introduction of novel techniques for direct transdifferentiation (Section 12.4) provide more straightforward alternatives. However, both of these techniques have one intrinsic disadvantage: Reprogrammed cells still carry the mitochondrial genome of the original cell. Thus, this approach is not suitable for the treatment of mitochondrial diseases (Section 3.1.4). In contrast, therapeutic cloning involves the transfer of the nucleus of a patient's cell into a donor oocyte that contains healthy mitochondria, thus permitting the development of a cell therapy for the correction of mitochondrial defects.

12.2
Adult Stem Cells

Adult stem cells are undifferentiated cells that exist throughout the postnatal organism. They are responsible for tissue renewal and for repair following damage. There is no consistent definition of the term "adult stem cell." At times, induced pluripotent stem cells (see

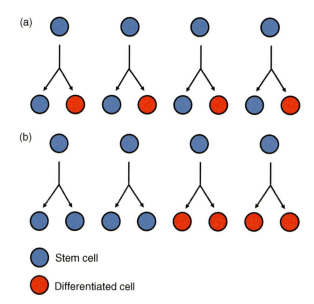

(a)

(b)

● Stem cell

● Differentiated cell

Fig. 12.5 Asymmetric and symmetric modes of stem cell division. (a) Asymmetric division of stem cells results in one daughter stem cell and one daughter cell that is destined to differentiate. (b) Symmetric division of stem cells generates two identical cells, which can be either two stem cells or two differentiated cells.

below) are considered adult stem cells. Nevertheless, here we will use the term to mean the tissue-specific, partially differentiated stem cells that are found throughout the human body after development. Adult stem cells share the common characteristics of all stem cells, that is, the ability to reproduce themselves (self-renewal) and to generate differentiated progeny cells. However, their molecular architecture, such as the expression of specific transcription factors and cell surface markers, is distinct from that of ESCs and iPSCs.

Stem cells can undergo two types of cell division (Figure 12.5): (i) Through asymmetric cell division, each stem cell generates one daughter stem cell and one progenitor cell with limited self-renewal potential that is destined to differentiate. (ii) Symmetric division gives rise to two identical daughter cells. These may be either two daughter stem cells or two differentiated cells. To maintain the stem cell population, symmetric division of one stem cell into two differentiated daughter cells requires another stem cell to divide and produce two stem cells. Most stem cells are currently believed to have the ability to switch between asymmetric and symmetric modes of division. Several signaling pathways including the Notch, Wnt, and TGF-β pathways control the self-renewal and differentiation of stem cells.

Adult stem cells are thought to exist in most of the body's tissues, including blood, brain, liver, intestine, and skin and are referred to by the tissue of their origin (e.g.,

mammary stem cells, dental pulp stem cells, and endothelial stem cells). Adult stem cells have the broad ability to differentiate into many types of cells, but they are usually committed to becoming a cell of the tissue they originate from. For example, hematopoietic stem cells give rise to all the blood cell types (Section 1.5), but they will not normally develop into a non-blood cell type. Adult stem cells are said to be restricted to certain *lineages*. However, more recent research has revealed that some types of adult stem cells can differentiate into cells of a different lineage under suitable conditions.

Among the most intensively studied adult stem cells are the hematopoietic stem cells. These cells are located in the bone marrow and generate all the cells of the blood in addition to immune cells. HSCs can divide to self-renew or to generate both common lymphocyte progenitor and common myeloid progenitor cells, which further differentiate to all types of lymphocytes (T cells, B cells, natural killer cells, and dendritic cells) and myeloid cells (including red blood cells, macrophages, granulocytes, and blood platelets) (Figure 1.26 in Section 1.5).

HSCs reside in a local tissue microenvironment called the bone marrow niche. The niche is perivascular and is created partly by mesenchymal stromal cells and endothelial cells. The cells creating the niche produce various factors that maintain the stem cells and regulate their fate. HSCs are particularly sensitive to radiation; they can be destroyed by a sufficiently high dose. Without further intervention, an irradiated animal would die of bone marrow failure, but it can be rescued by injecting HSCs into its bloodstream. The HSCs will engraft in the bone marrow niches and reconstitute the hematopoietic and immune systems of the treated animal. The ability of HSCs to recolonize an available niche is the basis for the use of HSCT to treat lymphomas, leukemias, and other blood diseases in human patients, as described in Section 12.6.1.

While the hematopoietic system is clearly self-renewing and contains stem cells, other organs do not renew. Rather, they consist of post-mitotic cells that are formed during embryogenesis or soon after birth, but never divide thereafter. The heart and the brain are two organs that would benefit from suitable stem cells to treat the debilitating diseases resulting from cell loss. However, the neurons of the brain and the cardiomyocytes in the heart belong to the category of post-mitotic cells. It is controversial if new neurons or cardiomyocytes can be formed during adult life. While new heart muscle is produced very slowly, if at all, the formation of new neurons has been observed in two locations in the brain, the subventricular zone that lines the lateral ventricle from where new neurons migrate to the olfactory bulb, and the dentate gyrus of the hippocampal formation.

These two regions contain neural stem cells that can differentiate into neurons and *glial* cells. The latter are nonneuronal cells that provide support and protection for neurons. They also maintain homeostasis and help insulate the neurons. Two types of glial cells exist: *astrocytes* and *oligodendrocytes*. In addition to these two areas of the brain, neurogenesis can be induced in other regions of the brain under certain conditions, such as tissue damage caused by ischemia.

When neuronal stem cells are cultured *in vitro*, they form what is called *neurospheres*. These are floating heterogeneous aggregates containing only a few neural stem cells and a larger number of cells with a finite division potential. Under suitable conditions, neurospheres differentiate into both neuronal and glial cells. There is great interest in neurospheres, because, unlike HSCs, they can be maintained in culture and may be useful for the cell therapy of neurodegenerative diseases.

Another type of multipotent cell that can proliferate in culture and has gained much attention in recent years is the mesenchymal stromal cell (MSC), originally referred to as a "mesenchymal stem cell." However, this designation is misleading, as MSCs develop only *in vitro* via cultivation. *In vivo* they do not express the specific markers found *in vitro* and no specific MSC niche is known. MSCs can be obtained from various sources. The earliest MSCs can be isolated from umbilical cord tissue, especially from Wharton's jelly and umbilical cord blood. These MSCs may be a particularly useful source for clinical applications, as they have more primitive properties than MSCs obtained later in life. Another rich source of MSCs is the developing tooth bud of the mandibular third molar (wisdom tooth). These stem cells eventually develop into at least 29 different unique end organs, including enamel, blood vessels, dental pulp,

and nervous tissues. Adipose tissue is another almost limitless source of MSCs. Although the exact function of MSCs in living organisms is not fully understood, MSCs are a valuable resource for biomedical research, since they can be easily obtained and grown in cell culture and can differentiate into various cell types, especially bone, cartilage, fat, or smooth muscle cells. MSCs obtained from the developing tooth bud have also been shown to produce hepatocytes.

Mammary stem cells are the source of cells for growth of the mammary gland and can give rise to both the luminal and myoepithelial types of breast cells. The mammary gland undergoes significant developmental changes during pregnancy, lactation, and involution, and the cells relevant to these processes are probably provided by stem cells. However, mammary stem cells may also play an important role in mammary carcinogenesis and in other types of tumors (Box 12.2).

Research in the past decade has revealed that stem cells reside in many adult tissues where they provide a reservoir of cells for reparative and regenerative processes. For biomedical purposes, adult stem cells are an attractive alternative to hESCs, as their use is devoid of the ethical dilemmas resulting from the destruction of an embryo. Research over the last several years has demonstrated the potential of adult stem cells to differentiate into numerous cell types. Although their potency is a source of controversy, it is unlikely that normal organisms contain pluripotent cells (except for sperm stem cells) after the earliest stages of embryonic development. It is also an open question if the multipotency of adult stem cells is sufficient to replace pluripotent hESCs for every application that can be envisioned. In addition, the more recently generated iPSCs discussed in the following section provide an attractive alternative source of pluripotent cells.

Box 12.2. Cancer Stem Cells

The existence of cancer stem cells (CSCs) has been a subject of debate for many years. However, after their initial discovery in myeloid leukemia cells (Section 4.7) in the late 1990s, CSCs have been convincingly demonstrated to be present in several solid tumors, including breast, brain, prostate and colon carcinomas and melanoma. CSCs possess the characteristics of self-renewal and differentiation into all the cell types found in a particular cancer. CSCs can seed tumors when transplanted into animal hosts and may account for resistance to therapy, relapse, and metastasis, although they often only comprise a small proportion of cells in a cancer (e.g., less than 1 in 10 000 cells in

acute myeloid leukemia). However, the proportion of CSCs seems to vary widely between tumor types.

Heterogeneity is a defining feature of many types of tumors. This heterogeneity is due to genetic variations, epigenetic changes, and other factors. According to the conventional clonal evolution theory (Figure 12.6a), a cell acquires a series of mutations that confers a growth advantage. These events produce a dominant clone that gives rise to tumor formation. All cells of the tumor arising from this clone have similar tumorigenic capacity. Tumor heterogeneity results from the diversity of cells present within the tumor. According to the more recently

(a) **Clonal evolution model**

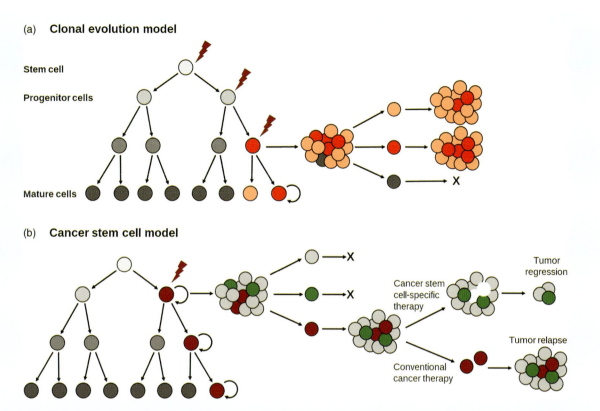

(b) **Cancer stem cell model**

Fig. 12.6   The clonal evolution and cancer stem cell (CSC) models. (a) According to the clonal evolution model, mutations accumulating in tumor cells confer a selective growth advantage. All tumor cells arising from the initial dominant clone have similar tumorigenic potential. Gray cells have lost their proliferative capacity due to stochastic events. (b) The CSC model is based on the assumption of a hierarchical organization of cells. Only a small subset of the cells, the CSCs, have the capacity for the self-renewal and generation of tumor cells. This model demonstrates the necessity of developing new cancer drugs that specifically target cancer stem cells. (Adapted from Ref. [1] with kind permission from Elsevier.)

proposed CSC model, only a small subset of the cells is responsible for sustaining tumorigenesis and establishing heterogeneity in the primary tumor. In the example shown in Figure 12.6b, a progenitor cell has acquired mutations and has become a tumor cell with stem cell-like properties. This progenitor cell can give rise to a wide range of tumor cells, thus accounting for tumor heterogeneity.

The origin of CSCs is still under debate and various sources of CSCs have been proposed: CSCs may originate by mutations from developing (adult) stem or progenitor cells resulting in dysregulation of proliferation and differentiation pathways. Alternatively, they may develop from differentiated cells that dedifferentiate, acquiring stem cell-like characteristics. Another theory states that the cancer stem cell is generated by a mutation in the stem cell niche population during development.

CSCs can be separated from other non-CSCs in a tumor sample based on their specific cell surface protein antigen profile and their ability to seed new tumors. Recent research has demonstrated that CSCs are not only important in the primary tumor but may also play a pivotal role in the metastatic process. There is growing evidence for the existence of two functionally distinct subsets of CSCs: Stationary CSCs embedded in the primary tumor and migrating CSCs that impart metastatic activity. An important step in the malignant transformation is the so-called epithelial mesenchymal transition (EMT). This is mediated by TGF-β, which converts stationary CSCs into mobile and migrating CSCs. The transition involves disruption of epithelial cell homeostasis, loss of membrane proteins in adherent junctions, and the acquisition of a migratory mesenchymal phenotype. Mobile CSCs divide asymmetrically, that is, one daughter cell begins to proliferate and differentiate, while the remaining CSC migrates a short distance before it undergoes another asymmetric division or begins to disseminate through blood and lymphatic vessels to produce metastasis.

The CSC model has several important implications for cancer treatment. A common characteristic of stem cells is their resistance to chemotherapeutic agents via the expression of efflux transporters (Section 9.1). CSCs pump cytotoxic substances, including chemotherapy, out of the cell. In addition, CSCs have a slow rate of cell turnover, which also confers resistance to chemotherapeutic agents that primarily target rapidly replicating cells. New insights into CSC biology may require a shift in how cancer therapeutics are developed. Most existing cancer treatments promote maximal tumor shrinkage (right part of Figure 12.6b). However, despite an efficient initial reduction in tumor mass, current treatments will not necessarily kill CSCs. Consequently, the remaining CSCs will produce new cancer cells, which can lead to tumor relapse and metastases. In contrast, the specific elimination of CSCs might result in tumor regression due to cell differentiation and/or death. Targeting pathways that maintain stem cell properties may be critical for the specific eradication of CSCs.

In 2009, a library of 16 000 compounds was screened for their ability to kill drug-resistant stem cell-like breast cancer cells. One of these compounds, the antibacterial ionophore salinomycin, which is currently used in veterinary medicine, was a 100-fold more efficient killer of CSCs than taxol, a commonly used breast cancer chemotherapeutic agent. Initial clinical studies were promising: Salinomycin was able to effectively eliminate CSCs and induce partial clinical regression of therapy-resistant cancers.

12.3
Induced Pluripotent Stem Cells

Induced pluripotent stem cells (iPSCs) are a special type of pluripotent stem cell and are obtained by reprogramming differentiated adult cells. These cells can propagate indefinitely or can be differentiated into other cell types in the body, for example, into neurons, heart, pancreatic, and liver cells. They are very similar to ESCs, but have some important advantages. Since they are produced directly from somatic cells, they bypass the need to destroy embryos, eliminating ethical considerations. In addition, they can be produced in a patient-specific manner, allowing the generation of autologous transplants without the risk of immune rejection. Thus, iPSCs hold great promise in the field of regenerative medicine. The iPSC technology was pioneered by Shinya Yamanaka and coworkers in Kyoto, Japan, who demonstrated in 2006 that the introduction of four specific genes converted differentiated adult fibroblast cells into iPSCs. For his pioneering work, Yamanaka was awarded the Nobel Prize for Physiology or Medicine together with Sir John Gurdon in 2012.

12.3.1
Generation of Induced Pluripotent Stem Cells

In their seminal experiment generating pluripotent stem cells from differentiated somatic cells, Yamanaka and coworkers used mouse fibroblasts. By retroviral gene transfer (Figure 12.7), they introduced cDNAs encoding four transcription factors: *Oct4* (*octamer-binding transcription factor 4*), *Sox2* (*sex-determining region Y (SRY)-box 2*), *Klf4* (*Kruppel-like factor 4*), and *c-Myc* (*c-myelocytomatosis oncogene*), dubbed the OSKM factors (as per the first letters of the four factors) or Yamanaka factors, into terminally differentiated fibroblast cells. The 4 factors were selected out of an initial set of 24 factors known to be important to maintain pluripotency in ESCs. This treatment produced ESC-like colonies that reactivate expression of the *Fbx15* gene, a specific marker for ESCs. The reprogrammed cells were named induced pluripotent stem cells. The spelling of iPSCs with a lower case "i" was chosen in the hope that the cells would spread across the world like the "iPod" that

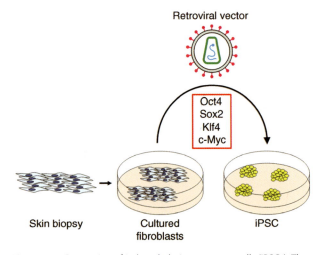

Fig. 12.7 Generation of induced pluripotent stem cells (iPSCs). The initial strategy to reprogram fibroblasts into pluripotent cells included the retroviral transfer of the four transcription factors Oct4, Sox2, Klf4, and c-Myc. These four factors reprogrammed differentiated cells into pluripotency.

became popular at about the same time. iPSCs had typical features of ESCs, that is, they possessed the capacity of unlimited self-renewal and pluripotency, as they could form embryoid bodies and differentiate into lineages from all three germ layers. In addition, they gave rise to teratomas and contributed to fetal chimeras when injected into blastocysts. However, due to several differences from real ESCs with respect to gene expression patterns and epigenetic markers, these cells failed to produce viable adult chimeras.

Shortly afterward, several groups independently developed procedures to reprogram fibroblasts into iPSCs that did produce viable chimeras after injection into blastocysts. The main difference between the approaches was the use of Nanog instead of Fbx15 as a marker for pluripotent cells.

The next logical step was to create iPSCs from adult human cells. Yamanaka's group and the group of James Thomson, who had also established the first hESC line, achieved this goal independently. While the Japanese group used the same four factors and retroviral system they had successfully applied to the murine cells, the US group used several different factors (Nanog and LIN28 instead of KLF4 and c-Myc) and a lentiviral system.

Although the procedures for the production of iPSCs were immediately recognized as being groundbreaking

for stem cell research, one limitation soon became obvious: 20% of the chimeric mice produced by injecting murine iPSCs into blastocysts developed cancer. This finding was not unexpected, as some of the transferred factors, for example, c-Myc and KLF4, were known to have oncogenic potential. In addition, retroviral vectors integrate into the host genome and may also induce tumor growth, as discussed in Section 11.2.1. The use in humans of iPSCs generated by the methods employed initially is still not possible. Another shortcoming of the reprogramming procedure is its slowness and low efficiency. Typically, it takes several weeks to generate iPSCs, with an efficiency below 1%. Stem cell researchers are, therefore, continuously developing improved methods to reduce the oncogenic potential of the iPSCs, to increase efficiency and to shorten the time required to produce iPSCs (Table 12.1).

Alternative approaches to generate iPSCs have employed different types of vectors to transfer the genes needed for reprogramming. As discussed in Section 11.2.1, the risk of insertional mutagenesis associated with retroviral vectors can be avoided by the use of vectors that remain episomal. In fact, adenoviral vectors were found to be suitable for the delivery of the reprogramming factors. The transient expression of the transgene by episomal vectors is sufficient as conversion of somatic

Table 12.1 Summary of major methods for reprogramming somatic cells into induced pluripotent stem cells (iPSCs).

Approach type	Vector	Factors	Advantages	Disadvantages
Integrating	Retroviral	OSKM	Reasonably efficient	Genomic integration, oncogenic
Nonintegrating	Adenoviral	OSKM	No genomic integration	Low efficiency
	Plasmid	OSNL	Only occasional genomic integration	Low efficiency, occasional genomic integration
Excisable	Transposon	OKSM	Reasonably efficient, no genomic integration	Labor-intensive screening of excised lines
DNA-free	Protein	OSKM	No genomic integration	Low efficiency, requirement of large quantities of pure proteins and multiple applications of proteins
	Modified mRNAs	OSKML	No genomic integration, faster reprogramming kinetics, controllable and highly efficient	Requirement for multiple rounds of transfection
	MicroRNAs	miR-200c, miR302 family, miR-369 family	No genomic integration, reasonably efficient, comparatively fast reprogramming kinetics, no exogenous transcription factors	Lower efficiency than other methods
Chemical	Retroviral	OSK + VPA	Reasonably efficient, omission of oncogenic factors	Genomic integration
	None	Seven chemicals	Reasonably efficient, omission of oncogenic factors, no genomic integration	Method needs improvement

O: Oct4; S: Sox2; K: Klf4; M: c-Myc; N: Nanog; L: LIN28; VPA: valproic acid.
Modified after Ref. [2].

cells into iPSCs only requires short-term activity of the reprogramming factors. Following modulation of the genetic program, iPSCs remain stably pluripotent without the need for further intervention. The four required genes can even be transferred by transfection of plasmid DNA, completely avoiding the use of a viral vector. However, an important drawback of these methods is that they still require cancer-promoting genes to accomplish reprogramming. Despite drastically reducing the risk of insertional mutagenesis, transfected plasmids have previously been found to integrate into the genome at low frequency. In addition, the efficiency of the non-retroviral approaches is very low. A transposon-based delivery system (Section 11.2.4) was shown to deliver the reprogramming factors at higher efficiency. Transposons initially integrate the delivered gene into the host cell genome, but eventually re-excise it so that the reprogramming procedure does not produce mutations. Although the system occasionally fails, the risk of insertional mutagenesis is still substantially reduced.

Additional strategies have been developed that completely avoid the transfer of DNA. In 2009, iPSCs were generated by treating somatic cells with recombinantly expressed proteins instead of transferring the encoding genetic material. To achieve cellular uptake of the four OSKM factors, the proteins were fused to a cell penetrating poly-arginine peptide. The cells generated were dubbed protein-induced pluripotent stem cells (piPSCs). Another method for reprogramming human cells was based on the transfer of *in vitro* transcribed mRNAs. A cocktail of mRNAs encoding the OSKM factors and Lin28 was transfected into human fibroblasts. The mRNAs were transiently translated into proteins, but did not produce permanent genetic changes. To improve the efficiency of this process, several chemically modified nucleotides were incorporated into the mRNAs to enhance their stability, and the interferon response was blocked to reduce cytotoxicity. Nevertheless, multiple rounds of transfection were needed to obtain pluripotent cells, termed RiPSCs (RNA-induced pluripotent stem cells). An additional RNA-based approach was based on the transfer of microRNAs (miRNAs) rather than mRNAs. As discussed in Section 13.4, miRNAs are short RNA molecules that regulate the expression of target mRNAs. Several miRNA families that are highly expressed in ESCs are capable of converting somatic cells into iPSCs (dubbed mi-iPSCs). Although the exact mechanism by which miRNAs mediate reprogramming of somatic cells is under investigation, they seem to inhibit pathways that limit reprogramming.

Other attempts to improve the reprogramming procedure aim to replace reprogramming factors by small molecules. In fact, valproic acid, a histone deacetylase inhibitor that is in medical use for the treatment of epilepsy, bipolar disorder, and the prevention of migraine headaches, was found to mimic the signaling that is usually caused by the transcription factor c-Myc. Hongkui Deng's stem cell group at Beijing University, China eventually succeeded in producing chemically induced pluripotent stem cells (CiPSCs) that were reprogrammed without transferring any genes. This group initially screened 10 000 small molecules and eventually identified a cocktail of seven compounds that was capable of converting cells at an efficiency comparable to that of the standard techniques. CiPSCs contributed to all major cell types when introduced into developing mouse embryos, demonstrating their pluripotency. These experiments were carried out with murine cells; the procedure must now be transferred to human cells. This may require adjustment of the chemical cocktail used for reprogramming.

More recently, it was claimed that subjecting cells to certain types of stress, such as low pH, resulted in the generation of pluripotent cells. However, these studies could not be independently reproduced and the claims were retracted.

Since the initial generation of iPSCs by Yamanka and coworkers, several additional procedures have been developed for this purpose. iPSCs have been obtained from various cell types, including fibroblasts, liver, skin, blood, prostate, and urinary tract cells. Nevertheless, protocols with improved efficiency and reprogramming kinetics are still needed for the production of iPSCs. For example, it has been shown that depletion of Mbd3, a core member of the nucleosome remodeling and deacetylation (NuRD) complex, results within 7 days in deterministic and synchronized iPSC reprogramming with near 100% efficiency in both mouse and human cells. In addition, the oncogenic potential of iPSCs still needs to be minimized prior to their use in humans.

12.3.2
Properties of Induced Pluripotent Stem Cells

As an increasing number of methods are being developed to generate iPSCs, strict criteria are required to demonstrate that a presumed iPSC line is, in fact, pluripotent. A crucial test to confirm pluripotency includes the differentiation of an iPSC into cells from all three germ layers. As stated previously, this can be performed *in vitro* by differentiating iPSCs into various cell types of the endoderm, mesoderm, and ectoderm, or by demonstrating *in vivo* teratoma formation by the iPSCs. In addition, iPSCs should give rise to cells from all the cell types of a newborn mouse after injection into an early embryo.

When these criteria are fulfilled, iPSCs have various advantages over classical hESCs. As already mentioned, their production does not require the destruction of an embryo, avoiding ethical considerations. In addition, they can be obtained in a patient-specific manner and are thus immunocompatible. hESCs produced by SCNT are also patient-specific, but their generation is technically challenging and currently requires numerous human oocytes. However, whether iPSCs are a suitable alternative to hESCs for use in research and therapy still requires detailed comparative analysis.

Initial characterization confirmed the pluripotency of iPSCs as they differentiate *in vitro*, form embryoid bodies, and develop into teratomas when injected into immuno-deficient mice. In addition, iPSCs express major markers of pluripotency at a level comparable to hESCs (e.g., *Oct4*, *Sox2*, and *Nanog*). However, additional analysis revealed several differences between iPSCs and hESCs. Several studies indicated that the type of the parental cell from which an iPSC was obtained can influence its capacity for differentiation. For example, iPSCs derived from bone marrow or B cells more efficiently differentiated along hematopoietic lineages than did fibroblast- or neural-pro-genitor-derived iPSC lines. Further investigation suggested that these limits were due to epigenetic modification patterns (Section 1.3.4). The iPSCs are thought to retain "epigenetic memory" (sometimes also called "somatic memory") of their past identity, which may influence patterns of global gene expression.

Several studies have thus addressed the global methylation status (also known as the methylome) of iPSCs. Overall, remarkable global similarities were observed between iPSCs and ESCs. However, several differentially methylated regions (DMRs) were identified; these may originate from incomplete reprogramming and may influence the differentiation of the iPSCs. However, DMRs are not consistently found and much is not understood about them. In addition, gene expression patterns and epigenetic markers seem to depend on the method by which the iPSCs were generated (e.g., viral versus nonviral approaches, see above). Despite intensive comparative studies, the question of whether iPSCs are equivalent to ESCs cannot be easily answered. However, ESCs themselves have considerable epigenetic hetero-geneity and differing propensities for differentiation. Thus, at the current stage of research, the various types of pluripotent cells are complementary to each other. Transcriptional and epigenetic variations within each class of cells and between iPSC and ESC populations will clearly affect their utility in research and in thera-peutic development. Overall, iPSCs appear to have several advantages over ESCs, including the relative ease of obtaining them in a patient-specific manner and the

absence of complex ethical issues associated with the generation of ESCs. In the future, iPSCs can be expected to become the researcher's choice of pluripotent cell.

12.4
Transdifferentiation and Direct Reprogramming

A more recently developed strategy has been designed to obtain a desired patient-specific cell type for regenerative or personalized medicine while bypassing the intermediate pluripotent state. This strategy is devoid of the risk of teratoma formation, which is one of the major problems of the therapeutic application of cells derived from ESCs or iPSCs (see below). Direct cell-type switching is usually achieved by the expression of lineage-specific transcription factors or miRNAs. Various studies have demonstrated that fibroblasts could, for example, be converted directly into neurons, cardiomyocytes, or hepatocytes. Exocrine pancreatic cells could be switched into insulin-secreting pancreatic β-cells and endothelial cells could be obtained from fibroblasts and amniotic cells.

Two slightly different approaches designed to achieve the cellular switch can be distinguished, despite their common feature that one cellular type can be converted to another without passing through a pluripotent stage. The first strategy, known as *transdifferentiation*, is usually the result of overexpression of lineage-specific factors that push a differentiated cell into the lineage of interest. The second strategy of *direct reprogramming* involves the short-term cellular conversion toward the pluripotent stage (without reaching it), before directing the cell toward a lineage-specific path.

In one of the early attempts at transdifferentiation, in 2008 exocrine pancreas cells were converted into insulin-secreting β-cells. This approach started with the screening of more than 1000 transcription factors and finally settling on 3 transcription factors that were sufficient for the desired cell conversion. The final cell product may be valuable for the long-term treatment of diabetes.

Following these experiments, several groups reported the transdifferentiation of fibroblasts into cardiomyocytes. This conversion can be achieved through overexpression of three factors, collectively referred to as the GMT factors (for GATA-binding protein 4, myocyte-specific enhancer factor 2C, and T-box 5). This reprogramming strategy was successful not only in cell culture but also *in vivo*. Retroviral transfer of the GMT factors into infarcted mouse hearts converted cardiac fibroblasts into cardiomyocyte-like cells with cardiac-specific gene expression and sarcomeric structures. Several groups reported that the *in vivo* induction of cardiomycytes resulted in significant functional recovery after myocardial infarction. However, it was also possible that the cardiomyocytes could have originated from immature cardiomyocytes or cardiac progenitor cells in the initial fibroblast pool.

Several methods have been developed to directly convert human skin cells into neurons. These approaches include combinations of different transcription factors and/or microRNAs. The transcription factors NeuroD1 and NeuroD2 played an essential role in the efficiency of the conversion process. The resulting neurons were functional and secreted the neurotransmitter dopamine. They became successfully integrated and functional when engrafted into neonatal mouse brains. However, the conversion of human fibroblasts from patients with Parkinson's disease (PD) proved to be both incomplete and less efficient. This problem must be solved before transdifferentiation can be used for the development of a cell-based treatment of neurodegenerative disorders.

The mechanism of transdifferentiation is not fully understood and the cells generated have several limitations. These include restricted proliferative capacity,

limited cell-type diversity, and the assumption of senescence. Therefore, the alternative approach of the so-called direct reprogramming has been developed, which involves the short-term opening of pluripotency-related pathways. Here, fibroblasts were reprogrammed with the four OSKM factors previously used to generate iPSCs. For the direct reprogramming procedure, overexpression of the four factors was restricted to 4 days compared to the standard treatment period of 8–12 days, used for the production of iPSCs. The cells, named partial-iPSCs, responded to different signaling environments and could be reprogrammed into specific cell types after an appropriate stimulus. For example, the procedure allowed the generation of vascular endothelial cells from fibroblasts. In contrast to iPSCs, these cells did not form tumors when injected into immunodeficient mice since they had not reached the pluripotent stage.

Figure 12.8 shows a model for the direct reprogramming of fibroblasts into cardiomyocytes. After a brief overexpression of the reprogramming factors, the cells reach an unstable intermediate stage. These intermediate

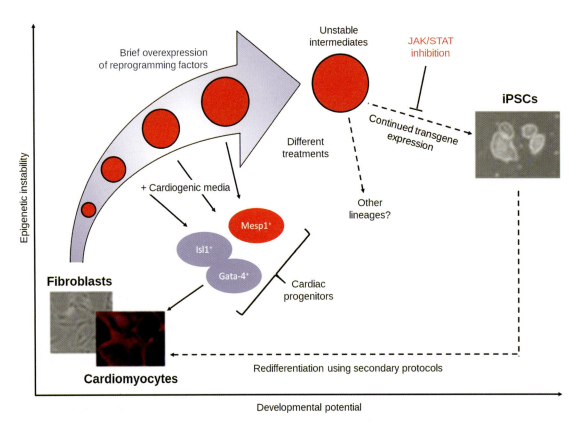

Fig. 12.8 Direct reprogramming of fibroblasts into cardiomyocytes. Overexpression of four factors (OKSM) reprograms fully differentiated fibroblasts to an unstable intermediate stage and eventually to the pluripotent iPSCs. Inhibition of the establishment of pluripotency (e.g., by inhibition of the JAK/STAT pathway) and the choice of suitable culturing conditions results in the generation of cardiac progenitors and eventually cardiomyocytes. (Adapted from Ref. [3] with kind permission from Macmillan Publishers Ltd., Copyright 2011.)

populations are characterized by an open-chromatin state marked by high epigenetic instability. The cells tend to relax back into more stable epigenetic states that may be either pluripotent or differentiated. Continued treatment of the cells with the reprogramming factors would result in the production of iPSCs. However, under suitable culturing conditions, cells proceed down a path that leads to cardiac progenitors and eventually to cardiomyocytes. This process can be supported by inhibition of the JAK/STAT pathway, which leads to pluripotency. The direct reprogramming procedure is not only faster than the iPSC methodology but is also nearly three times as fast as transdifferentiation by lineage-specific factors, as described above. Spontaneous cardiomyocyte contraction was observed as early as after 11 days after direct reprogramming, compared to the 4–5 weeks needed for transdifferentiation.

In an alternative approach, a combination of several miRNAs instead of transcription factors induced the direct cell reprogramming of fibroblasts to cardiomyocyte-like cells. The cells generated had properties typical of cardiomyocytes, such as spontaneous calcium oscillations and transients in response to depolarization. The miRNA approach was also successfully translated into an *in vivo* application: Injection of the miRNAs into the myocardium of ischemic mice resulted in direct conversion of fibroblasts to cardiomyocytes. However, human fibroblasts are usually more difficult to reprogram than mouse fibroblasts; improved methods need to be developed for medical applications.

Another problem that needs to be solved before any direct reprogramming strategy can be employed for regenerative medicine purposes in humans is in the use of retroviral vectors for the delivery of the reprogramming factors. As in iPSC research, the use of small molecules was evaluated to avoid potential oncogenesis. For example, the two small molecules forskolin and dorsomorphin enabled the neurogenin 2 transcription factor to convert human fetal lung fibroblasts into cholinergic neurons with high purity and efficiency. Neurons produced in this way demonstrated mature electrophysiological properties and exhibited motor neuron-like features. Practically speaking, direct reprogramming provides an alternative to the generation and differentiation of iPSCs that is less complicated, faster, and may have lower tumorigenic potential. The fact that differentiated cells can be directly reprogrammed into other cell types has challenged our views on the terminal nature of cellular differentiation and the integrity of lineage commitment. With proper strategies, all cells are potentially totipotent and may form any type of tissue.

12.5
Differentiation of Stem Cells

A major challenge in stem cell research lies in the differentiation of pluripotent stem cells, either ESCs or iPSCs, into a desired cell type, such as differentiated neurons, cardiomyocytes, or others. A straightforward approach is a lineage selection that starts with the undirected differentiation of pluripotent cells. As outlined above, pluripotent cells will form embryoid bodies and differentiate into virtually all types of cells after removal of the inhibitory factors that maintain pluripotency. The key challenge of this strategy is that it requires isolation of the desired cell type from the cell mixture. This can be achieved by antibody marking followed by fluorescence-activated cell sorting (FACS) (Section 2.5.2). Alternatively, a selection marker (e.g., a fluorescent protein or antibiotic resistance marker) can be expressed in cells under the control of cell-type specific promoters. However, it is questionable whether cells that include such genetic modifications can be employed for medical purposes.

Directed differentiation of pluripotent cells into a desired cell type has thus become a dominant strategy. This approach is based on the sequential application of extrinsic factors that direct the cellular phenotype in a specific direction. A conceptually simple idea is to expose the cells to the same sequence of inducing factors, at the same concentrations and times, as they would normally experience during embryonic development. Typically, four to six steps are required to differentiate a pluripotent cell into a desired cell type. Factors that direct the differentiation of pluripotent cells include growth factors, for example, transforming growth factor β (TGF-β), vascular endothelial growth factor (VEGF), fibroblast growth factor (FGF), or bone morphogenetic protein (BMP). Protocols for directed differentiation usually do not require embryoid body formation. In reality, however, the procedure is very complex and various factors, including the initial pluripotent cell line and the cultivation conditions influence the differentiation process. Consequently, multiple alternative protocols exist for the directed differentiation of pluripotent cells into certain cell types and it is not always possible to achieve a homogeneous cell population.

Neuronal cells are required for the treatment of traumatic or degenerative disorders of the central nervous system (CNS). Their generation has been intensively optimized and they are among the lineages most reliably obtained from pluripotent cell lines. A common approach to obtaining differentiated neural cells starts with so-called dual SMAD inhibition (dSMADi), which

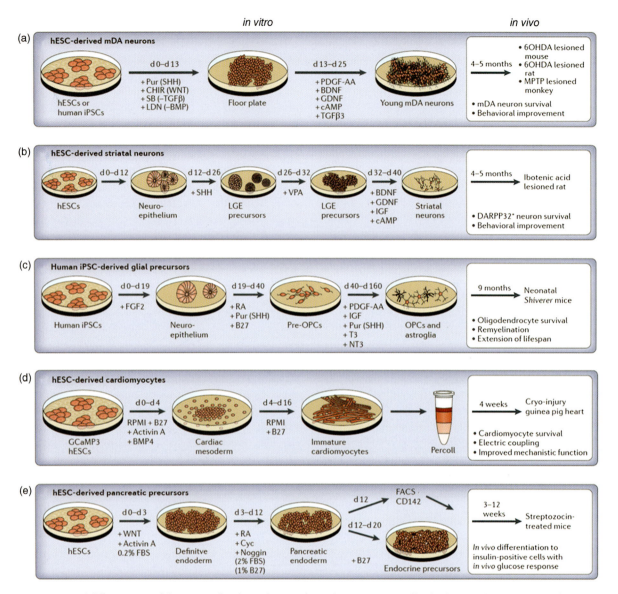

Fig. 12.9 Directed differentiation of therapeutically relevant lineages from pluripotent stem cells. The diagram depicts protocols for the generation of (a) midbrain dopamine (mDA) neurons, (b) striatal neurons, (c) glial precursors, (d) cardiomyocytes, and (e) pancreatic precursors. Small molecules and growth factors are used to direct cell fate. Intracellular factors that are induced or inhibited are shown in parentheses. (Adapted from Ref. [4] with kind permission from Macmillan Publishers Ltd., Copyright 2014.)

includes inhibition of TGF and BMP signaling. The dSMADi approach synchronizes the differentiation process and yields a nearly uniform population of early neural cells within 10 days of differentiation. Figure 12.9a–c summarizes protocols for the generation of three types of neuronal cell: (i) midbrain dopamine (mDA) neurons for the treatment of Parkinson's disease, (b) striatal neurons for the treatment of Huntington's disease (HD), and (c) glial precursors for the treatment of demyelinating disorders such as multiple sclerosis. Factors that direct the differentiation process may be small

molecules, such as LDN-193189, a cell-permeable inhibitor of BMP receptors, or proteins, for example, brain-derived neurotrophic factor (BDNF).

Parkinson's disease is a neurodegenerative disorder caused by the specific loss of midbrain dopamine neurons. The feasibility of replacing the dopaminergic neurons by cell therapy has been demonstrated by fetal tissue transplantation. However, since the clinical benefit of these approaches has been modest and some patients receiving fetal neurons experienced high-grade side effects, stem cell-based therapy is being developed as an alternative. The differentiation of pluripotent stem cells to reach the so-called floor plate intermediate, a transient structure along the midline of the developing embryo, is initiated by a combination of four small molecules (Figure 12.9a). Subsequently, the cells are exposed to a cocktail of growth factors that promote differentiation into young mDA neurons, which have successfully been tested in animal models of PD.

Huntington's disease is an autosomal dominant neurodegenerative disease caused by an expansion of CAG repeats within the *huntingtin* gene (Section 3.1.1). Striatal neurons are severely affected in patients with HD. Similar to PD, patients with HD have also been treated with fetal tissue, but the stem cell-based approach provides a more promising alternative. Differentiation of hESCs begins with induction into the neural rosette stage (the neuroepithelium) followed by sequential exposure to a signaling molecule called sonic hedgehog and then the small-molecule valproic acid (Figure 12.9b). Eventually, a mixture of growth factors and cAMP promote differentiation into striatal neurons.

Multiple sclerosis and some rare genetic disorders are caused by neuronal demyelination of neurons. The differentiation of pluripotent cells into myelinating oligodendrocytes, which produce myelin and insulate the neuronal axons, is under consideration for the treatment of these diseases. The differentiation process also begins with the isolation of neuroepithelial cells, which are then treated with factors to obtain oligodendrocyte precursor cells (OPCs) (Figure 12.9c). Engraftment of these cells into the neonatal CNS of myelin-deficient mice resulted in extensive remyelination *in vivo* and a significantly extended lifespan of the transplanted animals. Early glial precursors were the first hESC-derived products that were tested in humans, as outlined in the following section.

Neuronal cells are ectoderm derivatives. Cardiomyocytes are an example of the successful differentiation of pluripotent stem cells into mesoderm derivatives. One of the most successful strategies to generate cardiomyocytes is to expose pluripotent stem cells to activin A and BMP4 (Figure 12.9d). After continuous cultivation in a special medium (RPMI + a formulation consisting of numerous supplements known as B27), cardiomyocytes can be isolated by a physical separation technique called a Percoll density gradient. In animal models, these cells electrically coupled with the host heart cells and suppressed arrhythmias in injured hearts.

Pancreatic islet cells are endoderm derivatives that may become valuable for a cell-based therapy of type I diabetes. Past attempts at islet cell replacement include grafting fetal or cadaver tissue. To generate pancreatic precursors from hESCs, induction of the endoderm is achieved in the presence of molecules that trigger the WNT and the Nodal-activin-TGF-β signaling pathways (Figure 12.9e). Since current protocols for *in vitro* maturation do not yield cells with appropriate glucose responsiveness, pancreatic endoderm, which is not responsive, can be directly transplanted *in vivo* and will mature into insulin-positive cells with the appropriate glucose response.

Similar protocols to differentiate pluripotent stem cells, either ESCs or iPSCs, have been developed for virtually all types of clinically relevant cells. Figure 12.10 gives a summary of currently available differentiation protocols.

12.6
Medical Applications of Stem Cells

While therapy with adult, tissue-specific stem cells is clinically routine for the treatment of blood cancers, therapeutic applications based on pluripotent stem cells (hESCs or iPSCs) are still in their infancy. However, they already play an important role in biomedical research since patient-derived iPSCs can be used to create models for severe genetic diseases. The current status of various stem cell applications will be described in the following sections.

12.6.1
Adult Stem Cell Therapies

By far the most common type of stem cell therapy in current practice is hematopoietic stem cell transplantation (HSCT). This procedure is mainly used for the treatment of cancer, primarily lymphomas and leukemias (Section 4.7). Only about 5% of the HSCTs are done for the treatment of nonmalignant blood diseases and other conditions. Worldwide, about 50 000 HSCTs are performed every year. The term HSCT covers not only bone marrow transplantation, which is commonly associated with the treatment of leukemia, but also the transfer of hematopoietic stem cells obtained from nonmarrow sources such as peripheral blood or umbilical cord blood.

As described in Section 12.2, HSCs are particularly sensitive to radiation and can be selectively destroyed and replaced by new donor HSCs. This finding, which was based on observations made following the use of the atomic bombs in World War II, laid the basis for the treatment of leukemia. A sufficiently high dose of radiation can completely eradicate blood cell cancers; the patient is rescued from death by an infusion of healthy bone marrow. However, early attempts at doing so in the clinic failed due to the lack of knowledge of

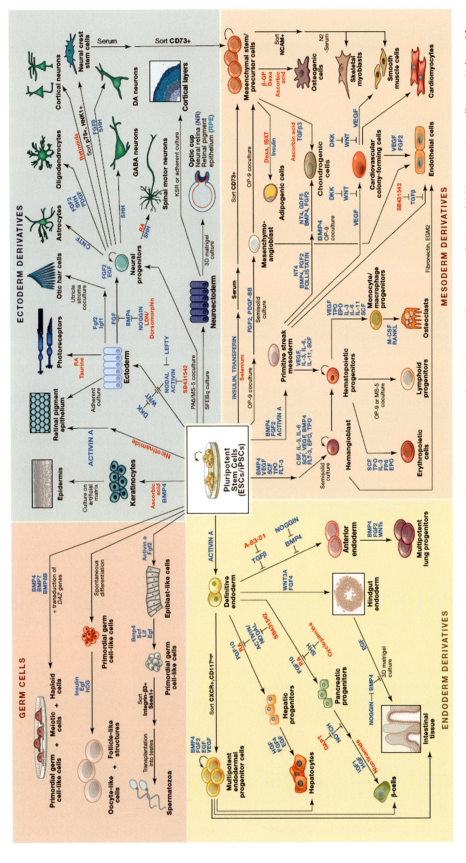

Fig. 12.10 Directed differentiation of pluripotent stem cells. The chart summarizes current strategies for the directed differentiation of ESCs and iPSCs into various cell types. (Adapted from Ref. [5] with kind permission from Elsevier.)

compatibility factors between the donor and the host. When incompatible, engrafted T-lymphocytes will recognize the new surroundings of the recipient as non-self and destroy it. This complication, called *graft-versus-host disease*, is the reverse of the graft rejection process observed when transplanting an incompatible organ. Elucidation of the complex human leukocyte antigen (HLA) (Section 1.5.1) system has helped to overcome these problems and to select patient–donor matches.

Unfortunately, there is no routine method to expand hematopoietic stem cells *in vitro*; transplantation must be carried out using freshly isolated cells. There are two main methods to obtain HSCs. In the traditional method, HSCs are harvested from bone marrow, typically by inserting a large-bore needle into the iliac bone of the pelvis. Since this procedure is performed under general anesthesia, there is a small associated risk to the donor. The recombinant production of granulocyte colony-stimulating factor (G-CSF, Section 10.2.3) permitted the development of a new procedure. This involves the mobilization of HSCs from the bone marrow into the peripheral blood and their collection by simple leukapheresis. However, since for unknown reasons cells so harvested carry a higher risk of producing graft-versus-host disease than those isolated directly from bone marrow, both methods remain in use. Another source of HSCs is blood from the umbilical cord of newborns. One method to purify HSCs from complex cell mixtures, that is, from collected bone marrow or blood, is fluorescence-activated cell sorting (FACS) (Section 2.5.2), which recognizes surface molecules on the cells of interest.

Modern therapy of leukemia by HSCT differs somewhat from the protocols used initially. The malignant HSCs are no longer completely destroyed by radiation or chemotherapy. Rather, the transplanted graft is now used to destroy residual tumor cells. This is called the *graft-versus-leukemia* effect and is mediated by engrafted lymphocytes that attack tumor cells.

A drawback of allogenic HSCT (from one individual to another) is the at least 10% mortality rate of the treatment, rendering it useful only for life-threatening diseases. More than half of all HSCTs are autologous, that is, the HSCs are harvested from the patient him- or herself and reinfused after treatment with cytotoxic chemotherapy. Autologous HSCT can also be combined with gene therapy. For example, HSCs can be harvested from individuals with severe combined immunodeficiency (SCID). After *ex vivo* insertion of a functional version of the deficient, disease-causing gene by a viral vector, the corrected HSCs are reinfused into the patient. Further details and problems with this procedure are described in Section 11.4.1.

Although HSCT is the dominant type of stem cell therapy in clinical practice, other applications have been developed in the last few decades. The first stem cell drug was approved in 2012 in Canada. Prochymal (produced by Osiris Therapeutics Inc.) is an allogenic stem cell therapy for the treatment of acute graft-versus-host disease following bone marrow transplantation in children. This devastating complication kills almost 80% of all affected children. The standard therapy for graft-versus-host disease is steroids, but the success rate is only 30–50%. Prochymal consists of MSCs purified from the bone marrow of adult donors. These cells can be infused through a simple intravenous line without the need for HLA-typing or immunosuppressing the recipient. In clinical trials, Prochymal produced clinically significant responses in approximately two thirds of the children with graft-versus-host disease who were unresponsive to steroids. Further trials are ongoing to confirm these results.

Another stem cell therapy in development attempts to regenerate skin of patients with severe burns. Stem cells from the basal layer of the epidermis can be grown in culture to form a tissue layer that can be used as a graft to cover burned areas. A shortcoming of this approach is that the grafted epidermis lacks hair follicles and sweat glands since they originate from other populations of stem cells. The lack of sweat glands is problematic, as they are important for regulating body temperature.

Many more potential applications of adult stem cell therapy are in early stages of clinical development. A field of intensive research is stem cell therapy of the heart, since the heart muscle lacks the ability to regenerate damaged tissues, for example, after a myocardial infarction. Some early studies indicated that grafts of autologous bone marrow will also repopulate the heart and other organs, in addition to niches in the bone marrow. Although the frequency of such events is probably not high enough to be of medical benefit, several clinical trials grafting autologous bone marrow into damaged hearts have been initiated. However, cardiac function improved only slightly and transiently. The supposed beneficial effects probably did not originate from repopulation of the heart with grafted cells, but rather from so-called paracrine effects. The transplanted cells probably secrete factors that modulate several processes, such as stimulation of blood vessel formation or the reduction of inflammation.

An impressive example of mature organ replacement in regenerative medicine was the development of a bioengineered tooth (Figure 12.11). The procedure began with the isolation of tissues and single-cell preparations from the epithelium and mesenchyme of tooth pulp. After organ culture *in vitro*, the bioengineered tooth pulp was transplanted to a hole in the alveolar bone of a murine tooth loss model, where it generated a functional

Natural tooth occlusion

Bioengineered tooth occlusion

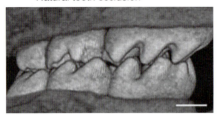 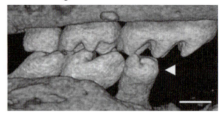

Scale bar: 500 μm

Fig. 12.11   Bioengineered tooth. The images compare natural and bioengineered tooth occlusion in a murine tooth loss model [6].

tooth. The engrafted bioengineered tooth was physiologically fully functional, suggesting that in the future this approach might be used to grow teeth in humans.

The first full transplant of a human organ grown from adult stem cells was performed to treat a female whose trachea had collapsed due to tuberculosis. A section of trachea from a donor was stripped of cells that could cause an immune reaction. The cartilage trunk was then seeded with stem cells from the patient's bone marrow and a new section of trachea was grown over 4 days in the laboratory. It was then transplanted into the patient's left main bronchus. Since the new trachea was reconstituted with cells from the recipient, the transplant was not rejected even though no immunosuppressive medication was used. After 5 years of follow-up, the patient had problems with the native trachea, but the tissue-engineered trachea itself remained open over its entire length, was well vascularized, and was completely recellularized with respiratory epithelium. It also retained normal function and no stem-cell-related teratoma or antidonor antibodies could be found.

Additional areas of active stem cell research include the therapy of eye damage, the repair of spinal cord injury, the treatment of neurodegenerative diseases and of diabetes, and the use of MSCs to increase cartilage formation in orthopedic conditions. For example, clinical trials have been reported on the surgical transplantation of MSCs into damaged knees to produce coverage of chondral defects. This procedure is also known as autologous chrondrocyte transplantation (ACT). An activity that received much media attention was the production of a burger from muscle stem cells (Box 12.3).

12.6.2
Pluripotent Stem Cells for Biomedical Research

There is great hope that pluripotent stem cells will open numerous new opportunities for the therapy of currently untreatable diseases. Basically, two alternative approaches can be followed (Figure 12.13). One widely recognized strategy is to differentiate the required type of cells from a source of pluripotent cells (which can either be iPSCs

Box 12.3. Stem Cell Burger

Current meat production methods impact the environment by contributing to the rising concentration of greenhouse gases such as carbon dioxide, methane, and nitric oxide. These problems will worsen with rising populations and the increasing demand for meat, particularly in developing countries. In addition, a growing number of people are uncomfortable with consuming meat produced by intensive mass animal farming. These issues have triggered attempts to grow meat *in vitro*. One approach was to isolate ESCs from farm animals. Since the differentiation of the pluripotent stem cells into muscle cells is not trivial, adult muscle stem cells can serve as an alternative source. These cells can be differentiated with adequate growth factor support

and the muscle cells grown on special grids to place tension on them and support their growth (Figure 12.12). Finally, the new muscle cells can be harvested and used for burger production.

The first stem cell burger was produced by the physiologist Mark Post of Maastricht University, The Netherlands. Its estimated cost was approximately €250 000. The burger was presented at a press conference in August 2013. It tasted like meat, but it was said to lack juiciness. It was suggested that the meat would be improved by allowing some of the stem cells to develop into fat cells. However, it is questionable whether consumers will eventually accept this artificially produced meat and even if they do, whether production will ever become economically viable.

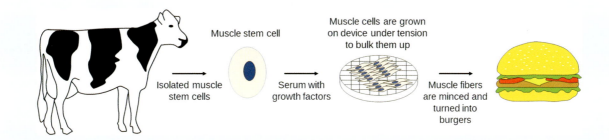

Fig. 12.12   Production of stem cell burgers. The production of stem cell burgers begins with the isolation of muscle stem cells from cattle. These cells are then differentiated into muscle cells and grown on special devices to place the appropriate tension on the muscle cells required to bulk them up. The newly grown muscle fibers can then be harvested and minced to become a burger.

or hESCs) and transplant the differentiated cells into patients to repair damaged tissues. This approach will be described in the following section. This section will focus on more advanced applications of the use of iPSCs to generate a disease model. In this case, patient-specific iPSCs are differentiated to the relevant cell type. The differentiated cell can then be used as a disease model, for example, to study disease mechanisms or to screen libraries for active compounds that eventually can be used for patient treatment.

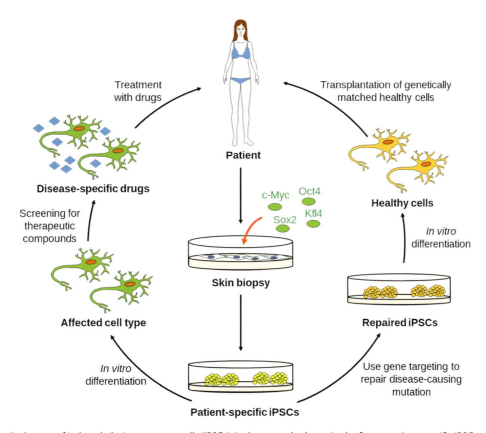

Fig. 12.13   Medical usage of induced pluripotent stem cells (iPSCs). In the example shown in the figure, patient-specific iPSCs are generated by reprogramming cells from a patient's skin biopsy with the four Yamanaka factors. In principle, any other method to generate iPSCs, as described in Section 12.3.1, can also be used. The example illustrates the use of iPSCs to model and treat a neurodegenerative disorder. Patient-specific iPSCs can be differentiated into neurons, which is then followed by screening of a library of compounds that can potentially be used for the treatment (left branch). Alternatively, the required cells can be differentiated from iPSCs and transplanted into the patient (right branch). In the example shown here, a disease-causing mutation needs to be repaired by gene-targeting approaches. (Adapted from Ref. [2] with kind permission from Macmillan Publishers Ltd., Copyright 2012.)

A great deal of knowledge about human physiology and the mechanisms underlying human genetic diseases has been derived from the study of mouse models. Genetic manipulations as described in Section 2.4 permit knocking out a gene-of-interest to study its physiological role. However, while this approach is often successful, it cannot model all human diseases due to species differences. Cystic fibrosis (Section 3.1.2) and the Lesch–Nyhan syndrome are two examples of genetic disorders that cannot be reproduced in animals, since mice do not express the relevant phenotype when the causative human mutations are introduced. In other cases, the mechanism of drug action differs substantially between mice and humans. For example, some neuroprotective drugs demonstrated promising results in preclinical studies with animal models but failed when tested in humans. These experiences underscore the importance of interspecies differences and suggest that the study of diseases and drug actions must be performed in human cells. However, the standard cell lines currently used in life science research often do not reproduce disease-relevant phenotypes.

Pluripotent human cells harboring causative mutations provide a novel source for the generation of cells of disease-relevant tissues and for the evaluation of pathophysiological processes and pharmacotherapy. Early attempts have used hESCs for these purposes. For example, embryos with genetic disorders were identified by preimplantation genetic diagnosis (PGD) (Section 8.1.2) and an hESC line derived from these embryos. Alternatively, a normal hESC line can be used and genetic modifications can be introduced, for example, by homologous recombination. However, both approaches are complicated and are associated with the ethical problems consequent to the need to destroy embryos in order to produce the hESC lines.

A much easier strategy is to use patient-specific iPSCs to study the pathomechanisms of genetic disorders. The reprogramming procedure developed by Yamanaka and coworkers and further developed by various groups, as described in Section 12.3.1, permits the generation of an iPSC line by the use of fibroblasts derived from a skin biopsy from an individual carrying a mutation. Other more recently developed strategies of generating iPSCs from the lymphocytes found in a blood sample make it even easier to produce patient-specific iPSCs.

In the past few years, numerous genetic diseases have been modeled with iPSCs. Among the early iPSC models were neurological diseases such as amyotrophic lateral sclerosis (ALS) (Box 12.4), Huntington's disease (HD), the fragile X syndrome, and multifactorial Parkinson's disease. An impressive example of the usefulness of iPSC-based disease models in biomedical research is found in drug screening for familial dysautonomia. This rare genetic disorder of the peripheral nervous system is caused by a point mutation in the gene encoding the inhibitor of nuclear factor-κB-kinase-complex-associated protein (IKBKAP). Patient-derived iPSCs were successfully differentiated in central and peripheral nervous system precursors that displayed the expected disease-related phenotypes. This *in vitro* model was used to screen multiple compounds, leading to the discovery that the plant hormone kinetin can partially normalize the disease phenotype.

Pathophysiological processes underlying cardiovascular disease have also been studied successfully in iPSC-based

Box 12.4. Amyotrophic Lateral Sclerosis and the Ice Bucket Challenge

Amyotrophic lateral sclerosis is a neurodegenerative disorder characterized by muscle spasticity and rapidly progressing muscle weakness. The average survival from onset to death is 3–4 years. The physicist Stephen Hawking is a well-known individual affected by ALS. His case is unusual, since he has survived with the disease for more than 50 years.

About 5–10% of all ALS cases are inherited. A defect in the *superoxide dismutase* (*SOD*) gene is associated with approximately 20% of all familial cases. The encoded enzyme acts as an antioxidant that protects the body from toxic free superoxide radicals produced in the mitochondria. Loss of SOD function leads to an accumulation of free radicals that cause cell damage.

In 2014, the wider public became aware of the disease through the ALS ice bucket challenge. The idea of this charitable activity was to dump a bucket of ice water on someone's head to promote awareness of the disease and encourage donations for research. Many celebrities took the challenge, including former US president George W. Bush, Bill Gates, Mark Zuckerberg, musicians (Eminem, Lady Gaga, and Shakira), actors (Harrison Ford, Ashton Kutcher, Tom Hanks, and Russell Crowe), and athletes (Tiger Woods, Usain Bolt, David Beckham, and Franz Beckenbauer). This activity raised more than $100 million for the ALS Association in a single month.

models. The long QT syndrome (LQTS) is a congenital disease associated with prolonged electrocardiographic QT intervals and a high risk of sudden cardiac death. Animal models of the disease have been hampered by distinct differences between the electrophysiological properties of human and murine cardiomyocytes. After generation of iPSCs from individuals with LQTS, cardiomyocytes that displayed prolonged action potentials in ventricular and atrial cells were differentiated. This model helped to elucidate additional details of the disease mechanism and permitted the evaluation of potential therapeutic agents.

Cardiomyocytes derived from iPSCs of healthy or diseased patients are generally valuable for drug testing. This is true not only for drugs that are intended to act on the heart but also for other drugs; many compounds have high-grade adverse cardiac side effects that cannot always be predicted from animal experiments.

Within only a few years, iPSC-based models for numerous other diseases have been established. These include severe combined immunodeficiency (ADA-SCID), hematological disorders (sickle-cell anemia, β-thalassemia), metabolic diseases (Lesch–Nyhan syndrome, type 1 diabetes), Duchenne and Becker muscular dystrophy, and cystic fibrosis. These models can be expected to help elucidate the disease-causing pathomechanisms and to identify new drugs. StemBANCC, an international academic–industry consortium, was initiated in 2012 with the aim of collecting and characterizing 1500 iPSC lines from 500 people. This collection will aid in the study of disease mechanisms and will also be used for drug and toxicology testing.

12.6.3
Therapeutic Applications of Pluripotent Stem Cells

Numerous conditions have been considered appropriate for cell therapy with pluripotent stem cells (Figure 12.14). These include neurodegenerative diseases and spinal cord injury, heart disease, blindness, deafness, diabetes, muscular dystrophy, and many other disorders caused by the malfunctioning or destruction of tissues. As an exhaustive discussion of all potential stem cell therapies is beyond the scope of this textbook, only a few areas can be discussed here as paradigms.

Despite the potential of stem cell therapy, some issues still hamper wider clinical applications of cells derived from pluripotent cells. The main issue associated with the use of both hESCs and iPSCs is the risk of teratoma formation. The ability to form these tumors is one of the defining features of pluripotency. For reasons that are not fully understood, iPSCs develop teratomas more

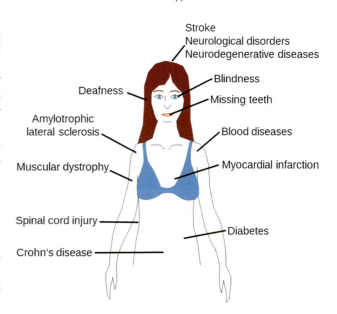

Fig. 12.14 Potential uses of cells differentiated from pluripotent stem cells.

efficiently and faster than hESCs. The direct transplanting of pluripotent cells thus cannot be justified. Rather, they will first have to be differentiated *in vitro* and the fully differentiated cells then implanted into the patient. However, a small risk will still remain that some of the transplanted cells will have not been differentiated and will retain their potential to form teratomas. It is hard to predict whether it will ever be possible to guarantee that 100% of the cells of a population will be fully differentiated. To avoid this problem, an alternative strategy is the transdifferentiation or direct reprogramming described in Section 12.4, since the pluripotent state is bypassed. As an increasing number of factors that define a certain cell type are discovered, it will soon be possible to reprogram fibroblasts (or other easily available cells) directly into the desired cell type, circumventing the risk of teratoma formation.

Another challenge in stem cell therapy is the delivery of the cells to the region where repair is required. This is comparatively easy for HSCs that can simply be injected into the bloodstream and will engraft into niches in the host bone marrow. In contrast, the repair of solid organs such as the heart or spinal cord is more challenging. Cells are usually directly injected into the damaged region of the organ, but their three-dimensional functional integration is not always achievable.

12.6.3.1　Diabetes
Diabetes is a metabolic disease in which patients have elevated blood sugar levels over a prolonged period. More than 200 million individuals are affected

worldwide; this number is expected to double over the next 25 years. The current standard treatment of diabetes includes the use of recombinantly produced insulin (Section 10.2.2). In the future, the therapeutic paradigm may shift to cell-based strategies, particularly for the treatment of type 1 diabetes, which is characterized by the lack of insulin production due to a loss of β-cells in the islets of Langerhans of the pancreas (Section 3.2). In the past decade, protocols have been developed to transplant islets taken from the pancreas of deceased human organ donors. The islets are usually infused into the liver, where they can detect blood glucose levels and secrete the appropriate amount of insulin. However, there are two disadvantages to this therapeutic approach. First, the supply of donor cells is insufficient to meet the increasing demand. Second, the procedure involves the transplantation of allogeneic grafts. Therefore, recipients must take immunosuppressive drugs for the rest of their lives, which creates the risk of contracting an infectious disease or certain cancers.

These problems underscore the potential of stem cell-based approaches. Several protocols have been developed to differentiate iPSCs into pancreatic β-cells. Animal models of diabetes have been produced by destroying the murine β-cells by injection of the drug streptozotocin. Immunodeficient mice tolerate grafts of human origin. It has been shown that β-cells derived from human pluripotent cells can ameliorate diabetes in diabetic animal models. While the β-cells initially generated still lacked many characteristics of *bona fide* β-cells, protocols developed more recently produced β-cells exhibiting the most important features. These include packaging of insulin into secretory granules and their secretion in a glucose-regulated manner.

Although this approach is among the most intensively discussed pluripotent stem cell therapies, one critical specific problem remains. The lack of immune rejection of autologously transplanted grafts is normally considered one of the major advantages of the iPSC technology. However, type 1 diabetes is often caused by autoimmune destruction of the β-cells. Therefore, cells derived from patient-specific iPSCs may rapidly be destroyed by the same autoimmune response. A solution to this problem may be the encapsulation of the cells within a material that permits release of insulin but protects the cells from cytotoxic lymphocytes.

Many people are concerned that the patient-specific preparation of iPSCs and their differentiation into grafts may be prohibitively expensive. This is true at the present time, but costs can be expected to decrease as stem cell technologies become more established. Diabetes is an example in which personalized stem cell therapy will not only improve the quality of life for the patient but

might eventually also be cost-effective, as type I diabetics currently require a lifelong supply of recombinantly produced insulin.

12.6.3.2 Heart Disease

As discussed in the previous section, stem cell therapies for heart disease are eagerly awaited as after a myocardial infarction, cardiomyocytes usually do not regenerate damaged areas of the heart muscle. Instead, the affected region fills with cardiac fibroblasts, which secrete extracellular matrix material and form a scar that cannot contract. Differentiation protocols for the generation of cardiomyocytes begin with the formation of mesoderm, then the anterior type of mesoderm followed by cardiac progenitor cells, and finally the cardiomyocytes themselves. Some animal experiments have indicated persistence of grafted cells and some improvement in cardiac function. However, the interpretation of the data is still under debate as it is not always clear whether the beneficial effects originate from functional engraftment of cardiomyocytes or from secondary effects, as discussed above.

12.6.3.3 Neurodegenerative Diseases

Another field that engenders hope for advancement based on pluripotent stem cell therapy is the treatment of neurodegenerative diseases. Parkinson's disease is one of the most common neurodegenerative conditions in elderly people. It primarily affects movement (rigidity and tremor), but can also lead to language and cognitive disabilities. This progressive disorder arises from a loss of neurons in the *substantia nigra*, a region of the midbrain, which use dopamine as a neurotransmitter. Current standard therapy is based on the administration of L-DOPA, a precursor substance that is converted into dopamine in dopaminergic neurons. This treatment temporarily diminishes the motor symptoms, but long-term use of the drug has severe side effects and response to treatment invariably diminishes.

For more than 30 years, cell therapy has been evaluated for the treatment of patients with PD. In several trials, grafts from aborted fetuses have been implanted into patients. In addition to the ethical issues inherent in using fetal tissue, the efficacy of the treatment has been controversial. Thus, human pluripotent cells have been considered an alternative source for the desired cells. In a landmark study, the group of Rudolf Jaenisch from the Whitehead Institute at the Massachusetts Institute of Technology (MIT) in Cambridge, Massachusetts, differentiated dopaminergic neurons from iPSCs. A PD animal model produced by the destruction of dopaminergic neurons by 6-hydroxydopamine was employed. Injection of the iPSC-derived dopaminergic neurons into the brain led to their functional integration and to an improvement of the condition of these PD rats. Cell replacement therapy in brain has since been considered one of the most promising areas of iPSC application. However, since patients with PD can survive for a comparatively long time, there is a significant risk that some pluripotent cells that did not differentiate may implant and develop into teratomas. In addition, the differentiation process may also produce some other types of neurons in addition to dopaminergic neurons. These neurons may make inappropriate connections and cause problems such as uncontrolled muscular movements.

12.6.3.4 Combinations of Stem Cell and Gene Therapy

The stem cell therapies discussed so far aim to regenerate tissue that was damaged for other than genetic causes, for example, ischemia in myocardial infarction or autoimmune-mediated dysfunction in type 1 diabetes. However, this approach cannot cure some diseases caused by a genetic defect. For example, Huntington's disease is caused by a CAG triplet expansion in the *huntingtin* gene (Section 3.1.1). The expansion of the encoded polyglutamine tract in the huntingtin protein results in massive cell death in the striatum of HD patients. Regeneration of the lost neurons with cells derived from patient-specific iPSCs that still carry the genetic defect will thus be futile. Therefore, therapeutic approaches require genetic correction as indicated in the right branch in Figure 12.13. It has been shown that replacement of the expanded CAG repeat with a normal repeat using homologous recombination permitted the production of normal neurons from patient-specific iPSCs and reversed the disease phenotype. Only genetically corrected cells will eventually be useful for the cell replacement therapy of HD.

In a similar approach, iPSCs derived from a humanized mouse model of sickle-cell anemia (Section 6.2) have been corrected by homologous recombination. Repaired iPSCs were differentiated into hematopoietic progenitors and transplanted into affected mice. The cells stably engrafted and corrected the disease phenotype. However, more recent research revealed that hematopoietic cells differentiated from pluripotent stem cells engraft poorly. This finding led to the idea of phenocopying the vascular-niche microenvironment of the hematopoietic cells, which enables direct reprogramming of human endothelial cells into engraftable hematopoietic cells without a transition through a pluripotent intermediate.

The CRISPR/Cas system briefly introduced in Box 11.3 of Section 11.4.3 may open new paths in combining genetic engineering and stem cell therapy. This system permits the specific introduction of base changes into the cellular genome and may be used to correct a genetic mutation in patient-specific iPSCs. The differentiated, corrected cells produced from the modified iPSCs may then be reimplanted into the patient.

This genetic correction strategy has been applied to various other diseases, for example, spinal muscular atrophy. However, it may even be possible to not only correct a genetic defect but also to improve upon basic cellular properties. For example, the combination of stem cell and gene therapy has been proposed for the treatment of patients with chronic liver disease from viral hepatitis (i.e., infection with HCV or HBV). Treatment of an infected patient with unmodified hepatocytes derived from his or her iPSCs would only be of short-lived benefit, as they would rapidly become infected by the hepatitis virus. This has led to the idea of transferring into hepatic cells a gene encoding a short hairpin RNA (shRNA) (Section 13.3.1) directed against the virus. The RNA interference (RNAi) response against the virus would render the hepatocytes virally resistant, allowing them to repopulate the liver over time and to restore normal liver function.

12.6.3.5 Clinical Trials

The first clinical trial incorporating cells derived from hESCs was approved by the FDA in 2009. The study used hESC-derived oligodendrocyte precursors for the treatment of spinal cord injury caused, for example, by a car accident. As outlined above, oligodendrocytes are a class of glial cells that produce the myelin sheaths of axons. The trial's aim was to remyelinate denuded axons at the site of injury, as preclinical studies had shown significant restoration of mobility in treated animals with spinal injuries. The clinical trial was conducted by Geron, a biotech company based near San Francisco, CA. At the end of 2010, the first patient received the stem cell-derived oligodendrocyte precursors. In total, four patients were treated. The phase I study focused on safety, and 1 year later the first results were presented. No adverse effects were observed. Despite these encouraging data, the company discontinued the trial for financial and strategic considerations.

Another trial carried out in the United States and the United Kingdom was initiated by Advanced Cell Technology, Inc. (now Ocata Therapeutics, Inc.) for the treatment of retinal degeneration. About 10% of all individuals over 65 are affected by age-related macular degeneration (AMD) that can lead to loss of central vision in severe cases. The wet form of the disease is characterized by overgrowth of blood vessels and can be treated by antibodies (Section 10.2.1) or aptamers (Section 14.3). However, there is no treatment for the dry form that is due to a defect in the retinal pigment epithelium (RPE), a layer of pigmented cells lying beneath the photoreceptors. RPE cells can be obtained relatively easily from pluripotent stem cells. In animal experiments, subretinal grafts of RPE cells improved visual function. An advantage of this approach is that RPE cell preparations are not thought to form teratomas. However, should a teratoma arise, the affected eye could be removed. The trial enrolled patients with AMD and Stargardt disease, a related juvenile form of macular degeneration. Follow-up 3 years after treatment initiation provided evidence for treatment safety, survival of the grafted cells, and improved visual acuity in at least some patients.

In late 2014, the biotech company Viacyte (San Diego, CA) received permission to initiate a phase I/II clinical trial in type 1 diabetes with pancreatic progenitor cells that were derived from hESCs. The cells are encapsulated in a specific semipermeable device that protects them from attack by immune cells. After implantation under the skin, the progenitor cells are expected to mature into insulin-producing β-cells that will regulate blood glucose.

While the first clinical trials used cells derived from hESCs, the first trial based on iPSCs commenced in Japan in late 2014. Similar to the above-mentioned trial, the iPSC trial examined the engraftment of RPE cells into a patient with AMD. A $1.3 \times 3.0 \, \text{mm}^2$ sheet of RPE cells was implanted into the patient's eye. This transplant was not necessarily intended to repair lost vision, but rather was an attempt to halt retinal deterioration and prevent further photoreceptor destruction. Many additional clinical trials using pluripotent stem cells are in the pipeline that can be expected to commence in the near future.

HSCT has been used successfully for the treatment of leukemia and other diseases for many years and it can be expected that more stem cell therapies will follow in the near future. Patient-specific iPSCs are now considered the most promising sources of pluripotent cells. They have already contributed to biomedical research by providing new disease models. iPSCs may also outperform hESCs in the clinic due to the ethical issues connected with hESCs and their induction of immune rejection. However, direct reprogramming of cells may eventually become the method of choice as it bypasses the pluripotent state and is thus devoid of the risk of teratoma formation.

References

1. Visvader, J.E. and Lindeman, G.J. (2012) Cancer stem cells: current status and evolving complexities. *Cell Stem Cell*, **10**, 717–728.
2. Robinton, D.A. and Daley, G.Q. (2012) The promise of induced pluripotent stem cells in research and therapy. *Nature*, **481**, 295–305.
3. Efe, J.A., Hilcove, S., Kim, J. *et al.* (2011) Conversion of mouse fibroblasts into cardiomyocytes using a direct reprogramming strategy. *Nat. Cell Biol.*, **13**, 215–222.
4. Tabar, V. and Studer, L. (2014) Pluripotent stem cells in regenerative medicine: challenges and recent progress. *Nat. Rev. Genet.*, **15**, 82–92.
5. Williams, L.A., Davis-Dusenbery, B.N., and Eggan, K.C. (2012) SnapShot: directed differentiation of pluripotent stem cells. *Cell*, **149**, 1174–1174 e1.
6. Oshima, M., Mizuno, M., Imamura, A. *et al.* (2011) Functional tooth regeneration using a bioengineered tooth unit as a mature organ replacement regenerative therapy. *PLoS One*, **6**, e21531.

Further Reading

Slack, J. (2012) *Stem Cells: A Very Short Introduction*, Oxford University Press, Oxford.

Adult Stem Cells

Morrison, S.J. and Scadden, D.T. (2014) The bone marrow niche for haematopoietic stem cells. *Nature*, **505**, 327–334.
Murphy, M.B., Moncivais, K., and Caplan, A.I. (2013) Mesenchymal stem cells: environmentally responsive therapeutics for regenerative medicine. *Exp. Mol. Med.*, **45**, e54.

Cancer Stem Cells

Visvader, J.E. and Lindeman, G.J. (2012) Cancer stem cells: current status and evolving complexities. *Cell Stem Cell*, **10**, 717–728.

Induced Pluripotent Stem Cells

Bao, X., Zhu, X., Liao, B. *et al.* (2013) MicroRNAs in somatic cell reprogramming. *Curr. Opin. Cell Biol.*, **25**, 208–214.
Hochedlinger, K. (2010) Your inner healers: a look into the potential of induced pluripotent stem cells. *Sci. Am.*, May: 46–53.
Robinton, D.A. and Daley, G.Q. (2012) The promise of induced pluripotent stem cells in research and therapy. *Nature*, **481**, 295–305.

Transdifferentiation

Kelaini, S., Cochrane, A., and Margariti, A. (2014) Direct reprogramming of adult cells: avoiding the pluripotent state. *Stem Cells Cloning*, 7, 19–29.

Differentiation of Pluripotent Stem Cells

Tabar, V. and Studer, L. (2014) Pluripotent stem cells in regenerative medicine: challenges and recent progress. *Nat. Rev. Genet.*, **15**, 82–92.

Medical Applications of Pluripotent Stem Cells

Robinton, D.A. and Daley, G.Q. (2012) The promise of induced pluripotent stem cells in research and therapy. *Nature*, **481**, 295–305.
Tiscornia, G., Vivas, E.L., and Izpisua Belmonte, J.C. (2011) Diseases in a dish: modeling human genetic disorders using induced pluripotent cells. *Nat. Med.*, **17**, 1570–1576.

Antisense, Ribozyme, and RNA Interference Strategies

13

Summary

- The common basis of all anti-mRNA strategies are the oligonucleotides that specifically bind to a target mRNA by Watson–Crick base pairing and prevent its translation into a protein. These technologies can be used to study the function of a gene by producing a loss-of-function phenotype, or to treat a disease by inhibiting the expression of a harmful gene.
- Antisense oligonucleotides (AS ONs) are usually approximately 14–18 nucleotides long and inhibit translation by blocking ribosomal translation of mRNA into protein or by inducing cleavage of the target RNA by RNase H.
- Ribozymes possess an intrinsic target RNA-cleaving catalytic activity independent of any cellular enzyme. However, due to their limited activity, their relevance as therapeutic agents is marginal.
- In contrast, over the last decade, RNA interference (RNAi) approaches have become a widely used technology. Short double-stranded RNA (dsRNA) molecules, either synthesized chemically or transcribed intracellularly from vectors, induce cleavage and subsequent degradation of a target RNA with high efficiency.
- MicroRNAs (miRNAs) are naturally occurring short, double-stranded RNA molecules that regulate cellular processes and are involved in much human pathology. They can also be employed as biomarkers for cancer or for cardiovascular disease. In addition, antisense molecules can inhibit microRNA function.
- Efficient oligonucleotide delivery and stabilization against nucleolytic degradation through chemical modification are major challenges for the development of any anti-mRNA approach. Therapeutic oligonucleotides have been tested in clinical trials for a large number of diseases, including viral infection, diabetes, and cancer. A growing number of these drugs may be approved in the near future.

Contents List

The vast majority of currently available chemically synthesized drugs work by modulating the activity of proteins, in many cases binding through steric interactions to the active centers of enzymes. Other small-molecule drugs block receptors on the surface of cell membranes. Frequently, however, these drugs are not specific and their interaction with other proteins can produce side effects. Also, conventional small-molecule approaches only address proteins that fold in a manner that permits interactions with such compounds. As a result, the so-called druggable genome is limited to about 3000 of the more than 20 000 human genes (Section 7.2.1). Anti-mRNA strategies act at an earlier level

Molecular Medicine: An Introduction, First Edition. Jens Kurreck and Cy Aaron Stein.
© 2016 Wiley-VCH Verlag GmbH & Co. KGaA. Published 2016 by Wiley-VCH Verlag GmbH & Co. KGaA.

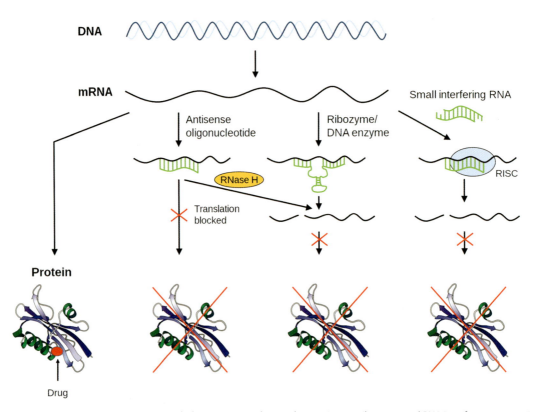

Fig. 13.1 Anti-mRNA strategies. Posttranscriptional silencing approaches, such as antisense, ribozyme, and RNA interference strategies, block gene expression at the level of the mRNA. Antisense oligonucleotides bind to the target RNA and prevent its translation, or induce cellular RNase H to degrade the mRNA. Ribozymes and DNA enzymes possess an intrinsic catalytic activity and can degrade the target RNA without the help of other enzymes. Short double-stranded RNA molecules called small interfering RNAs (siRNAs) are taken up by the RNA-induced silencing complex (RISC) and trigger the RNA interference pathway. The siRNAs guide the RISC to complementary RNAs, which are then degraded. In each case, the blockade or the degradation of the mRNA prevents the synthesis of the corresponding target protein. (Adapted from Ref. [1] with kind permission from John Wiley & Sons, Inc.)

in the course of gene expression; they thus bypass the problems of side effects due to unspecific interactions with proteins and the limitations imposed by unsuitable protein target folding. These approaches employ oligonucleotides that specifically bind to a target mRNA via Watson–Crick pairing and inactivate it (Figure 13.1). As a result, the encoded protein is no longer produced. Since the intervention is at the level of the mRNA, these technologies are collectively referred to as posttranscriptional gene silencing.

These silencing strategies are of great importance for both biomedical research and for the development of novel therapeutics. Sequencing projects that were initiated in the 1990s decoded the human genome (Chapter 7); however, the function of many encoded proteins is still unknown. Anti-mRNA strategies can switch-off single genes; analysis of the resulting loss-of-function phenotypes allows insights into the function of these genes. In addition, this approach can be used to validate targets, thus helping to determine if the potential target for a pharmacological substance has the

suspected significance in a disease process. For example, silencing strategies have been employed to shutoff newly discovered potential pain receptors to determine if an inhibitory substance directed against that target could reduce pain. There is also hope that anti-mRNA strategies will offer new therapeutic options in diseases for which there is currently no existing treatment. Theoretically, anti-mRNA technologies can inhibit the expression of any harmful gene. For example, oncogenes that trigger uncontrolled cell growth can be silenced. Similarly, genes that are essential for the replication of a harmful virus can be silenced. Anti-RNA strategies are thus complementary to gene therapy strategies (Chapter 11), in which a defective gene is replaced with an intact one.

Blocking the expression of a gene with an antisense oligonucleotide was first demonstrated in the late 1970s. AS ONs are short, single-stranded DNA molecules that bind to complementary mRNAs and either block their translation or induce their degradation via RNase H. The discovery of catalytically active RNA molecules, called ribozymes, dates back to the beginning of the

1980s, when it was demonstrated that ribozymes could degrade a target RNA without the aid of cellular (protein-) enzymes. This discovery was a scientific sensation at the time and inspired the idea of using these molecules as therapeutic agents. Since its discovery in the late 1990s, RNAi has revolutionized life sciences. Short double-stranded RNA molecules, called siRNAs (*small-* or *short-interfering RNA*), can be incorporated into a protein complex (RISC, *RNA-induced silencing complex*) and direct it to a target RNA, which is cleaved and subsequently degraded. Not only can artificially created double-stranded RNAs manipulate gene expression, but they can also occur naturally in eukaryotic cells and have many important functions. Other short RNA species, called miRNAs, regulate numerous cellular pathways and are involved in various pathological processes.

13.1
Antisense Oligonucleotides

In 1978, the American researchers Paul C. Zamecnik and Mary L. Stephenson first demonstrated the antisense mechanism by inhibiting the Rous Sarcoma Virus with AS ONs. Antisense molecules are usually short single-stranded DNA oligonucleotides or modified analogs, often 14–18 nucleotides in length. The AS ONs bind to the target RNA via complementary Watson–Crick base pairing (Figure 13.2). To avoid unintended side effects, a sequence that appears only once in the genome is the preferred target.

13.1.1
Mechanism of Action of Antisense Oligonucleotides

The inhibitory activity of AS ONs relies on two different modes of action (Figure 13.3). The first mechanism involves activation of the enzyme RNase H by the DNA oligonucleotide when it binds to RNA. RNase H plays an important role in replication by recognizing DNA and RNA duplex molecules and cleaving the RNA component of these hybrids. The mRNA is irreversibly inactivated

Fig. 13.2 Antisense oligonucleotides. An antisense oligonucleotide binds to a target RNA to inhibit gene expression. The AS ON is usually 15–20 nucleotides long. It consists of DNA or modified DNA analogs and binds to the target RNA via Watson–Crick base pairing.

(a) RNase H cleavage

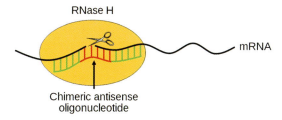

(b) Blocking of translation

Fig. 13.3 The mechanism of action of antisense oligonucleotides. (a) RNase H recognizes hybrids of DNA and RNA and degrades the RNA component of the duplex. In order to increase nuclease resistance, AS ONs often contain chemically modified nucleotides that do not activate RNase H. Chimeric AS ONs, known as gapmers, contain DNA bases in the center to recruit RNase H and nuclease protective analogs on both ends (for further details, see below). (b) The second mechanism of action of AS ONs consists of blocking the ribosome, thereby preventing translation. AS ONs that are directed against the 5′ end of the mRNA can impede the assembly of the ribosome. (Adapted from Ref. [1] with kind permission from John Wiley & Sons, Inc.)

and the AS ON is released to induce degradation of additional target mRNAs. The second mechanism leads to a blockade of translation without degrading the mRNA. In this case, the AS ON binds to the mRNA and prevents the ribosome from reading it. Since an active ribosome can sometimes displace a bound AS ON and continue to synthesize a protein, the AS ONs are often targeted against the 5′-region of the mRNA to prevent initial assembly of the ribosome. For this second mechanism, the AS ON has inhibitory activity only by virtue of binding to the target RNA. No catalytic destruction of the mRNA target occurs. Consequently, higher concentrations of the AS ONs are required than when RNase H activity is involved.

Another use of AS ONs is to correct an mRNA molecule without destroying it. For this purpose, AS ONs directed against splice sites can modulate maturation of the mRNA. Incorrect splice sites that otherwise would lead to false processing of the mRNA can be occluded. Figure 13.4 shows a situation where a mutation caused a new splice site in upstream exon X. The pre-mRNA is, therefore, incorrectly processed and part of the exon is lost during splicing. This aberrant splicing can either result in a shortened protein or it can cause a frameshift that leads to a completely nonfunctional protein. If an

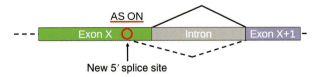

AS ON

Exon X ○ Intron Exon X+1

New 5′ splice site

Fig. 13.4 Correction of splicing by an antisense oligonucleotide (AS ON). In the example shown, a mutation has created a new splice site in exon X that leads to incorrect processing of the mRNA so that part of the exon is lost. This aberrant splicing can result in elimination of the protein or to a protein with impaired function. An AS ON can cover the aberrant splice site so that the exons can be correctly connected to each other.

AS ON occludes the new aberrant 5′ splice site, the spliceosome will not recognize it. Instead, the spliceosome will use the original splice site and exon X will correctly fuse with exon X + 1, repairing the defect. Alternatively, AS ONs can trigger exon skipping to eliminate exons containing harmful mutations or stop codons (Section 13.1.3). A requirement for the splicing modulation is the use of chemically modified AS ONs to avoid degradation of the target mRNA by RNase H cleavage.

13.1.2
Development and Stabilization of Antisense Oligonucleotides

The development of therapeutically useful AS ONs must overcome numerous challenges. Among others, these include the identification of regions of the target mRNA that are accessible for the AS ONs, the stabilization of the AS ONs against nucleolytic degradation, and the efficient delivery of the AS ONs into cells of the target tissue.

Since the interaction between the AS ONs and their target molecules takes place via Watson–Crick base pairing, they can theoretically be directed against any sequence of the complementary mRNA. However, this idea is based on a simplified concept of the structure of RNA. While textbooks discuss in detail the double helical structure of DNA and the complex folding of proteins, mRNA is often presented as a simple linear molecule. The fact is that RNA molecules also form complex secondary and tertiary structures (Figure 13.5a). In addition, many proteins bind to cellular RNAs. Therefore, not every region of a target RNA is suitable as a binding region for AS ONs.

As a result, the first important task in the development of an efficient AS ON is to identify an accessible region for target RNA binding. The secondary structure of RNA molecules can be predicted with bioinformatics (Figure 13.5b). Algorithms such as *mfold* (mfold.rna .albany.edu/?q=mfold) and the Unified Nucleic Acid Folding and hybridization package (UNAfold, mfold.rna .albany.edu) are available for public use.

However, messenger RNAs are often several thousand base pairs long. The consensus is that any calculation of the exact structure of such long RNA molecules is beyond the ability of bioinformatics, at least at the present time. The theoretically predicted binding sites for AS ONs thus require experimental validation. Typically, a series of AS ON candidates are screened for their knockdown efficiency to find one or several that are active.

The inhibition of the expression of a target gene is usually examined at both the RNA level by the use of quantitative RT-PCR (Section 2.2) or Northern blot and the protein level by Western blot. Of critical importance

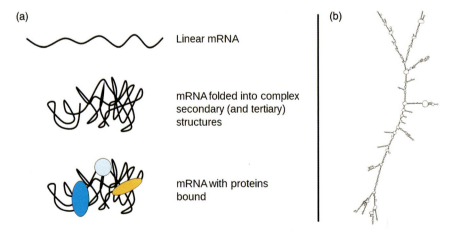

(a)

Linear mRNA

mRNA folded into complex secondary (and tertiary) structures

mRNA with proteins bound

(b)

Fig. 13.5 Structure of messenger RNAs. (a) mRNAs are often shown as linear molecules. In reality, however, they form complex secondary and tertiary structures, as proteins do. In addition, many proteins bind to mRNAs in cells. This limits the accessibility of large sections of the mRNA for the binding of AS ONs. (b) Secondary structure prediction of the mRNA of the Vanilloid Receptor TRPV1. The secondary structure was calculated with the *mfold* algorithm. The mRNA engages in base pairing, but there are also regions that are single-stranded and thus potentially more suitable for interaction with AS ONs.

is the use of suitable controls, for example, randomized AS ONs. These are used to determine whether the gene is silenced specifically due to an antisense mechanism or whether the observed effect is nonspecific. One well-documented example of the misinterpretation of experimental results involved the use of AS ONs containing "CpG" motifs ("CpG" represents a cytosine–phosphate–guanine sequence) in the early days of antisense technology. Several years passed until it was appreciated that any presumed antisense inhibition observed in human tumor xenografts was instead due to the activation of Toll-like receptors by the CpG motifs and subsequent immune activation (Section 14.4).

In addition to the identification of suitable binding sites for AS ONs, stabilization against nucleolytic degradation is critical for the efficacy of these molecules. In biological systems, nucleases usually degrade unmodified DNA oligonucleotides within a few hours. For this reason, chemically modified nucleotides are incorporated into AS ONs in order to extend their half-life in plasma and cells. All three components of a nucleotide (the base, sugar, and phosphate linkage) can be modified. However, analogs with modified bases have only seldom been used *in vivo*, and little is known about their pharmacological properties and potential toxicity. More frequently, AS ONs with modified sugar groups, in particular on the 2′-position of the ribose, or chemically modified internucleobase phosphates, have been clinically evaluated.

Phosphorothioates were among the first modified nucleotides and are still among the most commonly used ones. Due to the substitution by sulfur of a non-bridging oxygen atom at phosphorus, they are resistant to nucleolytic degradation (Figure 13.6). Like unmodified DNA, phosphorothioates induce the degradation of the target mRNA by RNase H. However, their affinity for complementary sequences is reduced relative to the iso-sequential phosphodiesters. Phosphorothioates bind to various proteins that can lead to numerous side effects, including activation of the complement cascade. On the other hand, the binding of phosphorothioates to plasma proteins, especially albumin, prevents their renal filtration and improves their half-life in serum.

Due to the disadvantages of employing phosphorothioates, a second generation of modified nucleotides has been developed for antisense applications. In addition to the phosphorothioate backbone, they also contain substitutions at the 2′-OH group bound to ribose. Important representatives of this generation of AS ONs in clinical development are the 2′-O-methyl-RNA (OMe) phosphorothioate as well as 2′-O-methoxyethyl-RNA (MOE) (Figure 13.6) phosphorothioate modifications. These variants improve on unmodified phosphorothioates by their

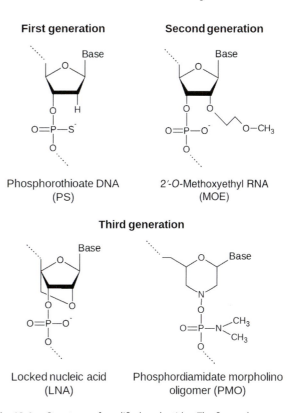

Fig. 13.6 Structures of modified nucleotides. The figure shows selected structures of modified nucleotides that are used in AS ONs under clinical investigation.

increased target affinity and lower toxicity. However, OMe- and MOE-modified AS ONs have a major disadvantage in that they are not recognized as substrates by RNase H and cannot induce degradation of target mRNA. To overcome this problem, so-called gapmers are employed. Gapmers usually consist of 3–5 modified nucleotides at each end as protection from nucleases and 6–10 unmodified phosphorothioate DNA elements in the center of the AS ONs to induce RNase H activity (Figure 13.7). Interestingly, the toxicity of the second- and third-generation building blocks combined with phosphorothioate linkages has been reported to be considerably reduced compared to base- or sugar-modified

Fig. 13.7 Design of a gapmer. In gapmers, modified building blocks are located on the 5′ and 3′ ends of the AS ONs (e.g., 2′-O-methyl-RNA (OMe) or locked nucleic acids (LNA)), which protect the oligonucleotide from exonucleases (marked in green). The monomers on a red background in the center are unmodified deoxyphosphodiester or phosphorothioate DNA, which induce activation of RNase H.

phosphorothioates. The reasons for this have not been determined.

Literally, hundreds of other DNA and RNA analogs have been developed to further improve the properties of the AS ONs. The first two modifications of these third-generation ONs to reach clinical trials were the locked nucleic acids (LNA) and the phosphorodiamidate morpholino oligomers (PMO) (Figure 13.6). In LNAs, a methylene bridge links the 2'-oxygen of the ribose to the 4'-carbon. LNAs are characterized by excellent resistance to nucleases and very high affinity for complementary RNA. They are thus effective at very low concentrations (e.g., in the low nanomolar range). Since LNAs do not activate RNase H, they are usually employed in gapmers. PMOs have an uncharged backbone, which minimizes their nonspecific interactions with cellular components. PMOs also lack the ability to trigger the degradation of the target mRNA by RNase H; thus, they are often employed to sterically block translation or to modulate splicing.

One of the greatest challenges to the clinical application of oligonucleotides, such as AS ONs, ribozymes, and siRNAs (see below), is their efficient delivery to the target tissue and across the cellular membrane. Due to the negative charges of the phosphate or phosphorothioate groups of their backbones, they cannot easily penetrate the hydrophobic cellular membrane. Transfection reagents are usually employed in cell culture experiments to achieve efficient oligonucleotide uptake into cells. Cationic lipids are most commonly used; they shield and neutralize the negative charges of the phosphate or phosphorothioate groups and permit passage of lipid–nucleic acid complexes through endosomal membranes after endocytosis.

Remarkably, in some organs *in vivo*, AS ONs are taken up spontaneously via an uncertain mechanism after repeated dosing or at high local concentrations. The AS ONs can be employed without transfection reagents in some animal models and for the occasional clinical indication. Figure 13.8 depicts the uptake of fluorescently labeled AS ONs in dorsal root ganglia (DRG) in rats after intrathecal (into the cerebrospinal fluid space) application of oligonucleotides in the absence of transfection reagents.

Moreover, it has also been discovered that short oligonucleotides are taken up *in vitro* by cultured cells. This leads to the knockdown of the target gene. However, in order for the short AS ONs to bind efficiently to their target mRNAs, they must be nuclease resistant and have a high affinity for the target; these are properties of phosphorothioate LNA gapmers. This technique is called *gymnosis*, which is defined as the silencing of gene expression in tissue culture in the absence of transfection agents or

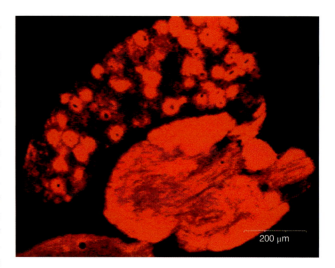

Fig. 13.8 Picture of antisense oligonucleotides (AS ONs) in the dorsal root ganglia (DRG) of rats. The fluorescently labeled AS ONs were injected intrathecally (into the spinal canal). Confocal fluorescence microscopy demonstrated that the oligonucleotides had been taken up into the DRG neurons. (Reprinted from Ref. [2] with kind permission from Elsevier.)

oligonucleotides conjugated with lipids, for example. Gymnosis has the advantage that the outcome of *in vitro* experiments more closely reflects results that can be expected *in vivo*.

13.1.3
Clinical Applications

Over 30 AS ONs have already been tested in clinical trials. The selection shown in Table 13.1 demonstrates the diversity of the diseases addressed by this technology. Since the antisense strategy can be applied to potentially inhibit the expression of any gene, its applications range from oncology to viral infection to inflammatory disease and diabetes. The majority of the AS ONs in advanced stages of clinical trials are phosphorothioates.

In most of these trials, the AS ONs are administered by intravenous (i.v.) injection. In some cases, application is local, for example, directly into the vitreous humor of the eye (intravitreal). With the newer modifications, oral delivery has been reported to lead to significant oligonucleotide uptake and bioavailability.

In 1998, the first AS ON was approved by the Food and Drug Administration (FDA) in the United States. Fomiversen was sold under the brand name Vitravene™ and was used for the treatment of cytomegalovirus-induced retinitis. Infections with cytomegalovirus are not a problem for healthy individuals, but can lead to blindness in HIV-infected patients. Treatment involves

Table 13.1 Selection of antisense oligonucleotides (AS ONs) in clinical trials or marketed.

Drug (company)	Indication	Target	Chemistry	Route	Status (2015)
Activation of RNase H					
Fomiversen (ISIS)	Cytomegalovirus	IE2 gene	PS ODN	Intravitreal	Approved
Mipomersen (Genzyme)	Cardiovascular	Apolipoprotein B	2′-MOE chimera	Systemic	Approved
Oblimersen (Genta)	Oncology	Bcl-2	PS ODN	Systemic	Phase III
Trabedersen (Antisense Pharma)	Oncology-glioblastoma	Transforming growth factor-β	PS ODN	Intratumoral	Phase III
GS-101 (Gene Signal)	Corneal neovascularization	Insulin receptor substrate-1	PS ODN	Topical	Phase III
LOR-2040 (Lorus)	Oncology	Ribonucleotide reductase	PS ODN	Systemic	Phase II
Archexin (Rexahn)	Oncology	AKT-1	PS ODN	Systemic	Phase II
TPI ASM8 (Pharmaxis)	Asthma	CCR3 and IL-5 receptors (two oligos)	PS ODN	Inhaled	Phase II
Alicaforsen (ISIS/Atlantic)	Colitis	Intercellular adhesion molecule-1	PS ODN	Enema	Phase II
Custirsen (ISISI)	Oncology	Clusterin	2′-MOE chimera	Systemic	Phase II
ATL1102 (ISIS)	Multiple sclerosis	CD49D	2′-MOE chimera	Systemic	Phase II
ISIS-PTP1B$_{RX}$ (ISIS)	Diabetes, type 2	Protein tyrosine	2′-MOE chimera	Systemic	Phase II
Monarsen (Ester Neurosciences)	Myasthenia gravis	Acetylcholine esterase	2′-O-Me chimera	Oral	Phase II
Apatorsen (OncoGeneX)	Oncology	Heat shock protein 27	2′-MOE chimera	Systemic	Phase II
ISIS-EIF4E$_{RX}$ (ISIS)	Oncology	eIF-4E	2′-MOE chimera	Systemic	Phase II
ISIS-CRP$_{Rx}$ (ISIS)	Cardiovascular/inflammation	C-reactive protein	2′-MOE chimera	Systemic	Phase II
ISIS-GCGR$_{Rx}$ (ISIS)	Diabetes	Glucagon receptor	2′-MOE chimera	Systemic	Phase II
ISIS-GCCR$_{Rx}$ (ISIS)	Diabetes	Glucocorticoid receptor	2′-MOE chimera	Systemic	Phase II
iCO-007 (ISIS/iCO)	Macular degeneration	C-Raf kinase	2′-MOE chimera	Intravitreal	Phase II
SPC2996 (Santaris)	Oncology	Bcl-2	LNA chimera	Systemic	Phase I/II
SPC2968 (Santaria)	Oncology	Hypoxia inducing factor1-α	LNA chimera	Systemic	Phase II
Translation inhibition					
AVI-7288 (Sarepta)	Marburg virus	Nucleocapsid protein	Morpholino	Systemic	Phase I
AVI-7100 (Sarepta)	Influenza virus	M1 and M2 matrix proteins	Morpholino	Systemic	Phase I
RNA binding					
GRN163L (Geron)	Oncology	Telomerase	Lipid-conjugated phosphoramidate	Systemic	Phase II
Splicing modulation					
Drisapersen/PRO-051 (Prosensa)	Duchene muscular dystrophy	Dystrophin	2′-O-Me	Systemic	Phase III
Eteplirsen/AVI-4658 (Sarepta)	Duchene muscular dystrophy	Dystrophin	Morpholino	Systemic	Phase II

PS: phosphorothioate; MOE: 2′-methoxyethyl; OMe: 2′-O-methyl; LNA: locked nucleic acid.
Selection based on Ref. [3] and company Web sites. Additional clinical trials with AS ON can be found on the Web site of ISIS pharmaceuticals: http://www.isispharm.com/Pipeline/index.htm).

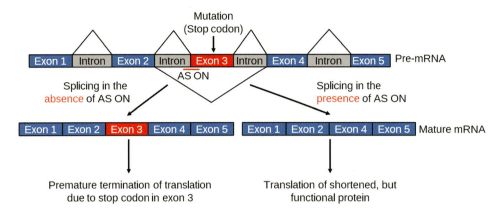

Fig. 13.9 Exon-skipping strategy using antisense oligonucleotides (AS ONs). The primary transcript consists of sequences corresponding to both exons and introns. In this example, exon 3 contains a stop codon. An AS ON directed against the splice site leads to this exon being skipped. The truncated mRNA now contains an open reading frame that can be translated into a slightly shortened but functional protein.

intravitreal injection of a phosphorothioate AS ON that is directed against an early gene of the virus. The therapy is well tolerated and effective, but the availability of alternative treatments and the reduction in the number of affected individuals due to the development of effective antiretroviral therapy has limited the sales and significance of Fomiversen in the clinical treatment of cytomegalovirus retinitis. Interestingly, it took another 15 years, until 2013, for the FDA to approve a second AS ON: Mipomersen (Kynamro™) is a 2′-MOE-modified AS ON directed against apolipoprotein B for the treatment of homozygous hypercholesterolemia (Section 3.1.1).

Many AS ONs have been directed against cancer-related targets. The AS ON Oblimersen (Genasense™) has been under very intense investigation. This AS ON is directed against the Bcl-2 mRNA and was investigated for the treatment of different types of cancer, including chronic lymphocytic leukemia and melanoma. Bcl-2 is an apoptosis inhibitor; silencing its expression would increase the proportion of cancer cells that undergo apoptosis, thus slowing tumor growth. The results from the first, but unfortunately not the second, large phase III clinical trial demonstrated a significant increase in survival of melanoma patients treated with Oblimersen. However, laboratory investigation demonstrated that the effects of Oblimersen on cells in tissue culture is not only due to specific antisense inhibition of Bcl-2 but also due to a nonspecific release of cytochrome c from the mitochondria, leading to apoptosis. It seems to be true that Oblimersen has a complex, multimodal mode of action instead of the single, simple, and specific mechanism initially thought to explain its activity.

As previously described, AS ONs can also be used to modulate splicing without leading to the degradation of the target mRNA. This approach is being used for the treatment of Duchenne muscular dystrophy (DMD), a recessive X-linked genetic disease that primarily affects males (Section 3.1.3). The muscle degeneration present in this disease frequently confines the affected patients to a wheel chair while they are still children. Death usually occurs before the age of 30. DMD is caused by various mutations in the dystrophin gene that prevent the production of a functional dystrophin protein. Other mutations in the dystrophin gene lead to a shorter but functional dystrophin protein, which results in the milder, Becker form of muscular dystrophy (BMD). The aim of antisense treatment of DMD is to skip over exons with the harmful mutation, and to thus lessen the symptoms of DMD in favor of those of the milder BMD. This strategy is known as exon skipping (Figure 13.9). 2′-O-methyl-modified AS ONs as well as PMOs have been used in clinical trials to skip over mutated exon 51 in DMD patients. Accumulation of dystrophin has been observed in muscle fibers around the injection site of patients who received the treatment. However, one limitation of this intervention is that a large number of different mutations can cause DMD. Skipping of exon 51 will be helpful for only 13% of DMD patients and the number that would benefit from other splice-switching AS ONs is even smaller. It is unclear if it will be one day possible to create personalized AS ONs for specific dystrophin mutations.

In addition to modulating splicing, PMOs have been employed to block translation. Two oligomers directed against hemorrhagic viruses (Ebola and Marburg virus) have been developed within the framework of the biological defense programs sponsored by the US government. These programs began after the successful treatment of penguins infected with West Nile Virus (Box 13.1).

In conclusion, two decades of clinical development have shown that AS ONs are usually well tolerated and

Box 13.1. The Penguins of Milwaukee

In 2002, the West Nile Virus, which can also be deadly to humans, infected penguins in the Milwaukee Zoo. The sick animals stopped eating and had to be force-fed and soon the first penguins died. AVI Biopharma, later renamed Sarepta Therapeutics, offered to develop antisense therapeutics to treat the infection. They quickly used the available sequence information to design a morpholino oligomer against the translation start site of the viral RNA that encodes a single polyprotein. A few days later, treatment began. All the animals that received the oligomer survived, while the untreated penguins died. This success not only delighted veterinarians but also caught the attention of the US Defense Threat Reduction Agency. An antisense strategy appears to be suited for the rapid development of countermeasures to bioterrorism threats. The US government subsequently asked AVI Biopharma to develop antisense medicines against possible biological weapons like the Ebola and Marburg viruses, which cause hemorrhagic fever.

are not associated with unacceptable toxic side effects. In most cases, unfortunately, the effectiveness of the AS ONs has been insufficient, which has made FDA approval difficult. To overcome the problem of limited efficacy, scientific attention initially turned to ribozymes, and later to more efficient RNAi approaches (see below). However, as discussed in Section 13.3, these strategies also have disadvantages.

13.2
Ribozymes

In the course of investigations in the early 1980s on introns of the *Tetrahymena thermophila* ciliate, Thomas Cech and coworkers discovered that these RNA molecules could splice themselves without the need for proteins. They gave these molecules the name "ribozyme" to indicate that they are *ribo*nucleic acids with en*zym*atic activity. These self-splicing molecules are not true enzymes in the narrow sense of the word. By definition, an enzyme must not only accelerate (catalyze) the rate of a reaction but must also not be consumed by the reaction it catalyzes. However, these introns can only process themselves once and then are no longer catalytically active. A short time later, Sydney Altman and colleagues described true ribozymes by demonstrating that the RNA of RNase P is active without a protein component. RNase P is a ubiquitous enzyme that processes the 5′ ends of tRNAs and emerges from the reaction intact. Because they fundamentally changed the scientific view of the world with their discovery of ribozymes, Cech and Altman jointly received the Nobel Prize for Chemistry in 1989. Previously, enzymatic activity was attributed exclusively to proteins, while nucleic acids were regarded solely as carriers of genetic information. The discovery of ribozymes demonstrated that RNA can carry genetic information and can also be catalytically active. These findings provided an important argument for the RNA world hypothesis (Box 13.2).

13.2.1
Classification of Ribozymes

Naturally existing RNA-cleaving ribozymes are roughly classified by their size. Large ribozymes consist of several hundred to a few thousand nucleotides, while small ribozymes are 30–150 nucleotides long. Among the large ribozymes are the self-splicing group I and group II introns in addition to RNase P, which consists of a catalytically active RNA and a protein component. The small ribozyme class includes the hammerhead, the hairpin, the hepatitis delta virus, and the Varkud satellite ribozymes. Artificial selection has also led to the isolation of ribozymes composed of DNA, also known as DNA enzymes or DNAzymes.

Besides the well-studied ribozymes already mentioned, other ribozymes have since been discovered that catalyze other important processes. For example, ribosomal RNA is primarily responsible for the peptidyl transferase activity of the ribosome (Box 13.2). In the spliceosome, ribozyme catalysis presumably plays a central role: A metabolically regulated ribozyme in the 5′ UTR of the mRNA of glutamine fructose 6-phosphate-amidotransferase in Gram-positive bacteria regulates gene expression. Another ribozyme is involved in the termination of transcription of the β-globin mRNA in human cells. Since new ribozymes, for example, the CPEB3 ribozyme, the CLEC2 ribozyme, and the twister ribozyme, are continuously being discovered, the repertoire of RNA-catalyzed reactions appears to be significantly greater than originally thought.

Box 13.2. RNA World Hypothesis

The RNA world hypothesis holds that before our present form of life, which is dominated by DNA and proteins, a time existed in which RNA was of primary importance. This helps to resolve the "chicken and egg" problem, which may be defined as whether the information carrier (DNA) was present first or whether proteins, which can amplify DNA and catalyze other processes but are not able to store and pass on information, emerged first. Since RNA can act as both a carrier of genetic information and a catalyst of biological reactions, it is possible that an RNA-dominated world filled the gap between the creation of the first biomolecules and the emergence of the cellular forms of modern life.

The RNA world hypothesis was first proposed in the 1960s and the discovery of ribozymes in the 1980s provided important arguments in its favor. For example, it was found that the peptidyltransferase activity of the ribosome is carried out primarily by ribosomal RNA and not by proteins (Figure 13.10). This activity could be a relic of the RNA world that survived the transition to the protein-dominated world. By means of artificial *in vitro* evolution, ribozymes have been selected that have many essential functions of a possible RNA world, for example, the ability to replicate RNA. However, the RNA world idea is not universally accepted.

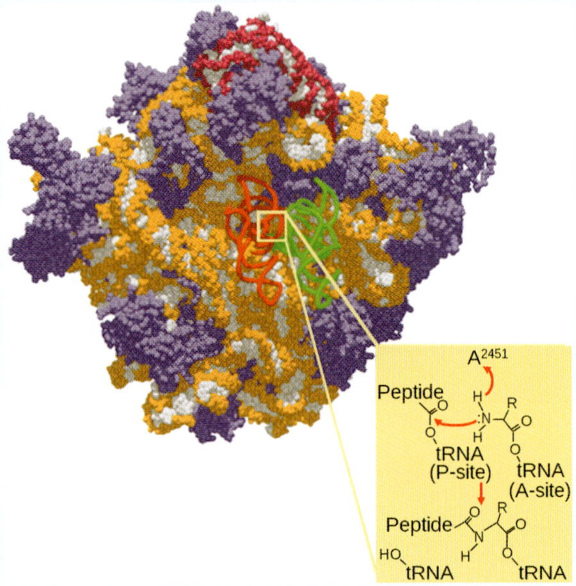

Fig. 13.10   Structure of the ribosome. The solution of the structure of the ribosome has revealed that there are only RNAs and no proteins in the immediate proximity of the peptidyl transferase catalytic center. This finding led Thomas Cech, the discoverer of ribozymes and a Nobel laureate, to publish a commentary entitled "The Ribosome is a Ribozyme." Since then, a discussion has continued as to whether proteins compose an essential component of the catalytic center. (Adapted from Ref. [4] with kind permission from American Association for the Advancement of Science.)

13.2.2
Development of Ribozymes for Medical Applications

Ribozymes can be employed for the degradation or correction of a target RNA. For example, as shown in Figure 13.11, the splicing activity of group I introns was used to replace a pathologically mutated section of the β-globin mRNA with the intact exon. Group I introns interact with the 5′ exon via an internal guide sequence (IGS). For the artificial repair ribozyme, a sequence complementary to the target mRNA replaced the natural IGS. The downstream exon segment of the group I intron was also modified to contain the intact version of the β-globin mRNA instead of the natural sequence. This approach succeeded in correcting the defect. The globin protein resulting from the translation of this construct no longer led to pathologic hemoglobin polymerization. All approaches using large ribozymes, however, are in an early stage of preclinical research and have not yet been tested in humans.

The therapeutic development of small ribozymes is more advanced than that of the larger type. Clinical trials have focused on hammerhead ribozymes, which were originally identified as plant pathogens. These pathogens replicate using a rolling circle mechanism, with the naturally occurring hammerhead variety acting on its own RNA. This is known as being active *in cis*. However, for medical use, it was necessary to develop ribozyme variants that are active *in trans* and can thus cleave a target molecule. A shortened variant of the hammerhead ribozyme fulfilled this requirement. It consists of two substrate-binding arms that bind to an mRNA complementary target and a catalytic center that cleaves the mRNA (Figure 13.12). Cleavage of the target RNA occurs in a metal-dependent catalytic cycle in which the ribozyme binds to the target mRNA by Watson–Crick base pairing. The catalysis typically involves Mg^{2+} as the cation. After the cleavage, the two product strands diffuse away and the ribozyme can process another target mRNA.

For any practical ribozymes use, similar challenges as those faced when using AS ONs must be overcome.

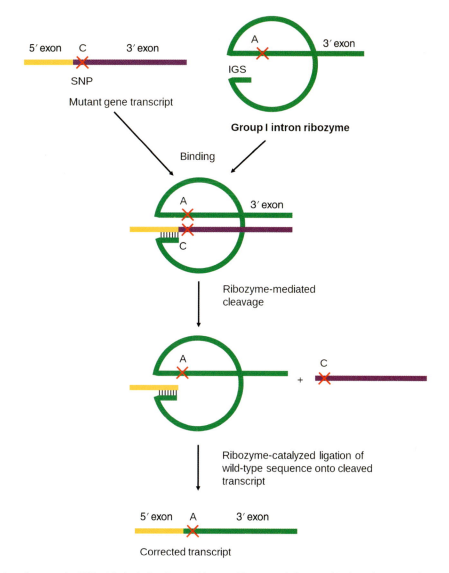

Fig. 13.11 Correction of mutated mRNA with the help of group I introns. The group I ribozyme binds to the mutated transcript, excises the defective locus, and replaces it with the correct sequence. SNP: single nucleotide polymorphism; IGS: internal guide sequence. (Adapted from Ref. [5] with kind permission from Macmillan Publishers Ltd., Copyright 2001.)

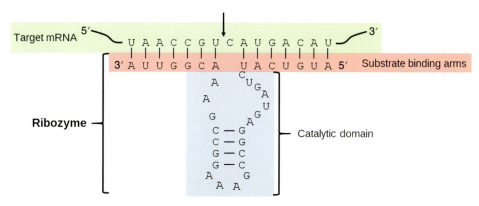

Fig. 13.12 Secondary structure model of the hammerhead ribozyme. The ribozyme consists of a catalytic domain (blue) and two substrate binding arms (red) targeting the complementary mRNA (green). The arrow indicates the cleavage site.

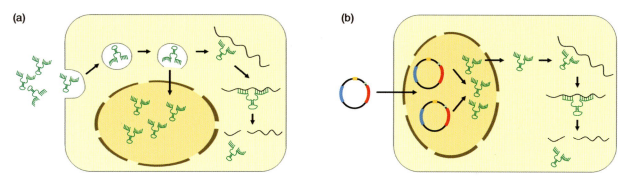

Fig. 13.13 Delivery of ribozymes. (a) For exogenous delivery, ribozymes are chemically synthesized and delivered to the cells where they cleave their target RNA. (b) For endogenous use, an expression cassette encoding the ribozyme is transduced into cells with the help of viral vectors. After transcription and transport to the cytoplasm, they cleave their target RNA.

First, a suitable region of the target RNA for binding and cleavage must be identified. This region should not only be single-stranded, as for antisense approaches, but it must also fulfill certain sequence criteria. For example, hammerhead ribozymes only cleave NUH triplets, where N can be any nucleotide, and H must be A, U, or C. The triplets AUC and GUC are cleaved with highest efficiency.

The fact that ribozymes are RNA molecules presents two options for cellular delivery (Figure 13.13). The ribozymes can either be (i) chemically synthesized and transfected into the target cells as are AS ONs, or (ii) viral vectors (Chapter 11) can transduce the target cells and generate ribozymes from the appropriate expression cassettes.

If not stabilized, chemically synthesized ribozymes are subject to rapid nucleolytic degradation. Since RNA is more sensitive to degradation than DNA, ribozymes must be protected against nucleases. However, incorporation of modified nucleotides into the catalytic center of ribozymes often drastically reduces their activity. In a systematic study using many modified ribozyme components, it was possible to develop a protected hammerhead ribozyme that was as active as the parental molecule. Only five unmodified ribonucleotides remained in the catalytic center, while all other positions were protected with $2'$-O-methyl- and $2'$-C-allyl-nucleotides and phosphorothioate linkages. An inverted thymidine at the $3'$ end of the molecule conferred additional nuclease resistance. These stabilized ribozymes had plasma half-lives of up to 10 days compared to less than a minute for the unmodified ribozyme. Several of the ribozymes that underwent clinical testing as described in the following section used this modification pattern.

The intracellular expression of a ribozyme requires a viral vector (Chapter 11) to deliver an expression cassette that contains a suitable promoter for intracellular expression. RNA polymerase III promoters are suitable for the strong expression of small RNAs; however, they are usually ubiquitously active. RNA polymerase II promoters can be inducible and tissue specific. Since viral transfer and the expression of short RNAs have advanced significantly during the development of RNAi technology, they will be discussed in more detail in Section 13.3.3.

13.2.3
Clinical Applications of Ribozymes

Several ribozymes have been evaluated in clinical studies for the treatment of viral infection or cancer. Hammerhead ribozymes, chemically synthesized ribozymes, and ribozymes expressed intracellularly after vector transfer have all been evaluated.

Ribozyme Pharmaceuticals Inc. used chemically stabilized ribozymes in their clinical trials. The candidate that advanced furthest in the clinic was the Angiozyme ribozyme. It was directed against the vascular endothelial growth factor receptor (VEGFR) and was intended to block tumor angiogenesis. Angiozyme inhibited tumor growth in several mouse models. In phase I studies, the ribozyme proved to be well tolerated and was advanced into a phase II study. Unfortunately, it was not sufficiently effective enough to achieve the intended clinical purpose. Other clinical studies in cancer patients were conducted with Herzyme, a hammerhead ribozyme directed against the human epidermal growth factor receptor type 2 (Her2). Overexpression of this receptor in breast cancer is associated with metastasis and a poor prognosis (Section 4.1). However, this ribozyme completed only a single tolerability study. In addition, a chemically synthesized ribozyme against the hepatitis C virus was in the early phases of clinical testing. None of these approaches was further developed.

The antiviral activity of ribozymes transferred with vectors was also clinically evaluated. The most advanced candidate was OZ1, a ribozyme directed against the

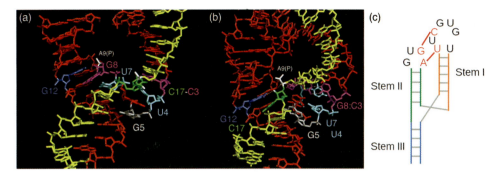

Fig. 13.14 Structure of the hammerhead ribozyme. Crystal structure of the minimal (a) and full-length hammerhead ribozyme (b). Several changes are induced by the interactions between stems I and II. The most important is the formation of a base pair between G-8 and C-3 (magenta) and the reorientation of the cleavage-site base, C-17 (green). (c) Modeled two-dimensional structure of the full-length hammerhead ribozyme. The catalytic core is in blue and the critical interactions between the loops of stems I and II (red) that cannot be formed in the minimal hammerhead ribozyme are indicated. (Parts (a) and (b) reprinted from Ref. [6] with kind permission from Elsevier.)

overlapping *vpr* and *tat* reading frames of HIV-1. A retroviral vector introduced the ribozyme expression cassette into hematopoietic progenitor cells. In a phase II study, it proved to be safe and effective in HIV-infected patients.

Interestingly, a laboratory result had a decisive influence on further clinical ribozyme development. A crystal structure demonstrated that an interaction between two loops of the natural hammerhead ribozyme is of critical importance for the enzyme to form the correct three-dimensional structure (Figure 13.14). This interaction is not possible in the shortened version of the ribozyme used for the *trans*-cleavage of target RNAs. The high concentration of magnesium ions used in the laboratory promoted a conformation that allowed target RNA cleavage by the truncated ribozyme. However, inside cells, the magnesium ion concentration is too low to support sufficiently high catalytic activity of the shortened hammerhead ribozyme required for clinical efficacy.

13.3
RNA Interference

RNA Interference has dramatically impacted basic biomedical research as laboratories worldwide adopted the technology shortly after its discovery. The ability to relatively easily silence the expression of almost any given gene opened up new frontiers for research. In addition, the therapeutic potential of RNAi was quickly recognized, accelerating its clinical development. In comparison to antisense and ribozyme strategies, RNAi is distinguished by its ease-of-use and high efficiency. It has thus become the most commonly used method of posttranscriptional gene silencing (PTGS).

Double-stranded RNA molecules trigger RNAi-mediated gene silencing. In the early 1990s, observations in plants indicated that the overexpression of genes that should have led to a specific pattern of flower coloring surprisingly led to an absence of color. The mechanism of what was called "cosuppression" was not fully understood at that time. At the end of the 1990s, experiments using the nematode *Caenorhabditis elegans* first elucidated the phenomenon now known as RNA interference. The American scientists Andrew Fire and Craig Mello demonstrated that double-stranded RNA molecules inhibit the expression of homologous genes (Figure 13.15). Interestingly, silencing is transitive in *C. elegans*, that is, RNAi-mediated inhibition of gene expression spreads from a single cell to the whole organism and from the organism to its progeny. Fire and Mello received a Nobel Prize for Medicine or Physiology in 2006 for their discovery.

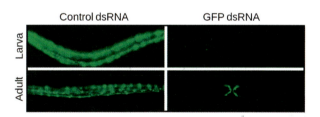

Fig. 13.15 Silencing of green fluorescent protein (GFP) in *C. elegans*. Due to a genetically engineered modification, the worms express highly fluorescent GFP. The progeny of animals injected with double-stranded RNA (dsRNA) molecules were analyzed by fluorescence microscopy. While adult animals and larva still exhibited strong fluorescence after administration of a control dsRNA (left), the fluorescent cell fraction markedly decreased after injection of dsRNA targeting GFP (right). The image was artificially colored. (Adapted from Ref. [7] with kind permission from Macmillan Publishers Ltd., Copyright 1998.)

The natural function of RNAi is still not completely understood, but it appears to protect cells from viruses, particularly in plants and invertebrates. However, the significance of RNAi as a mechanism of antiviral defense in mammals is still controversial since mammals possess an elaborate immune system. RNAi may thus be a relic, retained from lower organisms throughout evolutionary development. Nevertheless, RNAi appears to be important in higher eukaryotes to guarantee genetic stability and to keep mobile genetic elements, such as transposons, under control.

13.3.1
Mechanism of RNA Interference

As already mentioned above, dsRNA molecules are the trigger of RNAi. In lower eukaryotes such as the nematode *C. elegans*, dsRNAs are several hundred base pairs in length. An endonuclease called Dicer processes them into siRNAs that are approximately 21 nucleotides in length (Figure 13.16). The siRNAs are loaded into the actual effector complex, called RISC (RNA-induced

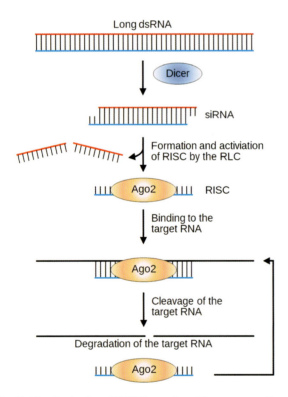

Long dsRNA

Dicer

siRNA

Formation and activiation of RISC by the RLC

Ago2   RISC

Binding to the target RNA

Ago2

Cleavage of the target RNA

Degradation of the target RNA

Ago2

Fig. 13.16 Mechanism of RNAi. The nuclease Dicer processes the long double-stranded RNA into 21mer siRNAs. One strand complexes with RISC while the other is degraded. The remaining strand directs RISC to the target mRNA, which is cleaved and degraded. RISC can then cleave additional target molecules. dsRNA: double-stranded RNA; siRNA: small interfering RNA; RISC: RNA-induced silencing complex; red: passenger strand of the siRNA; blue: guide strand of the siRNA.

silencing complex) with the aid of the RISC-loading complex (RLC). In the process of this loading and activation, one of the strands is presumably cleaved. The active strand found in RISC is termed the antisense or guide strand, and the discarded strand the sense or passenger strand. The antisense strand directs RISC to the target mRNA, to which it hybridizes by Watson–Crick base pairing. An important component of RISC, the protein Argonaute 2 (Ago2), possesses slicer activity and cleaves the target RNA. This creates two free ends unprotected by a cap or a poly A-tail that cellular nucleases rapidly degrade. RISC is released from the degraded target RNA and can cleave further target molecules in a multiple turnover process.

The use of RNAi as a genetic tool was restricted to lower organisms in the first few years after its discovery, since in mammals long double-stranded RNA molecules trigger a strong interferon response. Subsequently, two enzymes become activated, leading to a global termination of protein synthesis: Protein kinase R activation completely blocks translation and nonspecific RNase L activity destroys all RNAs. Protein synthesis is completely shut down and inhibition of individual target genes is no longer possible. A careful elucidation of the RNAi pathway enabled critical improvements in the technology, allowing its use in mammalian cells. It was known that only dsRNAs longer than 30 nucleotides trigger the interferon response, while siRNAs, which are shorter than 30 nucleotides, inhibited gene expression via the RNAi pathway. In 2001, Thomas Tuschl and colleagues demonstrated that siRNAs specifically and efficiently inhibit the expression of a target gene in mammalian cells. Standard siRNAs are 21 nucleotides long: 19 nucleotides of each strand form a helix while the two on the 3′ end do not base pair (Figure 13.17). After the introduction of siRNAs as RNAi triggers, it has been possible to determine the function of individual human genes and to develop novel therapeutic approaches for diseases hitherto deemed untreatable.

siRNAs are not only powerful tools for researchers; various organisms also express them endogenously. Originally, the natural occurrence of these so-called endogenous siRNAs (esiRNAs) was believed to be restricted to lower eukaryotes, which express an RNA-dependent RNA polymerase (RdRP). Eventually, mammals were also found to express esiRNAs as well. These esiRNAs appear to play an important role in both the control of mobile genetic elements and the regulation of protein-coding mRNAs.

In addition, in *C. elegans* and several other organisms, an RISC-independent mechanism exists in which the siRNA serves as a primer for the amplification of the RNA by RdRP. Since this mechanism is not present in mammals, it will not be discussed further.

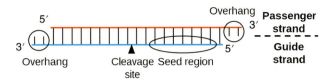

Fig. 13.17 Design of a small interfering RNA (siRNA). Standard siRNAs consist of two RNA strands, which are each 21 nucleotides long. Nineteen nucleotides form a double helix, while two nucleotides on each 3′ end form an overhang. Other designs also exist: siRNAs may be longer or have blunt ends. The passenger, or sense strand of the siRNA is discarded when RISC is loaded, while the guide or antisense strand directs RISC to the target mRNA and hybridizes with it, beginning with the seed region. RISC then cleaves the target RNA in the center of the duplex, 10 nucleotides from the 5′ end of the siRNA strand. (Adapted from Ref. [8] with kind permission from John Wiley & Sons, Inc.)

RNAi occurs in the cytoplasm. This is an important difference between the RNAi and antisense technologies: While AS ONs induce the degradation of the target RNA by RNase H in the nucleus (although more recent experiments have also demonstrated cytoplasmic antisense activity), the RNAi machinery is located in the cytoplasm (Figure 13.18). Colocalization studies have shown that the Argonaute proteins, which cleave the targeted mRNAs, are found in the processing bodies

(P-bodies). These distinct foci within the cytoplasm of the eukaryotic cell play fundamental roles in mRNA turnover. The different locations of silencing activity have implications for delivery: For siRNAs, it is sufficient that they pass through the cell membrane and enter the cytoplasm. AS ONs, however, may need to penetrate the nuclear membrane to reach their location of action.

The silencing activity of chemically synthesized siRNAs is transient. Depending on the rate of turnover of the target proteins, the inhibitory effect begins about 2 days after siRNA treatment and may persist for several days before protein levels returns to baseline. The target is usually only partially inhibited, which means that RNAi, like the antisense and ribozyme technologies, is a method to knockdown, or reduce, protein expression. This contrasts with the complete knockout of gene expression by homologous recombination, which is used for the generation of transgenic animals (Section 2.4).

An important advance in RNAi technology has been the development of strategies for the intracellular expression of dsRNAs. Expression systems permit the extension of the duration of RNAi-mediated silencing, the regulation of RNAi by reversible or irreversible inducers, and the use of viral transduction vectors. In

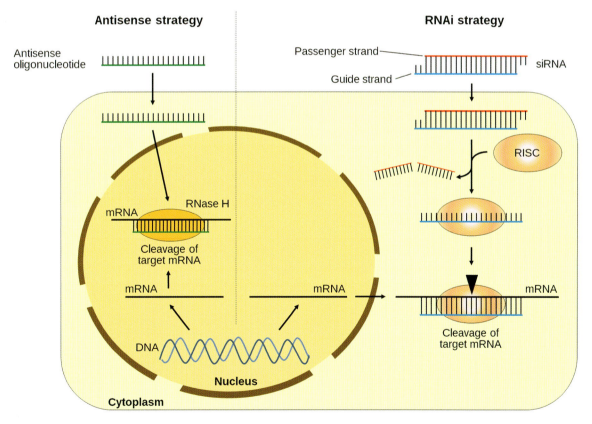

Fig. 13.18 Comparison of the mechanisms of antisense and RNAi technologies. As described in Section 13.1, most AS ONs activate RNase H. This enzyme is located in the nucleus. In contrast, the machinery of RNAi is located in the cytoplasm, where RNA cleavage occurs.

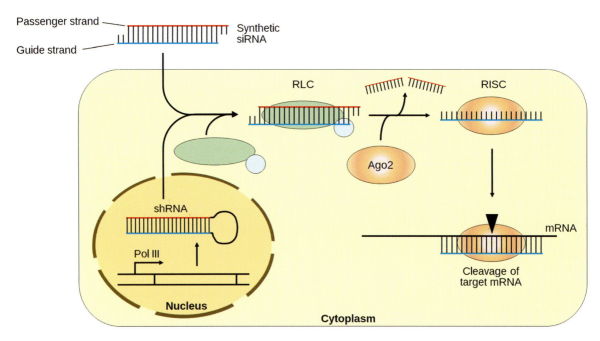

Fig. 13.19 Small interfering (siRNA)- and short hairpin (shRNA)-induced RNA interference (RNAi). Delivery agents can introduce chemically synthesized siRNAs into target cells. There they are loaded into RISC by RLC. The guide strand leads RISC to the target RNA and results in its cleavage by the Ago2 protein. Alternatively, an shRNA can be intracellularly expressed. After export from the nucleus into the cytoplasm, Dicer processes shRNAs into siRNAs so they can enter into the RNAi pathway. RISC: RNA-induced silencing complex; RLC: RISC-loading complex; Ago: Argonaute protein. (Adapted from Ref. [8] with kind permission from John Wiley & Sons, Inc.)

the expression approach, the dsRNA is usually transcribed in the nucleus as a self-complementary short hairpin RNA (shRNA), which is then exported into the cytoplasm (Figure 13.19). The Dicer enzyme then processes the shRNA to siRNA, which follows the previously described pathway. Expression of shRNAs is usually designed to be under the control of RNA polymerase III promoters. An example of this is the U6 promoter of the spliceosome, or the promoter of the H1-RNA of RNase P. Type III RNA polymerases generate large amounts of short RNAs lacking posttranscriptional modifications, such as a 5'-cap or a poly A-tail. Expression cassettes for shRNAs can also be employed to generate stable cell lines or transgenic animals for research purposes, where silencing is virtually permanent. Furthermore, elements like the tet-Operon positioned next to the promoter can reversibly turn silencing on and off with chemical inducers such as doxycycline (Section 2.4). Other systems, such as Cre-Lox, for example, can irreversibly activate RNAi.

A disadvantage of type III RNA polymerase promoters is that they do not restrict expression to particular cell type; they are active in almost all tissues. This problem has led to the development of other systems that can express short dsRNAs under the control of type II RNA polymerase promoters. For example, the shRNA may be placed in the sequence and processing context of a natural miRNA (Figure 13.20). The biology of miRNAs will be

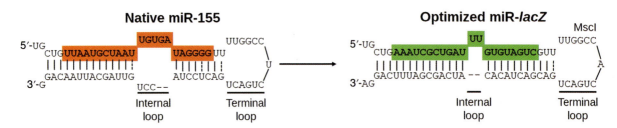

Fig. 13.20 Composition of an artificial miRNA. On the left side of the illustration, the sequence of the stem loop of the natural miR-155 is shown. The mature miR-155 is marked in orange. It can be replaced by the sequence of an siRNA against any given target mRNA, in this case lacZ (marked in green). A further sequence optimization is the introduction of a MscI restriction enzyme cleavage site. As described in the text, artificial miRNAs can be expressed under the control of RNA polymerase II promoters.

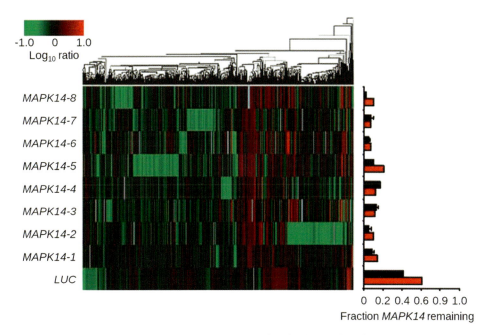

Fig. 13.21 Off-target effects of small interfering RNSa (siRNAs). A study analyzed global profiles of gene expression. siRNAs against mitogen-activated protein kinase 1 (MAPK14) regulated not only the target mRNA but also numerous other mRNAs. (Adapted from Ref. [9] with kind permission from Macmillan Publishers Ltd., Copyright 2003.)

covered in depth in Section 13.4. The shRNAs in a miRNA environment are also referred to as artificial miRNAs (amiRNAs). Since the natural microRNA pathway processes these amiRNAs, this approach is often highly efficient. The product of the processed amiRNA is a guide strand that is identical to the one generated from the corresponding shRNA, thus also permitting the silencing of any gene of interest. In addition to high efficiency, this approach permits the use of tissue-specific RNA polymerase II promoters. Thus, silencing can be restricted to certain organs, for example, the liver, to prevent the unintended side effects of silencing in other tissues.

The experience gained from gene therapy approaches has led to rapid progress in the vector-based delivery of RNAi. All approaches that trigger RNAi by the intracellular expression of short dsRNAs from expression cassettes profit from the availability of viral vectors as a means of delivery (Chapter 11 and Section 13.3.3).

13.3.2
Nonspecific Side Effects

Since siRNAs interact with their target RNA through Watson–Crick base pairing, they should in theory specifically inhibit the expression of their target only. Their high specificity has been viewed as a major advantage in comparison with most small-molecule protein inhibitors. However, under certain circumstances, siRNAs also cause nonspecific side effects.

Several years after the development of RNAi as a technology for gene silencing in mammalian cells, microarray analyses of global profiles of gene expression (Section 2.1) found the first hints of off-target effects, that is, the regulation of mRNA levels that were not the intended targets of the siRNA. These analyses showed that in addition to the target mRNA, numerous other mRNAs were inhibited (Figure 13.21). If these effects were produced by silencing of the target, all the siRNAs successfully inhibiting that target would be expected to generate the same pattern of downstream effects. However, every siRNA created a unique pattern, indicating that off-target effects were in fact being triggered by the siRNAs themselves. It is now clear that siRNAs can recognize not only 100% complementary target sequences but also other mRNAs that contain only partial sequence homology. The selection of sequences with minimal homology to other transcripts combined with the use of low concentrations of chemically modified siRNAs can minimize off-target effects.

But siRNAs do not only regulate nontargeted mRNAs, they can also trigger an interferon response. As previously described, the original assumption was that siRNAs would not trigger an interferon response due to their short length (less than 30 nucleotides). However, array analyses have found that in some cases siRNA treatment also induces interferon response genes. In other cases, siRNAs trigger cellular responses through Toll-like receptors. As will be described in Section 13.3.5, the first clinical trials based on RNAi were directed at inhibiting neovascularization in the eye. However, in subsequent investigations, doubts were raised that the siRNAs were producing any RNAi-mediated silencing of the target gene at all. Since the observed effects were independent of the siRNA sequence, activation of Toll-like receptor 3, which binds dsRNAs, may have indirectly been responsible for the observed antiangiogenic effects.

siRNAs can also trigger an interferon response through the cytosolic immunoreceptor RIG-I. Specific sequence motifs and other siRNA elements trigger this effect, and careful design can avoid it. However, for medical use such as treatment of viral infection or cancer, the induction of an interferon response might be desirable. This has led to the concept of bifunctional siRNAs, which both silence a target gene and stimulate an immune response. In a proof-of-concept study, an siRNA targeting the apoptosis inhibitor Bcl-2 that also contained sequence motifs that activate RIG-I was evaluated. In a mouse model, the antitumor activity of this bifunctional siRNA was higher than that of either an siRNA that silenced Bcl-2 only or a molecule that stimulated RIG-I, indicating synergy between the two modes of action.

Overdoses of short dsRNAs can have potent toxic side effects. A high dose of an adeno-associated virus (AAV) vector carrying an shRNA expression cassette led to severe liver toxicity when injected into mice. Presumably, high concentrations of shRNAs saturate the RNAi pathway and in the process block the endogenous miRNA pathway (Section 13.4). These results demonstrate that correct dosing is critical for the therapeutic use of RNA technology.

13.3.3
Delivery

As described for AS ONs and ribozymes, the hydrophobic cell membrane is a barrier to the transfer of negatively charged nucleic acids into cells. For *in vivo* use, reaching the target tissue is, of course, also critical. Achieving efficient delivery is the central challenge to the use of RNAi and extensive efforts are underway to overcome this hurdle.

There are two basic strategies for RNAi delivery: (i) the delivery of chemically synthesized siRNAs and (ii) the transfer of shRNA expression cassettes with the help of viral vectors. The delivery of siRNAs can be improved by targeting specific cell types with antibodies (Section 10.2.1), aptamers (Chapter 14), or receptor ligands. Unmodified siRNAs are very sensitive to nucleolytic degradation; incorporation of modified nucleotides is usually necessary to render them sufficiently resistant to nucleases for *in vivo* use. For this purpose, the modified elements previously described in the context of AS ONs, for example, phosphorothioate linkages,

2′-O-methyl-RNA and 2′-fluoro nucleotides or LNAs can be employed. However, the incorporation of unnatural nucleotides into siRNAs is challenging, since the modifications must not affect the silencing activity of the siRNA, which happens easily after modification of the guide strand. Therefore, careful evaluation is required to determine which positions tolerate modified nucleotides.

Cationic lipids are the most frequently used carriers for the transfer of siRNAs into cells. The positive charges of the lipids associate with the negatively charged siRNAs, forming what is referred to as a lipoplex (Figure 13.22a). Cationic lipids also interact with negatively charged groups on the surface of cells, which take up the complexes by endocytosis. Transfection reagents may also contain fusogenic lipids, which promote the intracellular release of the siRNAs from the endosomes. Coupling of the lipoplexes to polyethylene glycol (PEG) prevents renal clearance and extends the plasma siRNA half-life. For *in vivo* uses, less toxic liposomes are often used. These enclose the siRNAs in an inner vesicle (Figure 13.22b). Several clinical studies have used a class of liposomes known as "stable nucleic acid-lipid particles" (SNALPs). These consist of a cationic lipid, a fusogenic lipid, and a PEG coat on their surface. In addition to lipid-based systems, polymers of polyethyleneimine (PEI) are often used for the *in vivo* transfer of siRNAs.

An alternative strategy is to couple lipophilic molecules directly to the siRNA. A successful example is the use of siRNAs that carry cholesterol on the 3′ end of the sense strand (Figure 13.22c). Due in part to the lipophilic group, siRNAs injected intravenously efficiently reach the liver and the jejunum.

The targeted delivery of siRNAs allows the use of smaller doses and avoids side effects in nontargeted tissues. One of the first approaches to attempt the cell-type specific transfer of siRNAs employed antibodies recognizing specific proteins on the surface of the target cells. The antigen-binding fragments of the antibody (Fab) were fused to protamine, which bound several siRNAs noncovalently through ionic interactions (Figure 13.22d). The siRNAs entered only into cells that expressed the protein targeted by the antibody on their surface. As an alternative to antibody fragments, aptamers (Chapter 14) or the ligands of cellular receptors can specifically deliver complexes to cells. Furthermore, sophisticated nanoparticles can also efficiently deliver siRNAs. In some cases, the coupling of specific antibodies or receptor ligands to the surface of the nanoparticle can deliver siRNAs to specific cells. A nanoparticle that was successfully tested in clinical trials will be described in more detail in Section 13.3.5.

Vectors based on lenti-, adeno-, and adeno-associated viruses are commonly employed for the viral delivery of shRNA expression cassettes (Figure 13.23). Lentiviral

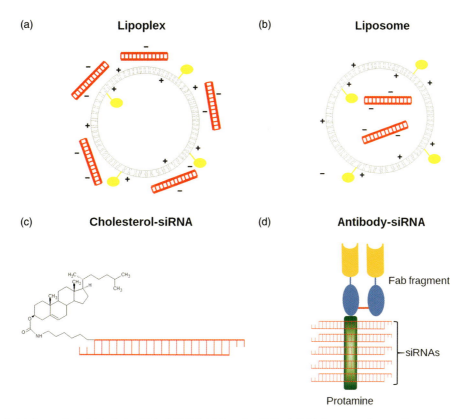

Fig. 13.22 Nonviral delivery of small interfering RNAs (siRNAs). (a) Lipoplex: Cationic lipids form complexes with negatively charged siRNAs. PEG (yellow) is frequently attached to improve pharmacokinetic characteristics. (b) Liposomes in which cationic lipids encapsulate the siRNA. (c) siRNA coupled to cholesterol to increase its lipophilicity. (d) Specific delivery by the coupling of siRNAs to the antigen-binding fragment of an antibody through positively charged protamine.

vectors integrate their genetic material into the host genome and lead to long-term expression of the shRNA. Adenoviral and AAV vectors remain episomal and are, therefore, suitable for transient RNAi. The choice of suitable serotypes can facilitate the transfer of the shRNA expression cassettes into specific target cells. Characteristics of the various types of vectors are described in detail in Section 11.2.

13.3.4
Preclinical Applications of RNA Interference

In only a few years, RNAi has become a standard tool in biomedical research. Every year, thousands of papers describe the use of RNAi to silence the expression of a gene, which can be accomplished relatively easily based on the sequence of the target mRNA. Employing the RNAi approach is invariably faster than attempting to find a small molecule that inhibits the target protein. RNAi is also capable of specifically silencing a selected member of a family of closely related genes whose protein products are too similar to be differentially targeted by chemical compounds.

Furthermore, RNAi has been used to study not only individual genes but also gene classes, and even the complete human genome. In one of the first published partial RNAi screens, the knowledge gained quickly led to a novel clinical use of a known drug. This screen involved the silencing of 50 genes that code for deubiquitinating enzymes and subsequently analyzing the data for those genes influencing the expression of the NF-κB transcription factor. It was discovered that the knockdown of the familial *cylindromatosis tumor suppressor* gene led to the induction of NF-κB. The importance of this gene for familial cylindromatosis, which produces an autosomal-dominant disposition for the development of benign skin appendages, was known, but its mechanism remained elusive. The RNAi screen established the connection between inherited mutations in the familial cylindromatosis gene and the modulation of NF-κB. Salicylic acid, a small molecule from which aspirin is derived, prevents the upregulation of NF-κB. Based on these findings, patients suffering from cylindromatosis were treated with a salicylic acid salve. The blockade of NF-κB resulted in both partial and, on occasion, complete remissions of appendage growth.

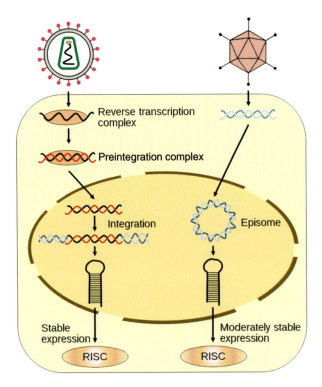

Fig. 13.23 Viral delivery of short hairpin RNA (shRNA) expression cassettes. Lentiviral vectors integrate into the host genome and allow long-term expression of the shRNA, while adenoviral and AAV vectors remain episomal and induce transient silencing of the target gene. (Adapted from Ref. [10] with kind permission from Macmillan Publishers Ltd., Copyright 2007.)

Although individual gene classes were investigated initially, libraries can also be employed to knockdown virtually every human gene. These libraries consist of either chemically synthesized siRNAs or viral vectors encoding the expression of shRNAs. Generally, the use of three to four siRNAs or shRNAs per target gene is required, since some are inactive and others trigger nonspecific effects. The actual screens utilize automated systems; the main areas of interest are cancer and virologic research.

In one experiment, a screen for HIV silenced every human gene, one gene at a time, in order to identify host factors that the virus requires for its replication. Only those genes whose knockdown led to a block of viral replication without decreasing cell viability were of interest. In addition to the known interaction partners in the viral life cycle, the screen revealed several hundred novel HIV dependency factors. These represent potential targets for therapeutics. Similar screens addressed other medically relevant viruses, such as influenza and HCV. In cancer research, RNAi screens have helped to identify factors that promote cell proliferation and are potential novel targets for cancer therapy.

RNAi can also target the virus directly. Efficient RNAi approaches exist for almost all medically relevant viruses, such as HIV-1, HBV, HCV, and influenza. The RNAi strategy employed heavily depends on the virus type. For RNA viruses, the siRNA targets the genomic RNA directly, while for DNA viruses the viral mRNAs are the target. Another great advantage of RNAi technology is its flexibility. On the appearance of a new variant of a known virus, such as the swine flu, or new types of viruses, such as the SARS Coronavirus, knowledge of the sequence only is required in order to design the siRNAs.

For various types of viruses, long-term studies have shown that a single siRNA or shRNA, like other antiviral approaches, is active for only a limited time. Depending on the error rate of the viral polymerase, mutants that are resistant to inhibition by a single RNAi trigger tend to emerge rapidly. Often, a single point mutation in the target site is sufficient to render the virus insensitive to inhibition by an siRNA or shRNA. By analogy with classical antiviral therapy in which several antiviral compounds are employed simultaneously (Section 5.4.1), multiple siRNAs or shRNAs can be used to suppress the emergence of resistant mutations, as it is unlikely that a virus can become resistant to multiple RNAi triggers simultaneously. For example, one of the first clinical studies of RNAi used four shRNAs against HBV.

13.3.5
Clinical Trials

The therapeutic potential of RNAi was quickly recognized. The first RNAi-based clinical studies began only 3.5 years after the initial proof-of-principle work that showed that RNAi can be used to silence genes in mammalian cells. The disease targeted was wet age-related macular degeneration (AMD), which is caused by abnormal blood vessel growth (neovascularization) induced by VEGF. AMD is a major cause of visual impairment and blindness in older adults. The first studies used local (intravitreal) injection directly into the eye, an organ where immune-mediated toxicity should be minimal. It was hypothesized that RNAi-mediated knockdown of VEGF or its receptor should stop the AMD process. However, as noted above, the most advanced trials were terminated when further research indicated that angiogenesis was not inhibited by specific knockdown of the target, but rather by activation of TLR3 activity (Section 13.3.2).

The selection of clinical studies using RNAi (Table 13.2), as previously discussed for antisense approaches, demonstrates the breadth of the use of silencing technology. In addition to eye disease, cancer, and viral infection, development of RNAi for many other diseases is underway. The technology may also be useful for cosmetic

Table 13.2 Selected RNAi therapeutics under clinical evaluation.

Candidate (company)	Indication	Target	Delivery
Bevasiranib/Cand5 (Opko Helath)	Wet AMD, diabetic macular edema	VEGF	Intravitreal needle injection (retina; local)
Sirna-027/AGN-745 (Sirna Therapeutics)	Wet AMD	VEGF-R1	Intravitreal needle injection (retina; local)
ALN-RSV01 (Alnylam)	RSV infection	Viral RNA	Inhalation of unformulated siRNAs (lung epithelium; local)
QPI-1002 (Quark/SBI Biotech)	Acute kidney injury, delayed graft function	p53	Intravenous naked siRNA (proximal tubule cells; systemic)
PF-4523655 (Quark/ Pfizer)	Wet AMD, diabetic macular edema	RTP801/REDD1	Intravitreal needle injection (retina; local)
QPI-1007 (Quark)	Chronic nerve atrophy, nonarteritic ischemic optic neuropathy	Caspase 2	Intravitreal needle injection
rHIV-shI-TAR-CCR5RZ (City of Hope)	HIV infection	Viral RNA and host factors	Lentiviral (hematopoietic stem cells; *ex vivo*)
NucB1000 (Nucleonics)	Hepatitis B viral infection	HBV RNAs	Liposomal plasmid (hepatocytes; systemic)
TD101 (TransDerm/ IPCC)	Pachyonychia congenita	Mutant keratin	Intradermal needle injection (skin; local)
Therapeutic vaccine (Duke University)	Metastatic melanoma	Immunoproteasome	Electroporation (autologous monocytes; *ex vivo*)
Excellair (ZaBeCor)	Asthma	Syk kinase	Inhalation of unformulated siRNAs (lung epithelium; local)
CALAA-01 (Calando)	Nonresectable or metastatic solid tumors	M2 subunit of ribonucleotide reductase	RONDEL (solid tumor cells; systemic)
ALN-VSP02 (Alnylam)	Liver cancer, cancer with liver involvement	VEGF, KSP	SNALP liposome (hepatocytes; systemic)
ALN-TTR01 (Alnylam)	Transthyretin amyloidosis	Transthyretin	SNALP liposome (hepatocytes; systemic)
ALN-TTR02 (Alnylam)	Transthyretin amyloidosis	Transthyretin	Improved SNALP liposome (hepatocytes; systemic)
ALN-TTRSC (Alnylam)	Transthyretin amyloidosis	Transthyretin	siRNA conjugated to a *N*-acetylgalactosamine (GalNAc) ligand (hepatocytes; subcutaneous)
ALN-PCS02 (Alnylam)	Hypercholesterolemia	PCSK9	SNALP liposome (hepatocytes; systemic)
Atu027 (Silence)	Advanced solid tumors	PKN3	AtuPLEX lipoplex (vascular endothelial cells; systemic)
SYL040012 (Sylentis)	Intraocular pressure and glaucoma	β-Adrenergic receptor 2	Eye drop (ciliary epithelial cells; local)
SYL1001 (Sylentis)	Ocular pain	TRPV1	Ophthalmic drops
PRO-040201/TKM-ApoB (Tekmira)	Hypercholesterolemia	Apolipoprotein B	SNALP liposome (hepatocytes; systemic)
TKM-PLK1 (Tekmira)	Solid cancers and lymphoma	Polo-like kinase 1	SNALP liposomal (solid tumor cells; systemic)
TKM-EBOLA (Tekmira)	Ebola infection	Viral RNA	SNALP liposome (hepatocytes and phagocytes; systemic)
FANG vaccine bi-shRNAfurin/GMCSF (Gradalis)	Ovarian cancer, colon cancer, advanced melanoma	Furin	Electroporation plasmid (autologous tumor samples; *ex vivo*)
siG12D LODER (Silenseed)	Operable pancreatic ductal adenocarcinoma	Mutated KRAS	LODER local drug elution
CEQ508 (Marina)	Familial adenomatous polyposis/colon cancer prevention	β-Catenin	Bacterial (mucosal layer of small and large intestine; oral)
RXI-109 (RXi)	Dermal scarring	CTGF	Intradermal needle injection (skin; local)
ND L02-s0201 (NDT)	Liver cirrhosis	HSP47	Vitamin A-coupled liposomes (hepatocytes, systemic)

Based on Refs [11,12] and company Web sites.

Box 13.3.   Hair Removal by Small Interfering RNAs (siRNAs)

As already mentioned, RNAi is not only of interest for medical uses but also for cosmetic purposes. The *Hairless* gene plays a critical role in the growth of hair. Through RNAi-mediated knockdown of *Hairless*, the hair follicle can be destroyed during the catagen stage (Figure 13.24). This leads to permanent changes in the hair follicle, so that hair growth ceases. Sirna Therapeutics initiated a program that attempted permanent hair removal by RNAi in a market estimated to be over $10 billion. The siRNA was delivered as a cream formulation. However, the skin proved to be too difficult a barrier to penetrate for any efficint siRNA delivery.

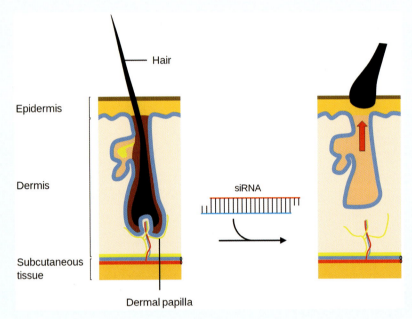

Fig. 13.24   Permanent hair removal by small interfering RNAs (siRNAs). The siRNA is targeted against the *Hairless* gene. Knockdown of this transcription factor destroys the hair follicle, which permanently prevents hair growth.

purposes (Box 13.3). Clinical studies have employed both chemically synthesized siRNAs and vector-expressed shRNAs. The siRNAs were either unmodified or were modified for greater stability. While the first clinical trials used naked siRNA molecules, later studies employed transfection reagents, most often liposomes or nanoparticles, to aid delivery. In many cases, this allowed systemic injection of the siRNAs rather than local application.

One of the most advanced studies for the therapeutic use of RNAi has involved treating respiratory syncytial virus (RSV) infection. For most adults, RSV produces mild symptoms that are often indistinguishable from the common cold. However, RSV can lead to serious illness in newborns and premature babies. Since the virus infects the respiratory tract, delivery of the siRNA by inhalation is easy and convenient. Clinical studies with infected adults have shown significant antiviral effects.

An approach to the treatment of HIV infection using a lentiviral delivery vector has been combined with several RNA technologies to prevent the emergence of escape mutants. Hematopoietic stem cells were collected from HIV-infected patients, treated with a lentiviral vector *ex vivo*, and then reinfused (Section 11.2.1). The lentiviral vector coded for an shRNA against the HIV-1 genes *rev* and *tat*, a hammerhead ribozyme against the coreceptor CCR5, and a decoy oligonucleotide against the transactivation response (TAR) element. The decoy oligonucleotide interacts with its target molecule like an aptamer (Chapter 14), relying on steric interactions instead of Watson–Crick base pairing. If successful, this triple approach could permanently protect hematopoietic cells from HIV.

The first direct proof that siRNAs actually function via an RNAi mechanism in humans was discovered during a

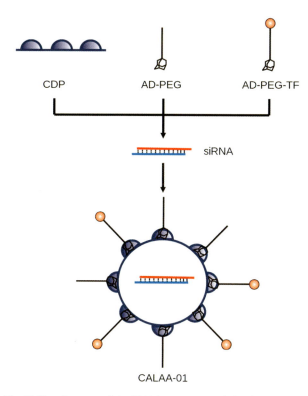

Fig. 13.25 Structure of the CALAA-01 nanoparticle for the targeted delivery of small interfering RNAs (siRNAs). For details, see text. AD: adamantine; CDP: cyclodextrin polymer; PEG: polyethylene glycol; TF: transferrin. (Adapted from Ref. [13] with kind permission from John Wiley & Sons, Inc.)

cancer therapeutic trial. In a phase I clinical trial, advanced melanoma patients received an siRNA targeted against the M2 subunit of ribonucleotide reductase incorporated into nanoparticles. These nanoparticles consisted of a sphere made from a cyclodextrin-based polymer with polyethylene glycol adamantine and transferrin protein attached to their surface (Figure 13.25). Many cancer cells overexpress the transferrin receptor on their surface, which mediates cellular entry of the nanoparticles. A special PCR technique was employed to demonstrate that the siRNAs led to the cleavage of the target mRNA at the expected position.

RNAi-based therapies have been found to be well tolerated in clinical studies. The development of RNAi, like the development of monoclonal antibodies, has its share of problems. The initial euphoria turned to skepticism after the discovery of the nonspecific effects caused by siRNAs. Nevertheless, a short time later, major pharmaceutical companies were again willing to invest large sums in biotech companies working in the RNAi field. As time went on, some of these same companies pulled out of the development of RNAi-based technologies, realizing that the technology required substantial improvement before

becoming commercially successful. In the interim, there is still hope that RNAi will lead to the development of innovative drugs. Ideally, progress in the delivery and chemical modification of antisense and siRNA molecules will mutually advance each other's therapeutic potential.

13.4
MicroRNAs

13.4.1
The Biology of MicroRNAs

MicroRNAs are the natural equivalent of the siRNAs used for research or therapeutic purposes. They are also approximately 21 nucleotides long, are present in the genomes of almost all eukaryotes, and are evolutionarily strongly conserved. The human genome contains more than 1500 different miRNAs that function as important posttranscriptional regulators of gene expression and control the activity of up to 60% of all protein-coding genes. Since each miRNA can regulate up to several hundred mRNAs, each of which can be controlled by multiple miRNAs, the network of regulation is highly complex. Usually, miRNAs repress each individual mRNA only modestly (30–50% downregulation). Regulation of gene expression by miRNAs can be viewed as a fine-tuning rather than as an on-off switch. Nevertheless, the extent of regulation is sufficient to induce significant phenotypic changes. miRNAs are involved in all important physiological processes studied to date, such as neurogenesis, heart development, and the maturation of blood and immune cells. They are of great significance for many diseases, including cancer, viral infection, and metabolic diseases. Like siRNAs, miRNAs regulate protein synthesis by base pairing with the target mRNA. Most miRNAs, however, form only partially complementary hybrids (Figure 13.26). In addition, they usually bind to the 3' UTR, while siRNAs are normally directed against the coding region. In general, miRNAs inhibit protein synthesis by either repressing translation and/or destabilizing their mRNA targets.

The biogenesis of miRNAs originates from either specific genes or mRNA introns (Figure 13.27). Initially, the long primary transcript, denoted as pri-miRNA, folds into a hairpin structure. It is processed by the enzyme Drosha and its cofactor DGCR8. Together, they form the so-called microprocessor that produces ~70-nucleotide precursor miRNA (pre-miRNA). Pre-miRNAs spliced directly out of introns that bypass the microprocessor complex are named "Mirtrons." Exportin-5 transports the pre-miRNA hairpin molecules from the nucleus into the cytoplasm, where Dicer processes the

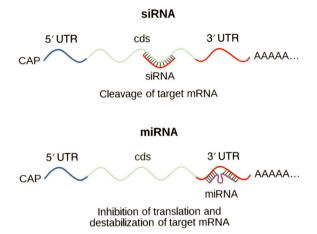

siRNA

5′ UTR   cds   3′ UTR

CAP   AAAAA…

siRNA

Cleavage of target mRNA

miRNA

5′ UTR   cds   3′ UTR

CAP   AAAAA…

miRNA

Inhibition of translation and
destabilization of target mRNA

Fig. 13.26 Comparison of small interfering RNAs (siRNAs) and microRNAs (miRNAs). siRNAs are usually fully complementary to the target mRNA and bind to the coding region, while most miRNAs bind to the 3′ UTR through partially complementary base pairing. siRNAs lead to cleavage of the target mRNA, while miRNAs inhibit translation and destabilize the target mRNA. UTR: untranslated region; cds: coding sequence.

pre-miRNA to yield a mature miRNA. After incorporation into RISC, the miRNA guides the complex to the target mRNA. The interaction between the miRNA and its target mRNA begins at what is called the seed region, nucleotides 2–8 of the miRNA (counting from its 5′ end).

Although the exact mechanism of miRNA-mediated gene silencing has not yet been completely elucidated, miRNAs appear to work primarily in two ways, either by inhibiting translation or by destabilizing the target mRNA. The interaction between miRISC and the cap on the 5′ end of the targeted mRNA prevents recognition by initiation factor eIF4E and the ribosome, thus blocking translation (Figure 13.28a). In addition, the assembly of the ribosome can be inhibited and sometimes postinitiation translation can be blocked. The degradation of

Fig. 13.27 MicroRNA pathway. (a) MiRNAs are initially generated as long primary transcripts (pri-miRNAs). They are processed while still in the nucleus to approximately 70mer pre-miRNAs, which are then exported into the cytoplasm. There they undergo the final processing step to mature miRNAs (b) and are loaded into RISC. RISC together with the miRNA is referred to as miRISC. (c) The miRNA guides the miRISC to the target RNA and usually binds to the 3′ UTR of an mRNA. An important element for the hybridization of the miRNA to its target mRNA is the seed region, which usually involves fully complementary and contiguous base pairing. Bulges or mismatches are often present in the central region of the miRNA–mRNA duplex. The binding of miRISC to the target mRNA represses its translation or triggers its degradation. DGCR8: DiGeorge syndrome critical region 8; Ago: Argonaute protein; TRBP: *trans*-activation response (TAR) RNA binding protein. (Adapted from Ref. [14] with kind permission from Macmillan Publishers Ltd., Copyright 2008.)

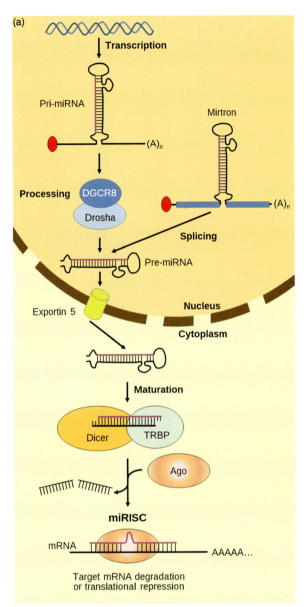

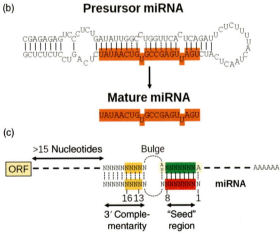

(a) Translational repression

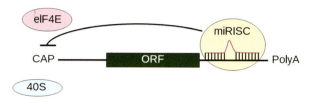

(b) mRNA decay

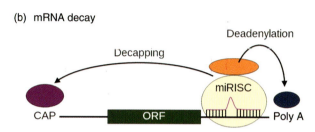

Fig. 13.28 Mechanisms of microRNA-mediated gene silencing. (a) Repression of translation is primarily due to the interaction between the miRNA-loaded RISC (miRISC) with the cap on the 5′ end of the target mRNA, so that the initiation factor eIF4E can no longer recognize it. This prevents recruitment of the small subunit of the ribosome. (b) In addition, the miRNA triggers the degradation of the mRNA. This begins with deadenylation of the poly A-tail. This process is followed by the removal of the 5′ end cap. The mRNA is then degraded by nucleases.

the target mRNA presumably begins with deadenylation of the poly A-tail (Figure 13.28b). Subsequent removal of the cap also leaves the mRNA unprotected against nucleases, which rapidly degrade it. It has not yet been determined to what extent each mechanism is responsible for the net degradation. It is possible that translation

is initially repressed, followed by deadenylation of the mRNA, followed by decapping and degradation of the destabilized mRNA.

Some miRNAs are present only for a limited period during certain stages of organismal development. They were thus originally named small temporal RNAs (stRNAs), though the name microRNAs is now used more commonly, since temporal restriction of activity is not a general feature of the short dsRNAs. Box 13.4 explains miRNA nomenclature. Numerous miRNAs are organ specific. The human miR-121 is a prominent example of a miRNA that is expressed only in liver. A single cell type may express less than 100 of the over 1500 different miRNAs. Levels of expression vary widely, from a few copies of a miRNA (<10) to over 10 000 transcripts.

The first physiological role described for miRNAs was in the release of insulin: miR-375 is highly expressed in the β cells of the pancreas. This miRNA binds to myotrophin mRNA, which regulates the release of insulin granules. Inhibition of miR-375 leads to an increase in the level of myotrophin, which causes an increased release of insulin granules. The overexpression of miR-375 leads to silencing of myotrophin and the release of less insulin.

13.4.2
MicroRNAs and Disease

As miRNAs are involved in the normal functioning of cells, it is not surprising that dysregulation of miRNAs has been associated with disease. miRNAs play a role in many disease processes, such as viral infection, cancer, and cardiovascular disease. As a result, their biological

Box 13.4. Nomenclature of miRNAs and Other Regulatory RNAs

The designation of a miRNA begins with the prefix "miR" followed by a dash and a number (e.g., miR-106), the latter often indicating the order of naming. Lower case "mir" refers to the pre-miRNA, while a capitalized "R" refers to the mature form. miRNAs with almost identical sequence (except for one or two nucleotides) receive an additional lower case letter; for example, miR-133a and miR-133b only differ in a single base position. In some cases, miRNAs identical in sequence are coded in multiple positions in the genome. They are designated by an additional dash-number suffix. Therefore, miR-133a-1 and miR-133a-2 have the same sequence, but they are coded on different chromosomes (18 and 20, respectively). Finally, the species of origin of a miRNA is coded in a three-letter prefix, for

example, hsa-miR-133a-1 designates a human (*Homo sapiens*) miRNA. If two mature miRNAs are processed from a single pre-miRNA, these receive an additional -3p or -5p suffix. If the amount of single-stranded miRNA of one arm of the pre-miRNA hairpin is significantly less than that of the opposite arm, it is indicated with an asterisk "∗." Several miRNAs appear in clusters. For example, the miRNA cluster miR-17-92 contains (at least) five miRNAs: miR-17, miR-18a, miR-19a/b, miR-20a, and miR-92a in a single transcript.

Besides miRNAs, there are other regulatory RNAs that do not code for proteins. These include the short piwi-interacting RNAs (piRNAs) that repress retrotransposition of repeat elements and longer noncoding RNAs (ncRNAs), which are also involved in gene regulation.

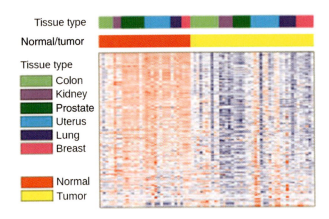

Tissue type

Normal/tumor

Tissue type

- Colon
- Kidney
- Prostate
- Uterus
- Lung
- Breast

- Normal
- Tumor

Fig. 13.29 Comparison of microRNA (miRNA) expression levels between normal and tumor tissue. Global analysis of miRNA expression levels by microarrays revealed that most of the miRNAs were expressed at lower levels in tumors compared with normal tissue. (Adapted from Ref. [15] with kind permission from Macmillan Publishers Ltd., Copyright 2005.)

function is under intense investigation to determine their suitability as biomarkers or therapeutic targets.

Experiments measuring the expression levels of every miRNA in cells have revealed altered miRNA regulation in disease states. For example, with the help of microarray analyses (Section 2.1), miRNA expression levels in healthy and diseased tissue can be compared (Figure 13.29). Tumors can be classified by microarray-based miRNA profiling or deep sequencing (Section 2.3). With the latter technology, it is possible to measure the expression level of known miRNAs, but also to discover new ones. Quantitative RT-PCR or Northern blots can confirm miRNA expression levels. An important step in the elucidation of miRNA function is the identification of the target mRNAs of the miRNAs. For this purpose, the 3′ UTRs of all human mRNAs have been bioinformatically examined for homologies to the miRNAs under investigation. This method has a level of uncertainty since, as previously described, miRNAs are not fully complementary to their target sequences. For this reason, reporter assays are often used to verify whether a 3′ UTR leads to miRNA-mediated regulation of gene expression.

To test the biological relevance of a cellular miRNA, it can either be overexpressed or inhibited. AS ONs are usually employed to inhibit a miRNA, as similar to mRNAs they bind to it and block its function, as described in Section 13.1. Like standard AS ONs, the miRNA inhibitors are chemically engineered to protect them from nucleolytic degradation. For *in vivo* applications, so-called antagomirs have been developed. These are single-stranded AS ONs that are partially modified with 2′-O-methyl nucleotides in addition to a cholesterol moiety on the 3′ end to improve cellular uptake. The

first clinically tested AS ONs targeted against miRNAs were stabilized with LNA nucleotides (see below).

The importance of miRNAs for the development of cancer has also been intensively studied. As is apparent in the heat map in Figure 13.29, tumors frequently contain reduced levels of mature miRNAs, but a few miRNAs are specifically upregulated. Depending on their role in cancer development, these miRNAs are referred to as *tumor suppressor miRNAs* or *oncomiRs*.

Tumor suppressor miRNAs control cell cycle progression, tumor invasion, and metastasis, and are similar to protein-encoding tumor suppressor genes (Box 4.1 in Chapter 4). Their loss promotes tumor development. For example, the let-7 miRNA family is downregulated in many cancer types and has been shown to target prominent oncogenes, such as *K-RAS*, *MYC*, and *HMGA2*. Therefore, a viable therapeutic strategy would be to overexpress the miRNA in cancer cells with, for example, the aid of a viral vector (Chapter 11). This strategy has reduced the rate of tumor growth in various animal models of cancer.

In contrast to tumor suppressor miRNAs, oncomiRs initiate or accelerate cancer formation or development when constitutively overexpressed. An example is miR-21, which is almost universally overexpressed in cancer. Its key downstream targets have been identified: One of these targets is the tumor suppressor *PTEN*, which is downregulated by the overexpression of miR-21. The suppression of *PTEN* ultimately leads to increased tumor cell proliferation, migration, and invasion. Inhibiting oncomiRs with AS ONs is thus a promising strategy to block the proliferation or metastasis of tumor cells.

miRNAs also play an important role in the cardiovascular system. Several cardiac-enriched miRNAs have been identified, including miR-1, miR-133, and miR-208, which are all important in cardiac development. Dysregulation of these miRNAs can lead to heart disease. For example, the downregulation of miR-1 has been correlated with hypertrophic cardiomyopathy. Myocardial ischemia modulates various miRNAs that regulate heat shock proteins and apoptotic factors. Further, since cells release at least some miRNAs into the circulation, they are also potential candidates as biomarkers (Box 13.5).

miRNAs are also of great significance in viral infection. Several hundred miRNAs have been identified that are encoded by viruses. Most of the viral miRNAs known today originate from members of the herpesvirus family, which relies on DNA for its genome. These viruses use cellular machinery to process their expressed miRNAs. Viral miRNAs serve in part to autoregulate viral gene expression. Several viral miRNAs have an important role in maintaining latent and persistent infection. Other viral miRNAs suppress host defenses by reducing specific host mRNAs.

Box 13.5.   Circulating miRNAs as Biomarkers

The discovery of miRNAs circulating in the bloodstream was a great surprise. The assumption had always been that miRNAs would be very sensitive to nucleolytic degradation and, thus, could not be present in extracellular fluids. Nevertheless, extracellular vesicles were isolated that were found to encapsulate miRNAs. These miRNAs did not appear to be randomly present in these so-called exosomes, but instead appeared to have been sorted in a nonrandom process. Current research is investigating the possibility that the exosome transport of miRNAs is a way for cells to communicate and regulate each other's genetic activity. As

mentioned, circulating miRNAs may serve as biomarkers, similar to circulating hormones. For example, during an acute myocardial infarction, the level of miR-208a in the bloodstream increases. This increase can be detected significantly earlier than the protein troponin I, which is often used for the diagnosis of a myocardial infarction. In a similar manner, circulating miRNAs can serve as biomarkers for cancer. Tumors increase the levels of specific miRNAs in the plasma. A diagnosis based on circulating miRNAs would be noninvasive, in contrast to an analysis of intracellular miRNA patterns, which would require a biopsy.

The role of RNAi as a part of the physiological defense of mammalian cells against viral infection has not yet been completely elucidated. Since mammals have developed an adaptive immune system and interferon-based defense mechanisms, it might appear that RNAi has lost importance as an antiviral mechanism. Nevertheless, numerous human miRNAs bind and repress the replication of viral RNAs (e.g., HBV, HIV, and coxsackievirus B3). Interestingly, viral suppressors of RNAi have been discovered that block the action of these host defenses.

A recent therapeutically relevant discovery was that HCV requires miR-122 for its replication. The virus has two binding sites for the miRNA in its 5′ UTR (Figure 13.30). Presumably, the interaction with the miRNA protects the viral RNA from nucleolytic degradation, thus promoting viral propagation. Blocking of miR-122 by an AS ON suppressed replication of HCV. The first miRNA inhibitor evaluated in a clinical study was an LNA-modified AS ON targeted against miR-122 (Miravirsen™). In a phase II study, long-lasting suppression of viremia in HCV-infected patients was achieved. Targeting a host cell miRNA instead of a viral RNA has two advantages. In studies with nonhuman primates, no resistant escape mutants emerged as is usually seen when a single siRNA against the virus is employed. This is because the virus cannot compensate for the suppressed miRNAs with mutations in the two binding sites in the 5′ UTR or by any other viral-specific mechanism. In addition, this approach is independent of the genotype of the virus, since all HCV variants appear to require miRNA binding. In parallel, an inhibitor of miR-122 that is composed of 2′-methoxyethyl nucleotides is also in clinical development for the treatment of

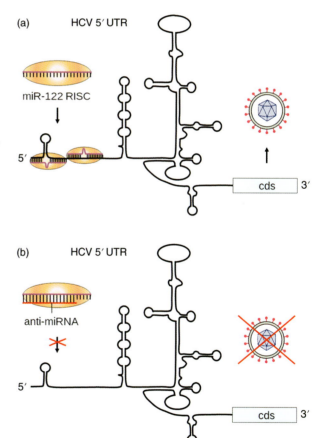

Fig. 13.30   Interaction of microRNA 122 with the 5′ untranslated region of the hepatitis C virus (HCV). (a) There are two binding sites for the liver-specific miR-122 in the 5′ UTR of HCV. The viral RNA is stabilized and protected via this interaction and viral propagation is promoted. (b) By sequestering mature miR-122 with LNA-modified AS ONs, the replication of HCV was suppressed.

HCV infection. This AS ON, named RG-101, is conjugated to *N*-acetylgalactosamine to improve hepatic uptake.

The principle of posttranscriptional silencing has been known for more than three decades: An oligonucleotide binds to a target RNA by Watson–Crick base pairing and blocks its function. The slow pace of clinical progress in the field is due in part to the fact that oligonucleotides differ significantly from traditional small-molecule drugs in their properties, size, production, and most importantly, their ability to be delivered to cells *in vitro* and *in vivo*. Their development is thus more comparable to that of monoclonal antibodies, which also required a substantial period of time before they became a class of successful new drug. The discovery of the RNAi pathway has immensely boosted gene silencing technologies. While ribozymes as therapeutic agents have faded into the background, AS ONs as well as siRNAs and shRNAs are in intensive development. Both strategies have advantages and disadvantages. RNAi approaches are usually extremely efficient. However, since the triggers of RNAi are double-stranded, production is more complex and regulatory approval more difficult. Single-stranded AS ONs are also taken up into cells more efficiently. Since the significance of miRNAs in many pathological processes has been demonstrated in recent years, a new field for the application of AS ONs as miRNA inhibitors has emerged. Further development of delivery technologies is the most important prerequisite for the successful clinical application of these silencing approaches.

References

1. Kurreck, J. (2003) Antisense technologies: improvement through novel chemical modifications. *Eur. J. Biochem.*, **270**, 1628–1644.
2. Christoph, T., Gillen, C., Mika, J. *et al.* (2007) Antinociceptive effect of antisense oligonucleotides against the vanilloid receptor VR1/TRPV1. *Neurochem. Int.*, **50**, 281–290.
3. Bennett, C.F. and Swayze, E.E. (2010) RNA targeting therapeutics: molecular mechanisms of antisense oligonucleotides as a therapeutic platform. *Annu. Rev. Pharmacol. Toxicol.*, **50**, 259–293.
4. Cech, T.R. (2000) Structural biology: the ribosome is a ribozyme. *Science*, **289**, 878–879.
5. Rice, M.C., Czymmek, K., and Kmiec, E.B. (2001) The potential of nucleic acid repair in functional genomics. *Nat. Biotechnol.*, **19**, 321–326.
6. Martick, M. and Scott, W.G. (2006) Tertiary contacts distant from the active site prime a ribozyme for catalysis. *Cell*, **126**, 309–320.
7. Fire, A., Xu, S., Montgomery, M.K. *et al.* (1998) Potent and specific genetic interference by double-stranded RNA in *Caenorhabditis elegans*. *Nature*, **391**, 806–811.
8. Kurreck, J. (2009) RNA interference: from basic research to therapeutic applications. *Angew. Chem., Int. Ed.*, **48**, 1378–1398.
9. Jackson, A.L., Bartz, S.R., Schelter, J. *et al.* (2003) Expression profiling reveals off-target gene regulation by RNAi. *Nat. Biotechnol.*, **21**, 635–637.
10. Kim, D.H. and Rossi, J.J. (2007) Strategies for silencing human disease using RNA interference. *Nat. Rev. Genet.*, **8**, 173–184.
11. Haussecker, D. (2012) The business of RNAi therapeutics in 2012. *Mol. Ther. Nucleic Acids*, **1**, e8.
12. Li, T., Wu, M., Zhu, Y.Y. *et al.* (2014) Development of RNA interference-based therapeutics and application of multi-target small interfering RNAs. *Nucleic Acid Ther.*, **24**, 302–312.
13. Kurreck, J. (2010) Proof of RNA interference in humans after systemic delivery of siRNAs. *Angew. Chem., Int. Ed.*, **49**, 6258–6259.
14. Filipowicz, W., Bhattacharyya, S.N., and Sonenberg, N. (2008) Mechanisms of post-transcriptional regulation by microRNAs: are the answers in sight? *Nat. Rev. Genet.*, **9**, 102–114.
15. Lu, J., Getz, G., Miska, E.A. *et al.* (2005) MicroRNA expression profiles classify human cancers. *Nature*, **435**, 834–838.

Further Reading

Burnett, J.C. and Rossi, J.J. (2012) RNA-based therapeutics: current progress and future prospects. *Chem. Biol.*, **19**, 60–71.
Sharma, V.K., Rungta, P., and Prasad, A.K. (2014) Nucleic acid therapeutics: basic concepts and recent developments. *RSC Adv.*, **4**, 16618–16631.
Watts, J.K. and Corey, D.R. (2012) Silencing disease genes in the laboratory and the clinic. *J. Pathol.*, **226**, 365–379.

Antisense Oligonucleotides

Bennett, C.F. and Swayze, E.E. (2010) RNA targeting therapeutics: molecular mechanisms of antisense oligonucleotides as a therapeutic platform. *Annu. Rev. Pharmacol. Toxicol.*, **50**, 259–293.
Chamberlain, J.R. and Chamberlain, J.S. (2010) Muscling in: gene therapies for muscular dystrophy target RNA. *Nat. Med.*, **16**, 170–171.
Crooke, S.T. (2004) Progress in antisense technology. *Annu. Rev. Med.*, **55**, 61–95.
Eckstein, F. (2014) Phosphorothioates, essential components of therapeutic oligonucleotides. *Nucleic Acid Ther.*, **24**, 374–387.
Kurreck, J. (2003) Antisense technologies: improvement through novel chemical modifications. *Eur. J. Biochem.*, **270**, 1628–1644.
Lightfoot, H.L. and Hall, J. (2012) Target mRNA inhibition by oligonucleotide drugs in man. *Nucleic Acids Res.*, **40**, 10585–10595.

Ribozymes

Doudna, J.A. and Cech, T.R. (2002) The chemical repertoire of natural ribozymes. *Nature*, **418**, 222–228.
Fedor, M.J. and Williamson, J.R. (2005) The catalytic diversity of RNAs. *Nat. Rev. Mol. Cell Biol.*, **6**, 399–412.
Mulhbacher, J., St-Pierre, P., and Lafontaine, D.A. (2010) Therapeutic applications of ribozymes and riboswitches. *Curr. Opin. Pharmacol.*, **10**, 551–556.
Schubert, S. and Kurreck, J. (2004) Ribozyme- and deoxyribozyme-strategies for medical applications. *Curr. Drug Targets*, **5**, 667–681.
Talini, G., Branciamore, S., and Gallori, E. (2011) Ribozymes: flexible molecular devices at work. *Biochimie*, **93**, 1998–2005.

RNA Interference

Haussecker, D. (2012) The business of RNAi therapeutics in 2012. *Mol. Ther. Nucleic Acids*, **1**, e8.

Kurreck, J. (2009) RNA interference: from basic research to therapeutic applications. *Angew. Chem., Int. Ed.*, **48**, 1378–1398.

Li, T., Wu, M., Zhu, Y.Y. *et al.* (2014) Development of RNA interference-based therapeutics and application of multi-target small interfering RNAs. *Nucleic Acid Ther.*, **24**, 302–312.

microRNAs

van Empel, V.P., De Windt, L.J., and da Costa Martins, P.A. (2012) Circulating miRNAs: reflecting or affecting cardiovascular disease? *Curr. Hypertens. Rep.*, **14**, 498–509.

Fabian, M.R., Sonenberg, N., and Filipowicz, W. (2010) Regulation of mRNA translation and stability by microRNAs. *Annu. Rev. Biochem.*, **79**, 351–379.

Fabian, M.R. and Sonenberg, N. (2012) The mechanics of miRNA-mediated gene silencing: a look under the hood of miRISC. *Nat. Struct. Mol. Biol.*, **19**, 586–593.

Grundhoff, A. and Sullivan, C.S. (2011) Virus-encoded microRNAs. *Virology*, **411**, 325–343.

Han, M., Toli, J., and Abdellatif, M. (2011) MicroRNAs in the cardiovascular system. *Curr. Opin. Cardiol.*, **26**, 181–189.

Hoshino, I. and Matsubara, H. (2013) MicroRNAs in cancer diagnosis and therapy: from bench to bedside. *Surg. Today*, **43**, 467–478.

Iorio, M.V. and Croce, C.M. (2012) MicroRNA dysregulation in cancer: diagnostics, monitoring and therapeutics. A comprehensive review. *EMBO Mol. Med.*, **4**, 143–159.

Jackson, A.L. and Levin, A.A. (2012) Developing microRNA therapeutics: approaching the unique complexities. *Nucleic Acid Ther.*, **22**, 213–225.

Jansson, M.D. and Lund, A.H. (2012) MicroRNA and cancer. *Mol. Oncol.*, **6**, 590–610.

Jeang, K.T. (2012) RNAi in the regulation of mammalian viral infections. *BMC Biol.*, **10**, 58.

Lindow, M. and Kauppinen, S. (2012) Discovering the first microRNA-targeted drug. *J. Cell Biol.*, **199**, 407–412.

Aptamers

14

Summary

- Aptamers are single-stranded nucleic acids that bind a ligand with high affinity and specificity. Target molecules can be metal ions, organic compounds, nucleotides, amino acids, peptides, and proteins, including proteins on the surface of viruses or cells.
- Aptamers are usually obtained from a large library of oligonucleotides by an *in vitro* selection procedure known as SELEX (Systematic Evolution of Ligands by EXponential enrichment).
- For uses in biological systems, aptamers must be stabilized against nucleolytic degradation. This is accomplished by the introduction of chemically modified nucleotides either during the selection procedure or incorporation of post-SELEX site-specific modification. Alternatively, aptamers consisting of the nonnatural L-RNA enantiomer, so-called Spiegelmers, have been developed and are highly resistant to nucleases.
- An advantage of aptamers compared to the antisense molecules discussed in Chapter 13 is the possibility of using them against extracellular targets, for example, coagulation factors or cytokines. Several aptamers have already been clinically evaluated.
- Decoy oligonucleotides are double-stranded DNA molecules that mimic the promoter of a target gene and bind transcription factors, preventing them from translocating into the nucleus and inducing transcription. The sequestration of proinflammatory transcription factors has been evaluated in clinical trials.
- Immunostimulatory oligonucleotides contain CpG motifs (cytosine–phosphate–guanosine). They bind to toll-like receptor 9 and activate an immune response. Immunostimulatory oligonucleotides are being used as vaccine adjuvants and as immunotherapy for allergy, cancer, and infectious diseases.

Contents List

Aptamers are single-stranded nucleic acids that bind to a target molecule with high affinity and specificity. The term aptamer is derived from the Latin word *aptus* meaning "to fit" and the Greek word *meros* for "region." Aptamers are usually created by selecting them from a large pool of oligonucleotides with random sequences. Aptamers can bind to a broad variety of targets ranging from metal ions, organic compounds, amino acids, and nucleotides to peptides, proteins, and other larger structures. Most aptamers are 15–50 nucleotides in length, while some aptamers are up to 100 nucleotides long. Depending on their size relative to the size of the targets, they can be thought of either as ligands for large molecules, as binding partners of equal size, or as receptors for small molecules (Figure 14.1).

While the antisense oligonucleotides, ribozymes, and siRNAs discussed in Chapter 13 bind to their target by Watson–Crick base pairing, aptamers fold into complex three-dimensional structures and recognize their targets by steric interactions. They can thus be regarded as the nucleic acid analog of a monoclonal antibody (Section 10.2.1), especially since binding affinities in the low nanomolar to picomolar range can be reached. Aptamers can also be thought of as an attractive alternative to antibodies, and may overcome some of their limitations. For example, the production of antibodies in animals is a difficult process and the screening of colonies for the production of monoclonal antibodies can be laborious and expensive. Furthermore, antibodies against targets with low immunogenicity or against toxic substances can be difficult if not impossible to produce. In contrast,

Molecular Medicine: An Introduction, First Edition. Jens Kurreck and Cy Aaron Stein.
© 2016 Wiley-VCH Verlag GmbH & Co. KGaA. Published 2016 by Wiley-VCH Verlag GmbH & Co. KGaA.

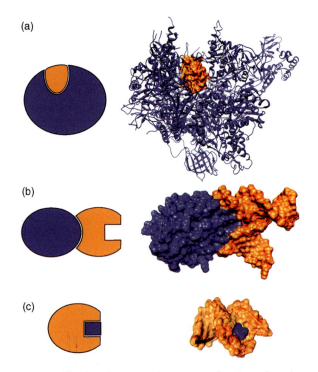

(a)

(b)

(c)

Fig. 14.1 The size of aptamers. Aptamers can function as ligands, binding partners of equal size or receptors for small molecules. (a) Binding of a 35 nt RNA aptamer binding to yeast RNA polymerase II. (b) Binding of a DNA aptamer to the approximately equally sized von Willebrand factor domain A1. (c) Binding of a 33 nucleotide long RNA to L-arginine. The aptamer is always depicted in orange, while the binding partner is drawn in blue. (Adapted from Ref. [1] with kind permission from the American Chemical Society.)

aptamers are created by *in vitro* selection and can be obtained for virtually any given target. Kinetic parameters, the temperature, and the conditions governing target interaction are factors that can be adjusted during the selection procedure. Aptamers can be synthesized chemically with reproducibly high purity. Less batch-to-batch variation can be expected than for the production of monoclonal antibodies. Finally, aptamers are free from biological contamination, are less sensitive to degradation than proteins, and have a longer shelf life.

Aptamers can be used for a variety of diagnostic and therapeutic purposes. For example, they can be employed for the *ex vivo* quantification of biomarkers such as C reactive protein (CRP), a general marker for inflammation and infection. An aptamer-based ready-to-use kit to determine the level of activated protein C, which plays an important role in regulating anticoagulation, is commercially available. In addition, a proof-of-principle demonstrating that aptamers can be employed for *in vivo* tumor diagnosis has been established. Finally, several aptamers have been developed for therapeutic purposes and are currently being evaluated in clinical trials. One aptamer named Macugen™ was approved by the FDA for the treatment of neovascular (wet) age-related macular degeneration (AMD).

Initially, only artificially selected aptamers were studied. It took more than a decade after the first *in vitro* selection of an aptamer until the first naturally occurring aptamers were discovered. These aptamers are cellular mRNAs and they are usually referred to as riboswitches (Box 14.1).

Box 14.1.   Riboswitches

Since the early 1990s, aptamers have been developed as artificially engineered tools for uses in biomedical research and for therapeutic purposes. In 2002, the first naturally occurring aptamers were discovered. These nucleic acid-based genetic regulatory elements were named riboswitches. Similar to aptamers, riboswitches bind small molecules through steric interactions. Riboswitches can be thought of as sensors for metabolites that are regulators of the activity of their own mRNAs. In contrast to previously known regulatory pathways, this mode of regulation does not require proteins. The finding that naturally occurring RNA can bind small molecules specifically and regulate gene expression in the absence of proteins supports the RNA world hypothesis. This states that RNA was the dominant molecule in early forms of life, while proteins and DNA emerged later (Box 13.2 in Section 13.2). Riboswitches may thus represent ancient regulatory systems.

Riboswitches are primarily located in the 5' UTR of certain mRNAs. They are generally composed of two domains: an aptamer and an expression platform (Figure 14.2). The ligand-binding domain interacts with a metabolite, which results in a structural change of the expression platform. Most expression platforms turn off gene expression in response to binding of the small molecule: This can be done at the transcriptional as well as at the translational level. In the latter case, riboswitch-mediated folding sequesters the ribosome-binding site (Shine–Dalgarno sequence), thus

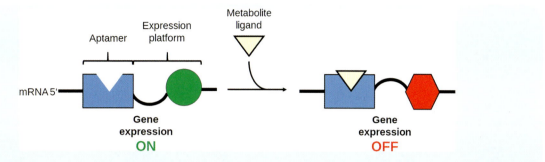

Fig. 14.2 Organization of a riboswitch. The riboswitch is usually located in the 5′ UTR of a bacterial mRNA. It consists of an aptamer and an expression platform. After binding of the metabolite ligand, the conformation of the expression platform is altered. In consequence, expression of the downstream gene is turned off. (Adapted from Ref. [2] with kind permission from John Wiley & Sons, Inc.)

inhibiting translation. In addition, premature termination of transcription can be the result of riboswitch-controlled formation of a transcription termination hairpin.

Most of the riboswitches known to date have been found in prokaryotes. For example, nearly 2% of the genes of *Bacillus subtilis* are under control of riboswitch RNAs. Since they regulate the expression of essential genes, riboswitches have been proposed as novel targets for antibiotics. In fact, the antibacterial drug pyrithiamine, whose mechanism of toxicity was previously unknown, was found to bind to the thiamine pyrophosphate riboswitch and to thus interfere with a vital bacterial process. These findings support the idea that riboswitches are attractive targets for the development of novel antibacterial drugs.

14.1
Selection of Aptamers

Aptamers are usually isolated from a large pool of oligonucleotides with different sequences by an *in vitro* selection procedure termed SELEX (Systematic Evolution of Ligands by EXponential enrichment). The SELEX process starts with the production of a library of oligonucleotides with high sequence diversity (Figure 14.3). The oligonucleotides consist of conserved sequences for primer binding separated by a random sequence of 20–40 bases. In theory, a pool of oligonucleotides with a random region of 40 nucleotides should contain $4^{40} = 1.2 \times 10^{24}$ different sequences. In practice, however, the library usually consists of 10^{12}–10^{15} different sequences.

For the selection of an RNA aptamer, the single-stranded DNA library is converted into a pool of double-stranded DNA molecules by PCR. This pool is subsequently transcribed into an RNA pool. This library, consisting of nucleic acids with high sequence and structural diversity, is allowed to interact with the target molecule, which is usually either bound to beads, to a column, or to a membrane. While nonbinding RNAs are washed away, the molecules that are retained are those that form a three-dimensional structure that is capable of binding tightly to the target molecules. Finally, the binding molecules are eluted, reverse transcribed, and amplified by PCR. The pool now enriched with target-binding nucleic acids is again transcribed into RNA and

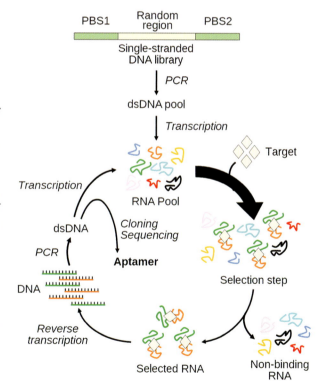

Fig. 14.3 Systematic Evolution of Ligands by EXponential enrichment (SELEX). Aptamers with high binding affinities can be obtained by *in vitro* selection. Initially, a library of DNA oligonucleotides with a randomized region is generated. This library is transcribed into an RNA pool with high diversity. Molecules that bind to the target are selected and amplified. After several rounds of selection, RNAs with high affinity are cloned and sequenced. PBS: Primer binding site.

used for another round of selection. After 5–15 cycles, an enriched pool of nucleic acids exists with high affinity for the target molecule. The sequences of these nucleic acids can be determined by cloning and sequencing. Aptamers are usually further modified after the SELEX process. For example, the total length of the aptamer, including the primer binding sites, may be up to 100 nucleotides, while the minimal ligand-binding motif may be as short as 15–20 nucleotides. Thus, aptamers can usually be truncated to facilitate synthesis without significant loss of binding affinity.

The SELEX methodology was developed independently in the laboratories of Larry Gold and Jack W. Szostak in 1990. While the initial procedures described the creation of RNA aptamers, protocols for the selection of DNA aptamers were published several years later. In general, DNA aptamers are selected from single-stranded DNA libraries after the strands are separated either during or after the PCR procedure. RNA and DNA aptamers have similar characteristics, although RNA aptamers possess an additional hydroxyl group that stabilizes intramolecular interactions and forms hydrogen bonds for high-affinity binding with the target molecule. However, they are intrinsically less (bio)stable than DNA molecules.

A great advantage of *in vitro* selection procedures is the opportunity to obtain highly specific molecules by adjusting the experimental protocol. This strategy, which is also known as counter-SELEX, was initially developed to create an aptamer that can discriminate between the alkaloids theophylline and caffeine, which differ only in a methyl group at N-7 (Figure 14.4). Theophylline is widely used as a bronchodilator in the treatment of asthma, bronchitis, and emphysema. To accurately determine serum levels of theophylline, an aptamer must be generated that efficiently discriminates it from caffeine and other chemically related substances present in serum samples. Accordingly, several rounds of a regular SELEX against theophylline were carried out, followed by additional rounds of a counter-SELEX when the enriched pool was first incubated with theophylline and then washed with caffeine. By doing so, all RNA

molecules that bound both alkaloids were removed, while those oligonucleotides that specifically bind theophylline were further enriched. The binding affinity of the resulting aptamer was 10 000-fold greater for theophylline than for caffeine.

Since the selection cycles are a reiteration of simple standard procedures (transcription, selection, reverse transcription, and PCR), the entire process can be automated. Using pipetting robots, several aptamers can be selected in parallel at high speed.

The SELEX procedure has also been combined with next-generation sequencing technologies (Section 2.3). The ability to determine several million sequences in a single run allows reducing the selection to as few as three rounds. Sequences that occur at high frequency after deep sequencing are those that originate from the RNAs that bind to the target molecule with high affinity.

It is very challenging to isolate aptamers against cell surface proteins, if only recombinant proteins are available, since their folding may not reflect the natural conformation in a membrane. To overcome this problem, the principle of cell-based SELEX (also known as cell-SELEX or whole-cell SELEX) has been developed (Figure 14.5). The idea behind this method is to first incubate the library with a nontarget cell to remove nonspecific binders (similar to a counter-SELEX step). The remaining oligonucleotides are then transferred to a cell line that expresses the particular target protein. Bound aptamers are subsequently recovered and further

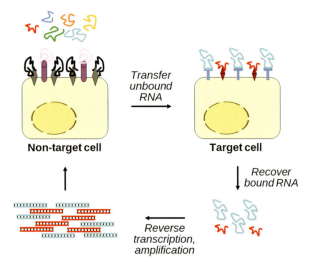

Fig. 14.5   Cell-based aptamer selection. For a cell-based SELEX, the oligonucleotide library is first incubated with a non-targeted cell as a counter-selection step. The unbound molecules are transferred to the target cell line where they bind to the target protein on the cell surface. Aptamers with high affinity are recovered, amplified, and enriched by several rounds of selection as already described for standard SELEX.

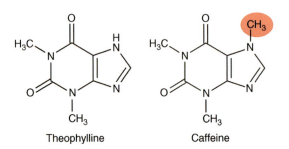

Theophylline   Caffeine

Fig. 14.4   Structure of theophylline and caffeine. Theophylline and caffeine are chemically closely related alkaloids. The latter carries an additional methyl group at N7.

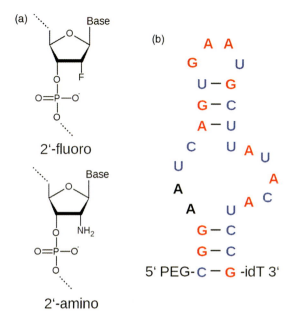

(a)

Base

F

O=P—O

2'-fluoro

Base

NH₂

O=P—O

2'-amino

(b)

A A

G   U

U — G

G — C

A — U

C   U A
U   U
            A
A   A
A   U   A   C
G — C
G — C

5' PEG-C — G -idT 3'

Fig. 14.6 Stabilization of aptamers by the introduction of modified nucleotides. (a) Chemical structures of 2′-fluoro and 2′-amino nucleotides. (b) Predicted secondary structure and modification pattern of the aptamer pegaptanib. 2′-O-methylated purines are shown in red, 2′-fluoro-modified pyrimidines are in blue, and two unmodified ribonucleotides are shown in black. A 40 kDa polyethylene glycol (PEG) moiety is also covalently attached to the aptamer. idT: inverted deoxythymidine.

library of modified RNA sequences was injected into tumor-bearing mice. RNA molecules that localized to tumor metastases were isolated and enriched by several rounds of selection. Finally, several aptamers that bound to the cancer cells with high affinity were isolated. Their target proteins were subsequently identified by mass spectrometry.

14.2
Modifications of Aptamers

As described for antisense molecules (Section 13.1.2), aptamers must be stabilized against nucleases for therapeutic applications. Conceptually, two strategies can be distinguished: The modified nucleotides can either be introduced during the selection procedure or can be incorporated by post-SELEX site-specific modification of the selected aptamer. A prerequisite for the inclusion of modified nucleotides during the SELEX procedure is their compatibility with the activity of the polymerases (RNA polymerase and reverse transcriptase) employed in the process. Modified pyrimidine nucleotides containing either a fluorine or an amino group at the 2′-position of ribose (Figure 14.6a) fulfill this requirement and have frequently been used for the selection of nuclease-resistant aptamers. In addition, mutated versions of the polymerases have been created that accept modified analogs of all four nucleotides.

Another sophisticated approach for the selection of aptamers with extraordinarily high nuclease stability is the "Spiegelmer" concept, which relies on the enantioselectivity of nucleases. Spiegelmers are synthesized from the unnatural L-form of nucleotides and have a very long half-life in biological fluids since they are not recognized by nucleases (Box 14.2).

enriched as in the standard SELEX method. This approach also allows discrimination between normal and cancer cells without any prior knowledge of the specific surface proteins expressed by the cancer cell and can lead to the discovery of new biomarkers.

An additional advance was the development of a protocol for the *in vivo* selection of aptamers. A nuclease-resistant

Box 14.2. Spiegelmers

RNA (as well as, but to a lesser extent, DNA) molecules are very unstable in biological fluids. Most RNA oligonucleotides are degraded by nucleases within a few minutes. While natural nucleotides contain the D-enantiomer of ribose at C2, their mirror image nucleotides containing L-ribose are not recognized as substrates by nucleases and are stable in human serum for many days. Aptamers synthesized from L-ribose monomers may have desirable pharmacological properties, including long plasma half-lives.

Though the non-natural stereochemical configuration protects L-RNA from nucleolytic degradation, it also precludes the use of the polymerases required for the selection procedure. This problem was

overcome by a variation of the SELEX approach that is based on the reciprocal specificities of binder pairs. If an aptamer binds a specified stereochemical configuration of a target molecule, the mirror image of the aptamer will bind to the enantiomer of the target with identical affinity (Figure 14.7, upper part).

The basic concept underpinning the Spiegelmer technology involves carrying out the SELEX with normal D-RNA against the mirror image of the target. The corresponding enantiomer of the D-RNA aptamer, the Spiegelmer, will then bind to the natural target with the same affinity (Figure 14.7, lower part). This Spiegelmer will also be resistant to nuclease degradation.

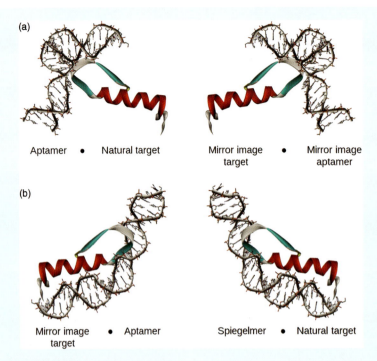

Fig. 14.7 Stereochemistry of Spiegelmers. Reciprocal specificities of aptamers composed of D-RNA and Spiegelmers composed of L-RNA and their natural and mirror image targets, respectively. (Adapted from Ref. [3] with kind permission from the Royal Society of Chemistry (http://www.rsc.org/shop/books/2008/9780854041169.asp).)

Since it is not possible to produce a large protein composed of unnatural D-amino acids, biologically relevant peptides or peptide epitopes of proteins are used as targets. A standard SELEX with natural D-RNA is then performed to identify aptamers with high affinity (Figure 14.8). The enantiomeric Spiegelmers of these aptamers will bind the natural target. Spiegelmers targeted against peptide chemokines are being evaluated in clinical trials, as will be outlined in Section 14.3.

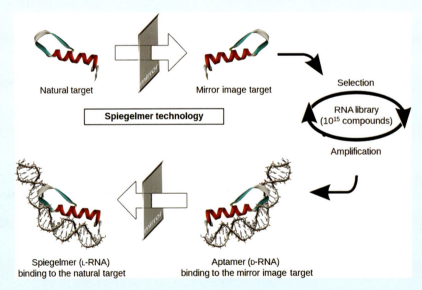

Fig. 14.8 Selection of a Spiegelmer. After production of the mirror image of the target, a standard SELEX is carried out. The aptamer binding to the mirror image target is then converted into its enantiomer, the Spiegelmer, which binds to the natural target and is nuclease resistant. (Adapted from Ref. [3] with kind permission from the Royal Society of Chemistry (http://www.rsc.org/shop/books/2008/9780854041169.asp).)

Post-SELEX modification of an aptamer is a challenging task, since modified nucleotides tend to alter its structure. The extent to which this happens must be measured experimentally, since the conformational changes induced by a modified nucleotide are almost impossible to predict.

Macugen (Section 14.3), the first aptamer to be approved by the FDA, was obtained by combining both strategies, that is, the introduction of modified nucleotides during the selection procedure and a post-SELEX modification. This aptamer was initially isolated from an RNA library comprised of oligonucleotides composed of 2′-ribopurines and 2′-fluoropyrimidines. After the selection procedure, the sequence motif was optimized and the aptamer further stabilized by the introduction of 2′-O-methyl nucleotides. As can be seen in Figure 14.6b, all purines except for two riboadenosines could be substituted without significant loss of binding activity.

Because aptamers are low-molecular weight molecules (5–15 kDa), they are rapidly cleared by the kidney. To increase circulation time and improve pharmacokinetic characteristics, high-molecular-mass moieties are commonly attached to them. Polyethylene glycol (PEG) polymers are frequently used for this purpose, as shown for Macugen in Figure 14.6b. Conjugation of PEG can increase the half-life of an aptamer in the mouse circulatory system from a few minutes to more than 1 day (Figure 14.9).

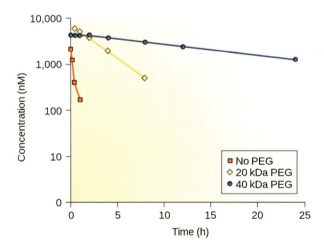

Fig. 14.9 Pharmacokinetics of polyethylene glycol (PEG)-coupled aptamers. Three different variants of a 39mer aptamer were administered intravenously to mice: The aptamer was either unconjugated or conjugated to 20 kDa or 40 kDa PEG. Attachment of PEG increased the half-life of the aptamer in the blood stream. (Adapted from Ref. [4] with kind permission from Macmillan Publishers Ltd., copyright 2010.)

14.3
Clinical Development of Aptamers

Since aptamers recognize a target with high specificity, they can be used for diagnostic purposes as well as for therapeutic applications. As agents for noninvasive imaging, aptamers combine a number of desirable properties. For example, they are much smaller than antibodies and can easily diffuse into dense tumor tissue. In addition, aptamers can easily be synthesized with functional groups on either of their molecular termini, for example, with chelators binding radioactive elements. Here, the rapid clearance from the bloodstream is an advantage for diagnostic purposes as it prevents extended exposure of the patient to radioactivity. As a proof-of-principle, U251 glioblastoma or MDA-MB-435 tumors were implanted into mice. An aptamer targeting Tenascin-C, a protein found in the extracellular matrix that can be used as a tumor marker, was labeled with radioactive technetium (^{99m}Tc). After intravenous injection of the aptamer, the tumors were localized. This aptamer entered clinical trials in Europe, but its further development was discontinued after a reorganization of the sponsoring company.

Efforts have also been made to develop aptamers as a new class of therapeutic agent. As described in Chapter 13, the major problem for the clinical application of anti-mRNA strategies is the inefficient delivery of the oligonucleotides into cells. Figure 14.10 illustrates that AS ONs, ribozymes, and siRNA must enter the cell to bind to their target mRNAs by Watson–Crick base pairing. Aptamers, however, can be used against extracellular targets such as signal peptides, extracellular proteins, or cell surface proteins, omitting the need for intracellular delivery. In fact, most of the aptamers tested in clinical trials are directed against extracellular targets (Table 14.1). Some examples of aptamers in therapeutic development will be described in more detail below.

In 2004, the first nucleic acid aptamer, pegaptanib (trade name Macugen), was approved by the FDA for the treatment of wet age-related macular degeneration (AMD), a major cause of visual impairment in older adults. Wet AMD causes vision loss due to abnormal blood vessel growth (choroidal neovascularization) leading to blood and protein leakage underneath the macula. Pegaptanib, a modified RNA aptamer described in Section 14.2, binds to vascular endothelial growth factor (VEGF)-165, the VEGF isoform primarily responsible for pathological neovascularization and vascular permeability. The drug is administered by intravitreal injection (directly into the vitreous humor of the eye). Treatment of patients suffering from AMD with pegaptanib resulted in a 45% relative benefit in mean change in vision after

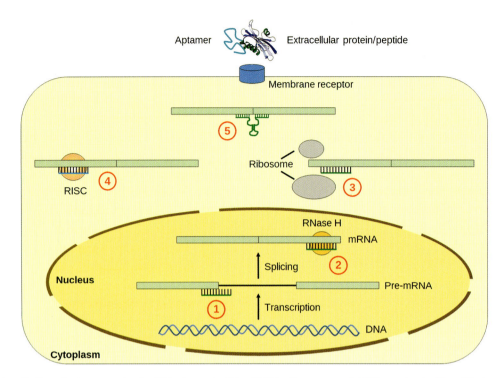

Fig. 14.10 Localization of aptamers and anti-mRNA oligonucleotides. AS ONs, ribozymes, and siRNAs must enter the cell to be active: AS ONs that modulate splicing (1) or induce RNase H-mediated degradation of a target mRNA (2) act in the nucleus, whereas AS ONs that block translation (3), siRNAs (4), and ribozymes (5) are functional in the cytoplasm. In contrast, aptamers can bind to extracellular peptides or proteins.

Table 14.1 Selection of aptamers that were or are still being tested in clinical trials.

Aptamer	Company	Target	Composition	Indication	Status
Pegaptanib (Macugen)	Eyetech Pharmaceuticals/ Pfizer	VEGF	OMe, 2′-F, PEG	Age-related macular degeneration	Approved
AS1411	Antisoma	Nucleolin	G-rich DNA	Acute myeloid leukemia	Phase III
REG1/2 (RB006/ RB007)	Regado Biosciences	Factor IXa	2′-F, PEG	Acute coronary syndrome	Phase III
ARC1779	Archemix	Von Willebrand factor	OMe, DNA, PEG	Thrombotic microangiopathies and carotid artery disease	Phase II
ARC1905	Archemix/ Ophthotech	Complement component 5	2′-F, PEG	Age-related macular degeneration	Phase I
E10030	Archemix/ Ophthotech	Platelet-derived growth factor	OMe, DNA, PEG	Age-related macular degeneration	Phase I
NU172	ARCA Biopharma	Thrombin	Unmodified DNA	Coronary artery disease	Phase II
NOX-E36	NOXXON Pharma	Chemokine (C-C motif) ligand 2	Spiegelmer, PEG	Type 2 diabetes, diabetic nephropathy	Phase II
NOX-A12	NOXXON Pharma	Chemokine (C-X-C motif) ligand 12	Spiegelmer, PEG	Multiple myeloma and non-Hodgkin's lymphoma	Phase II
NOX-H94	NOXXON Pharma	Hepcidin	Spiegelmer, PEG	Anemia of chronic disease	Phase II

2′-F: 2′-fluoro RNA; OMe: 2′-O-methyl RNA; PEG: polyethylene glycol.

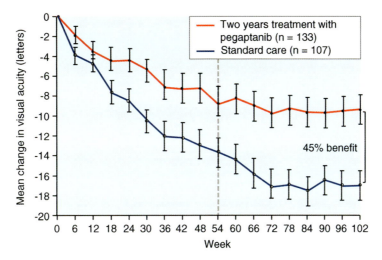

Fig. 14.11 Visual acuity of patients with wet age-related macular degeneration (AMD). The mean visual acuity of patients treated either with pegaptanib or receiving standard care was compared. (Adapted from Ref. [5] with kind permission from Macmillan Publishers Ltd., Copyright 2006.)

roughly 2 years, compared with those receiving standard care (Figure 14.11). Although clinically successful, pegaptanib has serious competition in the marketplace with antibody-based VEGF inhibitors such as ranibizumab (Section 10.2.1). Other aptamers against different targets are also in development to treat AMD.

Since aptamers do not necessarily have to be delivered into cells, intravenously injected aptamers are suitable agents to modulate blood clotting. Coagulopathy is an important problem in cardiovascular diseases and during certain clinical procedures. Anticoagulants (e.g., heparin, a highly sulfated glycosaminoglycan) are an important class of pharmaceutical agent. Treatment with heparin, however, can be associated with serious side effects such as spontaneous hemorrhage. To ameliorate this problem, aptamers targeting different factors of the clotting cascade (Table 14.1) are being developed as novel anticoagulants.

REG1/2 is a particularly innovative therapeutic system: Here, the anticoagulant activity of the aptamer can be regulated by an antisense-based antidote (Figure 14.12a). The aptamer, named RB006, binds to factor IXa and inhibits cleavage of factor X by the factor IXa–factor VIIa enzyme complex. As a result, the formation of fibrin clots is prevented. The second component of REG1/2, named RB007, is a fully 2′-*O*-methyl-modified AS ON that is complementary to the 5′ terminus of RB006. RB007 functions as an antidote because its binding to RB006 changes the conformation of the aptamer and inhibits its anticoagulant activity (Figure 14.12b). As a result, a treated patient's clotting parameters will rapidly return to normal. The controllable function of the aptamer/antidote system is a highly desirable property

during percutaneous coronary artery angioplasty and stenting, one that is not provided by any of the standard anticoagulant drugs. The system was evaluated in large phase III clinical trials. The aptamer was administered intravenously in the REG1 trial and subcutaneously in the REG2 trial. However, in late 2014, the REG1 trial was permanently terminated because of an unacceptable rate of serious anaphylactic adverse events. As anaphylaxis was thought to be the most likely induced by the PEG moiety of the aptamer, these events raised the question as to whether intravenously administered PEGylated oligonucleotides as a class can trigger allergic reactions.

Several L-RNA aptamers (Spiegelmers) (Box 14.2) are also being evaluated in clinical trials. Since the selection of the Spiegelmers requires the generation of the mirror image of the target, peptides and small proteins are the most suitable binding partners. Accordingly, the Spiegelmers listed in Table 14.1 are directed against extracellular peptides and small proteins. While NOX-A12 targets the chemokine CXCL12, a protein involved in tumor metastasis, NOX-E36 is being developed for the treatment of diabetic nephropathy. It does so by inactivating the chemokine CCL2, which plays a role in recruiting monocytes, T cells, and dendritic cells to sites of inflammation. A third Spiegelmer, NOX-H94, binds hepcidin, a master regulator of iron metabolism, and is being developed for the treatment of anemia.

In conclusion, in recent years, aptamers have been proposed as novel therapeutic agents. They have several advantageous properties, including a small size compared to antibodies, low immunogenicity and ease of production, and they are amenable to chemical

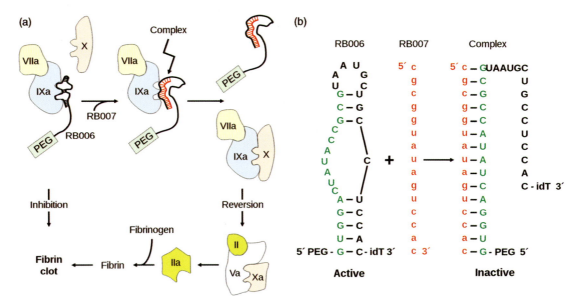

Fig. 14.12 Reversal of aptamer activity by an antidote. (a) The PEGylated aptamer RB006 binds to Factor IXa preventing cleavage of factor X and the subsequent formation of fibrin clots. The activity of the aptamer can be reversed by the addition of the AS ON RB007 antidote. (b) The REG1 anticoagulation system is composed of the aptamer RB006 and its antidote RB007. Base pairing of the antidote to the aptamer results in the neutralizing of its pharmacological effect. PEG, 40 kDa polyethylene glycol; idT: inverted deoxythymidine. ((a) adapted from Ref. [6] with kind permission from Macmillan Publishers Ltd., Copyright 2007.)

modification and conjugation to a variety of biomolecules. Furthermore, aptamers targeting cellular receptors can be employed for the targeted cellular delivery of therapeutic moieties, such as siRNAs (Section 13.3.3), drugs, toxins, and enzymes. These types of aptamers have been termed *escort aptamers*.

14.4
Decoy and Immunostimulatory Oligonucleotides

Decoy and immunostimulatory oligonucleotides are conceptually similar to aptamers as they function via steric interactions with the target, rather than by Watson–Crick base pairing with a cellular nucleic acid. Decoy oligonucleotides inhibit gene expression by blocking the activity of a specific transcription factor. Since transcription factors often control multiple genes, treatment with a decoy oligonucleotide will silence a group of genes simultaneously. Decoy oligonucleotides are short double-stranded DNA molecules that mimic the binding site of a transcription factor to the promoter(s) of its target gene(s). The concept underlying the decoy strategy is that the targeted transcription factor will recognize its binding sequence and thus interact with the decoy. Trapped by the oligonucleotide in the cytoplasm, the transcription factor will be prevented from

translocating to its promoter in the nucleus. As a result, every gene that is controlled by that transcription factor can in theory be silenced (Figure 14.13).

Several decoy oligonucleotides have been evaluated in clinical trials for the treatment of inflammatory diseases or neointimal hyperplasia. NF-κB and Stat-1 are proinflammatory transcription factors that have been considered as therapeutic targets for a decoy approach. The use of decoy oligonucleotides against either of these transcription factors diminished the expression of proinflammatory genes. Decoys were then evaluated in clinical trials to treat asthma and psoriasis (Stat-1) as well as atopic dermatitis (NF-κB). Although treatment with these decoy oligonucleotides, either by an inhalation route or topically is well tolerated, clinical development has not proceeded due to limited efficacy. A decoy oligonucleotide targeting the oncogenic transcription factor STAT3 was evaluated in a clinical trial for the treatment of head and neck tumors. Intratumoral injection of the decoy oligonucleotide reduced the activity of STAT3 and decreased the levels of several of its transcriptional targets. This is an example in which an oligonucleotide-based strategy succeeded in blocking a target that was considered undruggable by traditional small molecules.

Decoy oligonucleotides have also been designed to sequester the E2F transcription factor, which is known to play a role in the upregulation of genes involved in neointimal hyperplasia after coronary artery bypass graft surgery. In proof-of-concept experiments, the E2F decoy

(a)

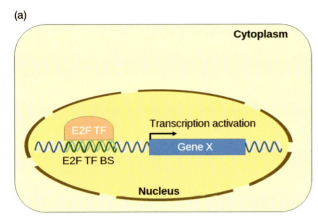

(b)

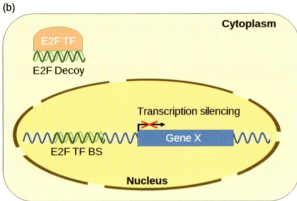

Fig. 14.13 Decoy oligonucleotide strategy. The figure shows an approach to silence transcription induced by the E2F transcription factor (E2F TF). (a) Interaction of the E2F TF with specific binding sites (E2F TF BS) upstream from multiple genes involved in cell cycle regulation activates transcription. (b) The decoy oligonucleotide (E2F Decoy) interacts with E2F TF, preventing its binding to the E2F TF BS. As a consequence, gene transcription is silenced. (Adapted from Ref. [7] with kind permission from Elsevier.)

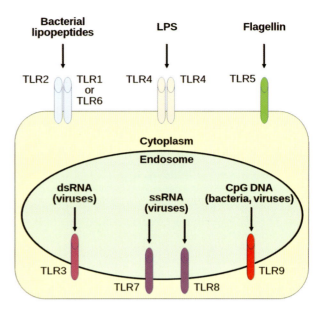

Fig. 14.14 Toll-like receptor (TLR) family. TLRs1, 2, 4, 5, and 6 are located at the cell surface, while TLRs3, 7, 8, and 9 are located in the endosomal compartment. The TLRs are activated by different ligands as indicated. CpG: cytidine–phosphate–guanosine; dsRNA: double-stranded RNA; LPS: lipopolysaccharide; ssRNA: single-stranded RNA. (Adapted from Ref. [8] with kind permission from the Royal Society of Chemistry (http://www.rsc.org/shop/books/2008/9780854041169.asp).)

oligonucleotide suppressed neointima formation in a rat model. However, the outcome of a large phase III clinical trial was disappointing. For progress to occur, more efficient methods for the organ-specific delivery of decoy oligonucleotides will be required.

An additional therapeutic approach is based on the immune-modulatory activity of certain oligonucleotides that bind to specific receptors. The most prominent immunostimulatory oligonucleotides are CpG oligonucleotides, which contain a cytosine followed by a guanosine. The "p" refers to the phosphodiester linkage connecting the nucleosides in DNA. When CpG motifs are unmethylated, they act as immunostimulants by activating Toll-like receptor 9 (TLR9). TLRs are a large family of pattern recognition receptors (PRRs). The members of the TLR family differ in their localization on the cell surface and in the endosomal compartment (Figure 14.14). Furthermore, they recognize different

ligands, such as bacterial lipopeptides, lipopolysaccharides (LPS), flagellin, double- or single-stranded RNA, and CpG oligonucleotides.

The immunostimulatory effects of CpG motifs were elucidated in the mid-1990s and oligonucleotides containing such motifs have been developed for therapeutic purposes. CpG oligonucleotides are recognized by TLR9, which is constitutively expressed in B cells and plasmacytoid dendritic cells (pDC). Depending on the number and location of the CpG motifs and the sequences that flank this motif, oligonucleotides containing CpG motifs can be grouped into different classes. The modes of action of the major classes are shown in Figure 14.15. Oligonucleotides containing a palindrome sequence and in which the cytidine and guanosine are joined by a phosphodiester linkage belong to class A and induce primarily pDC activation and secretion of INF-α. These oligonucleotides are poor activators of B cell proliferation. In contrast, oligonucleotides with a fully phosphorothioate backbone are class B oligonucleotides and are strong stimulants of B cell proliferation, but poor inducers of INF-α secretion. Class-C oligonucleotides have phosphorothioate backbones and form duplexes. They exert immune effects intermediate between the A- and B-classes.

CpG oligonucleotides have been evaluated in many clinical trials for a variety of applications ranging from vaccine

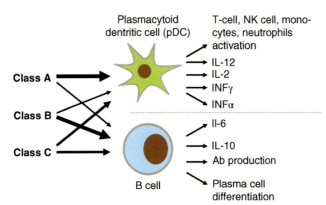

Fig. 14.15 Activation of plasmacytoid dendritic cells (pDC) and B cells by different classes of CpG oligonucleotides.

adjuvants to immunotherapies for allergy, cancer, and infectious diseases. However, the successful treatment of infectious diseases and cancer by CpG oligonucleotides has been challenging. The greatest potential for these compounds is probably in the enhancement of vaccine efficacy (Section 5.2). Among the most advanced is HEPLISAV^{TM}, a hepatitis B vaccine comprised of an immunostimulatory oligonucleotide and the hepatitis B surface antigen. This vaccine has been developed for individuals who respond poorly to standard hepatitis B vaccination. Large clinical trials have demonstrated the immunostimulatory activity of the CpG oligonucleotide when serving as an adjuvant. Use of an adjuvant can also help to preserve valuable vaccine doses if availability becomes limited in the event of a pandemic. For example, coadministration of a CpG oligonucleotide helped to reduce the dose of an influenza vaccine by 90% while maintaining the same level of IFNγ secretion induced with the vaccine at full dose. Additional CpG oligonucleotides are in development for potential use in allergic diseases and as an adjuvant for cancer vaccines. However, improving specificity is a major challenge for successful use.

Taken together, the development of aptamers, decoy oligonucleotides, and immunostimulatory oligonucleotides shows that nucleic acids can be used not only to bind to cellular RNA or DNA molecules by standard Watson–Crick base pairing, but can also function through spatial interactions with peptides and proteins.

References

1. Famulok, M. and Mayer, G. (2011) Aptamer modules as sensors and detectors. *Acc. Chem. Res.*, **44**, 1349–1358.
2. Winkler, W.C. and Breaker, R.R. (2003) Genetic control by metabolite-binding riboswitches. *ChemBioChem*, **4**, 1024–1032.
3. Eulenberg, D., Purschke, W., Anders, H.-J. *et al.* (2008) Spiegelmer NOX-E36 for renal diseases, in *Therapeutic Oligonucleotides*

(ed. J. Kurreck), Royal Society of Chemistry Publishing, Cambridge, pp. 200–225.
4. Keefe, A.D., Pai, S., and Ellington, A. (2010) Aptamers as therapeutics. *Nat. Rev. Drug Discov.*, **9**, 537–550.
5. Ng, E.W., Shima, D.T., Calias, P. *et al.* (2006) Pegaptanib, a targeted anti-VEGF aptamer for ocular vascular disease. *Nat. Rev. Drug Discov.*, **5**, 123–132.
6. Que-Gewirth, N.S. and Sullenger, B.A. (2007) Gene therapy progress and prospects: RNA aptamers. *Gene Ther.*, **14**, 283–291.
7. Fichou, Y. and Ferec, C. (2006) The potential of oligonucleotides for therapeutic applications. *Trends Biotechnol.*, **24**, 563–570.
8. Uhlmann, E. (2008) Immune stimulatory oligonucleotides, in *Therapeutic Oligonucleotides* (ed. J. Kurreck), Royal Society of Chemistry Publishing, Cambridge, pp. 142–162.

Further Reading

Aptamers

Burnett, J.C. and Rossi, J.J. (2012) RNA-based therapeutics: current progress and future prospects. *Chem. Biol.*, **19**, 60–71.
Famulok, M. and Mayer, G. (2014) Aptamers and SELEX in chemistry & biology. *Chem. Biol.*, **21**, 1055–1058.
Keefe, A.D., Pai, S., and Ellington, A. (2010) Aptamers as therapeutics. *Nat. Rev. Drug Discov.*, **9**, 537–550.
Li, Y., Wu, H., Niu, Y. *et al.* (2013) Development of RNA aptamer-based therapeutic agents. *Curr. Med. Chem.*, **20**, 3655–3663.
Ng, E.W., Shima, D.T., Calias, P. *et al.* (2006) Pegaptanib, a targeted anti-VEGF aptamer for ocular vascular disease. *Nat. Rev. Drug Discov.*, **5**, 123–132.
Pei, X., Zhang, J., and Liu, J. (2014) Clinical applications of nucleic acid aptamers in cancer. *Mol. Clin. Oncol.*, **2**, 341–348.
Sundaram, P., Kurniawan, H., Byrne, M.E. *et al.* (2012) Therapeutic RNA aptamers in clinical trials. *Eur. J. Pharm. Sci.*, **48**, 259–271.
Syed, M.A. and Pervaiz, S. (2010) Advances in aptamers. *Oligonucleotides*, **20**, 215–224.
Zhou, J., Bobbin, M.L., Burnett, J.C. *et al.* (2012) Current progress of RNA aptamer-based therapeutics. *Front Genet.*, **3**, 234.

Riboswitches

Blount, K.F. and Breaker, R.R. (2006) Riboswitches as antibacterial drug targets. *Nat. Biotechnol.*, **24**, 1558–1564.
Winkler, W.C. and Breaker, R.R. (2003) Genetic control by metabolite-binding riboswitches. *ChemBioChem*, **4**, 1024–1032.

Decoy Oligonucleotides

Fichou, Y. and Ferec, C. (2006) The potential of oligonucleotides for therapeutic applications. *Trends Biotechnol.*, **24**, 563–570.
Hecker, M., Wagner, S., Henning, S.W. *et al.* (2008) Decoy oligodeoynucleotides to treat inflammatory diseases, in *Therapeutic Oligonucleotides* (ed. J. Kurreck), Royal Society of Chemistry Publishing, Cambridge, pp. 163–188.
Koppikar, P. and Bromberg, J. (2012) State-of-the-art approach: using oligonucleotide decoys to target the "undruggable". *Cancer Discov.*, **2**, 670–672.

Immunostimulatory Oligonucleotides

Krieg, A.M. (2006) Therapeutic potential of Toll-like receptor 9 activation. *Nat. Rev. Drug Discov.*, **5**, 471–484.
Krieg, A.M. (2012) CpG still rocks! Update on an accidental drug. *Nucleic Acid Ther.*, **22**, 77–89.

Ethics in Molecular Medicine

15

Summary

- Philosophical ethicists have developed concepts to determine appropriate and inappropriate conduct and to justify moral principles. Some of the leading ethical theories focus either on the consequences of an action (consequentialism and utilitarianism) or on the adherence to rules (deontology).
- Advances in genetic testing have resulted in the claim of a right not to know to maintain individual autonomy. It has been argued that every person should have the right to decide whether or not he or she wants to obtain knowledge about his/her predisposition for late-onset (incurable) diseases. The right not to know one's genetic makeup is relevant to the relationship between applicants and employers or insurance companies, but may also involve the opposing interests of close relatives. The technological development of whole genome sequencing (WGS) poses new challenges to ensuring individual autonomy.
- The controversy over stem cells centers largely on the moral status of the embryo that is destroyed for the generation of human embryonic stem cell (hESC) lines. Some philosophers claim that early embryos are human beings that must not be instrumentalized for other's ends. Other bioethicists balance the protection of the embryo against the benefit of hESCs for therapeutic purposes. Strategies involving induced pluripotent stem cells or direct cell differentiation are ethically less controversial.
- Preimplantation genetic diagnosis (PGD) can be used to deselect embryos generated by IVF prior to implantation. Several ethical concerns have been raised about this technology. One major argument is that PGD may be misused not only to select against severe genetic disorders but also to choose a designer baby with specific desired properties. Sex selection by PGD for nonmedical reasons and the selection of a matching transplant donor (savior sibling) for a diseased child have also been criticized as ethically unjustifiable.

Contents List

The Basis of Bioethics
Fields of Application

- Genetic Testing and the Right "Not to Know"
- Stem Cell Research
- Preimplantation Genetic Diagnosis

Numerous ethical issues can arise from the application of biomedical research to clinical medicine, several of which were briefly addressed in previous chapters of this textbook. Some scientists have claimed that a sharp distinction exists between science and technology, stating that for the former the accumulation of knowledge is devoid of ethical content, while the latter invokes the practical application of scientific knowledge and hence generates ethical issues. However, this distinction is no longer valid. In the first half of the twentieth century, knowledge of physics directly led to the development of the atomic bomb. Several of the best physicists in the world were directly involved in the project. Biologists and physicians in Nazi Germany performed some of the most ethically indefensible acts ever, including unjustifiable experiments on humans and the lending of support to murderous eugenic programs. The latter included euthanasia and the sterilization of carriers of genetic diseases and members of certain races. It also included the mass extermination of Jews, the Holocaust, all carried out in the name of biology. These historic experiences clearly demonstrate that every scientific act has ethical implications.

However, not only do the failures of the past have an ethical dimension, but quotidian work in the life sciences also confronts us with profound ethical questions. Is it ethically justified to carry out animal experiments in the process of developing new drugs? What is the moral status of the human embryo? What reproductive medicine procedures, which can range from *in vitro* fertilization (IVF) to preimplantation diagnosis to the reproductive cloning of humans, cross ethical boundaries? These questions require every student and scientist working in molecular medicine to have at least some insight into the principles of bioethics.

This chapter will present a short general introduction to the philosophical basis of bioethics, and will address several selected topics in molecular medicine that are controversial: These include genetic testing, stem cell research, and preimplantation genetic diagnosis. The list of issues in bioethics is, of course, much longer, and also includes questions surrounding the patenting of human genes (or biological material in general), the release of genetically modified organisms, germline therapy, the reproductive cloning of humans, the pricing of new drugs (in developed and developing countries), the design of clinical trials, techniques assisting reproduction, assisted suicide, and many more. However, to adequately cover all these topics would be far beyond the scope of this book.

It is also important to note that the perspective from which ethical arguments are constructed are much more subjective than the scientific topics discussed so far. Here, however, we will attempt to sensitize the reader to the relevance of these arguments for biomedicine while not favoring any particular opinion.

15.1
The Basis of Bioethics

In general, ethics is the branch of philosophy that deals with moral principles. It systematizes, defends, and recommends concepts of proper conduct. An important claim of philosophical ethics is that it not only deals with appropriate and inappropriate behaviors but also produces reasons for what is morally right or wrong and justifies principles. In other words, philosophical ethics not only states that a behavior (e.g., lying) is wrong, but also gives reasons why it is wrong. In this chapter, we will use the terms *ethics* and *morals* synonymously, although morality is sometimes considered to refer to general attitudes and standards of behavior, while ethics is usually taken as

the systematic treatise of good and bad behavior as a philosophical discipline.

Bioethics belongs to the field of *applied ethics*, which draws upon ethical theory to ask what a person (or a community) is supposed to do in very specific situations or within a particular domain of action. Specifically, bioethics deals with the moral and social implications of the consequences resulting from advances in the biological sciences (including molecular medicine).

An exhaustive overview of ethical theories is beyond the scope of this book. Instead, we will describe only one basic distinction that can be used to classify the various ethical theories: While some ethicists focus on the consequences of our actions and view them as the basis for any judgment about the rightness of that conduct (*consequentialism*), others base the morality of an action on the action's adherence to a rule (this philosophical tradition is known as *deontology*, from the Greek word *deon* for *duty*).

The theory of consequentialism may be simplified by the saying that the ends justify the means. Any method of achieving a goal is acceptable if the goal is sufficiently morally important. The most prominent form of consequentialism is *utilitarianism*. The central aim of this ethic is the production of the greatest good for the greatest number of people. The theory was originally introduced by Jeremy Bentham in the eighteenth century. According to Bentham, good is what tends to maximize pleasure and at the same time minimize pain. In other words, the right action is the one that produces the greatest happiness.

One of the major proponents of contemporary utilitarianism is Peter A. D. Singer. Based on his ethical principles, Singer became an animal rights activist and garnered popularity for his book *Animal Liberation*, first published in 1975. He argued that the well-being of all sentient beings ought to be considered seriously. Rights should be conferred according to the level of a creature's self-awareness, regardless of their species. In his view, the boundary between human and animal is completely arbitrary. Humans, he says, tend to be *speciesist* in ethical matters, discriminating against nonhumans. Singer argues that while animals have lower intelligence than the average human, severely intellectually challenged humans may have lower mental capacity than some animals. This view has led not only to increased animal protection regulations but also has had implications for newborns (and human individuals with severely diminished intellectual abilities) who lack the essential characteristics of personhood such as rationality, autonomy, and self-consciousness. According to Singer's view, the murder of a newborn baby is, therefore, never equivalent to the killing of a person, that is, a being who wants to

Fig. 15.1 Immanuel Kant (1724–1804). Kant was a German philosopher and is often considered to be a central figure in modern philosophy. Portrait by G. Doebler. (Image courtesy of Museum Stadt Königsberg in Duisburg, Germany.)

continue to live. Singer has been accused of supporting euthanasia, as parents may decide to terminate the life of a newborn who will never perceive happiness because of a severe disease.

In contrast to utilitarianism, which derives rightness or wrongness from the outcome of the act itself, deontology emphasizes adherence to the rules. Accordingly, an act may be considered right even if the act produces bad consequences. The most prominent representative of this philosophical tradition is the German philosopher Immanuel Kant (Figure 15.1). Kant's aim was to find ethical principles that were based on rational procedures and were universally applicable. His moral law, which he calls the *Categorial Imperative*, can be regarded as the philosophically justified version of the so-called Golden Rule, that is, "treat others as you wish to be treated yourself." Another important obligation of Kantian ethics is to treat every person as an end in himself or herself and not as a means to one's own ends. This argument becomes relevant, for example, when discussing the ethical question of whether a *savior sibling* may be selected by preimplantation genetic diagnosis to provide a matching transplant donor for an already existing child with a severe disease. This is discussed in Section 15.2.3.

Kant's moral universalism has had a wide impact on discussions of the legal and political concepts that surround questions of human rights and equality and still influences philosophical thinking today.

In the 1970s, the German-born, Jewish philosopher Hans Jonas (1903–1993), who was forced to emigrate from Germany in 1933, questioned the traditional principles of ethics in light of human actions in the modern, technological age. He considered that nature has become vulnerable to man's technological intervention to the point that even the preservation of our species is questionable. For example, it is possible that the present generation could sacrifice the future of mankind for its own happiness by living an environmentally polluting luxurious lifestyle. In addition, the existence of humans could be terminated by technological catastrophes such as an atomic war or the outbreak of an uncontrollable biohazard. According to Jonas, this new type of human action requires a new version of the categorical imperative: "Act so that the effects of your action are compatible with the permanence of genuine human life." Jonas' main work, *The Imperative of Responsibility*, helped to catalyze the environmental movement. In later works, he addressed various ethical issues pertinent to the modern life sciences, such as genetic testing (see below), the reproductive cloning of humans, and organ transplantation.

John Rawls is one of the most important Anglo-American philosophers of the late twentieth century. His *Theory of Justice* can be seen as a modern form of Kantian theory and has become one of the most influential works in modern moral and political philosophy. Rawls' central argument is that the most reasonable principles of justice are those everyone would accept and agree on from a fair position. In his theory, he aimed to demonstrate that the seemingly competing claims of freedom and equality can be integrated into a seamless unity, justice as fairness. The Theory of Justice includes a famous thought experiment: A group of people has the task of reaching agreement about the political and economic structure of a society. However, each individual deliberates behind a *veil of ignorance*, that is, each lacks knowledge of his or her gender, race, age, intelligence, and so on. Each individual will then design a social structure that will secure himself or herself maximal advantage and, for example, hold discrimination against blacks or women to be unjust. Based on this scenario, Rawls deduces two basic principles: the liberty principle, which establishes equal basic liberties for all citizens, and the principle of equality. Rawls' Theory of Justice has not remained only a pure philosophical study, but has attracted the attention of jurists, economists, and politicians.

15.2
Fields of Application

The following sections will discuss several paradigmatic ethical issues posed by biomedical research and its applications. This discussion is by necessity incomplete; rather, it attempts to provide examples of some aspects of biomedicine that are currently intensively discussed. As already mentioned, we will try not to draw conclusions, but rather will depict the most important arguments raised for or against a particular approach.

15.2.1
Genetic Testing and the Right "Not to Know"

The "right to know" is an important legal principle. Individuals have the right to obtain information concerning their lives, such as a chemical to which they may have been exposed, data stored about themselves, or a medical diagnosis. However, with the development of genetic testing in the early 1980s, this view changed: The claim for a right to know was accompanied by the claim of a "right not to know." Together, both principles protect an individual's autonomy and the ability to determine his or her own fate.

Hans Jonas (Figure 15.2) was among the first to introduce the right not to know into philosophical discussion. In his original version, Jonas argued against the reproductive cloning of human beings, a procedure that was futuristic at that time but has become increasingly realistic since (though not yet realized; Box 12.1 in Section 12.1.2). As per his argument, the cloned person would know too much about himself or herself. In addition, others would also have too much information about this individual. This knowledge is likely to lead to a kind of determinism. For example, the clone of a famous tennis player would be trained to become a tennis player and the clone of a musician would be educated to become a musician. Only the right not to know will permit the individual to live free from a knowledge burden, allowing every day to come as a surprise.

This argument, originally raised against the reproductive cloning of humans, was rapidly adopted in the field of genetic testing. As discussed in detail in Chapter 8, several categories of genetic testing can be distinguished. The claim of a right not to know is particularly relevant for tests in an asymptomatic individual that are predictive for a severe genetic disorder that may appear later in life. For example, mutations in the two *BRCA* genes are associated with a 36–85% risk of developing breast cancer. Presymptomatic tests, for example, for mutations in the *huntingtin* gene, are indicative of a 100% probability of developing Huntington's disease (HD) at a later age. However, some diseases, such as HD, have no

Fig. 15.2　Hans Jonas (1903–1993). The philosopher Hans Jonas was among the first to introduce the right "not to know" into ethical discussion. (Image courtesy of Marketing Gesellschaft Mönchengladbach mgh, Germany.)

therapeutic options. For a patient to have knowledge of carrying the predisposition may thus produce severe psychological consequences, preventing an unburdened life. However, while some individuals will feel their life plans destroyed by knowledge of their genetic information, others will prefer knowledge about their genetic predispositions as they will feel freed from the burden of uncertainty. Thus, an autonomous and self-determined decision includes both the right to know and the right not to know.

The early discussion on the right to know and not to know centered on insurance and employment relationships. Employers and insurance companies have an interest in their applicant's genetic makeup to minimize absence from work due to illness and to minimize their own financial risk. Two scenarios must be distinguished. In the case in which an individual does not have any information about his or her genetic predispositions, an employer or insurer might ask for genetic testing to evaluate the applicant for late onset diseases. However, a broad consensus has been reached that an individual's autonomy, including his or her right not to know, has higher ethical value than the company's business interests. The prohibition on requesting the genetic testing of an applicant also prevents discrimination. Only in rare, specific situations in which a particular genetic predisposition might pose a health risk in the work place would genetic testing be considered a legitimate protection of an employee. The situation is different if the individual already knows his or her genetic predisposition. In this case, an applicant for a life insurance policy will have to disclose this knowledge to prevent its misuse, for example, to procure a high pay-out policy knowing that he or she will die early from a genetic disease (*adverse selection*). The insurer may also be exempt from providing indemnification if the insurant did not disclose a known genetic predisposition at the time of the issuance of the policy.

More recently, the discussion has shifted to potential conflicts between an individual's interests and those of affected close relatives. For example, a young adult whose grandfather or grandmother died of HD might want to have a genetic test performed for HD predisposition. If the result is positive, the respective parent will automatically know that he or she will be affected as well. The right to know of the young adult conflicts with the right not to know of his or her parent. Similar cases can occur in other constellations of relatives. For example, another scenario involves a woman whose mother died of breast cancer. She might want to be evaluated for mutations in the *BRCA1* gene that are responsible for the majority of hereditable breast cancer cases. However, her test result would be valid for her twin sister as well (or with a 50% probability for a nontwin sister). Again, the right to know and not to know of the two sisters may conflict. In many cases, the nondisclosure of test results to close relatives will not be a viable option. Knowledge of a predisposition for HD will usually cause a change in the affected individual's plans for the future; a carrier of cancer-related mutations in the *BRCA1* gene might decide to undergo mastectomy, and so on.

A variant of the problem arises when a genetic test reveals a predisposition for a severe, late-onset disease and the affected individual does not desire to disclose this information to his relatives. Because of laws protecting medical secrecy, the health care professional is not allowed to disclose the information to third parties, even though this information might help to prevent disease. To date, no satisfactory solution has been found to resolve the conflict between the right to know and not to know in personal relationships.

More recently, ethical considerations resulting from genetic testing have taken into account the new genetic knowledge produced by next-generation sequencing (NGS). As discussed in detail in Chapters 7 and 8, this technology will increasingly be applied for WGS in personalized medicine. However, WGS usually reveals several hundred disease-relevant variations for each individual. These may include, for example, predispositions for the risks of developing myocardial infarction or neurodegenerative diseases. In some cases, this knowledge may encourage changes in lifestyle or the prophylactic prescription of drugs such as statins. However, in other cases, WGS may reveal a genetic predisposition to severe, incurable diseases as secondary findings unrelated to the primary medical question. For example, employing NGS to optimize drug treatment of a specific disease may reveal a variation causing HD in the individual's genome. Therefore, some clinical laboratories are exploring informed consent models that permit patients to select the information to be disclosed *prior* to the WGS. The choice should also include the question of whether all data from their genetic analysis should be stored for potential analysis in the future. The ethical dilemma becomes even more complicated in pediatric genetic testing. Here, an additional problem arises as to when to disclose the results for adult-onset genetic conditions. Thus, it may not be advisable to test for any late-onset disorders in minors (or at least not to disclose them), unless such diseases can be ameliorated or prevented by specific treatments or lifestyle changes.

This discussion of the right not to know is not merely a philosophical issue; it has also been recognized by legal bodies addressing biomedical issues. For example, the UNESCO *Universal Declaration on the Human Genome and Human Rights* states: "The right of every individual to decide whether or not to be informed of the results of genetic examination and the resulting consequences should be respected."

There is also broad agreement that genetic tests should be carried out by qualified health care professionals. However, this consensus has been undermined by direct-to-consumer genetic testing (Box 15.1). It should also be noted that the right not to know is not an absolute right, as it is valid only under the condition that there is no risk of serious harm to other persons. This issue may become relevant with respect to infectious disease. For example, in a discussion in Germany in the 1980s and 1990s, some health care professionals and politicians called for mandatory HIV tests with every blood test performed in a clinical laboratory. However, this measure never became law as it was blocked by the right not to know. HIV testing still requires the informed consent of the tested individual. However, after

Box 15.1. Direct-to-Consumer Genetic Testing

In recent years, the market for so-called direct-to-consumer genetic testing has been steadily increasing. These tests are marketed by commercial companies directly to consumers without involving a physician, and can provide information about the risk of developing hundreds of disorders. Array technology (Section 8.3.3) and, more recently, NGS (Section 8.3.2), are commonly used. The tests may help people adjust their lifestyles or take prophylactic medical measures. However, as many individuals may not have the ability to interpret the data correctly, information about increased disease susceptibility may cause significant psychological stress. Therefore, several genetics societies have recommended that genetic testing should always be carried out by health care professionals who can offer pretest counseling, can order the appropriate tests, can accurately interpret the results, and can perform professional counseling when discussing the results, offering psychosocial follow-up if necessary. Most countries still do not extensively regulate direct-to-consumer testing.

having given informed consent, the individual cannot decline to know the test result; in case of a positive result, the physician must explain the risks for his or her sexual partner(s) to the infected person.

15.2.2
Stem Cell Research

In recent years, no field of life science research has been discussed as intensively and has been as controversial as stem cell research. The scientific background of this topic is covered in detail in Chapter 12. There are great hopes that this research will lead to the development of new therapeutic options for severe and currently untreatable diseases, such as Parkinson's or Huntington's disease. Stem cells may also one day help to repair damaged heart muscle after a myocardial infarction; they may also allow diabetics to dispense with lifelong insulin injections. However, not only are there technical hurdles that must be overcome to achieve these goals, but there are also ethical concerns about the research itself. The discussion centers primarily on the use of human embryonic stem cells (hESCs), which relies on the destruction of a developing embryo. The most significant ethical issue is thus relevant to the moral status of the embryo. This debate is profoundly influenced by religious traditions, although philosophical ethics claims to be independent of religious beliefs.

A central point of the stem cell controversy is the question of when human life begins. In the Christian tradition, human life begins when a sperm cell fertilizes an egg cell to form a zygote. This argument has also been adopted by many secular philosophers. The view that the early embryo is a human being possessing full human rights logically leads to the rejection of the morality of embryonic stem cell research. If the embryo has human dignity, it must not be instrumentalized for research purposes or medical applications. Destruction of the embryo for the purpose of the generation of a stem cell line violates the sanctity of life and is ethically illegitimate. Proponents of this position usually also argue that the development of the human embryo proceeds as a continuum.

In contrast, it has been suggested that various points in embryonic development, only after which will the embryo possess inviolable human dignity, are the true beginning of human life. In humans, the fertilized ovum typically implants after about 9 days. Prior to this time, a large fraction of fertilized eggs fail to implant and to initiate a pregnancy. Therefore, implantation of the embryo into the womb has been suggested as marking the point at which the developing human being deserves protection. This view is supported by the argument that intrauterine contraceptive coils also permit fertilization of the egg but prevent implantation. Another opinion states that up to approximately day 13 of gestation, the embryo is capable of *twinning*, that is, of separating into two or more embryos. It has, therefore, been suggested that the embryo obtains full rights from this stage forward, when the individuality of the developing human being is confirmed. Alternatively, the beginning of human life can be viewed as the inverse of the end of life, which is legally defined as the brain death. This would mean that life begins with the onset of brain activity, which occurs approximately on day 70. As embryos are incapable of surviving on their own outside the womb, some have not considered them as equivalent to human life. In this view, they have only the potential for life. Using modern medical technologies, fetuses are viable outside the womb after about 22 weeks. However, although there are good arguments for these different viewpoints, each of these definitions of when life "begins" has been criticized as arbitrary.

Nevertheless, it has been argued that while human life begins with fertilization and develops continuously, it does not initially possess inviolable rights. Rather, rights and the extent of protection are graded and become more comprehensive as development proceeds. This view has been illustrated by a thought experiment: A nurse is about to escape from a burning hospital. On her way out, she has the choice of either saving a rack of dishes with 10 000 embryos prepared for *in vitro* fertilization or a single newborn lying helplessly in the next room. She cannot carry both. If an embryo has full human dignity and possesses the sanctity of human life, the ethically correct choice would be to save the 10 000 embryos and abandon the baby. However, it is difficult to believe that many humans would decide against the baby: This indicates that we intuitively assign graded rights depending on the stage of development.

One of the main arguments of the proponents of stem cell research is the fact that hESC lines are (usually) generated from embryos created for *in vitro* fertility treatments and that will no longer be used. It has been estimated that several hundred thousand such embryos have been stored in the United States alone. The vast majority of them will be destroyed at some point. Therefore, it has been argued that these superfluous embryos may also be used to generate new hESCs lines that may help treat patients and save lives, rather than throwing them away. At the same time, there are suspicions that granting permission to generate hESC lines might lead to the purposeful generation of large numbers of surplus IVF embryos.

Proponents of hESC research have also pointed out that abortion is legal in many countries. Therefore, a double standard is being applied if the destruction of an early embryo for the generation of an hESC line is forbidden, while abortion is legal when the fetus is much further developed.

Even if the generation of an hESC line (which causes destruction of an embryo) is considered unethical, the question remains as to whether moral restrictions should apply to research that uses preexisting hESC lines that have been generated by others. One ethical concern is that it was immoral to generate the hESC line, as the destruction of the embryo violated its human rights and instrumentalized the embryo, which should be considered as an end in itself. The use of existing hESC lines must then be banned since their generation was unethical. In this view, a moral double standard would be present if one considers the generation of an hESC line immoral but the use of preexisting hESC lines acceptable.

Nevertheless, allowing the use of existing hESC lines coupled with a ban on the generation of new hESC lines has become the legal standard in many countries (see below). It has been argued that hESC lines are pluripotent, that is, they are no longer totipotent as the embryo itself is, and cannot develop into a viable human. This view permits the conferring of full human dignity and rights on the embryo, including the sanctity of its life, while regarding hESC lines as material that can be used for research purposes. Another perspective argues that the destruction of the embryo had happened anyway; therefore, it would be wasteful not to use the generated hESC lines for future biomedical research purposes and applications. Proponents of stem cell research argue that we have the duty to develop new therapies to cure currently untreatable diseases and help suffering patients. However, the existence of such an obligation has been denied by others.

Practically speaking, it has been stated that prohibiting research with existing hESC lines in one country would not prevent this type of research in other countries. Thus, the argument goes, it does not make sense to forbid such experiments. It would also be difficult for countries that do not permit hESC research to handle the question of what to do if an hESC-based therapy was successfully developed in another country. Would it be justifiable to withhold a successful therapy in one country that was created by a strategy that is considered unethical in another? Even if withholding the treatment is ethically justifiable, can it be practically implemented in law? There is little doubt that patients will travel to other countries that permit hESC-based therapies if it is in their interest to do so.

Furthermore, there is also disagreement as to whether or not permission to use existing hESC lines would increase the number of newly generated hESC lines by those who are not affected by a ban on embryo destruction. Permission to import existing hESC lines has sometimes been coupled to a cutoff date prior to the implementation of the regulation. This is intended to prevent the stimulation of the creation of new hESC lines in hESC exporting countries.

The debate about research involving embryos or hESCs is accompanied by controversy regarding therapeutic cloning (Section 12.1.2). Although this discussion cannot be extensively recapitulated here, the leading arguments will be discussed. For two major reasons, even some supporters of hESC research disapprove of therapeutic cloning: First, methods developed for therapeutic cloning might be misused for the reproductive cloning of humans. Both processes use the identical first step, somatic cell nuclear transfer (SCNT). As reproductive cloning is commonly disapproved of except for a few outliers (Box 12.1 in Section 12.1.2), many people claim that prohibiting therapeutic cloning is the only way to ensure that reproductive cloning remains impossible. The second major reason is the need for human egg cells. Harvesting of egg cells carries health risks, and widespread use of SCNT technology might also produce ethical conflicts (e.g., an illegal human egg market, pressure on women to become egg donors, etc.).

These disagreements are also reflected by a very diverse worldwide legal situation. Some countries, including China, have permissive policies and allow the use of human embryos and even therapeutic cloning. Israel also has permissive policies for stem cell research that is based on a (not uncontroversial) interpretation of the Jewish Talmud that the embryo should only receive protective rights at later stages of development. The legal situation in Europe is inconsistent. Some countries, including Finland, Greece, The Netherlands, Sweden and the United Kingdom, allow the production of hESC lines from surplus IVF embryos. However, the United Kingdom is the only country that permits cloning of embryos to produce stem cell lines for therapeutic purposes. Some countries, for example, Ireland, Austria, Denmark, Germany, and France, prohibit any production of hESC lines. Germany, however, even though it prohibits the production of hESC lines, permits the import of already existing hESCs for promising research. Initially, permission for the import of hESCs was restricted to cell lines that were created before January 2002. Since all hESC lines existing at that time were contaminated by murine feeder cells, the cutoff date was later pushed up to May 2007. This decision was heavily criticized as it was argued that making a regulation based on a movable cutoff date was meaningless. The stem cell policy in the United States is complex. Although the production and research on hESC lines are not regulated (and thus not forbidden) by federal law, federal funding of such research is regulated. In 2001, the conservative Bush administration chose not to allow taxpayer funding for research on hESC lines that were not already in existence at that time. In 2009, President Obama removed the restrictions on federal funding for the newer hESC lines. Although federal funding can still not be used to create new hESC lines, the new policy allows applying such funding to research involving any hESC line that was created using private funds (or was generated abroad).

It is likely that scientific developments will eventually render obsolete the ethical debate on research involving embryonic stem cells. As described in Section 12.1.1, researchers at Advanced Cell Technology have developed a method to obtain an hESC line without destroying the embryo. Furthermore, the main focus of stem cell research has shifted to the creation and use of induced pluripotent stem cells (iPSCs) (Sections 12.3.1 and 12.6). These cells are not only patient specific and do not induce an immune response, but they also do not create ethical issues since their generation does not require the destruction of an embryo. Another future option is the direct reprogramming strategy described in Section 12.4, which bypasses the pluripotent state and also does not create ethical questions. However, there is still broad consensus among stem cell researchers that experiments that will include hESC lines are essential in comparing the properties (differentiation potential, epigenetic markers, teratogenic potential, etc.) of various cell types.

15.2.3
Preimplantation Genetic Diagnosis

The term preimplantation genetic diagnosis refers to genetic profiling of embryos generated by IVF prior to implantation (Section 8.1.2). The analysis requires a biopsy of the early embryo (Figure 15.3) to obtain genomic material for evaluation. The selection of suitable embryos can improve the success rate of a pregnancy following IVF. The approach can also be used to discriminate against embryos with genetic disorders, which will be viable but diseased. In addition, PGD can be employed to select embryos of one sex in the context of *family balancing*. PGD has also been used to choose an embryo with a required human leukocyte antigen (HLA) so that it can donate cells for transplant for a diseased sibling. All of these applications of PGD raise specific ethical issues that will be discussed briefly in the following section.

PGD is a selection procedure; deselected embryos may be discarded or donated for scientific experiments. An ethical assessment of the technology, therefore, strongly

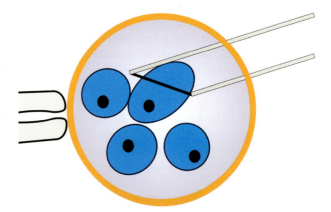

Fig. 15.3 Embryo biopsy. At day 3 of its development, the embryo consists of approximately eight cells called blastomeres. Each of these cells has the potential to develop into a viable human. It is thus possible to take one or two blastomeres from the embryo for genetic testing, a procedure called embryo biopsy.

depends on the moral status of the embryo, which was discussed in more detail in the previous section. Those who regard embryos as complete human beings and refer to the sanctity of the life of the embryo will consider the technology generally unethical. In contrast, if the rights of the embryo are graded depending on its developmental stage, the moral legitimacy of embryo selection depends on the aim pursued by the procedure. For example, the utilitarian perspective would argue that the selection of an embryo with the potential to become a happy human being, and which also makes its parents happy, would outweigh the negative aspects of destroying damaged embryos.

The rate of an IVF resulting in a live birth strongly depends on the age of the woman. IVF is successful approximately 30–40% of the time for younger women and decreases to a few percent for women above 45. PGD can help to deselect embryos with severe chromosomal abnormalities and thus may prevent reproductive failure. This approach is associated with the lowest level of ethical issues, as PGD is used only to increase the success rate of an already existing technology. Further, the deselected embryos are unlikely to be viable anyway. (Some organizations, for example, the Roman Catholic Church, generally oppose assisted reproductive technologies such as IVF, as they separate the goal of reproduction from marital values and natural procreation. However, this is a different general ethical controversy that will not be extensively discussed here.)

The PGD controversy centers around the identification of (mono)genetic disorders. PGD may be used to evaluate IVF embryos for a disease-relevant mutation that is known to occur in a family, for instance, after the loss of a child due to a genetic disorder. To prevent a recurrence, parents may ask to deselect embryos with this mutation in a new pregnancy. The situation becomes more complex if PGD is used not only for early-onset fatal diseases but also for late-onset diseases (e.g., Huntington's disease) or for risk predictors such as cancer susceptibility genes (e.g., *BRCA1* and *BRCA2*). Nevertheless, the American Society for Reproductive Medicine has considered such uses of PGD ethically justifiable even if there is no disease therapy available.

Most countries permit the use of PGD for the diagnosis of disabling or lethal genetic disorders. This regulation also acknowledges the pragmatic argument that selection by PGD is ethically less problematic than a trial pregnancy, in which a woman requests a prenatal genetic test several weeks after initiation of a pregnancy (Section 8.1.2) and then decides to abort if the fetus carries a harmful mutation.

While many people and regulatory bodies consider PGD ethically justifiable in the setting of a severe genetic disorder, the question remains of how to define severe disease. It has been proposed that a government body should assemble a list of severe disorders that qualify for PGD. However, it is unclear how such a commission would be composed: Would political figures or religious representatives be included? It would also be necessary to continuously update this list to take into account newly developed diagnostic and therapeutic procedures. An alternative suggestion was to consider the potential burden on the parents. This would be analogous to the abortion question in some countries, where abortion is justifiable if the pregnant woman feels incapable of raising the child.

However, leaving the decision to the affected parents-to-be has been criticized as promoting eugenics. There is no doubt that infantile Tay–Sachs disease, which usually leads to death by 3–4 years of age, is a severe disease. However, opera singers may consider a cleft lip and palate in a child to be an unbearable handicap, whereas others would regard this as a tolerable inconvenience. Some critics suspect a *slippery slope*: A technology such as PGD may initially be used to prevent severe disease, but later it may also be employed to avoid minor annoyances. Finally, PGD may be misused to select for positive properties such as intelligence or beauty. The term *designer baby* has become a catchphrase that reflects the expectation that PGD may select a child with the best possible characteristics. Fears also exist that extensive selection by PGD may lower the tolerance for existing disabled people and promote discrimination. However, only approximately 4% of all disabling disorders are caused by a genetic defect. It has also been stated that the decision to deselect embryos with a specific genetic disorder represents a decision about the value of life with a specific handicap: It has been argued that this is discriminatory to those who live with this handicap.

Another conflict may arise between the technological options of PGD and the right not to know, as discussed in Section 15.2.1 in the case of a family history of HD. Parents may wish not to know their own carrier status, but may at the same time want to ensure that they have offspring free of the disease. This may place the physician in an ethical dilemma if no unaffected embryos are available for transfer. In this case, a mock transfer would have to be carried out to prevent the parents-to-be from learning they are carriers of the defect.

The situation has become even more complex with the advent of WGS technologies. While PGD has mainly been employed to test for mutations that have previously occurred in a family, WGS will increasingly be used in PGD (and in prenatal genetic testing in general). While genetic testing of an embryo has typically been used to avoid a suspected monogenic disorder, WGS will reveal genetic risk factors for each embryo tested. Moreover, selection will be based on general health profiles rather than on the deselection of a specific risk factor. WGS will also reveal a child's genetic variations that are currently of unknown significance, but may become so in the future. Finally, it may disclose previously unknown genetic disorders in the prospective parents, giving rise to another conflict with their right not to know.

A survey in 2006 revealed another unexpected application of PGD: 3% of American PGD clinics admitted having intentionally used the technology to select an embryo for the presence of a disability. In some cases, deaf couples want to have a deaf child. Another similar example is dwarfism. Embryos of two parents with dwarfism have a 25% chance of normal height, a 50% chance of dwarfism, and a 25% risk of a double dominant mutation, which is usually fatal soon after birth. PGD can be used to eliminate embryos with the lethal mutation, thus avoiding doomed pregnancies. However, it may also be used to select either against dwarfism or for it, if the parents so wish.

The inheritance of some genetic disorders is linked to a sex chromosome; thus, sex selection may help prevent the disease. This was the intent of the first PGD ever carried out, which occurred in 1990. However, according to a 2006 survey, 42% of the IVF-PGD clinics in the United States provided PGD for nonmedical sex selection. This application has been referred to as *family balancing*, and often happens when a couple already has several children of the same sex. In some cultures, however, sex selection for nonmedical reasons may support the discriminatory view that male progeny are more valuable. At the same time, it may prevent the mass killing of female firstborns in some countries. Still, the selection of embryos on the basis of a parental preference for a boy or a girl may lead to a shift in the sex ratio, which may have severe social implications (e.g., a fraction of the population may never have a chance of finding a mate) as can already be seen in some societies. In addition, sex selection has been criticized as the first step in the direction of selection for non-health-related traits.

Another controversial application of PDG is in the selection of an embryo to provide a transplant for a sibling who is affected with a fatal disease, such as cancer or Fanconi's anemia. PDG can be used to test for genetic compatibility (human leukocyte antigen typing); only those embryos that are compatible with the existing sick child will be implanted. Although this procedure can save the life of a diseased child by hematopoietic stem cell transplantation (HSCT), it is highly controversial. One argument against this approach involves the welfare of the donor child. This concern is not relevant as long as umbilical cord blood is taken at birth and used for the HSCT. However, an issue arises if peripheral blood or even bone marrow has to be harvested from the savior sibling, since these procedures involve a risk, although a minor one, for the sibling.

An important argument against the selection of a matching transplant donor is the ethical demand to regard every human being as an end in itself and not as a means to another person's end. The selection of a matching transplant donor would thus be unethical, as the savior sibling would be misused as a means to save the diseased child. However, it would make an ethical difference if the parents wanted to have another child anyway and selected one who, in addition, can save its sibling, versus if the parents decided to have another child only to save the already existing one. In the latter case, the savior sibling would be instrumentalized in an illegitimate manner. It may become psychologically distressing for the savior sibling to know that it owes its existence only to the fact that it was supposed to save its sibling. In practice, it is impossible for the attending physician to ascertain the parental intentions. The issue became popular when both the houses of the UK Parliament recommended that screening should be permitted to provide a sibling match for children suffering from serious and life-threatening conditions. PGD has been used to select matched donors since 2001.

Further Reading

Mepahm, B. (2005) *Bioethics – An Introduction for the Biosciences*, Oxford University Press, Oxford, UK.

Talbot, M. (2012) *Bioethics – An Introduction*, Cambridge University Press, Cambridge, UK.

The Right Not to Know

Andorno, R. (2004) The right not to know: an autonomy based approach. *J. Med. Ethics.*, **30**, 435–439; including commentary 439–440.

Ayuso, C., Millan, J.M., Mancheno, M. *et al.* (2013) Informed consent for whole-genome sequencing studies in the clinical setting: proposed recommendations on essential content and process. *Eur. J. Hum. Genet.*, **21**, 1054–1059.

Helgesson, G. (2014) Autonomy, the right not to know, and the right to know personal research results: what rights are there, and who should decide about exceptions? *J. Law Med. Ethics*, **42**, 28–37.

Preimplantation Genetic Diagnosis

Hens, K., Dondorp, W., Handyside, A.H. *et al.* (2013) Dynamics and ethics of comprehensive preimplantation genetic testing: a review of the challenges. *Hum. Reprod. Update*, **19**, 366–375.

Abbreviations

AAV	Adeno-associated virus
ABC	ATP-binding cassette
Abl	Abelson kinase
aCGH	Array comparative genomic hybridization
acK	Acetyl lysine
ACT	1. Artemisinin combination therapy
	2. Autologous chondrocyte transplantation
AD	Alzheimer's disease
ADA	Adenosine deaminase
ADC	Antibody-drug conjugate
ADCC	Antibody-dependent cellular cytotoxicity
ADR	Adverse drug reaction
AFP	Alpha fetoprotein
Ago	Argonaute protein
AHS	Abacavir hypersensitivity syndrome
AI	Aromatase inhibitor
AIDS	Acquired immunodeficiency syndrome
AKR	Aldo-keto-reductase
ALCL	Anaplastic large cell lymphoma
ALK	Anaplastic lymphoma kinase
ALL	Acute lymphoblastic leukemia
ALS	Amylotrophic lateral sclerosis
AMACR	Alpha-methyl coenzyme A racemase
AMD	Age-related macular degeneration
amiRNA	Artificial microRNA
AML	Acute myeloid leukemia
Amp	Ampicillin
AP/MS	Affinity purification/mass spectrometry
APAF1	Apoptotic protease activating factor 1
APC	Antigen-presenting cell

APL	Acute promyelocytic leukemia
Apo	Apolipoprotein
AR	Androgen receptor
ARE	Androgen response element
AREG	Amphiregulin
ARMS	Amplification refractory mutation system
ARS	Autonomously replicating sequence
AS	Angelman syndrome
AS ON	Antisense oligonucleotide
ATM	Ataxia-telangiectasia mutated
ATRA	All-trans retinoic acid
AZT	Azidothymidine
BAC	Bacterial artificial chromosome
BBB	Blood–brain barrier
Bcr	Breakpoint cluster region
BDNF	Brain-derived neurotrophic factor
BER	Base-excision repair
BH_2	Dihydrobiopterin
BH_4	Tetrahydrobiopterin
BHD	Birt–Hogg–Dube
BHK	Baby hamster kidney
BL	Burkitt lymphoma
BLG	Beta-lactoglobulin
BLyS	B-lymphocyte stimulator
BMD	Becker muscular dystrophy
BMP	Bone morphogenetic protein
BRCA	Breast cancer
BRE	TFIIB recognition element
BSE	Bovine spongiform encephalopathy
BTK	Bruton tyrosine kinase
cAMP	Cyclic AMP
CAPB	Carcinoma prostate brain
CBP	CREB-binding protein
CCAAT	CCAAT box

Molecular Medicine: An Introduction, First Edition. Jens Kurreck and Cy Aaron Stein.
© 2016 Wiley-VCH Verlag GmbH & Co. KGaA. Published 2016 by Wiley-VCH Verlag GmbH & Co. KGaA.

cccDNA	Covalently closed circular DNA	CYP	Cytochrome P450
CCD	Charge-coupled device	DAG	1,2-Diacylglycerol
CCM	Chemical cleavage of mismatch	DAPI	4′,6-diamidine-2-phenylindole
CCR5	C-C chemokine receptor 5	DBD	DNA-binding domain
CD	Cluster of differentiation	ddNTP	Dideoxynucleotide triphosphate
CDC	1. Complement-dependent cytotoxicity	DG	Dystroglycan
		DGC	Dystrophin–glycoprotein complex
	2. Centers for Disease Control and Prevention	DGGE	Denaturing gradient gel electrophoresis
Cdk	Cyclin-dependent kinase	DHF	Dihydrofolic acid
cDNA	Complementary DNA	DHODH	Dihydroorotate dehydrogenase
CDS	Coding sequence	dHPLC	Denaturing high-performance liquid chromatography
CeGaT	Center for Genomics and Transcriptomics	DHT	Dihydrotestosterone
CEN	Centromere	DIGE	Differential in gel electrophoresis
Cer	Cerebroside	DMD	Duchenne muscular dystrophy
CF	Cystic fibrosis	DME	Drug-metabolizing enzyme
Cff-DNA	Cell-free fetal DNA	DMR	Differentially methylated region
CFTR	Cystic fibrosis transmembrane conductance regulator	DNMT	DNA methyltransferase
		dNTP	Deoxynucleotide triphosphate
CGH	Comparative genomic hybridization	Dox	Doxycycline
CHO	Chinese hamster ovary	DPE	Downstream promoter element
CIMP	CpG island methylator phenotype	DRD	Dopamine responsive dystonia
CiPSC	Chemically induced pluripotent stem cell	DRG	Dorsal root ganglia
		DSB	Double-strand break
CJD	Creutzfeldt–Jakob disease	dSMADi	Dual SMAD inhibition
CLL	Chronic lymphocytic leukemia	dsRNA/dsDNA	Double-stranded RNA/DNA
cM	Centimorgan	EGF	Epidermal growth factor
CMD	Congenital muscular dystrophy	EGFR	Epidermal growth factor receptor
CML	Chronic myeloid leukemia	ELISA	Enzyme-linked immunosorbent assay
CNS	Central nervous system	ELSI	Ethical, legal, and social issues
CNV	Copy-number variation	EM	Extensive metabolizer
COX	Cyclooxygenase	EMA	European Medicines Agency
CPE	Cytopathic effect	EML4	Echinoderm microtubule-associated protein like-4
CREB	cAMP-response element-binding protein	EmPCR	Emulsion PCR
CRISPR	Clustered regulatory interspaced short palindromic repeats	ENCODE	Encyclopedia of DNA elements
		Env	Envelope protein
cRNA	Complementary RNA	EpCAM	Epithelial cell adhesion molecule
CRPC	Castration-resistant prostate cancer	EPO	Erythropoietin
CSC	Cancer stem cell	ER	1. Endoplasmic reticulum
CSF	Colony stimulating factor		2. Estrogen receptor
C_T	Threshold cycle	EREG	Epiregulin
CTLA	Cytotoxic T-lymphocyte antigen	eRF1	Eukaryotic release factor
CVS	Chorionic villus sampling	Erk	Extracellular signal-regulated kinase
CXCR4	C-X-C chemokine receptor type 4	ESA	Erythropoiesis-stimulating agents

ESC	Embryonic stem cell
ESI	Electrospray ionization
esiRNA	Endogenous siRNA
ETS	E26 transformation specific
Fab	Antigen-binding fragment (of an antibody)
FACS	Fluorescence-activated cell sorting
FAP	Familial adenomatous polyposis
FBSN	Familial bilateral striatal necrosis
Fc	Crystallizable fragment (of an antibody)
FDA	US Food and Drug Administration
FGF	Fibroblast growth factor
FGFR	Fibroblast growth factor receptor
FH	1. Familial hypercholesterolemia
	2. Fumarate hydratase
FISH	Fluorescence *in situ* hybridization
FMR	Fragile X mental retardation
FRET	Fluorescence resonance energy transfer
GABA	Gamma aminobutyric acid
Gag	Group-specific antigen
Gal	Galactose
Gal NAc	*N*-acetylgalactosamine
GAPDH	Glyceraldehyde-3-phosphate dehydrogenase
G-CSF	Granulocyte colony-stimulating factor
GEF	Guanosine exchange factor
GET	Genomes environments traits
GFP	Green fluorescent protein
GGCX	γ-Glutamyl carboxylase
GGI	Genomic grade index
Glc	Glucose
GLP-1	Glucagon-like peptide-1
GM-CSF	Granulocyte macrophage colony-stimulating factor
gp	Glycoprotein
GRAS	Generally regarded as safe
GSH	Glutathione
GST	Glutathione-S-transferase
GTA	GATA transcription factor
GTF	General transcription factor
HA	Hemagglutinin
HAART	Highly active antiretroviral therapy
hAAT	Human alpha antitrypsin
HAR	Human accelerated region

HATs	Histone acetyltransferases
HBcAg	Hepatitis B core antigen
HBsAg	Hepatitis B surface antigen
HBV	Hepatitis B virus
HCC	Hepatocellular carcinoma
HCG	Human chorionic gonadotropin
HCM	Hypertrophic cardiomyopathy
HCV	Hepatitis C virus
HD	Huntington's disease
HDAC	Histone deacetylase
HDACI	Histone deacetylase inhibitor
HDF	Host dependency factors
HEK	Human embryonic kidney cells
Her	Human epidermal growth factor receptor
HFE	High for Fe
HGH	Human growth hormone
HGP	Human Genome Project
HIF	Hypoxia-inducible factor
HIV	Human immunodeficiency virus
HLA	Human leukocyte antigen
HLRCC	Hereditary leiomyomatosis and renal cell cancer
HMG	3-Hydroxy-3-methyl-glutaryl
HMG-CoA	Hydroxymethylglutaryl-CoA
HNPCC	Hereditary nonpolyposis colorectal cancer
HNSCC	Head and neck squamous cell carcinoma
HPP	Human Proteome Project
HPRT	Hypoxanthine-guanine phosphoribosyltransferase
HPV	Human papilloma virus
HSC	Hematopoietic stem cell
HSCT	Hematopoietic stem cell transplantation
HSP	Heat shock protein
HSV	Herpes simplex virus
HTS	1. High-throughput screening
	2. High-throughput sequencing
HUPO	Human Proteome Organization
IC	Imprinting center
ICAT	Isotope-coded affinity tagging
ICM	Inner cell mass
IDL	Intermediate-density lipoprotein
IEF	Isoelectric focusing

Ig	Immunoglobulin	MAP	MUTYH-associated polyposis
IGS	Internal guide sequence	MAPK	Mitogen-activated protein kinase
IHC	Immunohistochemistry	MASP	MBL-associated serine protease
IKBKAP	Inhibitor of nuclear factor κB-kinase complex-associated protein	Mat	Maternal
		Mb	Megabases
IKMC	International Knockout Mouse Consortium	M-BCR	Major breakpoint cluster region
		MBL	Mannose-binding lectin
IL	Interleukin	MBP	Maltose-binding protein
IM	Intermediate metabolizer	MCL	Mantle cell lymphoma
Indels	Insertions and deletions	MCN	Minichromosome maintenance
INF	Interferon	mDA	Midbrain dopamine
Inr	Initiator element	MDR	Multidrug resistance
IP_3	Inositol-1,4,5-trisphosphate	MDS	Myelodysplastic syndrome
IPASS	Iressa Pan-Asia Study	MDSC	Myeloid-derived suppressor cells
iPSC	Induced pluripotent stem cell	meK	Methyl lysine
IPV	Inactivated polio vaccine	MELAS	Mitochondrial encephalomyopathy, lactic acidosis, and stroke-like episodes
IRES	Internal ribosome entry site		
ITR	Inverted terminal repeats		
iTRAQ	Isobaric tags for relative and absolute quantification	meR	Methyl arginine
		MERRF	Myoclonic epilepsy with ragged-red fibers
IVF	*In vitro* fertilization		
Jak	Janus kinase	M-FISH	Multiplex fluorescence *in situ* hybridization
kb	Kilobases		
KSS	Kearns–Sayre syndrome	MHC	Major histocompatibility complex
LABA	Long-acting β-adrenoceptor agonist	mi-iPSC	MicroRNAs-induced pluripotent stem cell
LAR	Luminal androgen receptor		
LAT	ʟ-Amino acid transporter 1	MILS	Maternally inherited Leigh syndrome
LBD	Ligand-binding domain	MIQE	Minimum information for publication of quantitative real-time PCR experiments
LCA	Leber's congenital amaurosis		
LDL	Low-density lipoprotein		
LFS	Li–Fraumeni syndrome	miRISC	miRNA-loaded RISC
LGMD	Limb girdle muscular dystrophy	miRNA	MicroRNAs
LHON	Leber's hereditary optic neuropathy	MIT	Massachusetts Institute of Technology
LIF	Leukemia inhibitory factor		
LINE	Long interspersed nuclear element	MLL	Mixed lineage leukemia
LNA	Locked nucleic acid	MLPA	Multiplex ligation-dependent probe amplification
LOI	Loss of imprinting		
LPL	Lipoprotein lipase	MMR	Measles–mumps–rubella (vaccine)
LQTS	Long QT syndrome	MMRD	Mismatch repair protein deficiency
LS	Leigh syndrome	MMRV	Measles–mumps–rubella–varicella (vaccine)
LSD	Lysosomal storage disease		
LTR	Long terminal repeat	modENCODE	Model organism ENCODE
mAb	Monoclonal antibody	MODY	Maturity onset diabetes of the young
MAC	Membrane attack complex		
MALDI	Matrix-assisted laser desorption/ionization	MOE	2′-methoxy-ethyl RNA
		MRC	Medical Research Council
		MRI	Magnetic resonance imaging

MRSA	Methicillin-resistant *Staphylococcus aureus*	ORR	Objective response rate
		OS	Overall survival
MS	1. Multiple sclerosis	PABP	Poly(A)-binding protein
	2. Mass spectrometry	PAC	P1-derived artificial chromosome
MS/MS	Tandem mass spectrometry	PAH	Phenylalanine hydroxylase
MSC	Mesenchymal stroma cell	PAMP	Pathogen-associated molecular pattern
MSI	Microsatellite instability		
MSL	Mesenchymal stem-like subtype	PAP	Prostatic acid phosphatase
MSS	Microsatellite stable	PAR	Poly-ADP ribose
mtDNA	Mitochondrial DNA	PARP	Poly-ADP ribose polymerase
MTE	Motif ten element	Pat	Paternal
mTOR	Mammalian target of rapamycin	PBP	Penicillin-binding protein
Myc	Myelocytomatosis viral oncogene	PCNA	Proliferating cell nuclear antigen
NA	Neuraminidase	PCR	Polymerase chain reaction
NAD	Nicotinamide adenine dinucleotide	pCR	Pathological complete response
NAPQI	*N*-Acetyl-*p*-benzoquinone imine	PD	1. Programmed death
NARP	Neuropathy, ataxia, and retinitis pigmentosa		2. Parkinson's disease
		pDC	Plasmacytoid dendritic cell
NCI	National Cancer Institute	PDE	Phosphodiesterase
ncRNA	Noncoding RNA	PDGF	Platelet-derived growth factor
NER	Nucleotide excision repair	PDGFR	Platelet-derived growth factor receptor
Neu NAc	*N*-Acetylneuraminic acid		
NGS	Next-generation sequencing	PDL	Programmed death ligand
NHL	Non-Hodgkin lymphoma	PDR	Pan drug resistant
NIPD	Noninvasive prenatal diagnosis	PDX	Patient-derived xenograft
NIPT	Noninvasive prenatal testing	PEG	Polyethylene glycol
NK cell	Natural killer cell	PEP	Phosphoenolpyruvate
NME	New molecular entity	PEPT	Peptide transporter
NMR	Nuclear magnetic resonance	PFS	Progression-free survival
nNOS	Neuronal nitric oxide synthase	PGD	Preimplantation genetic diagnosis
NNRTI	Nonnucleoside reverse-transcriptase inhibitor	PGP	1. P-glycoprotein
			2. Personal Genome Project
NRTI	Nucleoside analog reverse-transcriptase inhibitor	PgR	Progesterone receptor
		PGS	Preimplantation genetic screening
NSCLC	Non-small cell lung cancer	PGx	Pharmacogenetics/pharmacogenomics
NTC	No-template control	PhChr	Philadelphia chromosome
NuRD	Nucleosome remodeling and deacetylation	PI	Phosphoinositide
		PI3K	Phosphatidylinositol 3-kinase
OATP	Organic anion transporting polypeptide	PIC	Preinitiation complex
		PIN	Prostatic intraepithelial neoplasia
OCT	Organic cation transporter	PIP$_2$	Phosphatidylinositol-4,5-bisphosphate
OMe	2′-*O*-Methyl RNA		
OMIM	Online Mendelian inheritance in man	piPSC	Protein-induced pluripotent stem cell
OPC	Oligodendrocyte precursor cell	piRNA	Piwi-interacting RNA
OPV	Oral polio vaccine	PKA	Protein kinase A
Ori	Origin of replication	PKD	Polycystic kidney disease

PKU	Phenylketonuria	SABA	Short-acting β-adrenoceptor agonist
PLC	Phospholipase C	SARS	Severe acute respiratory syndrome
PM	Poor metabolizer	SBS	Sequencing by synthesis
PMO	Phosphorodiamidate morpholino oligomer	scAAV	Self-complementary AAV
PNS	Peripheral nervous system	scFv	Single-chain variable fragment
PPi	Pyrophosphate	SCID	Severe combined immunodeficiency
PrP	Prion protein	SCLC	Small cell lung cancer
PRR	Pattern recognition receptor	SCNT	Somatic cell nuclear transfer
PS	1. Phosphorothioate	SCT	Stem cell transplantation
	2. Phosphoryl serine	SDS-PAGE	Sodium dodecyl sulfate-poly-acrylamide gel electrophoresis
PSA	Prostate-specific antigen	SEER	Surveillance, epidemiology, and end results program
PSCA	Prostate stem cell antigen		
PSI	Protein structure initiative	SELEX	Systematic evolution of ligands by exponential enrichment
PTEN	Phosphatase and tensin homolog		
PTGS	Posttranscriptional gene silencing	SER	Smooth endoplasmic reticulum
PUBS	Percutaneous umbilical cord blood sampling	SERM	Selective estrogen receptor modulator
PWS	Prader–Willi syndrome	SHARP	Sorafenib hepatocellular carcinoma assessment randomized protocol
QF-PCR	Quantitative fluorescence PCR		
qPCR	Quantitative PCR	SHBG	Sex hormone-binding globulin
qRT-PCR	Quantitative reverse-transcription PCR	shRNA	Short hairpin RNA
		SILAC	Stable isotope-labeled amino acids in cell culture
RA	Rheumatoid arthritis		
rAAV	Recombinant AAV	SINE	Short interspersed nuclear element
RARA	Retinoic acid receptor alpha	siRNA	Small (or short) interfering RNA
Ras	Rat sarcoma	SKY	Spectral karyotyping
Rb	Retinoblastoma	SLE	Systemic lupus erythematosus
RCC	Renal cell carcinoma	SM	Smooth muscle
RdRP	RNA-dependent RNA polymerase	SMRT	Single molecule real time
RER	Rough endoplasmic reticulum	SNALP	Stable nucleic acid lipid particle
RF	Release factor	snoRNA	Small nucleolar RNA
RFLP	Restriction fragment length polymorphism	SNP	Single-nucleotide polymorphism
		snRNA	Small nuclear RNA
RiPSC	RNA-induced pluripotent stem cell	snRNP	Small nuclear ribonucleoprotein complex
RISC	RNA-induced silencing complex		
RLC	RISC loading complex	SOD	Superoxide dismutase
RNAi	RNA interference	SOLiD	Sequencing by oligo ligation detection
RNAP	RNA polymerase		
RNA-Seq	RNA sequencing	SPR	Surface plasmon resonance
RPA	Replication protein	SRP	Signal recognition particle
RPE	Retinal pigment epithelium	SSB	Single-strand break
RSV	Respiratory syncytial virus	SSCP	Single-stranded conformation polymorphism
RT	Reverse transcriptase		
RTK	Receptor tyrosine kinase	SSR	Short (or simple) sequence repeat
RT-PCR	Reverse-transcription PCR	ssRNA/ssDNA	Single-stranded RNA/DNA

STAT	Signal transducer and activator of transcription		TSE	Transmissible spongiform encephalopathy
STR	Short tandem repeat		tSMS	True single-molecule sequencing
STS	Sequence tagged site		TTP	Thrombotic thrombocytopenic purpura
T1DM	Type 1 diabetes mellitus			
T2DM	Type 2 diabetes mellitus		uK	Ubiquitinated lysine
TAF	TBP-associated factor		UM	Ultrarapid metabolizer
TAR	Trans-activation response element		UPD	Uniparental disomy
			UTR	Untranslated region
TATA	TATA box		VAPP	Vaccine-associated paralytic poliomyelitis
TB	Tuberculosis			
TBP	TATA-binding protein		VDPV	Vaccine-derived poliovirus
T_C	Cytotoxic T cell		VEGF	Vascular endothelial growth factor
TCR	T-cell receptor		VEGFR	Vascular endothelial growth factor receptor
TEL	Telomere			
TERT	Telomerase reverse transcriptase		V_{HH}	Variable domain of the H chain of heavy-chain antibodies
Tet	Tetracycline			
TetO	Tet operator		vHL	Von Hippel–Lindau
TF	Transcription factor		VKOR	Vitamin K epoxide reductase
TGF	Transforming growth factor		VKORC1	VKOR complex subunit 1
T_H	T helper cell		VLP	Virus-like particle
THF	Tetrahydrofolic acid		vRNA	Viral RNA
TKI	Tyrosine kinase inhibitor		VUS	Variants of unknown significance
TLR	Toll-like receptor		VZV	Varicella-zoster virus
TNF	Tumor necrosis factor		WADA	World Anti-Doping Agency
TOF	Time of flight		WES	Whole exome sequencing
tPA	Tissue-type plasminogen activator		WGS	Whole genome sequencing
TPMT	Thiopurine-*S*-methyltransferase		WHO	World Health Organization
TRBP	Trans-activation response (TAR) RNA-binding protein		Xist	X-inactive specific transcript
			XP	Xeroderma pigmentosum
Tre	Tetracycline response element		YAC	Yeast artificial chromosome
T_{reg}	Regulatory T cell		ZNF	Zinc finger nuclease
TSC	Tuberous sclerosis complex		ZMW	Zero-mode waveguide

Glossary

Adaptive immune response A response highly specific for a given pathogen that prevents its spread. It comprises cellular and humoral immunity, which produces pathogen-specific antibodies.

Allele One of a number of alternative forms of the same gene or genetic locus in the chromosome. A diploid organism contains two alleles for each gene, which may or may not be identical.

Allogenic The transplantation of cells, tissues, or organs from one individual to another of the same species.

Aneuploidy The finding of an abnormal number of chromosomes in a cell.

Antibody (also called immunoglobulin). A molecule produced by the immune system that specifically binds to a foreign agent (the antigen). An antibody has a Y-shaped structure and neutralizes pathogens or marks them for destruction by other components of the immune system.

Apoptosis The process of programmed cell death in multicellular organisms.

Aptamer Usually, a single-stranded oligonucleotide that binds to a target molecule, for example, a protein, by steric interactions.

Autologous When cells, tissues, or organs are grafted into the same individual from whom it was derived.

Caspase A cysteinyl aspartate-specific protease that hydrolyzes cellular proteins. Caspases are the main mediators of apoptotic processes.

CD (Cluster of differentiation) Molecules located on the cell surface that are commonly used to characterize and define immune cells.

Chromatin The complex of DNA and proteins in the nucleus of eukaryotic cells.

Chromosome A linear DNA molecule associated with proteins in the nucleus of eukaryotic cells. It comprises a large part of the organism's genome.

Compound heterozygosity The condition of having two heterogeneous recessive alleles at a particular locus that cause genetic disease in combination. Commonly, both alleles are mutated, but at different locations.

Concordance The probability that a pair of genetically identical individuals (twins) will have a certain phenotypic characteristic, given that one of the pair has the characteristic.

Contig A contiguous sequence read resulting from the assembly of numerous overlapping small DNA fragments.

Copy number variation An alteration of the genomic DNA, defined as relatively large deletions or duplications ranging from 1 kb to several megabases in size.

Coverage (also known as read depth). The average number of reads representing a given nucleotide in a reconstructed sequence.

Dicer An endoribonuclease belonging to the RNase III family that cleaves longer double-stranded RNAs and pre-microRNAs into short double-stranded RNAs and mature microRNAs, respectively.

Embryonic stem cells Cells generated from the inner cell mass of a blastocyst, an early-stage embryo. They can be grown in culture without limit and can be induced to differentiate into all types of tissues.

Epigenetics The study of heritable changes in gene activity that are not caused by changes in the DNA sequence. For example, changes in DNA methylation and histone modification alter gene expression without changing the underlying DNA sequence.

Euchromatin Relatively lightly packed chromatin that is transcriptionally active.

Expressivity Variations in a phenotype among individuals carrying the same genotype. For example, variations in the severity of a disease phenotype in individuals carrying the same mutant allele.

ex vivo Indicates that an experiment or a treatment is performed outside an organism in or on tissue/cells in an artificial environment with the minimum alteration of natural conditions.

Molecular Medicine: An Introduction, First Edition. Jens Kurreck and Cy Aaron Stein.
© 2016 Wiley-VCH Verlag GmbH & Co. KGaA. Published 2016 by Wiley-VCH Verlag GmbH & Co. KGaA.

Gene expression The process by which information stored in a gene is used for the synthesis of a functional gene product, usually a protein.

Genome The complete set of genetic (hereditary) material in an organism. The study of the entire set of DNA within a cell or an organism is called genomics.

Genomic imprinting An epigenetic phenomenon in which the expression of certain genes depends on the parent of origin. Imprinted genes are silenced so that the respective genes are expressed only from the nonimprinted allele inherited from the mother or from the father.

Genotype The genetic makeup of an organism or an individual. It usually refers to a specific characteristic, often the genetic constitution of alleles at a specific locus.

Green fluorescent protein (GFP). A protein that exhibits bright green fluorescence when exposed to light in the blue to ultraviolet range. It is frequently used as a reporter in cell and molecular biology.

Haploinsufficiency This occurs when an organism has only one functional copy of a gene, while the other copy is inactivated by mutation. The functional copy cannot compensate for the loss of the second allele to produce the wild type, leading to a disease state.

Haplotype A set of linked genetic variants that are inherited as a unit.

Heterochromatin Nonexpressed eukaryotic chromatin that is highly condensed.

Heteroplasmy The presence of a mixture of more than one type of mitochondrial genome within a cell, usually wild-type and mutated mtDNA.

Homoplasmy A cell in which all copies of the mtDNA are identical. The homoplasmic mtDNA copies may be normal or mutated.

Hormone A molecule that mediates intercellular signaling, usually at a distance. It is secreted by one cell type and often induces a physiological response in another cell, usually at a distance.

Immunoglobulin See antibody.

Imprinting See genomic imprinting.

Indel Short insertions and deletions of nucleotides in the DNA.

Induced pluripotent stem cells A type of pluripotent stem cell that originates from differentiated normal somatic cells. They can propagate indefinitely or differentiate into other cell types.

Innate immune response The first line of defense of mammals against invading pathogens. It consists of cells and macromolecules that detect and destroy the infectious agent in a nonspecific manner.

in silico Indicates that a biological question is being addressed by computer simulation. The phrase was coined as an analogy to the terms *in vivo* and *in vitro*.

in situ Indicates that a phenomenon is being studied in the place where it occurs without removing it to artificial surroundings.

in vitro Indicates that an experiment is being carried out with components of an organism that have been isolated from their usual biological surroundings. These "test tube" experiments allow a very detailed and more convenient analysis than can be done with a whole, living organism. However, the experimental setup is artificial and the results do not always reflect the *in vivo* situation.

in vivo Indicates that an experiment is being carried out in a whole living organism as opposed to experiments in an artificial environment. *In vivo* experimentation usually refers to the testing of an approach in animal models.

Karyotype The set of chromosomes in a cell or an individual. The term also refers to the number and appearance of chromosomes in the nucleus.

Kinase An enzyme that transfers a phosphoryl group from high-energy donor molecules, for example, ATP, to specific substrates. The process is called phosphorylation. Protein kinases transfer the phosphate to a target protein.

Lyonization (also known as X-inactivation). A process by which one of the two copies of the X chromosome present in females is inactivated.

microRNAs A short endogenous RNA of ~21 nucleotides that regulates gene expression through RNA interference.

Mitochondriopathies Also known as mitochondrial diseases, these are disorders produced by dysfunctional mitochondria caused either by mutations in the nuclear or mitochondrial DNA.

Mosaicism The presence of two or more cell lines in an individual that differ in genotype or chromosomal constitution, although they have been derived from a single zygote.

Mutein A genetically modified variant of a therapeutically relevant protein produced with the aim of improving the pharmacokinetic and/or pharmacodynamic properties of the protein.

Nanobody A recombinant, single-domain antibody derived from a camelid heavy-chain antibody that consists only of a dedicated antigen-binding V_HH domain.

Oncogene Usually refers to mutated forms of normal cellular genes, so-called proto-oncogenes. An oncogene is capable of transforming a cell into a tumor cell.

PEGylation The process of covalently attaching polyethylene glycol (PEG) to another molecule, for example, a protein or an aptamer.

Penetrance The proportion of individuals carrying a particular genetic variation that develop the disease phenotype.

Personalized medicine Refers to the tailoring of medical treatment to the individual characteristics of each patient. The use of genetic information plays a large role in the personalization of treatment.

Pharmacodynamics The study of the effects of a drug on the human organism, including investigations into the mechanisms of drug action and the relationship between drug concentration and effect.

Pharmacogenetics Deals with the effect that the genotype has on the individual's drug response. It identifies inherited differences between individuals in their reaction to drugs.

Pharmacogenomics Synonym for pharmacogenetics.

Pharmacokinetics The description of the fate of a drug after administration to a human. This includes the substance's tissue distribution, the rate of metabolism, and degradation and the elimination from the body.

Phosphatase An enzyme that removes a phosphate group from its substrate (e.g., a protein) by hydrolyzing phosphoryl ester groups into a phosphate ion, leaving the remaining molecule with a free hydroxyl group. This reaction results in the dephosphorylation of the substrate molecule.

Plantibodies A portmanteau for an antibody produced in genetically engineered plants.

Pleiotropy Occurs when a genetic variation influences multiple, seemingly unrelated phenotypic traits.

Pluripotency The ability of a cell to differentiate into any cell of the three germ layers (endoderm, mesoderm, and ectoderm) and the germline.

Prenatal diagnosis Testing for diseases in a fetus or embryo before it is born.

Preimplantation genetic diagnosis Genetic profiling of embryos after *in vitro* fertilization and prior to implantation.

Prion protein A type of disease-causing protein. The prion protein may occur in its properly folded form, denoted PrP^C (for cellular), or in the disease-linked, misfolded form PrP^{Sc} (for scrapie).

Prodrug A pharmacologic substance that is administered to the patient in an inactive form, which then is converted into its active form through metabolic processes.

Proenzyme See zymogen.

Proofreading An additional error-correcting process that is carried out by some enzymes. For example, DNA polymerases recognize incorrectly incorporated bases and excise the nucleotide so that the correct base can be inserted.

Proprotein See zymogen.

Proteome The entire set of proteins in a cell, tissue, or organism at a given time. Proteomic studies include the quantitation, localization and analysis of modifications, interactions, and activities of all of a cell's proteins.

Pseudogene A dysfunctional relative of a normal gene that has lost its activity through mutation. It is usually derived by duplication of the functional gene.

Replication The process of producing an identical copy from one original molecule. Each strand of the double-stranded original DNA molecule serves as template for the production of a complementary strand. The process is carried out by a DNA-dependent DNA polymerase.

Reproductive cloning The production of an entire organism that is a genetically identical copy of an animal (or human).

Restriction enzyme Also known as restriction endonuclease, an enzyme that recognizes a specific (often palindromic) nucleotide sequence of double-stranded DNA and cleaves it endonucleolytically.

Ribonucleoprotein A complex composed of protein and RNA.

Riboswitch A regulatory element of an mRNA that can modulate the synthesis of an encoded protein as a consequence of the binding of a small molecule.

Ribozyme An RNA molecule with catalytic activity.

RNA interference A mechanism of posttranscriptional gene regulation. Short double-stranded RNA molecules trigger the degradation of a homologous RNA by the RNA-induced silencing complex (RISC).

RNase H A ribonuclease that cleaves the RNA moiety of a DNA/RNA double-strand. The H indicates that the RNase recognizes this molecule as a hybrid.

Second messenger Signaling molecules that transfer extracellular events such as binding of a hormone to a receptor on the cell surface to a target molecule inside the cell. Cyclic AMP is an example of a second messenger molecule.

Short hairpin RNA (shRNA) A self-complementary double-stranded RNA molecule that folds back on itself in a tight hairpin turn. shRNAs are generated after the intracellular expression of triggers for RNAi-mediated silencing. The cellular protein Dicer processes the shRNA into small interfering RNAs (siRNAs).

Short interfering RNA See small interfering RNA.

Single-nucleotide polymorphism A variation of a single nucleotide in the DNA sequence between members of a biological species.

Small interfering RNA (siRNA) A short double-stranded RNA molecule that can be used to silence the expression of a gene by RNA interference.

Spiegelmer An aptamer consisting of L-RNA, that is, the enantiomers (mirror images) of the naturally occurring nucleotides.

Stem cells Cells that reproduce themselves and whose offspring can differentiate into specialized cell types.

Therapeutic cloning The creation of genetically identical human cells for medical applications, for example, to produce a transplant for regenerative medicine.

Therapeutic window The range of dosages that allow efficient treatment in the absence of adverse effects.

Totipotency The ability of a cell to divide and differentiate into all cells of an organism and to construct a complete and viable organism.

Transcription Synthesis of RNA molecules with DNA as a template. Transcription is carried out by a DNA-dependent RNA polymerase.

Transcriptome The entire collection of all RNAs that a cell transcribes.

Transdifferentiation The procedure of converting fully differentiated cells into other cell types while bypassing an intermediate pluripotent state.

Translation The process by which proteins are synthesized based on the information contained in a messenger RNA as specified by the genetic code. Protein biosynthesis is carried out by the ribosome and requires transfer RNAs to deliver amino acids and protein factors in addition to the messenger RNA.

Tropism The ability of a virus/pathogen to infect a specific cell type or cells of a specific organ and replicate in them. Host tropism refers to the ability to infect a certain species.

Tumor suppressor gene Genes that are able to protect a cell from uncontrolled proliferation. When the gene is mutated or loses its function, the cell can progress to a cancer cell.

Uniparental disomy Occurs when a person receives two copies of a chromosome from one parent and no copy from the other parent. This mode of inheritance can, for example, result in phenotypical abnormalities in the case of imprinted genes.

Virulence factors Molecules (e.g., proteins) that enable a microorganism to establish itself within a host and enhance its potential to cause disease.

Zymogen An inactive precursor of a protein (enzyme) that must undergo limited proteolysis to become fully active (also known as a proprotein or proenzyme). In the process, its propeptide is excised.

Index

Molecular Medicine: An Introduction, First Edition. Jens Kurreck and Cy Aaron Stein.
© 2016 Wiley-VCH Verlag GmbH & Co. KGaA. Published 2016 by Wiley-VCH Verlag GmbH & Co. KGaA.

National Center for Biotechnology Information (NCBI) 181
Nazi Germany 345
NCBI (National Center for Biotechnology Information) 181
Neanderthal Genome Project 179, 180
necrotic cells 9
neointimal hyperplasia 342
neovascularization, choroidal 339
NER (nucleotide excision repair) system 73, 74
nested PCR 141
neural stem cells 284
neuraminidase inhibitors 150
neurodegenerative diseases, genetic disorders 68, 69
– Huntington's disease, *see* Huntington's disease
– Parkinson's disease, *see* Parkinson's disease
– stem cell therapies 284, 290–293, 296–300
neurons, cholinergic 27
– dopaminergic, *see* dopaminergic neurons
– synapses 26
neuropeptides 26
neurotransmitters 26, 27
neutral mutations 13
neutral protamine Hagedorn (NPH) 247
"never smoker" lung cancers 93
nevirapine 144
new molecular entity (NME) 3
newborn screening 200
next-generation sequencing (NGS) 45–51, 182–184, 210, 211
– bioethics 349, 350
– Neanderthal Genome Project 179
– Venter, Craig 186
nilotinib 216
nivolumab 107
NKX3.1 110
no-template control (NTC) 43
nomenclature, antibodies 243
– miRNAs 327
non-Hodgkin lymphoma (NHL) 115, 116
non-small cell lung cancer (NSCLC) 93–96
nonfunctional protein products 14,310
noninherited genetic disorders 61
nonliving vaccines, bacterial 160
– viral 133–139
nonviral delivery, siRNAs 320, 321
nonviral gene transfer 266, 267
Northern blot 38–40, 306, 328
NPH (neutral protamine Hagedorn) 247
nuclear envelope 5
nuclear membrane 4
nucleocapsid 125
nucleolytic degradation 307, 314, 337
nucleosides 18, 19, 142–146, 343
– DNA viruses 149, 150
– nucleoside analog reverse-transcriptase inhibitor (NRTI) 142–146, 226
– second-generation analogs 142
nucleotides, modified 18, 19, 307, 308, 337
– NER system 73, 74

– reversible terminator 49
nucleus 4–6

o
oblimersen 310
off-target effects 107, 319
Okazaki fragments 11
oligodendrocyte precursor cells (OPCs) 292, 293
oligonucleotides, antisense (AS ONs), *see* antisense oligonucleotides
– aptamers 333–344
– decoy 342–344
– immunostimulatory 342–344
– siRNAs 316–326, 329, 330
oligopotent stem cells 278, 279
oncogenes 87
– sporadic prostate cancer 111–114
oncology, molecular, *see* molecular oncology
– oncogenesis "driver" mutations 93
– oncolytic viruses 263, 264
– paradigm shift 2
– Personal Genome Project 186
oncolytic viruses 263, 264
oncomiRs 328
oncoretroviral vectors 260–262
Oncotype DX® test 99, 213
Online Mendelian Inheritance in Man (OMIM) 64
opportunistic pathogens 154
opsins 75
opsonization 136
Optaflu 137
oral bioavailability 2, 143
oral polio vaccine (OPV) 134, 135
Orbitrap mass analyzer 193
organelles 77
– human cells 4–8
organic anion transporting polypeptides (OATP) 217
organic cation transport (OCT) proteins 217
organs, artificial 53, 54
Orthomyxoviridae 130
orthopoxviruses 150
Orthoretrovirinae 260
oseltamivir 149, 150
OSKM factors 286, 288
overexpressing, genes 52
oxaliplatin-based adjuvant therapy 104
oxazolidinones 166

p
p53 87, 110, 263, 272
P-glycoprotein (PGP) 218
packaging signal (Ψ) 262, 263
PAH (phenylalanine hydroxylase) 72, 73
pairing, Watson–Crick 303–306
palivizumab 243
pan-drug resistance (PDR) 166
pancreas, β-cells 300